HYDROCARBON BIOREMEDIATION

Edited by

Robert E. Hinchee and Bruce C. Alleman
Battelle, Columbus, Ohio

Ron E. Hoeppel
Naval Civil Engineering Laboratory, Port Hueneme, California

Ross N. Miller
USAF Center for Environmental Excellence, Brooks AFB, Texas

Library of Congress Cataloging-in-Publication Data

Catalog record is available from the Library of Congress

Direct all inquiries to CRC Press, Inc., 2000 Corporate Blvd., N.W., Boca Raton, Florida 33431.

Lewis Publishers is an imprint of CRC Press

International Standard Book Number 0-87371-984-0
Printed in the United States of America 1 2 3 4 5 6 7 8 9 0
Printed on acid-free paper

CONTENTS

Foreword ix

Articles

REMOVAL OF GASOLINE VOLATILE ORGANIC COMPOUNDS VIA AIR BIOFILTRATION: A TECHNIQUE FOR TREATING SECONDARY AIR EMISSIONS FROM VAPOR-EXTRACTION AND AIR-STRIPPING SYSTEMS 1
A. G. Saberiyan, M. A. Wilson, E. O. Roe, J. S. Andrilenas, C. T. Esler, G. H. Kise, and P. E. Reith

BIOFILTRATION FOR TREATMENT OF GASOLINE VAPORS 12
J. S. Devinny, V. F. Medina, and D. S. Hodge

DEVELOPMENT OF A BIOREACTOR TO PURIFY GROUNDWATER 20
J. H. Verheul, J. Heersche, W. de Bruin, G. Schraa, P. Vis, and A. Rinzema

A VADOSE COLUMN TREATABILITY TEST FOR BIOVENTING APPLICATIONS 32
R. S. Baker, J. Ghaemghami, S. Simkins, and L. M. Mallory

COST EFFECTIVENESS AND FEASIBILITY COMPARISON OF BIOVENTING VS. CONVENTIONAL SOIL VENTING 40
H. J. Reisinger, E. F. Johnstone, and P. Hubbard, Jr.

THE EFFECTS OF SOIL TYPE, CRUDE OIL TYPE AND LOADING, OXYGEN, AND COMMERCIAL BACTERIA ON CRUDE OIL BIOREMEDIATION KINETICS AS MEASURED BY SOIL RESPIROMETRY 58
M. H. Huesemann and K. O. Moore

BIOSTIMULATION VERSUS BIOAUGMENTATION—THREE CASE STUDIES 72
M. E. Leavitt and K. L. Brown

NITRATE-BASED BIOREMEDIATION OF PETROLEUM-CONTAMINATED AQUIFER AT PARK CITY, KANSAS: SITE CHARACTERIZATION AND TREATABILITY STUDY 80
S. R. Hutchins and J. T. Wilson

APPLICATION OF NITRATE AS ELECTRON ACCEPTOR AT AN IN SITU BIOREMEDIATION OF AN ABANDONED REFINERY SITE: PILOT STUDY AND LARGE-SCALE OPERATION 93
G. Battermann, R. Fried, M. Meier-Löhr, and P. Werner

APPLYING FIELD-EXPEDIENT BIOREACTORS AND LANDFARMING IN ALASKAN CLIMATES 100
C. M. Reynolds, M. D. Travis, W. A. Braley, and R. J. Scholze

BIOREMEDIATION OF THE *EXXON VALDEZ* OIL SPILL: MONITORING SAFETY AND EFFICACY 107
R. C. Prince, J. R. Clark, J. E. Lindstrom, E. L. Butler, E. J. Brown, G. Winter, M. J. Grossman, P. R. Parrish, R. E. Bare, J. F. Braddock, W. G. Steinhauer, G. S. Douglas, J. M. Kennedy, P. J. Barter, J. R. Bragg, E. J. Harner, and R. M. Atlas

IN SITU BIOREMEDIATION: AN INTEGRATED SYSTEM APPROACH 125
C. H. Nelson, R. J. Hicks, and S. D. Andrews

IN SITU BIOREMEDIATION OF A GASOLINE AND DIESEL FUEL CONTAMINATED SITE WITH INTEGRATED LABORATORY SIMULATION EXPERIMENTS 133
A. Robertiello, G. Lucchese, C. Di Leo, R. Boni, and P. Carrera

EFFECTS OF HYDROGEN PEROXIDE ON THE IN SITU BIODEGRADATION OF ORGANIC CHEMICALS IN A SIMULATED GROUNDWATER SYSTEM 140
C. J. Lu

THE EVOLUTION OF A TECHNOLOGY: HYDROGEN PEROXIDE IN IN SITU BIOREMEDIATION 148
R. A. Brown and R. D. Norris

BIOREMEDIATION OF OIL-CONTAMINATED SHORELINES: THE ROLE OF CARBON IN FERTILIZERS 163
P. Sveum, L. G. Faksness, and S. Ramstad

LIMITATIONS ON THE BIODEGRADATION RATE OF DISSOLVED BTEX IN A NATURAL, UNSATURATED, SANDY SOIL: EVIDENCE FROM FIELD AND LABORATORY EXPERIMENTS 175
R. M. Allen-King, K. E. O'Leary, R. W. Gillham, and J. F. Barker

HETEROGENEITY IN CONTAMINANT CONCENTRATION AND MICROBIAL ACTIVITY IN SUBSURFACE SEDIMENTS 192
C. M. Aelion and S. C. Long

NATURAL BIORECLAMATION OF ALKYLBENZENES (BTEX) FROM A GASOLINE SPILL IN METHANOGENIC GROUNDWATER 201
J. T. Wilson, D. H. Kampbell, and J. Armstrong

THE USE OF INTERNAL CHEMICAL INDICATORS IN PETROLEUM AND REFINED PRODUCTS TO EVALUATE THE EXTENT OF BIODEGRADATION 219
G. S. Douglas, R. C. Prince, E. L. Butler, and W. G. Steinhauer

Technical Notes

THE ROLE OF SURFACTANTS IN ENHANCED IN SITU BIOREMEDIATION 237
J. Ducreux, D. Ballerini, and C. Bocard

VENTING AND BIOVENTING FOR THE IN SITU REMOVAL OF PETROLEUM FROM SOIL 243
J. van Eyk

OPTIMIZING IN SITU BIOREMEDIATION AT A PILOT SITE 252
H. Bonin, M. Guillerme, P. Lecomte, and M. Manfredi

DISSOLUTION KINETICS OF TOLUENE POOLS IN SATURATED POROUS MEDIA 258
M.-F. Yeh and E. A. Voudrias

ENHANCED AEROBIC BIOREMEDIATION OF A GASOLINE-CONTAMINATED AQUIFER BY OXYGEN-RELEASING BARRIERS 262
C.-M. Kao and R. C. Borden

MAXIMIZATION OF *METHYLOSINUS TRICHOSPORIUM* OB3b SOLUBLE METHANE MONOOXYGENASE PRODUCTION IN BATCH CULTURE 267
J. P. Bowman and G. S. Sayler

BIODEGRADATION OF NAPHTHENIC ACIDS AND THE REDUCTION OF ACUTE TOXICITY OF OIL SANDS TAILINGS ... 274
D. C. Herman, P. M. Fedorak, M. D. MacKinnon, and J. W. Costerton

OXIDATION CAPACITY OF AQUIFER SEDIMENT 278
G. Heron, J. C. Tjell, and T. H. Christensen

BIOREMEDIATION: A SOUTH AFRICAN EXPERIENCE 285
Z. M. Lees and E. Senior

NATURAL BIOREMEDIATION OF A GASOLINE SPILL 290
R. C. Borden, C. A. Gomez, and M. T. Becker

AGGREGATION OF OIL- AND BRINE-CONTAMINATED SOIL TO ENHANCE BIOREMEDIATION 296
D. H. McNabb, R. L. Johnson, and I. Guo

EVALUATION OF LIPOSOME-ENCAPSULATED CASEIN AS A NUTRIENT SOURCE FOR OIL SPILL BIOREMEDIATION 303
R. T. Herrington, G. D. Sayles, C. E. Furlong, R. J. Richter, and A. D. Venosa

BIOREMEDIATION AND PHYSICAL REMOVAL OF OIL ON SHORE 311
P. Sveum and C. Bech

THE ROLE OF BIOSURFACTANTS IN BIOTIC DEGRADATION OF HYDROPHOBIC ORGANIC COMPOUNDS 318
W. P. Hunt, K. G. Robinson, and M. M. Ghosh

IN SITU BIOREMEDIATION IN ARCTIC CANADA 323
J. G. Carss, J. G. Agar, and G. E. Surbey

MONITORING AN ABOVEGROUND BIOREACTOR AT A PETROLEUM REFINERY SITE USING RADIORESPIROMETRY AND GENE PROBES: EFFECTS OF WINTER CONDITIONS AND CLAYEY SOIL 329
R. Samson, C. W. Greer, T. Hawkes, R. Desrochers, C. H. Nelson, and M. St-Cyr

BIOTREATMENT OF PETROLEUM HYDROCARBON-CONTAINING SLUDGES BY LAND APPLICATION: A CASE HISTORY AND PROSPECTS FOR FUTURE TREATMENT 334
N. A. Persson and T. G. Welander

MATRIX EFFECTS ON THE ANALYTICAL TECHNIQUES USED TO MONITOR THE FULL-SCALE BIOLOGICAL LAND TREATMENT OF DIESEL FUEL-CONTAMINATED SOILS 343
M. A. Troy and D. E. Jerger

INITIAL RESULTS FROM A BIOVENTING SYSTEM WITH VAPOR RECIRCULATION 347
D. C. Downey, O. A. Awosika, and E. Staes

FIELD DEMONSTRATION OF NATURAL BIOLOGICAL ATTENUATION 353
H. S. Rifai and P. B. Bedient

ASPECTS OF SOIL VENTING DESIGN 362
N. V. Mark-Brown

ADVANCES IN VACUUM HEAP BIOREMEDIATION AT A FIXED-SITE FACILITY: EMISSIONS AND BIOREMEDIATION RATE 368
G. R. Hater, J. S. Stark, R. E. Feltz, and A. Y. Li

BIODEGRADATION OF BTEX COMPOUNDS IN A BIOFILM SYSTEM UNDER NITRATE-REDUCING CONDITIONS 374
J. P. Arcangeli and E. Arvin

BIOREMEDIATION OF WASTE OIL-CONTAMINATED GRAVELS VIA SLURRY REACTOR TECHNOLOGY 383
M. A. Wilson, A. G. Saberiyan, J. S. Andrilenas, R. S. Miller, C. T. Esler, G. H. Kise, and P. DeSantis

EVALUATION OF AN IN SITU BIOREMEDIATION USING HYDROGEN PEROXIDE 388
J. Heersche, J. Verheul, and H. Schwarzer

BIOREMEDIATION OF HYDROCARBON-CONTAMINATED SOIL 398
D. S. Hogg, M. R. Piotrowski, R. P. Masterson, M. R. Jorgensen, and C. Frey

THE USE OF MULTIPLE OXYGEN SOURCES AND NUTRIENT DELIVERY SYSTEMS TO EFFECT IN SITU BIOREMEDIATION OF SATURATED AND UNSATURATED SOILS 405
R. D. Norris, K. Dowd, and C. Maudlin

AROMATIC HYDROCARBON DEGRADATION SPECIFICITY OF AN ENRICHED DENITRIFYING MIXED CULTURE 411
B. K. Jensen and E. Arvin

BACTERIAL DEGRADATION UNDER IRON-REDUCING CONDITIONS 418
H.-J. Albrechtsen

LABORATORY FEASIBILITY STUDY FOR BIOREMEDIATION OF A KEROSENE-CONTAMINATED SITE 424
M. E. Watwood and D. L. Carr

DISTRIBUTION OF NONAQUEOUS PHASE LIQUID IN A LAYERED SANDY AQUIFER 431
C. D. Johnston and B. M. Patterson

REVIEW OF BIOREMEDIATION EXPERIENCE IN ALASKA 438
B. L. Kellems and R. E. Hinchee

COLD CLIMATE APPLICATIONS OF BIOVENTING 444
S. K. Ong, A. Leeson, R. E. Hinchee, J. Kittel, C. M. Vogel, G. D. Sayles, and R. N. Miller

A SIMPLE METHOD FOR DETERMINING DEFICIENCY OF OXYGEN DURING SOIL REMEDIATION 454
H. Würdemann, M. Wittmaier, U. Rinkel, and H. H. Hanert

Author List 459

Index 469

FOREWORD

Bioremediation as a whole remains an emerging and rapidly changing field. Since the first symposium in 1991, significant advances have been made. In 1991, only one paper was devoted to biofilters, but in 1993 biofiltration was of significant interest to generate an entire session. Natural attenuation, also little discussed in 1991, was of great interest in 1993. Increased interest in surfactant-enhanced biodegradation, bioventing, air sparging, and metals bioremediation also was apparent. Interest in chlorinated solvent bioremediation remained steady but strong, and a trend toward field application has developed. The ratio of laboratory to field studies has clearly moved toward the field. This ratio is an indication of a technology that is beginning to mature, but by no means indicates that bioremediation is a mature technology. Clearly, many knowledge gaps and needs exist. Well-developed relationships between laboratory bench-scale testing and field practice frequently are lacking. Although much encouraging laboratory research work has been done with chlorinated and many other recalcitrant organics, field practice of bioremediation is still largely directed at petroleum hydrocarbon-contaminated sites. The number of well-documented field demonstrations of bioremediation is increasing; however, before bioremediation can mature into a readily accepted and widely understood technology area, many more field demonstrations will be necessary.

This book and its companion volumes, *Bioremediation of Chlorinated and Polycyclic Aromatic Hydrocarbon Compounds* and *Applied Biotechnology for Site Remediation*, represent the bulk of the papers arising from the Second International Symposium on In Situ and On-Site Bioreclamation held in San Diego, California, in April 1993. Two other books, *Air Sparging* and *Emerging Technology for Bioremediation of Metals*, also contain selected papers on these topics.

The symposium was attended by more than 1,100 people. More than 300 presentations were made, and all presenting authors were asked to submit manuscripts. Following a peer review process, 190 papers are being published. The editors believe that these volumes represent the most complete, up-to-date works describing both the state of the art and the practice of bioremediation.

The symposium was sponsored by Battelle Memorial Institute with support from a wide variety of other organizations. The cosponsors and supporters were:

Bruce Bauman, *American Petroleum Institute*
Christian Bocard, *Institut Français du Pétrole*
Rob Booth, *Environment Canada, Wastewater Technology Centre*
D. B. Chan, *U.S. Naval Civil Engineering Laboratory*
Soon H. Cho, *Ajou University, Korea*
Kate Devine, *Biotreatment News*
Volker Franzius, *Umweltbundesamt, Germany*
Giancarlo Gabetto, *Castalia, Italy*
O. Kanzaki, *Mitsubishi Corporation, Japan*

Dottie LaFerney, *Stevens Publishing Corporation*
Massimo Martinelli, *ENEA, Italy*
Mr. Minoru Nishimura, *The Japan Research Institute, Ltd.*
Chongrak Polprasert, *Asian Institute of Technology, Thailand*
Lewis Semprini, *Oregon State University*
John Skinner, *U.S. Environmental Protection Agency*
Esther Soczo, *National Institute of Public Health and Environmental Protection, The Netherlands*

In addition, numerous individuals assisted as session chairs, presented invited papers, and helped to ensure diverse representation and quality. Those individuals were:

Bruce Alleman, *Battelle Columbus*
Christian Bocard, *Institut Français du Pétrole*
Rob Booth, *Environment Canada, Wastewater Technology Center*
Fred Brockman, *Battelle Pacific Northwest Laboratories*
Tom Brouns, *Battelle Pacific Northwest Laboratories*
Soon Cho, *Ajou University, Korea*
M. Yavuz Corapcioglu, *Texas A&M University*
Jim Fredrickson, *Battelle Pacific Northwest Laboratories*
Giancarlo Gabetto, *Area Commerciale Castalia, Italy*
Terry Hazen, *Westinghouse Savannah River Laboratory*
Ron Hoeppel, *U.S. Naval Civil Engineering Laboratory*
Yacov Kanfi, *Israel Ministry of Agriculture*
Richard Lamar, *U.S. Department of Agriculture*
Andrea Leeson, *Battelle Columbus*
Carol Litchfield, *Keystone Environmental Resources, Inc.*
Perry McCarty, *Stanford University*
Jeff Means, *Battelle Columbus*
Blaine Metting, *Battelle Pacific Northwest Laboratories*
Ross Miller, *U.S. Air Force*
Minoru Nishimura, *Japan Research Institute*
Robert F. Olfenbuttel, *Battelle Columbus*
Say Kee Ong, *Polytechnic University, New York*
Augusto Porta, *Battelle Europe*
Roger Prince, *Exxon Research and Engineering Co.*
Parmely "Hap" Pritchard, *U.S. Environmental Protection Agency*
Jim Reisinger, *Integrated Science & Technology*
Greg Sayles, *U.S. Environmental Protection Agency*
Lewis Semprini, *Oregon State University*
Ron Sims, *Utah State University*
Marina Skumanich, *Battelle Seattle*
Jim Spain, *U.S. Air Force*
Herb Ward, *Rice University*
Peter Werner, *University of Karlsruhe, Germany*
John Wilson, *U.S. Environmental Protection Agency*
Jim Wolfram, *Montana State University*

The papers in this book have been through a peer review process, and the assistance of the peer reviewers is recognized. This typically thankless job is essential to technical publication. The following people peer-reviewed papers for the publication resulting from the symposium:

Jens Aamand, *Water Quality Institute*
Nelly M. Abboud, *University of Connecticut*
Daniel A. Abramowicz, *GE Corporate R&D Center*
Dan W. Acton, *Beak Consultants Ltd., Canada*
William Adams, *Monsanto Company U4E*
Peter Adriaens, *University of Michigan*
C. Marjorie Aelion, *University of South Carolina*
Robert C. Ahlert, *Rutgers University*
David Ahlfeld, *University of Connecticut*
Hans-Jorgen Albrechtsen, *Technical University of Denmark*
Bruce Alleman, *Battelle Columbus*
Richelle M. Allen-King, *University of Waterloo*
Sabine E. Apitz, *NCCOSC RDTE DIV 521*
John M. Armstrong, *The Traverse Group*
Boris N. Aronstein, *Institute of Gas Technology*
Mick Arthur, *Battelle Columbus*
Erik Arvin, *Technical University of Denmark*
Steven D. Aust, *Utah State University*
Serge Baghdikian
M. Talaat Balba, *TreaTek - CRA Company*
D. Ballerini, *Institut Francais du Pétrole*
N. Bannister, *University of Kent, England*
Jeffrey R. Barbaro, *University of Waterloo*
James F. Barker, *University of Waterloo*
Morton A. Barlaz, *North Carolina State University*
Denise M. Barnes, *Ecosystems Engineering*
Edward R. Bates, *U.S. Environmental Protection Agency*
Tad Beard, *Battelle Columbus*
Cathe Bech, *SINTEF Applied Chemistry, Norway*
Pamela E. Bell, *Hydrosystems, Inc.*
Judith Bender, *Clark Atlanta University*
James D. Berg, *Aquateam Norwegian Water Technology Centre A/S*
Christopher J. Berry, *Westinghouse Savannah River Company*
Sanjoy K. Bhattacharya, *Tulane University*
Jeffery F. Billings, *Billings & Associates*
James N. P. Black, *Stanford University*
Joan Blake, *U.S. Environmental Protection Agency*
Robert Blanchette, *University of Minnesota*
Bert E. Bledsoe, *U.S. Environmental Protection Agency*
Christian Bocard, *Institut Français du Pétrole*
Gary Boettcher, *Geraghty & Miller, Inc.*
David R. Boone, *Oregon Graduate Center*
James Borthen, *ECOVA Corporation*
Edward J. Bouwer, *Johns Hopkins University*
John P. Bowman, *University of Tennessee*
Joan F. Braddock, *University of Alaska*

A. Braun-Lullemann, *Forstbotanisches Institut der Universität Göttingen, Germany*
Susan E. Brauning, *Battelle Columbus*
Alec W. Breen, *U.S. Environmental Protection Agency*
James A. Brierley, *Newmont Metallurgical Services*
Fred Brockman, *Battelle Pacific Northwest Laboratories*
Kim Broholm, *Technical University of Denmark*
Thomas M. Brouns, *Battelle Pacific Northwest Laboratories*
Edward Brown, *University of Northern Iowa*
Guner Brox, *EIMCO Process Equipment*
Gaylen R. Brubaker, *Remediation Technologies, Inc.*
Wil P. de Bruin, *Wageningen Agricultural University, The Netherlands*
Robert S. Burlage, *Oak Ridge National Laboratory*
David Burris, *Tyndall Air Force Base*
Timothy E. Buscheck, *Chevron Research and Technology Company*
Larry W. Canter, *University of Oklahoma*
Jason A. Caplan, *ESE Biosciences*
Peter J. Chapman, *U.S. Environmental Protection Agency*
Abe Chen, *Battelle Columbus*
G. O. Chieruzzi, *Keystone Environmental Resources*
Soon Hung Cho, *Ajou University, Korea*
Patricia J. S. Colberg, *University of Wyoming*
Edward Coleman, *MK Environmental*
Ronald L. Crawford, *University of Idaho*
Steven L. Crawford, *DPRA Inc.*
Craig Criddle, *Michigan State University*
Jon Croonenberghs, *Coors Brewing Company*
Scott Cunningham, *DuPont Central Research and Development*
Mohamed F. Dahab, *University of Nebraska*
Lois Davis, *Sybron Chemicals, Inc.*
Wendy J. Davis-Hoover, *U.S. Environmental Protection Agency*
Peter Day, *Rutgers University*
Sue Markland Day, *University of Tennessee*
Mary F. DeFlaun, *Envirogen*
Richard A. DeMaio, *Mycotech Corporation*
Dave DePaoli, *Oak Ridge National Laboratory*
Allen Deur, *Polytechnic University*
Kate Devine, *Biotreatment News*
L. Diels, *Vlaamse, Instelling voor Technologisch Onderzoek, Belgium*
Greg Douglas, *Battelle Ocean Sciences*
Douglas C. Downey, *Engineering-Science, Inc.*
David Drahos, *SPB Technologies*
Murali M. Dronamraju, *Tulane University*
Jean Ducreux, *Institut Français du Pétrole*
James Duffy, *Occidental Chemical Corporation*
Ryan Dupont, *Utah State University*
Geraint Edmunds
Elizabeth A. Edwards, *Beak Consultants Ltd., Canada*
Richard Egg, *Texas A&M University*
David L. Elmendorf, *University of Central Oklahoma*
Mark Emptage, *DuPont Company*
Burt D. Ensley, *Envirogen*
Michael V. Enzien, *Westinghouse Savannah River Site*

David C. Erickson, *Harding Lawson Associates*
Richard A. Esposito, *Southern Company Services*
J. van Eyk, *Delft Geotechnics, The Netherlands*
Brandon J. Fagan, *Continental Recovery Systems, Inc.*
Liv-Guri Faksness, *SINTEF Applied Chemistry, Norway*
John Ferguson, *University of Washington*
J. A. Field, *Wageningen Agricultural University, The Netherlands*
Pedro Fierro, *Geraghty & Miller, Inc.*
Stephanie Fiorenza, *Amoco Corporation*
Paul E. Flathman, *OHM Remediation Services Corporation*
John Flyvbjerg, *Water Quality Institute, Denmark*
Cresson D. Fraley, *Stanford University*
W. T. Frankenberger, *University of California, Riverside*
James Fredrickson, *Battelle Pacific Northwest Laboratories*
David L. Freedman, *University of Illinois*
Ian V. Fry, *Lawrence Berkeley Laboratory*
Clyde W. Fulton, *CH2M HILL*
Kathryn Garrison, *Geraghty & Miller, Inc.*
Edwin Gelderich, *U.S. Environmental Protection Agency*
Richard M. Gersberg, *San Diego State University*
John Glaser, *U.S. Environmental Protection Agency*
Fred Goetz, *Mankato State University*
C. D. Goldsmith, *EnvironTech Mid-Atlantic*
James M. Gossett, *Cornell University*
Peter Grathwohl, *University of Teubingen, Germany*
Charles W. Greer, *Biotechnology Research Institute, Canada*
Christian Grøn, *Technical University of Denmark*
D.R.J. Grootjen, *DSM Research BV, The Netherlands*
Matthew J. Grossman, *Exxon Research & Engineering*
Ipin Guo, *Alberta Environmental Centre, Canada*
Haim Gvirtzman, *The Hebrew University of Jerusalem*
Paul Hadley, *California Environmental Protection Agency*
John R. Haines, *U.S. Environmental Protection Agency*
Kenneth Hammel, *U.S. Department of Agriculture*
Mark R. Harkness, *GE Corporate R&D Center*
Joop Harmsen, *The Winand Staring Center for Integrated Land, Soil and Water Research, The Netherlands*
Zachary Haston, *Stanford University*
Gary R. Hater, *Chemical Waste Management, Inc.*
Tony Hawke, *Groundwater Technology Canada Ltd.*
Caryl Heintz, *Texas Tech University*
Barbara B. Hemmingsen, *San Diego State University*
Stephen E. Herbes, *Oak Ridge National Laboratory*
Gorm Heron, *Technical University of Denmark*
Ronald J. Hicks, *Groundwater Technology, Inc.*
Franz K. Hiebert, *Alpha Environmental, Inc.*
E. L. Hockman, *Amoco Corporation*
Robert E. Hoffmann, *SiteRisk Inc.*
Desma Hogg, *Woodward-Clyde Consultants*
Brian S. Hooker, *Tri-State University*

Kevin Hosler, *Wastewater Technology Centre, Canada*
M. Akhter Hossain, *Atlantic Environmental*
Perry Hubbard, *Integrated Science & Technology, Inc.*
Michael H. Huesemann, *Shell Development Company*
Scott G. Huling, *U.S. Environmental Protection Agency*
Jasna Hundal, *CH2M HILL*
Peter J. Hutchinson, *The Hutchinson Group, Ltd.*
Aloys Huttermann, *Forstbotanisches Institut der Universität Göttingen, Germany*
Mary Pat Huxley, *U.S. Naval Civil Engineering Laboratory*
Charles E. Imel, *Ecosystems Engineering*
Danny R. Jackson, *Radian Corporation*
Peter Jaffe, *Princeton University*
Trevor James, *Woodward Clyde Ltd., New Zealand*
D. B. Janssen, *University of Groningen, The Netherlands*
Minoo Javanmardian, *Amoco Oil Company*
Ursula Jenal-Wanner, *Western Regional Hazardous Substance Research Center, Stanford University*
Bjorn K. Jensen, *Water Quality Institute, Denmark*
Douglas E. Jerger, *OHM Remediation Service Corporation*
Randall M. Jeter, *Texas Tech University*
Richard L. Johnson, *Alberta Environmental Centre, Canada*
George Johnson, *Stillwater, Inc.*
C. D. Johnston, *CSIRO, Australia*
Donald L. Johnstone, *Washington State University*
E. Fraser Johnstone, *Exxon Company*
K. C. Jones, *Lancaster University*
Warren L. Jones, *Montana State University*
E. de Jong, *Wageningen Agricultural University, The Netherlands*
Linda de Jong, *University of Washington*
Miryan Kadkhodayan, *University of Cincinnati*
Don Kampbell, *U.S. Environmental Protection Agency*
Yacov Kanfi, *Israel Ministry of Agriculture*
Chih-Ming Kao, *North Carolina State University*
Leslie Karr, *U.S. Naval Civil Engineering Laboratory*
Keith Kaufman, *RESNA Industries, Inc.*
S. Keuning, *Bioclear Environmental Biotechnology, The Netherlands*
T. Kent Kirk, *U.S. Department of Agriculture*
Michael D. Klein, *EG&G Rocky Flats, Inc.*
Calvin A. Kodres, *U.S. Naval Civil Engineering Laboratory*
Simeon J. Komisar, *University of Washington*
Raj Krishnamoorthy, *Keystone Environmental Resources, Inc.*
M. Kuge, *Industrial and Fine Chemicals Division, Environmental Technology*
Debi Kuo, *University of Tennessee*
Bruce E. LaBelle, *California Environmental Protection Agency*
William F. Lane, *Remediation Technologies, Inc.*
Margaret Lang, *Stanford University*
Robert LaPoe, *U.S. Air Force*
Barnard Lawes, *DuPont Company*
Maureen E. Leavitt, *IT Corporation*
Clifford Lee, *DuPont Environmental Remediation Services*
Kun Mo Lee, *Ajou University, Korea*

Michael D. Lee, *DuPont Environmental Remediation Services*
Richard F. Lee, *Skidaway Institute of Oceanography*
Paul LeFevre, *Coors Brewing Company*
Robert Legrand, *Radian Corporation*
Terrance Leighton, *University of California*
Sarah K. Leihr, *North Carolina State University*
M. Tony Lieberman, *ESE Biosciences*
Carol D. Litchfield, *Chester Environmental*
Kenneth H. Lombard, *Bechtel Savannah River, Inc.*
Sharon C. Long, *University of North Carolina*
Charles R. Lovell, *University of South Carolina*
Ja-Kael Luey, *Battelle Pacific Northwest Laboratories*
J. S. Luo, *Center for Environmental Biotechnology*
Stuart Luttrell, *Battelle Pacific Northwest Laboratories*
John Lyngkilde, *Technical University of Denmark*
Ian D. MacFarland, *EA Engineering, Science, and Technology, Inc.*
Joan Macy, *University of California*
Andzej Majcherczyk, *Forstbotanisches Institut der Universität Göttingen, Germany*
David Major, *Beak Consultants Ltd., Canada*
Pryodarshi Majumdar, *Tulane University*
Leo Manzer, *E. I. DuPont, De Nemours & Co., Inc.*
Nigel V. Mark-Brown, *Woodward-Clyde International, New Zealand*
Donn Marrin, *InterPhase Environmental, Inc.*
Dean A. Martens, *University of California, Riverside*
Michael M. Martinson, *Delta Environmental Consultants, Inc.*
Perry L. McCarty, *Stanford University*
Gloria McCleary, *EA Engineering*
Linda McConnell, *Logistics Management Institute*
Mike McFarland, *Utah Water Research Laboratory, Utah State University*
Ilona McGhee, *University of Kent, England*
David H. McNabb, *Alberta Environmental Centre, Canada*
Sally A. Meyer, *Georgia State University*
Kathy Meyer-Schulte, *Computer Science Corporation*
Robert Miller, *Oklahoma State University*
Ali Mohagheghi, *Solar Energy Research Institute*
Peter Molton, *Battelle Pacific Northwest Laboratories*
Ralph E. Moon, *Geraghty & Miller, Inc.*
Jim Morgan, *The MITRE Corporation*
Frederic A. Morris, *Battelle Seattle Research Centers*
Pamela J. Morris, *University of Florida*
Klaus Müller, *Battelle Europe*
Julie Muolyta, *Stanford University*
H. S. Muralidhara, *Cargill, Inc.*
Reynold Murray, *Clark Atlanta University*
Karl W. Nehring, *Battelle Columbus*
Christopher H. Nelson, *Groundwater Technology, Inc.*
Per H. Nielsen, *Technical University of Denmark*
Dev Niyogi, *Battelle Marine Research Laboratory*

Robert Norris, *Eckenfelder, Inc.*
John T. Novak, *Virginia Tech*
Evan Nyer, *Geraghty & Miller*
Joseph E. Odencrantz, *Lavine-Fricke Consulting Engineers*
Laurra P. Olmsted, *Brown & Root Civil, England*
Brian O'Neill, *Dearborn Chemical Company, Ltd., Canada*
Richard Ornstein, *Battelle Pacific Northwest Laboratories*
David Ostendorf, *University of Massachusetts*
Donna Palmer, *Battelle Columbus*
Anthony V. Palumbo, *Oak Ridge National Laboratory*
Sorab Panday, *HydroGeologic Inc.*
Joel W. Parker, *The Traverse Group*
John H. Patterson, *Continental Recovery Systems*
Richard E. Perkins, *DuPont Environmental Biotechnology Program*
James N. Peterson, *Washington State University*
Erik Petrovskis, *University of Michigan*
Brent Peyton, *Battelle Pacific Northwest Laboratories*
Frederic K. Pfaender, *University of North Carolina at Chapel Hill*
S. M. Pfiffner, *University of Tennessee*
George Philippidis, *National Renewable Energy Laboratory*
Peter Phillips, *Clark Atlanta University*
C.G.J.M. Pijls, *TAUW Infra Consult B.V., The Netherlands*
Keith R. Piontek, *CH2M HILL*
Michael Piotrowski, *Biotransformations, Inc.*
Augusto Porta, *Battelle Europe*
Roger C. Prince, *Exxon Research and Engineering Co.*
Parmely "Hap" Pritchard, *U.S. Environmental Protection Agency*
Jaakko A. Puhakka, *University of Washington*
Santo Ragusa, *CSIRO Division of Water Resources, Australia*
Ken Rainwater, *Texas Tech University*
Svein Ramstad, *SINTEF Applied Chemistry, Norway*
Grete Rasmussen, *University of Washington*
Mark E. Reeves, *Oak Ridge National Laboratory*
Roger D. Reeves, *Massey University, New Zealand*
H. James Reisinger, *Integrated Science & Technology, Inc.*
Charles M. Reynolds, *U.S. Army Cold Regions Research and Engineering Laboratory*
Hanadi S. Rifai, *Rice University*
Derek Ross, *ERM Inc.*
J. J. Salvo, *GE Corporate, R&D Center*
Réjean Samson, *Biotechnology Research Institute, Canada*
Erwan Saouter, *Center for Environmental Diagnostics and Bioremediation*
Bruce Sass, *Battelle Columbus*
Eric K. Schmitt, *ESE Biosciences, Inc.*
Gosse Schraa, *Wageningen Agricultural University, The Netherlands*
Alan G. Seech, *Dearborn Chemical Co., Ltd., Canada*
Robert L. Segar, *University of Texas at Austin*
Douglas Selby, *Las Vegas Valley Water District*
Patrick Sferra, *U.S. Environmental Protection Agency*
Daniel R. Shelton, *U.S. Department of Agriculture*
Tatsuo Shimomura, *Ebara Research Co. Ltd., Japan*
Mark Silva, *American Proteins, Inc.*

Thomas J. Simpkin, *CH2M HILL*
Judith L. Sims, *Utah State University*
Rodney S. Skeen, *Battelle Pacific Northwest Laboratories*
George J. Skladany, *Envirogen*
Marina Skumanich, *Battelle Seattle Research Centers*
Lawrence Smith, *Battelle Columbus*
Gregory Smith, *ENSR Consulting and Engineering*
Darwin Sorenson, *Utah State University*
Jim Spain, *Tyndall Air Force Base*
Gerald E. Speitel, *University of Texas at Austin*
D. Springael, *Vlaamse, Instelling voor Technologisch Onderzoek, Belgium*
Thomas B. Stauffer, *Tyndall Air Force Base*
Robert J. Steffan, *Envirogen*
H. David Stensel, *University of Washington*
Jan Stepek, *EA Engineering Science and Technology*
David Stevens, *Utah Water Research Laboratory, Utah State University*
Gerald W. Strandberg, *Oak Ridge National Laboratory*
Janet Strong-Gunderson, *Oak Ridge National Laboratory*
John B. Sutherland, *U.S. Food & Drug Administration*
C. Michael Swindoll, *DuPont Environmental Remediation Services*
Robert D. Taylor, *The MITRE Corporation*
Alison Thomas, *U.S. Air Force*
Francis T. Tran, *Diocese Loire-Atlantique, Seminaire Des Carmes*
Mike D. Travis, *RZA-AGRA Engineering and Environmental Services*
Sarah C. Tremaine, *Hydrosystems, Inc.*
Jack T. Trevors, *University of Guelph, Canada*
Marleen A. Troy, *OHM Remediation Services Corporation*
Mark Trudell, *Alberta Research Council, Canada*
Michael J. Truex, *Battelle Pacific Northwest Laboratories*
Samuel L. Unger, *Groundwater Technology, Inc.*
J. P. Vandecasteele, *Institut Français du Pétrole*
Ranga Velagaleti, *Battelle Columbus*
Albert D. Venosa, *U.S. Environmental Protection Agency*
Stephen J. Vesper, *University of Cincinnati*
Bruce Vigon, *Battelle Columbus*
John S. Waid, *La Trobe University, Australia*
Terry Walden, *BP Research*
Mary E. Watwood, *Idaho State University*
Lenly Joseph Weathers, *University of Iowa*
Marty Werner, *Washington State University*
Mark Westray, *Remediation Technologies Inc.*
David C. White, *University of Tennessee*
Patricia J. White, *Battelle Marine Research Laboratory*
Jeffrey Wiegand, *Alton Geoscience*
J. W. Wigger, *Amoco Corporation*
Peter Wilderer, *Technische Universität München, Germany*
Barbara H. Wilson, *Dynamac Corporation*
John T. Wilson, *U.S. Environmental Protection Agency*
Roger M. Woeller, *Water & Earth Science Associates, Ltd., Canada*
Arthur Wong, *Coastal Remediation*

Jack Q. Word, *Battelle Marine Research Laboratory*
Darla Workman, *Battelle Pacific Northwest Laboratories*
Brian A. Wrenn, *University of Cincinnati*
Lin Wu, *University of California*
Robert Wyza, *Battelle Columbus*
Andreas Zeddel, *Forstbotanisches Institut der Universität Göttingen, Germany*
Gerben Zylstra, *Rutgers University*

The editors wish to recognize some of the key contributors who have put forth significant effort in assembling this book. Lynn Copley-Graves served as the text editor, reviewing every paper for readability and consistency. She also directed the layout of the book and production of the camera-ready copy. Loretta Bahn worked many long hours converting and processing files, and laying out the pages. Karl Nehring oversaw coordination of the book publication with the symposium, and worked with the publisher to make everything happen. Gina Melaragno coordinated manuscript receipts and communications with the authors and peer reviewers.

Special thanks are also due to Robert Taylor and Jim Morgan of MITRE Corporation. In addition to providing numerous peer reviews, they provided substantial technical input in support of the coediting process.

None of the sponsoring or cosponsoring organizations or peer reviewers conducted a final review of the book or any part of it, or in any way endorsed this book.

Rob Hinchee
June 1993

REMOVAL OF GASOLINE VOLATILE ORGANIC COMPOUNDS VIA AIR BIOFILTRATION: A TECHNIQUE FOR TREATING SECONDARY AIR EMISSIONS FROM VAPOR-EXTRACTION AND AIR-STRIPPING SYSTEMS

A. G. Saberiyan, M. A. Wilson, E. O. Roe, J. S. Andrilenas, C. T. Esler, G. H. Kise, and P. E. Reith

ABSTRACT

A study was performed to determine both the maximum loading rate and the maximum concentration of gasoline volatile organic compounds (VOCs) per unit volume of packing material in model air biofilters. Three air biofilters were constructed containing 1 ft^3, 2 ft^3, and 4 ft^3 (0.028 m^3, 0.057 m^3, and 0.113 m^3) of packing material (Sphagnum moss), to establish the linearity of the relationships. During Phase I of the study, the system was inoculated with hydrocarbon-degrading *Pseudomonas* spp. and was allowed to equilibrate in the presence of 50 ppmv gasoline vapors, at a constant flowrate of 8.5 ft^3/min (0.241 m^3/min). Phase I results indicated about a 25% VOC removal rate for the smallest biofilter, up to about 85% VOC removal for the largest biofilter. Phase II of the study consisted of incrementally increasing influent VOC concentrations while maintaining a constant flowrate of 8.5 ft^3/min (0.241 m^3/min). Initial data, at various VOC concentrations, indicate that the 4 ft^3 (0.113 m^3) biofilter is capable of removing between 70 and 90% of the VOCs, whereas the 1 ft^3 (0.028 m^3) biofilter removes about 30% of the VOCs. Phase III of this study consisted of maintaining a constant VOC loading rate while varying the flowrate. The combination of data from Phases II and III provided adequate information to show the linearity of the relationship between the flowrate and volume of packing material used. This should allow for effective sizing of field air biofilter systems.

INTRODUCTION

Biological treatment of waste gases is well established in Europe, principally in the Netherlands and Germany, where it has been accepted as the best available control technology (BACT) (Leson & Winer 1991). Biological filters are commonly used in the treatment of wastewater, and various biological methods exist for treating solid wastes impacted with volatile organic compound (VOC) contaminants. VOCs typically are generated by two widely used remedial processes: vapor extraction and air stripping systems (Vasconcelos & Smith 1991). Both operations entail media transfer of the contaminants from the soil or groundwater to an airstream, without any contaminant destruction. As a result, their use as a BACT, without the inclusion of a secondary emissions control technology such as carbon adsorption, thermal treatment, and/or condensation, often is prohibited. These secondary measures are expensive and may produce toxic by-products. However, there are indications that an air biofilter may provide an inexpensive, reliable, low-energy consuming technology for secondary airstream treatment (Leson & Winer 1991). Further examination of this technology is required for development of field-scale air biofilters.

An air biofilter consists of one or more beds of packing material, such as compost, peat, wood chips, or soil. These beds provide a high surface area support material for the microorganisms. Waste gases and oxygen are passed through the inoculated packing material, where the microbial degradation of the contaminants takes place (Douglass 1991). Maximum treatment efficiency is obtained when optimum environmental conditions are present within the system. The major environmental parameters affecting treatment efficiency include oxygen and nutrient availability, moisture, and a suitable range of temperature and pH (Leson & Winer 1991).

Microorganisms are attached to the surface area of the packing materials. In the presence of enough moisture, a wet biolayer (known as a "biofilm") is produced (Pomeroy 1982). Hydrocarbon-degrading bacteria in the wet layer of the biofilter are capable of using VOCs as a source of carbon and energy. The final by-products of this usage are carbon dioxide, water, and biomass. The two basic processes that occur within the biofilm are the diffusion of the contaminant from the gaseous phase into the biofilm, and the metabolism of the contaminant by the bacteria (Meunier et al. 1981). The contaminant will diffuse from the gaseous phase into the biofilm. The biofilm is in a state of equilibrium according to Henry's law when the rate of contaminant diffusion to the film is equal to the rate of contaminant consumption by the bacteria (Blenkinsopp & Costerton 1991).

The removal rate in biofilm kinetics is limited by both the diffusion through the biolayer and the level of microbial activity. It is predicted that biodegradation is a zero-order kinetic reaction (Diks & Ottengraf 1991). When the influent contaminant concentrations are high, diffusion is not a limiting factor. In this case, the biofilm will be fully saturated with contaminant and microbial activity is the limiting factor. In other words, the rate of biodegradation depends on the efficiency of the microorganisms that metabolize the contaminant. At lower contaminant concentrations, where the biofilm is not saturated, diffusion is the rate-limiting factor (Meunier et al. 1981).

Due to the lack of practical and theoretical design information regarding the application of this technology, a feasibility study was executed prior to full-scale system design. The objective of this study was to assess the applicability of using a microbial air filter to treat gasoline VOCs in contaminated airstreams generated by a typical remediation system. The targeted field trial site is a service station where soils and groundwater are impacted with gasoline. The feasibility study was designed to provide data suitable for the design of a field-scale unit and to simulate conditions that might be limiting factors in the operation of the field-scale unit.

BIOFEASIBILITY STUDY

The biofeasibility study, conducted on a pilot scale, assessed the practicality of treating impacted air from gasoline remediation sites with an air biofiltration system. The study was performed to determine basic engineering design requirements and to ensure repeatable results. Three pilot filters were constructed, each containing a different quantity of the same packing material. The various amounts were chosen to establish the linearity between contaminant degradation and the packing volume used. The packing selected was Sphagnum moss inoculated with hydrocarbon-degrading bacteria, primarily *Pseudomonas* spp. In an effort to simulate actual field operating conditions, BP brand unleaded gasoline (benzene 2%, toluene 7%, xylene 8%, 1,2,4-trimethylbenzene 2%, ethylbenzene 2%, cyclohexane 1%) was chosen as the contaminant. The use of a single constituent vapor as the contaminant may have made quantification of maximum loading concentrations more precise, but the goal of this study was to emulate actual contaminants expected in field applications. VOC readings were measured with a common field instrument, the GasTech 1314. Results were obtained in parts per million (mg/L) and are represented as the percent of contaminant reduction in the effluent airstream as compared to the influent air supply. Additional test data were used to correlate GasTech readings to actual gasoline vapor concentrations, for use in future field implementation and monitoring.

Pilot-Scale Design

The first step involved setting the basic pilot filter design parameters. Of particular interest were the basic geometry of the system, the transfer of contaminants from the airstream to the biofilm, the packing material selected, and the method of airflow introduction. The independent variables of the test included humidity, pH, and temperature. Dependent variables included air flowrates and contaminant (gasoline vapor) loading concentrations.

The selection of packing material is important in developing a highly efficient air biofilter. Parameters considered when choosing a good packing material include a large surface area to support effective microbial distribution and high biomass production, moisture retention to provide optimal biological activity, and high porosity to minimize airflow resistance. For this design, Sphagnum moss was selected for its ability to meet these prerequisites (Douglass et al. 1991). The benefits of using this packing material outweighed the potential drawbacks associated with eventual compaction and microbial decomposition of the Sphagnum moss. The packing volumes selected for the units were 1 ft^3, 2 ft^3, and 4 ft^3 (0.028 m^3, 0.057 m^3, and 0.113 m^3) of packing. Space limitations and easily obtainable materials resulted in a vertical orientation of the filters (Figure 1).

The issue of contaminant transfer from the airstream to the biofilm also had to be addressed. Contaminant diffusion and metabolism were previously described in this text. Gasoline vapors were introduced to the airstream by bubbling feed air into a 2 L flask containing 1 L of gasoline and piping the generated vapor into the influent air supply. Fresh gasoline was added on a regular basis to keep a fixed volume of 1 L in the generator flask. GasTech readings were taken approximately 8 hours after these additions were made.

Additionally, the issue of vacuum versus pressure flow had to be considered. The filters could be pressurized, pushing the airstream and contaminant through the system, or the filters could be operated under a vacuum, drawing the airstream and contaminant through the system. The latter process was selected primarily for safety reasons, as a leak would result in air being drawn into the system under vacuum, rather than the release of gasoline vapors to the laboratory environment

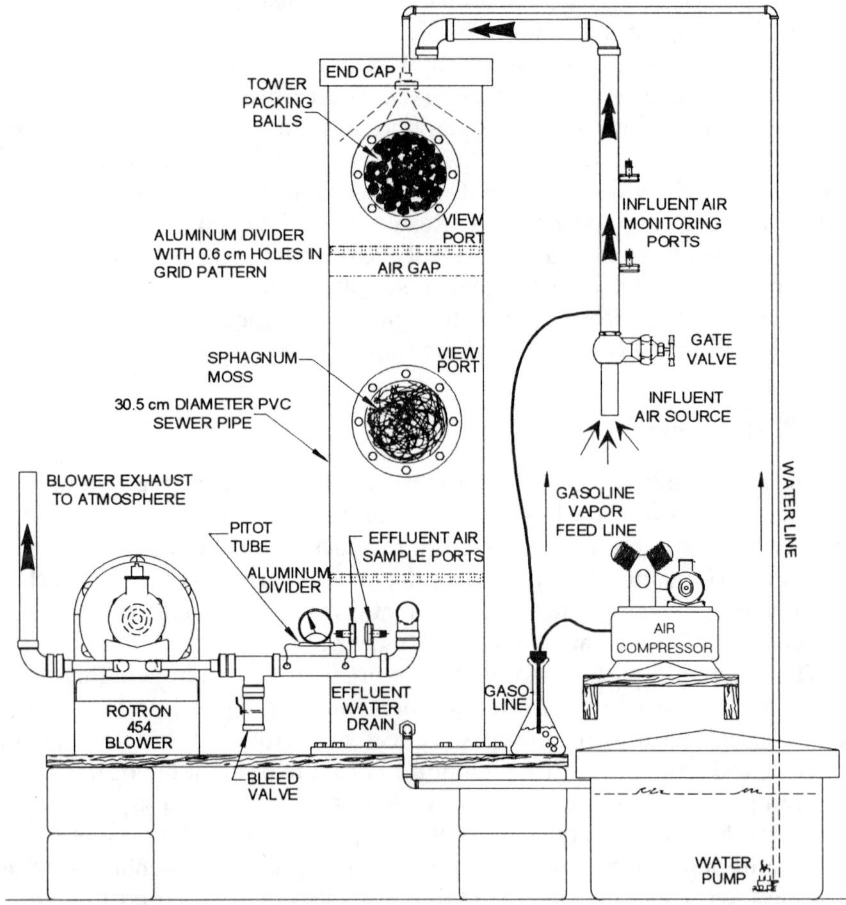

FIGURE 1. Pilot-scale air biofilter.

under pressure. The influent air line of each filter was open to the atmosphere; air was drawn through the filters by a Rotron 454 blower connected to the air effluent line. The loading rates selected were in the range of 4.9 to 16.4 ft/min (1.5 to 5 m/min) (Leson & Winer 1991). Airflow through each filter was measured in the effluent air line using a Pitot tube flowmeter to allow for continuous monitoring and to indicate serious flow-channeling problems. Unless adjusted, the flow meters maintained a relatively constant value. If any plugging or channeling problems had occurred, the resulting change in the air flow value would have been detected. Filter differential pressure was also occasionally measured, again to monitor for the occurrence of flow channeling. Ideally in a system of this nature, filter media pressure losses should be kept to a minimum.

Also, a steady amount of water should be maintained within the system. A relative humidity of approximately 98% was chosen, this is a ratio of roughly 7.5 gallons (28.4 L) of water per 100,000 ft^3 (2,832 m^3) of air (Leson & Winer 1991). In addition, water was passed through the system to keep the packing material saturated. A pump, timer, and spray-heads were used for water delivery. The filters were designed to have a closed water-recirculation system to keep the packing moist and to ensure distribution of inorganic nutrients to the bacterial community. The closed water system also prevented generation of contaminated water. Concurrent air and water flows were chosen both to facilitate the transfer of the contaminant from the air to the water, and to help humidify the incoming air. Air stripping packing balls were placed at the top of the filters to further this effort. During the course of the study, daily physical inspection of the packing materials was relied upon to ensure maintenance of a reasonable moisture content. On average, 5 gallons (18.9 L) of fresh water were added to the water reservoir on a weekly basis, to compensate for losses due to evaporation.

Pilot-Scale Operation

Before collecting the operational data, the filter units were run until a period of equilibrium had been established, to produce usable results. Three phases were executed, each to obtain data with as many variables held constant as possible. Due to the indoor location of the pilot-scale filters, it was possible to maintain a relatively constant temperature (18 to 25°C), except during the final stages when average daily temperatures dropped to 11°C. In the initial stages, the acidic nature of the packing material required the addition of a Tris(Hydroxymethyl)aminomethane pH buffering solution, to maintain a uniform pH of 7. This became less problematic as the filter units became stable and more neutralized. The entire pilot test spanned roughly 5 months. No rigid time frame was set, and each phase was allowed to proceed as required to attain a 80% removal efficiency. This study was performed to allow better field application of air biofiltration technology.

Phase I. Phase I involved system start-up and an equilibration period for all filter units. The system was inoculated with a consortium of hydrocarbon-degrading bacteria, primarily *Pseudomonas* spp. The inoculum was introduced via the water distribution system. The nutrients urea [CH_4N_20] and ammonium phosphate [$(NH_4)_2HPO_4$] were added to the system in the ratio of 10:1 nitrogen to phosphorus (Riser-Roberts 1992). This translated into 2.75 g of urea and 0.6 g of ammonium phosphate dissolved into the 30-gallon (112.5-L) water reservoir every 2 weeks.

After inoculation, the system was allowed to equilibrate for 1 month in the continual presence of 50 ppmv influent gasoline vapors and a flowrate of 8.5 ft^3/min (0.241 m^3/min), to apply selective pressure on the bacteria present. Influent and effluent air concentrations of VOCs were monitored daily with a GasTech 1314. To ensure that the contaminants were actually being metabolized rather than being dissolved into the water, the water recirculating through the system was tested periodically, and the maximum contaminant concentration found was 1 mg/L.

Phase II. Phase II of the experiment involved maintaining a constant flowrate of 8.5 ft^3/min (0.241 m^3/min) in each filter unit while gradually increasing the influent contaminant concentrations. An initial loading concentration of 100 ppmv was increased on a periodic basis until testing stopped and data were collected at 1,000 ppmv.

After the first part of phase II was completed (described above), the largest filter, 4 ft^3 (0.113 m^3), was selected for maximum influent contaminant concentration testing. The influent load for this filter was reset to 500 ppmv and, over a 55-day time period, was gradually increased to 2,000 ppmv, where failure occurred. Failure is defined as the point at which field-scale operation is not cost effective i.e., < 40% reduction of VOCs. Initially, the influent loads were increased weekly, but later as the influent concentrations reached higher levels, the filters required a longer adjustment period. The time allowed for stabilization was 5 days for the 500-ppmv and 1,000-ppmv influent concentrations, and was increased to 22 days for the 1,500-ppmv and 2,000-ppmv concentrations.

Phase III. Phase III of the study involved only the smaller filter units, 1 and 2 ft^3 (0.028 and 0.057 m^3) of packing. This phase consisted of incrementally increasing the flowrate (decreasing residence time), while maintaining a constant contaminant-loading rate of 500 ppmv. A 4.3 ft^3/min (0.122m^3/min) airflow was chosen as the lowest air flowrate to be tested, and 11.5 ft^3/min (0.326 m^3/min) was chosen for the highest (Leson & Winer 1991). To determine maximum flowrates, the initial flowrate of 4.3 ft^3/min (0.122 m^3/min) was increased periodically until the filters failed (< 40% reduction of VOCs).

Results of Biofeasibility Study

Phase I. For the first 2 weeks of the equilibration period, the maximum achievable contaminant removal rate for all three filters was 50%. By the end of the fourth week, removal rates stabilized at approximately 70%. The efficiency failed to improve beyond this limit due to the low concentration of influent contaminants

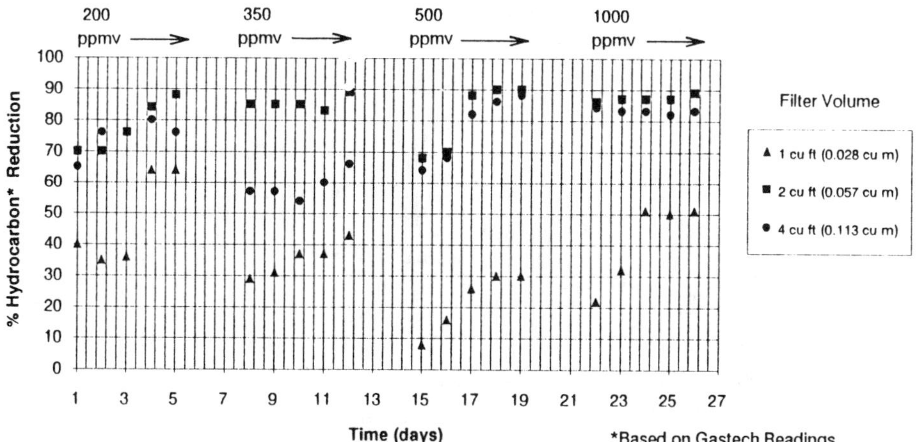

FIGURE 2. Phase II, comparison of removal efficiency of three packing volumes at 8.5 cfm (0.241 m^3/min) airflow.

(insufficient carbon supply). Plate count analyses of the recirculated water performed after the first 2 weeks revealed populations on the order of 4.2×10^3 colony forming units per milliliter of water (CFU/mL). At the end of the equilibration phase, population numbers were on the order of 1.5×10^6 CFU/mL.

Phase II. The results from keeping the flow constant at 8.5 ft^3/min (0.241 m^3/min) and varying the influent contaminant concentrations are presented below (Figure 2). The data indicated a 36% VOC average removal rate from the biofilter with 1 ft^3 (0.028 m^3) of packing material, an 83% average removal rate for the biofilter with 2 ft^3 (0.057 m^3) of packing, and a 73% average VOC removal rate for the biofilter containing 4 ft^3 (0.113 m^3) of packing material. Plate counts performed during this phase indicated that population numbers in the water ranged from 6.4×10^7 to 2.5×10^8 CFU/mL.

The results from the continued testing of the 4 ft^3 (0.113 m^3) biofilter (Figure 3) indicated a maximum removal rate of 88%, with an influent concentration of 500 ppmv. A 84% maximum removal rate, at an influent concentration of 1,000 ppmv, was obtained over a 5-day operational period. A 98% maximum removal rate, at a 1,500 ppmv influent, was obtained by allowing the filter to operate at this level for 22 days. The 4-ft^3 (0.113 m^3) filter experienced failure at an influent contaminant concentration of 2,000 ppmv and constant flowrate of 8.5 ft^3/min (0.241 m^3/min). Even after 23 days of operating time, the filter was unable to improve efficiency and reached a maximum removal rate of 38%. This drastic decrease in microbial activity can be attributed to both the increase in contaminant concentrations and the 10°C decrease in the ambient air temperature. Average daily temperatures at this stage of the experiment were approximately 11°C.

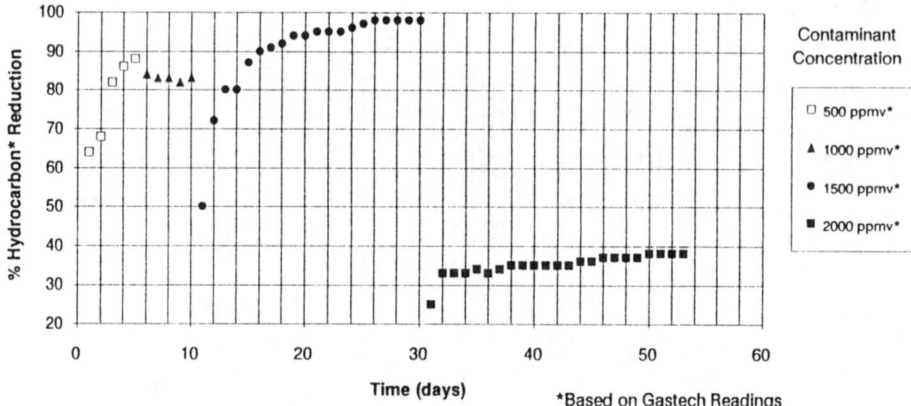

FIGURE 3. Phase II, removal efficiency for the 4-ft³ (0.113-m³) filter at a constant flowrate of 8.5 cfm (0.241-m³/min) and at various contaminant concentrations.

Interestingly, the 4-ft³ (0.113-m³) biofilter required minimum influent concentrations of 500 ppmv, as measured by the Gastech 1314, before optimal removal rates could be attained. This finding suggests that in addition to a maximum concentration capability, a minimum concentration requirement should also be considered during the design of microbial air filter systems.

Phase III. The results of testing various loading rates for the two smaller filters are summarized below (Figures 4 and 5). An initial loading rate of 4.3 ft³/min (0.122 m³/min) at 500 ppmv influent concentration was chosen and resulted in maximum efficiency of 90% removal in the 1-ft³ (0.280-m³) filter and a 96% removal rate in the 2-ft³(0.057-m³) filter. After 1 week, the flowrate was

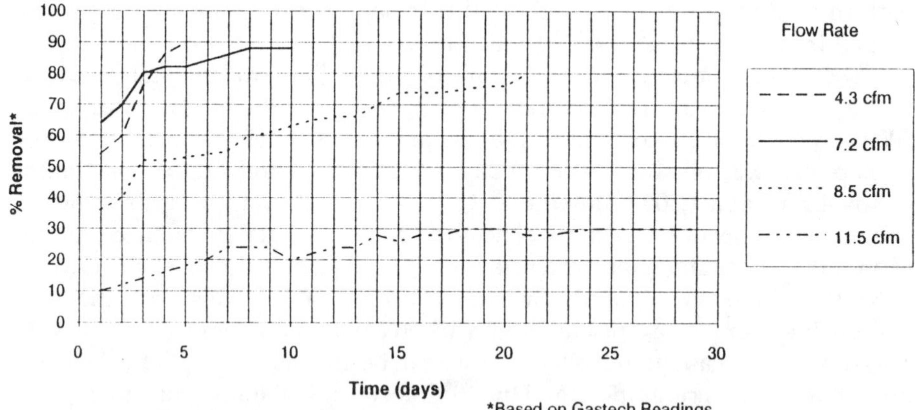

FIGURE 4. Phase III, removal efficiency for the 1-ft³ (0.028-m³) filter at a constant contaminant concentration of 500 ppmv and at various flowrates.

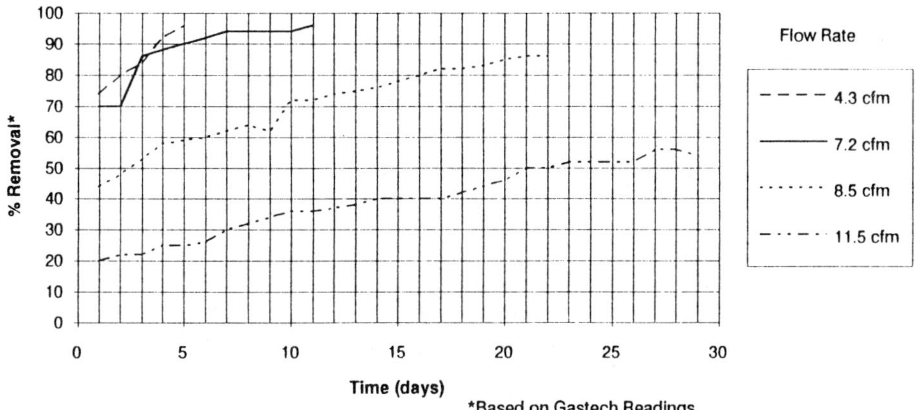

FIGURE 5. Phase III, removal efficiency for the 2-ft^3 (0.057-m^3) filter at a constant contaminant concentration of 500 ppmv and at various flowrates.

increased to 7.2 ft^3/min (0.204 m^3/min), again maintaining a 500-ppmv influent concentration, and the 1-ft^3 (0.280-m^3) filter achieved a maximum efficiency of 88% removal, while the 2-ft^3 (0.057-m^3) filter again reached 96% removal.

For the remainder of this experiment, the stabilization period for each stage was increased to 22 days. As a result, the filters were allowed more time to adjust to the new flowrates after increases were made. At 8.5 ft^3/min (0.241 m^3/min), the 1-ft^3 (0.028-m^3) filter reached 80% removal after 22 days of operation, and the 2-ft^3 (0.057-m^3) filter reached 86%. At 11.5 ft^3/min (0.326 m^3/min), the filters were allowed to operate for 22 days, but the performance of the 1-ft^3 (0.028-m^3) filter dropped significantly. It was only able to reach a removal rate of 30%, and this was considered to be failure of the system. The 2-ft^3 (0.057-m^3) filter performed slightly better, achieving a maximum removal rate of 56%. Plate counts performed throughout this phase reflected relatively stable populations in the water, on the order of 3.8×10^7 to 5.3×10^8 CFU/mL.

The relationship between removal efficiency and retention time (Figure 6) indicates that the rate of contaminant reduction increases as the retention time increases. It levels off when the bacteria are operating at their maximum removal efficiency at a given loading rate. Invariably, after increasing the loading rate, the efficiency of each filter dropped significantly for the first few days. Over the following days, however, the efficiency was found to gradually increase, eventually reaching consistent contaminant reduction levels. The data indicated that the filters, when operating at a less than maximum loading rate, can rapidly acclimate to changing loading rates and can obtain better than 95% VOC reduction under test conditions for extended periods of time.

These points indicate that this technology will certainly have practical field applications. The relationship between the filter volume and the degradation capacity (g/hr) is linear (Figure 7). This plot should allow conservative sizing of

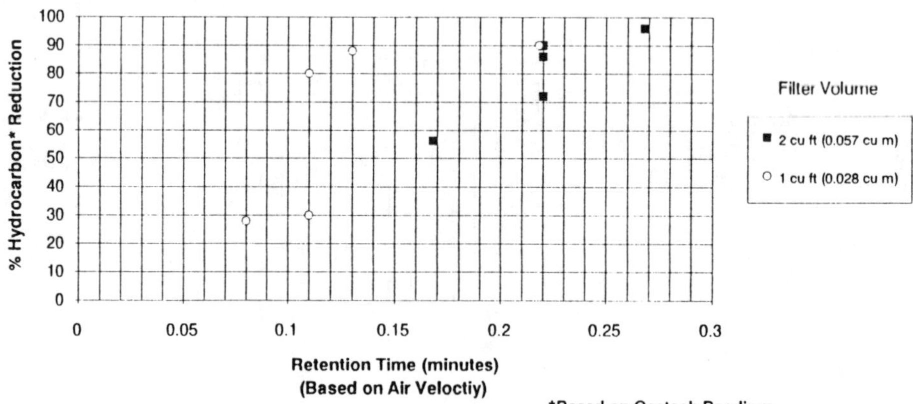

FIGURE 6. Removal efficiency vs retention time for a 500 ppmv constant contaminant concentration.

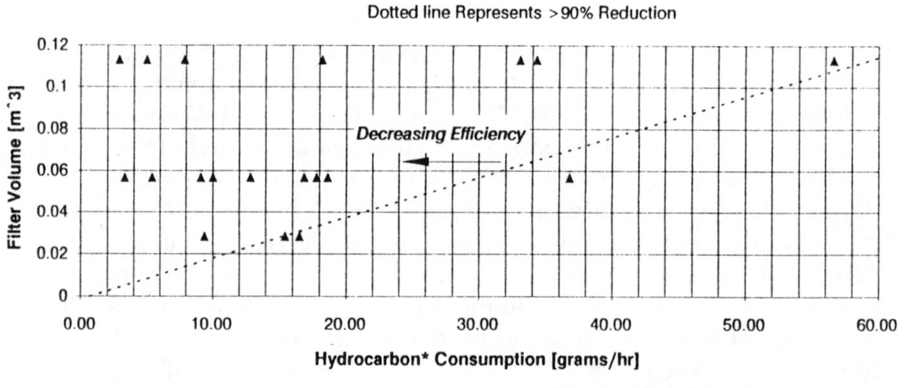

FIGURE 7. Contaminant loading rate vs consumption rate per cubic meter of packing material.

a biofilter given the loading rate and assuming that other variables are held constant as per the pilot study.

CONCLUSION

Volatile organic compound contaminants can be removed effectively from secondary air streams using air biofilters as a treatment process. Much interest in biofilters has recently been generated due to the demonstrated performance levels, and relatively low operational costs, of this type of technology.

The goal of this study was to determine both the maximum contaminant concentrations and maximum contaminant loading rates possible per unit volume of packing material. Based on calculated retention times and the data obtained from this study, the maximum possible contaminant loading rate was determined for 90% removal efficiency. The approximate value obtained was 16 g/hr of gasoline vapors per 1 ft^3 (0.028 m^3) of packing material. Based on this information, a full-size unit will be designed for field implementation.

ACKNOWLEDGMENT

We thank BP Oil Company (Environmental Resource Management) for funding this project and for technical assistance in the development of the air biofilters used in this study.

REFERENCES

Blenkinsopp, S. A., and J. W. Costerton. 1991. *Understanding Bacterial Biofilms.* Elsevier Science Publishers Ltd., UK, 0617-9430/91.

Diks, R. M. M., and S. P. P. Ottengraf. 1991. "A Biological Treatment System for the Purification of Waste Gases Containing Xenobiotic Compounds." In R. E. Hinchee and R. F. Olfenbuttel (Eds.), *On-Site Bioreclamation: Processes for Xenobiotic and Hydrocarbon Treatment*, pp. 452-463. Butterworth-Heinemann, Stoneham, MA.

Douglass, R. H. 1991. "Biofiltration Shows Potential as Air Pollution Control Technology." *The Air Pollution Consultant*, November/December, pp. 1.1-1.2.

Douglass, R. H., J. M. Armstrong, and W. M. Korreck. 1991. "Design of a Packed Column Bioreactor for On-Site Treatment of Air Stripper Off Gas." In R. E. Hinchee and R. F. Olfenbuttel (Eds.), *On-Site Bioreclamation: Processes for Xenobiotic and Hydrocarbon Treatment*, pp. 209-225. Butterworth-Heinemann, Stoneham, MA.

Leson, G., and A. M. Winer. 1991. "Biofiltration: An Innovative Air Pollution Control Technology For VOC Emission." *J. Air Waste Manage. Assoc. 41*(8): 1045.

Meunier, A. D., K. J. Williamson, and A. M. ASCE. 1981. "Packed Bed Biofilm Reactors: Simplified Model." *Journal of Environmental Engineering, 107*(EE2): 307-317.

Pomeroy, R. D. 1982. "Biological Treatment of Odorous Air." *Journal WPCF 54*(2): 1541.

Riser-Roberts, E. 1992. *Bioremediation of Petroleum Contaminated Sites*, pp. A-202. CRC Press, Inc., Boca Raton, FL.

Vasconcelos, J. J., L. Y. C. Leong, and J. E. Smith. 1991. "VOC Emissions and Associated Health Risks." *Water Environment & Technology*, May: 47-50.

BIOFILTRATION FOR TREATMENT OF GASOLINE VAPORS

J. S. Devinny, V. F. Medina, and D. S. Hodge

ABSTRACT

The use of biofilters offers a promising technology for treating point source air discharges contaminated with organic vapors. Their performance is best analyzed by measuring contaminant concentrations at several points through the support medium, establishing a profile of concentration vs. depth. Comparisons were made of granular activated carbon (GAC) and compost as filter media for treatment of air contaminated with ethanol. The biodegradation rate constant and mass partition coefficient were identified as important parameters for optimization. Bench-scale and pilot-scale reactors filled with GAC were used to treat gasoline vapors. Biofiltration provided high rates of treatment in all cases.

INTRODUCTION

Biological treatment of contaminant discharges from point sources shows substantial promise as a pollution abatement technology (Leson & Winer 1991). Successful implementation of biofiltration allows on-site treatment without creation of hazardous residue, carbon regeneration, or fuel costs for combustion.

Development of efficient biofiltration requires a fundamental knowledge of the process and optimization of filter design. The characteristics of the support medium that contribute to rapid contaminant biodegradation must be analyzed.

METHODS

Analytical

All of the experiments reported here used previously described apparatus (Hodge et al. 1991b; Medina et al. 1993; Medina et al. 1992). For the bench-scale columns, artificial airstreams were contaminated with ethanol or gasoline vapors and treated in the laboratory. The pilot-scale apparatus was used to treat off-gases from a soil vapor extraction site contaminated with gasoline. In each case, gas samples were analyzed in the laboratory using gas chromatography (GC) with a flame ionization detector.

Column Profiles

An important aspect of the experiments was that contaminant concentrations were determined at many points along the length of the biofilters. Concentration profiles were measured, rather than limiting data to input and output concentrations. This technique has several advantages, as described in the following paragraphs.

The shape of the profile allows evaluation of biodegradation kinetics. A straight line indicates zero-order biodegradation, for which degradation rates are not dependent on substrate concentrations. An exponential decline suggests first-order or diffusion-limited processes.

During startup of a new column, much of the pollutant removal will be a result of adsorption on the medium. While this is occurring, profiles will show the s-curve typical of adsorption breakthrough, rather than the simple decline of biodegradation, allowing ready identification of the adsorption phase.

Input spikes can provide valuable insight if profile data are collected, and may be intentionally created in laboratory experiments (Hodge & Devinny 1993a; Hodge et al. 1993a). A sudden change in concentration propagates through the biofilter much more slowly than the air moves, because most of the contaminant is adsorbed at any time. The effect is analogous to the movement of a peak through a chromatography column. The ratio of peak velocity to air velocity is the retardation factor. It is equal to one plus the ratio of the mass of contaminant in the solids/water phase to the mass in the air (the mass partition coefficient). A defined peak can be followed through the biofilter, and its rate of movement can be determined. Measurement of the retardation factor allows calculation of the mass partition coefficient. This parameter can also be measured by removing material from the biofilter and performing a batch test, but removal and testing may alter the material, changing the partition coefficient from the value in the active biofilter.

Profiles also allow meaningful interpretation of results where input concentrations are not constant. Concentration fluctuations can be troublesome in experimental work. Output concentrations reflect the contaminants remaining after biodegradation of vapors that entered the biofilter many hours previously, making simultaneous measurements of input and output misleading. Poor performance may be falsely indicated, for example, if a low input concentration is compared with an output concentration that reflects a previous period of higher input concentration.

Finally, full-scale biofilters are likely to be operated under conditions of variable load. The ability to evaluate the response of the biofilter to such loading is valuable.

CHARACTERISTICS OF BIOFILTER PACKING MATERIAL

The selection of biofilter packing material is an important design choice. The elimination capacity (grams of contaminant removed per cubic meter of biofilter per hour) is largely a function of packing medium characteristics. The material must provide a favorable environment for the microbial ecosystem. Inorganic nutrients are necessary, and must be added to materials such as granular activated carbon (GAC) or diatomaceous earth. Other filter materials, such as compost or soil, may provide adequate nutrient concentrations without amendment. The buffer capacity of the medium must be sufficient to maintain optimum pH, or the pH must be controlled (Ottengraf 1986).

The porosity of the medium is an important factor. Interparticle porosity, the spaces between the support particles, allows passage of air through the biofilter. For a given volumetric flowrate, interstitial velocity in the biofilter is inversely proportional to the interparticle porosity. Intraparticle porosity, within the particles, is taken as part of the solids/water phase. It provides much of the adsorption capacity of the medium.

MODELING

Mathematical models of biofilter processes have been developed (Devinny et al. 1991; Hodge et al. 1991a; Hodge & Devinny 1993b). The results of this study are well described by a first-order model, with the assumptions of steady state, negligible dispersion, and rapid contaminant transfer between phases. The contaminant partition coefficient, interstitial air velocity, medium porosity, and biological degradation rate constant are used:

$$C = C_o \exp\left\{-\frac{bk_h (1-\theta)\, x}{V\theta}\right\} = C_o \exp\left\{-\frac{bK_m\, x}{V}\right\} \qquad (1)$$

where
- C = contaminant concentration in the air phase, mg/cc
- C_o = influent air phase contaminant concentration, mg/cc
- b = first-order biological rate constant, /hour
- k_h = concentration partition coefficient, equilibrium value for ratio of contaminant concentration in the solids/water phase to the contaminant concentration in the air phase, unitless
- K_m = mass partition coefficient, equilibrium value for ratio of contaminant mass in the solids/water phase to the contaminant mass in the air phase, unitless
- x = distance of travel in biofilter, cm
- V = interstitial velocity of air, cm/hr
- θ = interparticle porosity of packing material, unitless

The biological degradation rate term in equation 1 is first order. The assumption is that degradation occurs at a rate proportional to the concentration of contaminant found in the solids/water phase. The values of bK_m are a fundamental measure of the effectiveness of a biofilter medium and its biomass.

RESULTS

Bench-Scale Treatment of Ethanol

Results for an experiment treating ethanol on GAC and compost are typical of those obtained in this study (Table 1). The contaminant removal efficiency at steady state was measured, and the biodegradation rate constant was calculated from that result (Hodge et al. 1992). The other parameters were measured independently. GAC provided better removal efficiency overall, although compost had a higher biodegradation rate constant.

The biodegradation rate constant has a substantial effect on the contaminant removal efficiency. The higher value for compost suggests that it offers a better environment for microbial growth and activity. The structure of a GAC particle allows biological growth only on the outer surface, or in the pores larger than a bacterial cell. Much of the intraparticle porosity is not available to microorganisms. The compost particles may have larger intraparticle pores, allowing the microorganisms to use a greater fraction of their internal volume. It may also be that the chemical nature of compost makes it a better environment for microorganisms.

The GAC was artificially seeded, while the compost biofilter used indigenous organisms resulting from the composting process. It may be that the GAC microbial ecosystem was less fully developed.

Typical values for model parameters may be combined with various values of the biodegradation rate to illustrate its effect on biofilter efficiency (Figure 1). Predicted biofilter performance is sensitive to the biodegradation rate constant, and better treatment will result if the modifications can increase its value.

It is also notable that the GAC had a lower overall interparticle porosity than the compost used in this experiment. This means that more solids/water phase is available per unit biofilter volume. Because degradation occurs only in this phase, decreasing the interparticle porosity increases biofilter effectiveness. The degree to which this can be done is limited, however, because the pressure required to drive the air through the biofilter increases, and clogging will occur when porosities are too low.

Biofilter efficiency is also affected by the mass partition coefficient. This is the greatest advantage of GAC over compost (Table 1). As the partition coefficient increases, the mass of contaminant on the medium increases. First-order kinetics predict that an increase in contaminant concentration increases the elimination rate. The effect is demonstrated for a wide range of values in a model calculation (Figure 2). The typical range for mass partition coefficients (1000-9000) is in the steep portion of the efficiency curve, so that increases in the coefficient will produce significant improvements in biofilter performance.

The model assumes that all of the contaminant in the solids/water phase is available for microorganisms. Reversibly adsorbed molecules have average

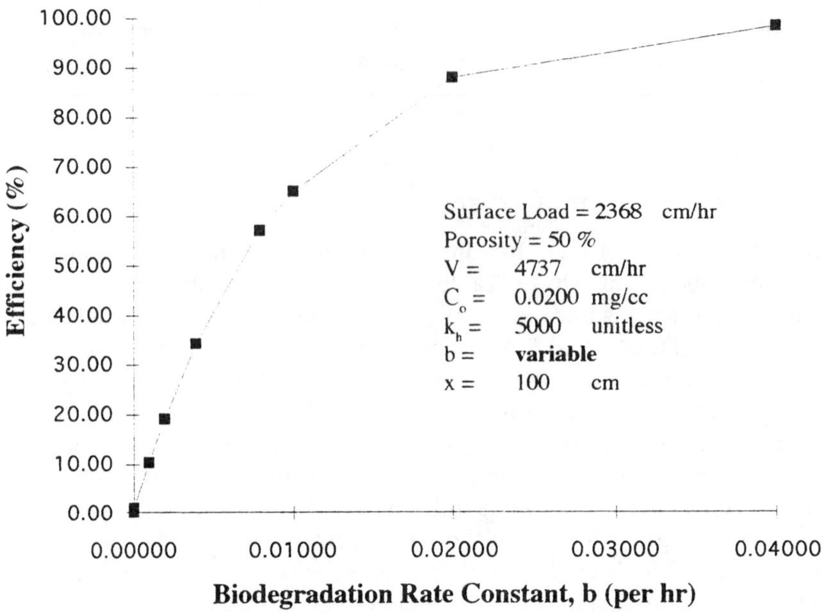

FIGURE 1. Effect of biodegradation rate constant on removal efficiency.

residence times on the carbon surface of 10-3 to 10-7 seconds (Graham, 1992). Because they are being rapidly transferred in and out of the water phase, they are available to the microorganisms. Irreversibly adsorbed molecules, however, are more strongly bonded. This may or may not prevent degradation, depending on whether the bonding interferes with the action of exoenzymes secreted by the microorganisms. The irreversibly adsorbed portion of the contaminant may be from 12% to 50% of the total (Graham, 1992; Medina et al. 1993). Thus the amount of adsorbed material available for degradation is between 50% and 100% of the total.

TABLE 1. Comparison of parameters for treatment of ethanol on GAC and compost.

Parameter	GAC	Compost
Biodegradation rate constant	0.00320 hr^{-1}	0.0062 hr^{-1}
Contaminant partition coefficient	8,674	2,937
Porosity	25	45
Interstitial velocity	9,474 cm/hr	5,263 cm/hr
bK_m	83 hr^{-1}	22 hr^{-1}
Contaminant removal efficiency	58%	34%

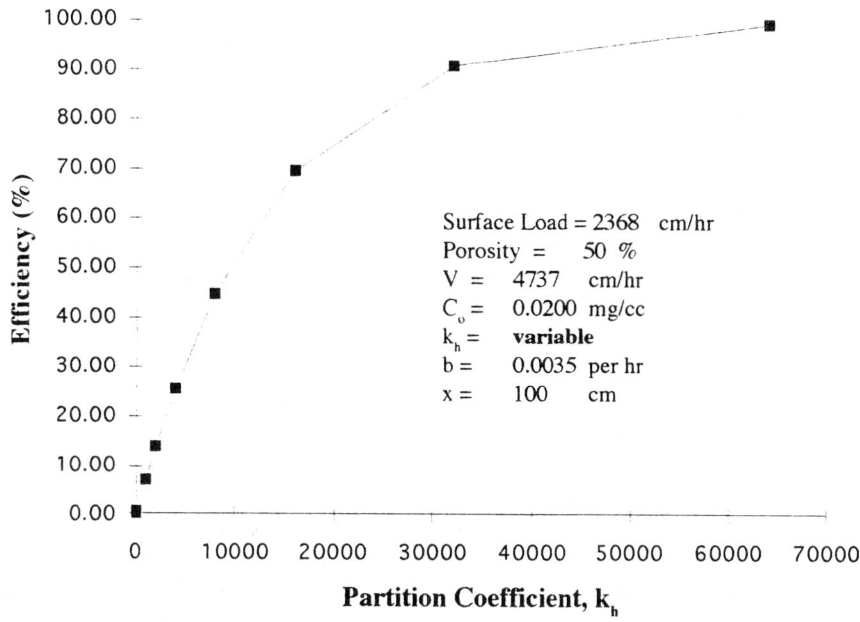

FIGURE 2. Effect of partition coefficient on removal efficiency.

Overall, GAC adsorbs organics well, producing high partition coefficients. It may also have a lower interparticle porosity. This keeps the mass of contaminants in the solids/water phase high and produces a high elimination capacity. Compost, on the other hand, provides an ideal environment for microbial growth, and possibly benefits from a greater surface area available to microorganisms.

Comparison of Pilot-Scale and Bench-Scale Systems

Bench-scale and pilot-scale biofilters for treatment of gasoline vapors were compared. The bench-scale unit was 5.1 cm in diameter and 90 cm long, and was operated in the laboratory using an artificially contaminated airstream. The pilot-scale reactor was 61 cm in diameter and 90 cm deep, and was operated in the field, treating off-gases from a soil vapor extraction project (Medina et al. 1992). Both systems used GAC (type KP 601, provided by Westates Carbon) as the support medium for the microorganisms. Both biofilters were seeded with mixed cultures of microorganisms (from petroleum-contaminated soils) and nutrients. Both columns used prehumidification as the primary means of moisture control.

There were also some differences. The prehumidification system for the bench-scale system was less efficient, allowing some decline in the water content of the GAC. Microorganisms may be inhibited if they are too dry. While contaminant loads were similar for the pilot-scale system, they were obtained by lower input concentrations and higher air flowrates.

Data were taken describing concentration profiles in the columns. The profiles were fitted to an exponential curve, as described by equation 1. Values for bK_m (eqn. 1) were determined. Values for K_m can be determined separately, using either batch measurements or spike tests. Spike tests were not successful for gasoline, however, because the various components moved through the column at various rates.

The profiles were chosen to avoid the startup period, when adsorption is the primary cause of contaminant removal (Medina et al. 1992). In addition, for the bench-scale unit, a period of time was identified during which biodegradation was dominant and the biomass was well developed. The average value of bK_m for the bench-scale reactor (12 observations) was 0.30/hour. For the pilot-scale reactor (30 observations), it was 0.32/hour.

Data were gathered for each system showing the average amount of gasoline degraded per unit filter volume (elimination capacity) as a function of pollutant load (Figure 3). (Complete data for the pilot-scale biofilter appear in Medina et al., 1993. The data shown here are for loads comparable to those on the bench-scale biofilter.) The bench-scale biofilter was more effective, with approximately double the elimination capacity at the higher loads. This occurred, in spite of the bench-scale biofilter's lower value for bK_m, because the interstitial velocity was lower.

CONCLUSIONS

The simple model developed was useful in understanding and predicting biofilter performance. The bench-scale results produced reasonably accurate estimates of pilot-scale performance.

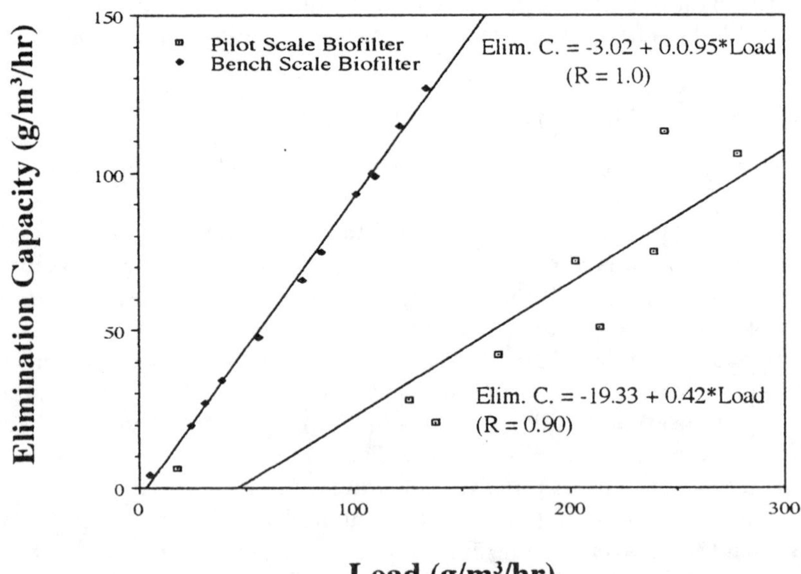

FIGURE 3. Elimination capacity vs. load for bench and pilot-scale biofilters.

Biofilters were effective for treatment of both ethanol and gasoline vapors, in the sense that degradation rates per unit of biofilter volume were high. Ethanol was treated more efficiently, reflecting its greater degradability.

Improvements in biofilter performance may be achieved by improving environmental conditions for the microbial ecosystem, and thereby improving the biodegradation rate constant. Media with high adsorptivities, producing high mass partition coefficients, are likely to be beneficial. Media with lower porosities will be valuable, but may be limited by pressure drop across the biofilter and biological clogging.

ACKNOWLEDGMENTS

Funding for some of the work reported here was provided by Westates Carbon. Valuable help was provided by Dr. James R. Graham and Mukund Ramaratnam.

REFERENCES

Devinny, J. S., V. F. Medina, and D.S. Hodge. 1991. "Bench Testing of Fuel Vapor Treatment By Biofiltration." *National Research and Development Conference on the Control of Hazardous Materials*, Hazardous Materials Control Research Institute, Anaheim, CA, February 20-22.

Graham, J. R. 1992. "Introduction to Activated Carbon and Its Properties", In *Activated Carbon: Properties, Applications and Regeneration Methods*, Westates Carbon Inc., Emeryville, CA, June 12.

Hodge, D. S., and J. S. Devinny. 1993a. "Biofilter Treatment of Ethanol Vapors." Accepted with requests for revision, *Journal of Environmental Engineering*, American Society of Civil Engineers.

Hodge, D. S., and J. S. Devinny. 1993b. "Modeling Biofilter Performance." Submitted for publication. Accepted with requests for revision, *Environmental Progress.*

Hodge, D. S., J. S. Devinny, V. F. Medina, and Y. Wang. 1992. "Biofiltration: Application in Controlling of VOC Emissions." Presented at 47th Annual Purdue University Industrial Waste Conference, West Layfayette, IN, May 11-13.

Hodge, D. S., V. F. Medina, and J. S. Devinny. 1991a. "Biological Methods for Air Decontamination: Modeling Hydrocarbon Removal." *Environmental Engineering. Proceedings of the 1991 Specialty Conference*, pp. 352-357, American Society of Civil Engineers, Reno, NV July.

Hodge, D. S., V. F. Medina, R. L. Islander, and J. S. Devinny. 1991b. "Treatment of Hydrocarbon Fuel Vapors in Biofilters." *Environmental Technology* 12:655-62.

Leson, G., and A. M. Winer. 1991. "Biofiltration: An Innovative Air Pollution Control Technology for VOC Emissions." *Journal of the Air and Waste Management Association* *41*(8):1045-1054.

Medina, V. F., T. Webster, M. Ramaratnam, D. S. Hodge, and J. S. Devinny. 1992. "Treatment of Soil Vapor Extraction Off-Gases by GAC Based Biological Filtration: Bench and Pilot Studies." *Emerging Technologies for Hazardous Waste Management*, American Chemical Society I&EC Division Special Symposium, Atlanta GA, September 21-23.

Medina, V. F., T. S. Webster, M. Ramaratnam, J. R. Graham, and J. S. Devinny. 1993. "Treatment of Gasoline Vapors from a Soil Vapor Extraction Site Using a Pilot-Scale GAC-Based Biofilter." Submitted for publication.

Ottengraf, S. P. P. 1986. " Exhaust Gas Purification," In H. J. Rehm and G. Reed (Eds.), *Biotechnology* 8:425-452.

DEVELOPMENT OF A BIOREACTOR TO PURIFY GROUNDWATER

J. H. Verheul, J. Heersche, W. de Bruin, G. Schraa, P. Vis, and A. Rinzema

ABSTRACT

A bioreactor based on the principle of dry filtration, a process used to purify drinking water, was developed to purify polluted groundwater. Successful laboratory experiments resulted in an optimization study and a pilot study where it was demonstrated that the elimination capacity of the reactor could reach 100% for organic loads of 50 to 60 $g/m^3.h$. The highest measured elimination capacity was 60 $g/m^3.h$, which is high compared to other biofilm reactors. A backflush cycle once every 3 days was sufficient for loads up to 50 $g/m^3.h$, as demonstrated by the optimization study. The mean residence time during the pilot study was 6 minutes, at a flowrate of 3 m^3/h. The reactor can compete with other comparable groundwater purification techniques with regard to technology and business economics.

INTRODUCTION AND OBJECTIVES

Groundwater released during remediation generally holds such high concentrations of contaminating substances that it cannot be direct directly discharged into sewage systems or surface waters. Benzene, toluene, ethylbenzene and different isomers of xylene (BTEX), as well as tetrachloroethene (PCE) and trichloroethene (TCE), are among the compounds and combinations of compounds most frequently found in polluted soil and groundwater. These components usually are removed by withdrawal of the groundwater and purification through physical/ chemical methods. The residual products left by such processes unfortunately shift the contamination from one environmental compartment to another, which makes the application of a biological alternative desirable.

The development of a biological remediation method, was aimed at developing a reactor in which, through a biotechnological process groundwater contaminated by the substances listed above could be purified. The strategy followed three objectives: (1) to collect data on the microbial degradation process of the compounds to be tested; (2) to apply this knowledge in a reactor to be designed and optimized; and (3) to customize and test the developed reactor concept at a contaminated site.

The project comprised two phases: Phase a: (a.1) implementation of a feasibility study (Schraa et al. 1989a); (a.2) laboratory research (Schraa et al. 1989b); and (a.3) realization of a reactor concept (Schraa et al. 1989b). Phase b: (b.1) laboratory research investigating the degradation of aromatics; (b.2) preliminary technological study; (b.3) optimization study concerning a bioreactor; (b.4) pilot study concerning a bioreactor; and (b.5) investigation of the anaerobic degradation of PCE.

The planning and results of phases b.2 through b.4 are presented in this paper. The investigation focused on eliminating volatile monoaromatics and naphthalene. The description and results of this investigation are preceded by a summarized explanation of some theoretical aspects. The research concerning anaerobic degradation of tetrachloroethene is published elsewhere (de Bruin et al. 1992, de Bruin & Schraa 1992).

THEORETICAL BACKGROUND

A biotechnological process for elimination of aromatic hydrocarbons from groundwater has to meet three criteria: (1) it must be able to reach a low concentration of effluent; (2) elimination must be effected mainly through degradation and not through volatilization or adsorption; and (3) the mean rate of transformation in the reactor must be high.

Various constituent processes play a role in eliminating aromatic hydrocarbons from groundwater in a three-phase reactor: (1) transformation of aromatic hydrocarbons and oxygen in the biofilm; (2) transfer of aromatic hydrocarbons from the water phase to the air phase (volatilization); (3) transfer of oxygen from the air phase to the water phase; and (4) adsorption of aromatic hydrocarbons to the biofilm and the carrier material.

Monocyclic aromatic hydrocarbons adsorb very little on biomass, as they have a low octanol-water coefficient. No adsorption processes were taken into account. The volatilization of aromatic hydrocarbons can be limited by (1) a low value of the of mass transfer coefficient and the specific interface between air and water; (2) a low absorption capacity of the airflow; and (3) the passing of air and water downflow through the reactor. The resistance against substance transfer, for both oxygen and aromatic hydrocarbons, mainly concerns the liquid. Limiting the volatilization of aromatics automatically results in insufficient aeration. Therefore volatilization should preferably be minimalized by limiting the flowrate of the air.

Degradation rate depends on kinetic parameters describing the growth and death of biomass, the thickness of the biofilm, and the concentration of biomass in the biofilm. Lower substratum concentrations on the biofilm surface slow down the degradation process, as the intrinsic reaction rate is lowered or the diffusion in the biofilm is limited. Consequently, the mean degradation rate in the reactor can be raised by (1) a large specific surface of the biofilm; (2) a high concentration of biomass in the reactor; and (3) high substratum concentrations at the surface of the biofilm. The mean concentration of aromatic hydrocarbons can be raised by restricting axial mixing of the water in the reactor. The concentration of oxygen can be raised by ensuring a high aeration capacity.

Few experimental data on the conditions required to reach a high biomass concentration per m^2 carrier surface are available, and very little theorizing has taken place so far. The available data seem to indicate that the biomass concentration per m^2 carrier surface is proportional to the concentration of the limiting substratum in the water. Sustaining the biofilm requires a certain minimum concentration. Below this minimum, the biofilm wears out and dies off faster than it can grow. In a plug flow reactor this can result in a threshold concentration, a residual concentration that is not degraded.

A dry filter is a packed column in which both air and water move downflow through the reactor. Such a filter has several advantages: (1) a large specific surface of the carrier material (1,500-3,000 m^2/m^3), resulting in a high concentration of biomass; (2) plug flow features, resulting in high substratum concentrations at the surface of the biofilm; (3) air and water flow in the same direction through the sandbed, limiting the volatilization of aromatic hydrocarbons; and (4) a high aeration capacity at a low air:water ratio, resulting in a low absorption capacity of the airflow.

PRELIMINARY LABORATORY INVESTIGATION WITH A DRY FILTER

Description of Dry Filter

The dry filter, which was kept at a temperature of 10°C, consisted of a Perspex™ tube with an inner diameter of 20 mm. The column was 1 m high, with the level of the sandbed at 93 cm. Sampling taps were placed in the filter at varying levels. The filtering material of the bed consisted of sand grains of 1 to 2 mm, with a mean of 1.5 mm, thus having a large specific surface. The pores in the filterbed ensured good oxygen transfer and a microscaled turbulence flow. Groundwater, with toluene, *ortho*-xylene, ethylbenzene, and naphthalene added, was sprinkled downflow through the dry filter, together with an airflow. Table 1 presents the schedule for the dry filter investigation. The flowrate of the air amounted to 0.33 L/h during the entire period.

Results of Preliminary Laboratory Investigation with a Dry Filter

After inoculation and recirculation of the groundwater, the microbic elimination of the aromatic hydrocarbons could be observed within some days. Figure 1 presents the organic loading and elimination capacity of the dry filter during the entire investigation period. With inoculation and recirculation, a dry filter can be started within 2 weeks and will reach an elimination capacity of 20 $g/(m^3.h)$ within a few weeks. The filter smoothly absorbed widely fluctuating influent

TABLE 1. Schedule for the dry filter investigation.

Period (days)	Phase	Flowrate water (m/h)	Organic load (g/(m³.h)
0 -10	start	1.3	0.82
10-23	continuous flowthrough	1.3	0.82
24-42		1.3	9.6-13.7
43-50		7.0	9.6-13.7
51-69		7.0	25-50

The reactor was inoculated with an enriched culture of bacteria grown on toluene, *ortho*-and *meta*-xylene, and naphthalene.

concentrations at a constant flowrate. The elimination capacity remained complete up to 35 g/(m^3.h), but beyond that it became less than adequate. The highest measured elimination capacity amounted to about 60 g/(m^3.h). This parameter was measured under varied processing conditions at 10°C.

Next, variations were applied to the parameters (1) hydraulic load, (2) organic load, and (3) air:water ratio. The effects of these parameters on the efficiency and stability of the transforming process were studied. The largest elimination capacity that could be reached was around 60 g of aromatic hydrocarbons per m^3

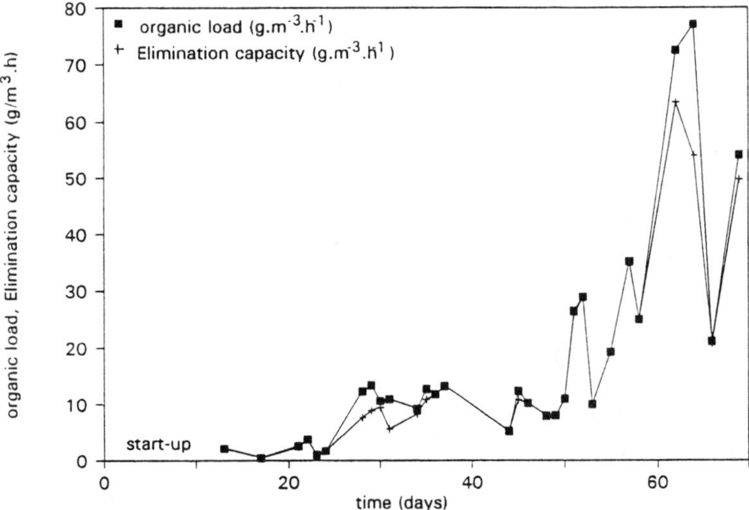

FIGURE 1. Organic load and elimination capacity of the dry filter during the preliminary laboratory investigation.

reactor per hour. Varying concentrations of aromatic hydrocarbons in the influent had no adverse effect on elimination performance.

Volatilization of aromatic hydrocarbons could be observed only at inadequate transformations and at air:water ratios exceeding 1. The oxidation of iron and manganese and the growth of the bacterial mass caused an increased pressure drop across the dry filter. When this increase grew too strong, the material that was added or deposited had to be removed by backflushing the filter with water, at superficial liquid rates between 15 and 20 $m.h^{-1}$. The frequency of backflushing depended on the loading level. Backflushing did not adversely effect the transformation capacity of the dry filter (Vis et al. 1992).

OPTIMIZATION INVESTIGATION

Description of Dry Filter

Figure 2 presents a diagram of the dry filter used in the optimization investigation. Spring water was passed through a closed buffer tank (0.1 m^3) and onto the dry filter with a mono-pump. Toluene and *ortho*-xylene (at a 1:1 ratio) and nutrients were added to the stirred contents of the buffer tank. Water was passed downflow through the dry filter, with a sprinkler ensuring even distribution across the filter surface. Air entered simultaneously downflow by means of a mass flow control (mfc) unit. The volumes of the reactor and the filterbed amounted to 20.6 and 18.3 L respectively, and their heights were 2.25 and 1.95 m. Inside the

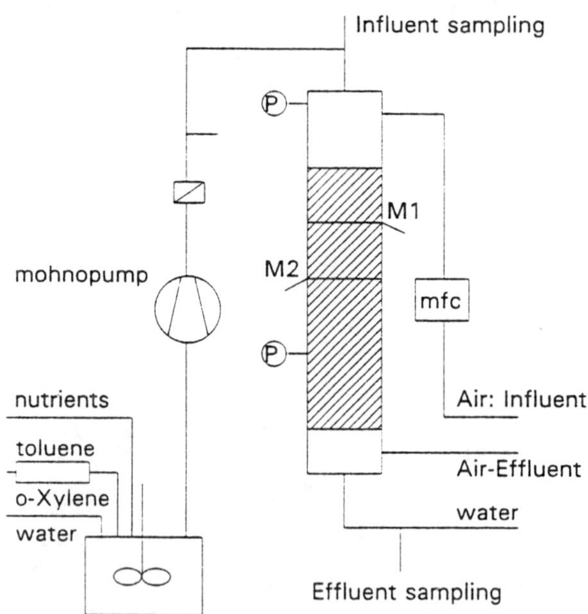

FIGURE 2. Diagram of the dry filter used during the optimization study.

column, sampling points were mounted at various distances in the bed. The column was fitted with pressure gauges above the bed. The course of the pressure drop through the filterbed was monitored constantly. The packing material consisted of sand (1.5 to 2.5 mm). The reactor was inoculated with a enrichment culture of toluene and *ortho*-xylene degrading bacteria.

Results of Optimization Investigation

Figure 3 shows the elimination capacity plotted against the influent concentrations of toluene, *ortho*-xylene, and total aromatics (sum of toluene and *ortho*-xylene) for a hydraulic surface load of 3.8 $m^3/(m2.h)$ (Q_{water} = 35 L/h, Q_{air} = 30 L/h). The dry filter was loaded with both aromatics. Figure 3 indicates that the elimination capacity of the total aromatics increased in direct proportion to the concentration of influent up to a value of 60 to 70 $g/(m^3.h)$. The elimination of aromatics was complete up to this value with an effluent concentration below 0.5 µg/L. When the organic load was increased, the elimination capacity grew to 90 to 100 $g/(m^3.h)$, but the aromatics were not totally eliminated, nor did the effluent concentration (>100 µ/L) meet the standard required for discharge into the sewage system. The elimination capacity of toluene and *ortho*-xylene was proportionally increased with an increase of influent concentration. The elimination capacity for *ortho*-xylene reached a maximum of around 40 $g/(m^3.h)$, after which it started decreasing. The maximum for toluene was double this amount. An explanation for this difference may be found in the fact that the microorganisms degrading toluene and the ones degrading *ortho*-xylene compete for oxygen.

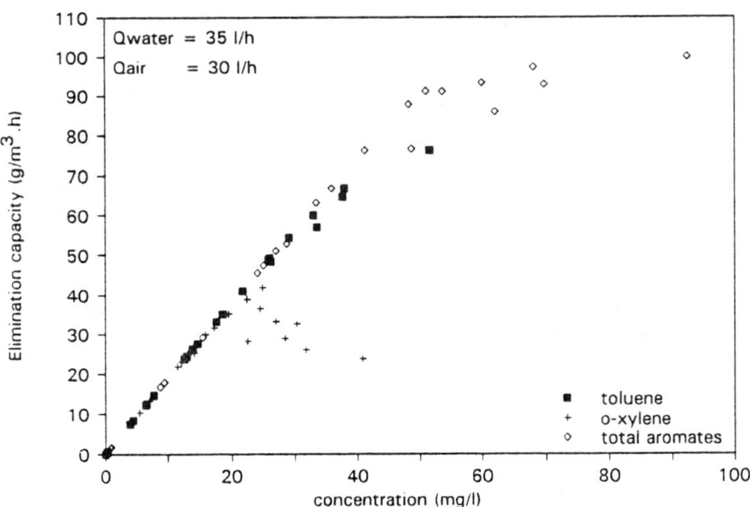

FIGURE 3. Elimination capacity vs. influent concentration of toluene, O-xylene, and total aromatics during the optimization study. Hydraulic load 3.8 $m^3/(m^2.h)$.

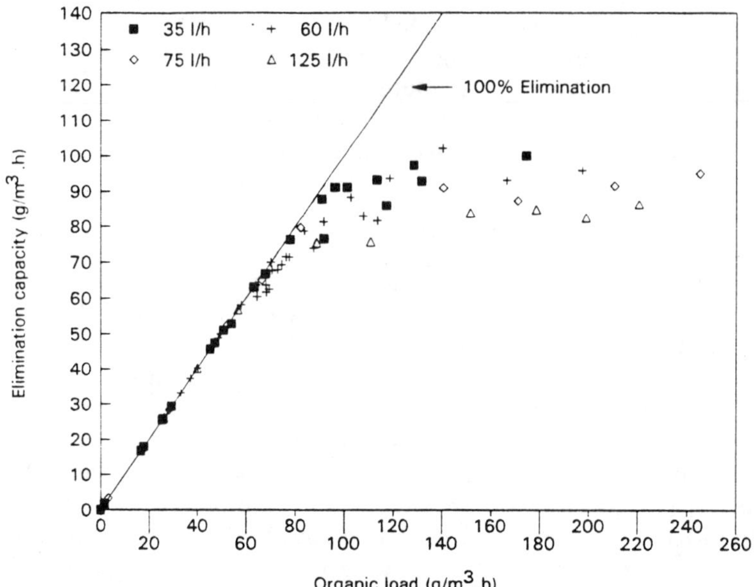

FIGURE 4. Elimination capacity vs. organic load at different hydraulic loads as measured during the optimization study.

Figure 4 shows the elimination capacity plotted against the total organic load at different hydraulic loads, for which the elimination of aromatics was 100% at an aromatics load of 55 g/(m^3.h). With regard to the maximum elimination capacity, Figure 4 indicates that this capacity showed a slight reduction at increasing hydraulic surface loads with equal aromatics loads. It was necessary to provide water poor in N/P with a continuous supply of nutrients (NH_4Cl and K_2PO_4 were added) to reach a high elimination capacity, both during the starting phase and during the continuous-operation phase. Also the dry filter needed backflushing every three days at an organic load of 50 g/(m^3.h). Backflushing was done at a pressure drop of 5 to 8 m water column. The elimination of free biomass during backflushing could result in a reduction of the elimination capacity. The elimination capacity reached with a mixture of toluene and *ortho*-xylene corresponded to the elimination capacity reached with a mixture of benzene, toluene, ethylbenzene, and *meta-para*-xylene. The components benzene and *ortho*-xylene were hardest to eliminate when the filter was overloaded with a BTEX-mixture (Vis et al. 1992).

PILOT STUDY WITH A DRY FILTER

After studying the successful elimination of aromatics in a small laboratory-scale dry filter, the performance of a larger reactor was studied under field conditions. Trial runs focused on the standards set in the first phase of the investigation and on establishment of the elimination capacity concerning volatile aromatics and naphthalene. The management assessment looked at its simplicity, reliability, short starting period, sensitivity with regard to fluctuating loads, and frequency of backflushing related to pressure drop. The elimination capacity was assessed with regard to elimination capacity concerning volatile aromatics and naphthalene, residual concentrations in effluent, elimination capacity of mixed substances, and influence of backflushing behavior on elimination capacity. Volatilization was measured by sampling the effluent air on tenax followed by gas chromatographic (GC) analyses.

Description of Pilot Installation

Figure 5 shows a schematic representation of the pilot installation, for which the same principles hold as for the optimization investigation. The installation was placed near an operating groundwater purification plant (GPP). Part of the water destined for this GPP was pretreated and diverted to the pilot installation. The groundwater was dosed with BTEX and naphthalene in a mixing tank, offering the opportunity of increasing the organic load to the values required for the tests. The aromatics were dosed through a pump from a chemical storage barrel. Table 2 shows the dimensions of the dry filter. The sand bed rested on a bottom plate with ducts for the transport of air and water. A backflush blower with a capacity of 40 m^3/h and a backflush pump with a capacity of 34 m^3/h were installed.

Testing Program

The testing program was aimed at starting the reactor, and establishing the elimination capacity and the backflushing behavior.

Starting the Reactor. The filter was started in three different ways: (1) by recirculating the groundwater across the reactor after dosing it with the aromatics; (2) as in (1), but after inoculation with an enriched culture of microorganisms, and (3) by submerging the reactor in groundwater with aromatics added.

Elimination Capacity. The elimination capacity for aromatics (toluene, ethylbenzene, *ortho*-xylene, and naphthalene) was studied as a function of hydraulic and organic loads. The hydraulic load was varied between 1 and 3.4 m^3/h, and the organic load was varied between around 2 and 100 $g/(m^3.h)$. During the experimental stage it became clear that a compressor would have to be installed to provide the system with sufficient air. The pilot installation was aimed at reaching an effluent concentration for total contamination between 10 and 50 µg/L.

The flowrates of air and water, the oxygen content and Ph of influent and effluent, and the pressure drop through the filter were monitored daily. Influent and effluent were analyzed three times a week for TEX and naphthalene.

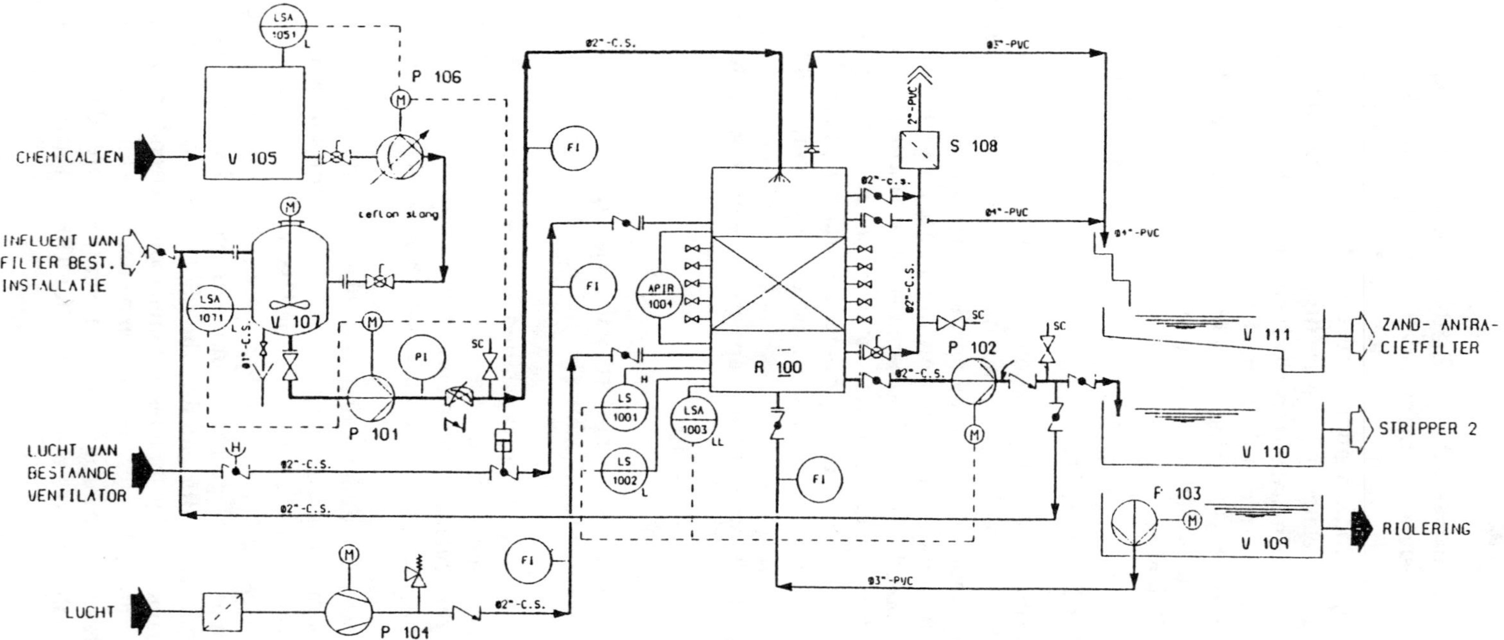

FIGURE 5. Schematic representation of the pilot installation.

TABLE 2. Dimensions of the dry filter.

diameter	0.915 m
total height	2.9 m
height of filterbed	1.5 m
total volume reactor	1.9 m^3
volume filterbed	1.0 m^3
surface area	0.66 m^2
capacity influent pump	5.0 m^3/h
capacity effluent pump	5.5 m^3/h

During the testing period the hydraulic load was varied from 1 to 2.3 up to 3 to 3.4 m^3/h, with the load of aromatics varying from 2 to 25 mg/L.

Backflushing Behavior. The reactor was backflushed manually, based on pressure drop. The frequency of backflushing and the influence of this process on the elimination capacity were studied.

Results

Starting the Reactor. The results of the starting experiments could not be unambiguously interpreted. It was be concluded, however, that the load of aromatics should be kept small at the start. Both recirculation and submerging are viable alternatives, but they should be accompanied by a good oxygen supply.

Elimination Capacity. The elimination capacity may vary widely for each component. *Ortho*-xylene generally proved the hardest to eliminate, but with naphthalene 100% elimination could be reached as a rule. Figures 6 and 7 summarize the results of the pilot study. The elimination capacity for the total contaminants is presented as a function of the organic load for hydraulic loads between 1 to 2.3 m^3/h and 3 to 3.4 m^3/h.

These results lead to the conclusions that: an elimination capacity of 100% is feasible for loads up to around 80 g/(m3.h). An adequate oxygen supply is crucial in this respect. A significantly lower elimination capacity could generally be traced to an insufficient oxygen supply or to problems with the flowthrough of the reactor bed. An effluent concentration between 10 and 50 µg/L appears quite feasible, and will allow the effluent of the bioreactor to be discharged directly onto the sewage system. Volatilization was generally less than 0,5%. From these results it may be concluded that the bioreactor is highly usable for the elimination of aromatics from groundwater.

Backflushing Behavior. Backflushing of the reactor did not pose any problems provided the compressor was large enough. The process readily set the sand bed

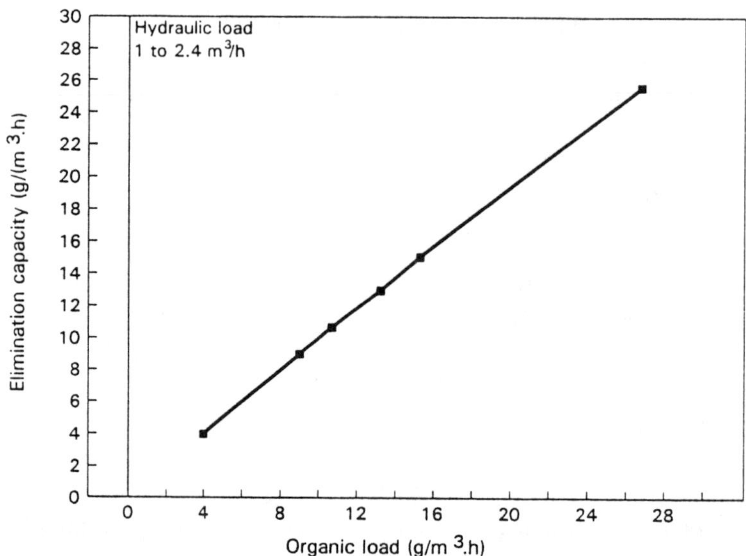

FIGURE 6. Elimination capacity vs. organic load (total contaminants) at a hydraulic load of 1 to 2.3 m^3/h as measured during the pilot investigation.

floating, thereby freeing biomass and Fe deposits. The organic load also appeared to be an influencing factor. In practice, the reactor was backflushed every 2 to 3 days on average; increasing to every day adversely affected the elimination capacity, as it probably inadvertently flushed out biomass as well (Verheul & Doelman 1992).

CONCLUSIONS

The "dry filtration" bioreactor in this study performs with high organic loads, high elimination capacities (both up to 50 to 60 g/(m^3.h) and low effluent concentrations. The effluent concentrations match the standards for discharge on the sewage system. In some cases the effluent concentrations were even lower, so the reactor system might even be used in situations where will be discharged into surface water. The system has been demonstrated to be usable to the purify groundwater contaminated with volatile aromatics. The bioreactor described is technologically and economically competitive with conventional and comparable techniques.

ACKNOWLEDGMENTS

The fundamental research on microbiological processes, process conditions, and optimization of the reactor concept was conducted by the Microbiology and Environmental Technology sections of the Agricultural University of Wageningen.

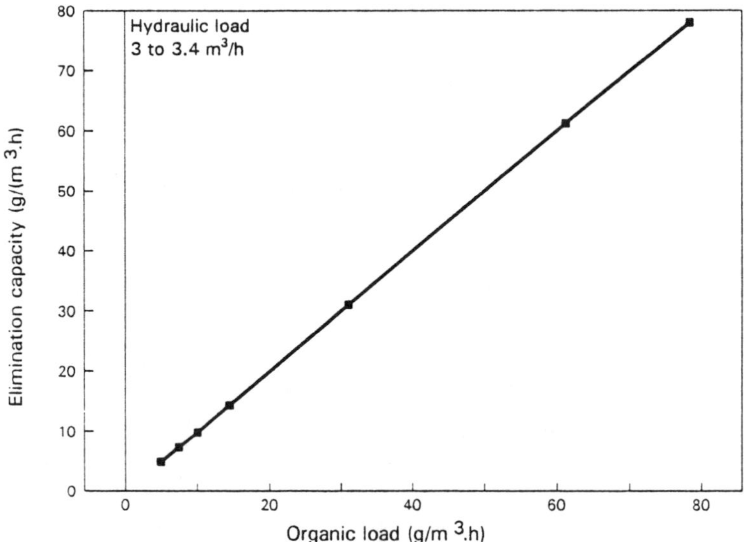

FIGURE 7. Elimination capacity vs. organic load (total contaminants) at a hydraulic load of 3 to 3.4 m^3/h as measured during the pilot investigation.

DHV Environment & Infrastructure executed the field work, customizing and testing the developed reactor under field conditions.

This project was cofunded by The Netherlands Agency for Energy and the Environment as part of the Environmental Technology Program, The Netherlands Integrated Soil Research Program, and the Institute for Inland Water Management and Waste Water Treatment.

REFERENCES

Bruin de, W. Kotterman, J. J., Posthumus, M. A., Schraa, G., and Zehnder, J. B. 1992a. "Complete Biological Reductive Transformation of Tetrachloroethene to Ethane." *Applied and Environmental Microbiology 58*(6):1996-2000.

Bruin de, W., and Schraa G. 1992b. "Reductieve dechlorering van tetrachlooretheen in een bioreactor." Landbouwuniversiteit, Wageningen, Nederland.

Schraa, G., W. Bruin de, J. Sluis van, and H. Rozema. 1989a. "De ontwikkeling van bioreactoren voor de reiniging van grondwater. Rapportage haalbaarheidsstudie (fase a.1)." DHV Milieu & Infrastructuur, Amersfoort, Nederland.

Schraa, G., W. Bruin de, J. Sluis van, and H. Rozema. 1989b. "De ontwikkeling van bioreactoren voor de reiniging van grondwater. Rapportage laboratoriumonderzoek en uitwerking reactorconcept (fase a.2)." DHV Milieu & Infrastructuur, Amersfoort, Nederland.

Verheul, J., and P. Doelman. 1992. "De ontwikkeling van bioreactoren voor de reiniging van grondwater. Rapportage veldproef." DHV Milieu & Infrastructuur, Amersfoort, Nederland.

Vis, P., W. Bruin de, A. Rinzema, and G. Schraa. 1992. "De ontwikkeling van bioreactoren voor de reiniging van grondwater. Biodegradatie van aromaten: laboratorium- en optimalisatieonderzoek." Landbouwuniversiteit, Wageningen, Nederland.

A VADOSE COLUMN TREATABILITY TEST FOR BIOVENTING APPLICATIONS

R. S. Baker, J. Ghaemghami, S. Simkins, and L. M. Mallory

ABSTRACT

A vadose column method designed to simulate the dynamics of air and water flow in the vadose zone under unsaturated conditions above the capillary fringe was developed to assess the treatability of material contaminated with volatile, biodegradable organic compounds under conditions representative of in situ bioventing. Preliminary testing using Plexiglas™ columns and gas chromatographic (GC) analyses indicated that losses due to volatilization and aqueous transport accounted for a range of 0.2 (ethylbenzene) to 28.1% (benzene) of added contaminant masses. For studies described here, the biotreatment apparatus was rebuilt using Pyrex™ columns, and additional data were obtained using [*U*-ring-^{14}C]toluene with gamma-irradiated sterile controls. More than 25% of the ^{14}C added as toluene was oxidized to $^{14}CO_2$ without any addition of inorganic nitrogen. A small addition (10 mg/kg) of combined N increased the amount of toluene mineralized to more than 30%. A larger addition of 40 mg N per kg of vadose material reduced the mineralized fraction of toluene to less than 10%. A loss of nearly 8% of added ^{14}C-toluene through outgassing occurred only from a column filled with subsoil initially sterilized by gamma irradiation. In all other cases, volatile losses were less than 0.15%.

INTRODUCTION

The use of airflow to remove volatile organic contaminants from soil and subsoil and the biodegradation of contaminants by indigenous microbiota are two attractive potentially low-cost alternatives for the remediation of contaminated vadose material. These two technologies often are combined and presented as a bioventing treatment technique (Dupont et al. 1991, Ely & Heffner 1988). It is highly desirable to be able to predict the successful treatment of a contaminated site prior to undertaking costly field studies. Conventional laboratory tests for organic contaminant biodegradation often use a fully mixed slurry treatment regime. These procedures dramatically change the physical, chemical, and biological properties of the contaminated environment. Slurry procedures often ignore loss of material by volatilization and may change the distribution of needed nutrients including the availability of oxygen as a terminal electron acceptor. The method presented here is designed to simulate the dynamics of air and water flow in the vadose zone under unsaturated conditions just above the capillary fringe.

A number of studies have used both unsaturated and saturated columns of contaminated subsurface soil as test material to investigate the biodegradation and volatilization of organic compounds including petroleum mixtures. Some of the more recent reports that focus on bioventing and soil vapor extraction include the work of Hosler et al. (1992), Miller et al. (1992), and Russell et al. (1992).

Biodegradation of petroleum compounds has been extensively studied. For such highly reduced organic substrates to be effectively mineralized (converted to CO_2 and H_2O), a number of conditions must be met. Temperature, pH, and salinity must be appropriate, the microorganisms capable of degradation must be present, the target compounds must be accessible, and adequate levels of a terminal electron acceptor and inorganic nutrients must be available. Degradation of simple aromatics, including benzene, toluene, ethylbenzene, and the xylenes (BTEX), is often thought to proceed at a most rapid rate when oxygen is used as the terminal electron acceptor (Gibson & Subramanian 1984). The more soluble components of petroleum normally are available as dissolved substrates. However, petroleum-contaminated material often has a very high concentration of carbon compared to other nutrients, particularly nitrogen. Aelion and Bradley (1991) demonstrated that an increase in available N enhanced the degradation rate of jet fuel by a microbial community active at a JP-4 spill site. Nitrogen limitation is not unique to the subsurface. Addition of N (and P) to heavily contaminated shoreline sediments apparently enhanced petroleum biodegradation following the grounding of the *T/V Exxon Valdez* (Lindstrom et al. 1991). In brief, suboptimal conditions, especially nutrient limitations, may serve to limit BTEX biodegradation (Leahy & Colwell 1990).

A primary goal of this study was to determine both the loss of target compound through vapor transport and the loss due to mineralization by active microbial metabolism. Preliminary tests (R. Baker, unpublished results) with Plexiglass columns™ and GC analyses indicated that losses due to volatilization and aqueous transport accounted for a range of 0.2 (ethylbenzene) to 28.1% (benzene) of added contaminant masses. Biodegradation might have accounted for the destruction of the balance of the added compounds. However, this could not be confirmed because microbial activity in the azide-poisoned control, although somewhat suppressed, was not effectively inhibited. Proof of biodegradation requires a loss of target compound directly attributable to microbial activity (Madsen et al. 1991).

We report here the results of a vadose column testing method designed to allow the continuous monitoring of vented ^{14}C-toluene and respired $^{14}CO_2$, and the application of this method to an examination of the effect of adding combined nitrogen on target compound loss. Results using vadose material collected from a BTEX-contaminated site are to be followed by field validation of the method using pilot-scale treatment data.

METHODS AND MATERIALS

Experimental Design

A ^{14}C tracer experimental design was used to monitor the fate of toluene, added in an aqueous solution of toluene, ethylbenzene, and *o*- and *p*-xylenes (TEX) to vadose material samples. These samples were maintained under conditions approximating those found in subsoil located above the capillary fringe. Respired $^{14}CO_2$ and ^{14}C-toluene lost through vapor transport were measured for 40 days following the addition of TEX. The system was studied under a low vacuum to prevent loss of target compound through leaks directly to the laboratory atmosphere. All liquid and gases were passed through traps to ensure a continuous accounting of added contaminant.

Vadose Zone Soil

Subsoil previously contaminated with TEX was a sandy loam in texture and was obtained from a depth of 4.5 m beneath the ground surface at an industrial solvent tank farm. Other relevant data for this subsoil include total organic carbon content [0.2% (w/w)], pH (6.5), saturated hydraulic conductivity ($1.2 \cdot 10^{-3}$ $cm \cdot s^{-1}$), and inorganic nitrogen content as NH_4^+-N (2.9) and NO_3^--N (2.0 $mg\text{-}N \cdot kg^{-1}$). The groundwater table at the site was encountered at a depth of approximately 5.2 to 5.8 m below grade. GC analyses of the freshly excavated vadose material revealed the presence of TEX but no benzene. The site is considered a suitable candidate for in situ treatment by bioventing, because the presence of operating industrial facilities precludes excavation.

Substantial losses of TEX were unavoidable during drying, sieving, and thorough mixing of the subsoil prior to packing. This pretreatment, however, did afford uniformity of initial subsoil conditions among the four columns. Use of undisturbed cores would have required far more experimental units to account for spatial heterogeneity.

Incubation Apparatus

Four Pyrex™ (Corning, Inc., Corning, New York) columns, 30.5 cm × 7.5 cm inner diameter (i.d.), were packed with 2 kg each of contaminated vadose material. The room housing the apparatus was maintained between 21 and 23°C. A humidified, organic-free airstream of 5 mL/min entered and exited the columns at depths of 5 and 25 cm, respectively. The air outlet was 5 cm above a porous plate in contact with a 102-cm hanging water column. In this way the soil moisture was equilibrated to a level corresponding to a matric suction of 0.1 bar. Tensiometers were placed opposite the air entry and exit ports. Small Tenax™ resin columns, capable of trapping nonpolar compounds, were placed at the bottom of the hanging water column. A pair of larger (10 mm i.d. × 80 mm) Tenax™-filled columns were connected in tandem at the air exit ports. After passing through the Tenax™ traps, exhaust gases were bubbled through a 1N NaOH solution to trap respired CO_2. Airflow was induced by a vacuum pump attached downstream of the carbon dioxide trap. Sorbed compounds were eluted from the Tenax™ traps with hexane. Aqueous subsamples from the NaOH traps and hexane extracts were assayed for radioactivity using Beckman liquid scintillation counter model LS3801 that was programmed to automatically correct for quenching.

Experimental Treatments

A single packed column was sterilized by exposure to 4 megarad gamma irradiation (Radiation Services, Univ. of Massachusetts, Lowell, Massachusetts). The remaining three columns were amended with 0, 10, or 40 mg-N·kg^{-1}, added as a mixture of NH_4NO_3 and $(NH_4)_2SO_4$ to provide ammoniacal and nitrate nitrogen at a ratio of 3:2 approximating the ratio of NH_4^+-N and NO_3^--N in the freshly excavated soil. Each of the columns was amended with 45 mL of water containing toluene, ethylbenzene, and *p*- and *o*- xylenes in a 6:2:1:1 ratio by weight. A total of 7.34 mg of ^{12}C-toluene was added to each column. Then, an additional 1.20 µg [*U*-ring-^{14}C]toluene (Sigma Chemical Corp, St. Louis, Missouri) having a specific activity of 53.5 mCi/mmole was added to each column. Radiolabeled toluene represented less then 0.02% of added toluene, thus allowing a true tracer experimental design.

RESULTS

Airflow through the columns was initiated 2 days after addition of ^{14}C-toluene and nonlabeled hydrocarbons to the soil. Prior to day 7, the vacuum pumps were operated only intermittently while the four systems were verified to be free of leaks from the outside. Continuous operation of the pumps began on day 7.

Evidence of Biodegradation

Counts from $^{14}CO_3^{2-}$ in samples withdrawn from the NaOH solutions on day 7 (Figure 1) revealed that substantial amounts of $^{14}CO_2$ were produced before large amounts of air passed through the columns. Evolution of $^{14}CO_2$ from the column containing sterilized subsoil was evidence that hydrocarbon-degrading microorganisms, possibly introduced with the in-flowing air, were able to colonize the soil in sufficient numbers for their activities to be detected. Nevertheless, a smaller amount of $^{14}CO_2$ was evolved from the column filled with sterilized vadose-zone subsoil than from any other column. The nonsterilized column that received no addition of combined nitrogen released over 4 times more $^{14}CO_2$ than did the otherwise identical sterilized column (Figure 1). An even larger amount of $^{14}CO_2$ was evolved from the column that was amended with 10 mg of combined nitrogen per kg of nonsterile soil (Figure 1). Far less mineralization occurred in the column that received 40 mg of combined N per kg of soil than in any of

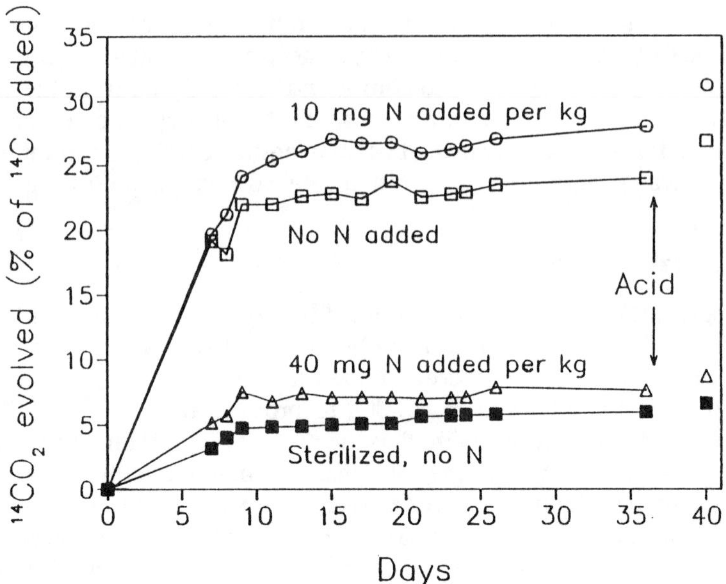

FIGURE 1. Cumulative amounts of CO_2 evolved from three nonsterile, vadose-zone soil columns amended with indicated amounts of combined nitrogen, and from one column containing soil initially sterilized by gamma irradiation. Vacuum pumps were turned off after 32 days of incubation, and the columns were flooded with 0.1 N HCl as shown by the arrows to recover $H^{14}CO_3^-$. After the acid was removed, the vacuum pumps were run for an additional 24 h prior to the final sampling (on day 40).

the other nonsterilized columns (Figure 1). The $^{14}CO_2$ evolved from this column was little in excess of that released from the sterilized control.

After 26 days of incubation, evolution of $^{14}CO_2$ from the columns appeared to have ceased (Figure 1). Because this subsoil had a nearly neutral pH, the columns were flooded with HCl (after sampling on day 36) to drive any dissolved bicarbonates into the gas phase. Although some additional counts were expelled from the columns by acidification (Figure 1), greater amounts of $^{14}CO_2$ were released before day 26 for all columns.

Volatilization

Nearly 8% of the ^{14}C-toluene added to the sterilized column (Figure 2, right axis) was recovered on the Tenax™ traps in the exhaust line. In contrast, no more than 0.13% of the ^{14}C added was ever recovered from any of the nonsterilized columns as untransformed ^{14}C-toluene in the exhaust gas stream (Figure 2, left axis). The column amended with 40 mg of N per kg lost about 2.5 times more ^{14}C as untransformed toluene in the exhaust gases than did either of the other two nonsterilized columns (Figure 2). Thus, an inverse relationship appeared

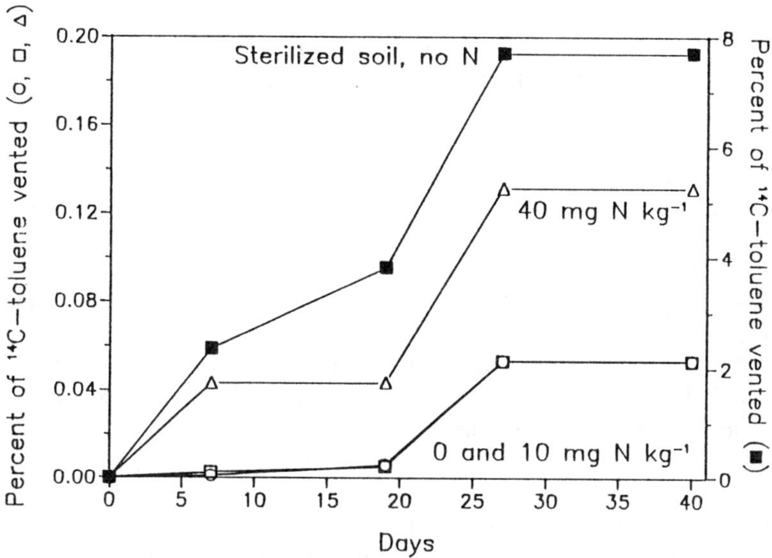

FIGURE 2. Cumulative percentages of added ^{14}C-toluene trapped on Tenax™ resin from gases exhausted from four vadose-zone soil columns. Nearly identical amounts of toluene were vented from nonsterile soil amended with 0 (open squares) and 10 (open circles) mg of combined nitrogen per kg of soil, as indicated by the superimposed curves.

to exist between the mineralized and outgassed amounts of ^{14}C-toluene. The two columns evolving the most $^{14}CO_2$ (Figure 1, open circles and squares) lost the smallest amounts of ^{14}C as vented toluene (Figure 2, same symbols). The sterilized column released the least $^{14}CO_2$ but the most ^{14}C-toluene. The more heavily fertilized (40 mg-N·kg^{-1}) column was intermediate in the amounts of ^{14}C-toluene lost through mineralization and outgassing.

Efforts to recover ^{14}C from the soil at the end of the study using extractions with methanol and hexane have so far yielded very few counts. Only about 1.0% of the ^{14}C added to the sterilized column has been recovered using these solvents. Identical amounts of ^{14}C (about 0.7% of added label) have been recovered from the columns fertilized with either 10 or 40 mg-N·kg^{-1}, and approximately 0.9% of the ^{14}C has been recovered from the soil that was neither fertilized nor sterilized. No ^{14}C was detected in the small volume of water draining from the hanging water column for any soil core.

DISCUSSION

It is clear from Figure 1 that microbial activity is responsible for the mineralization of toluene added to contaminated vadose samples. It is therefore reasonable to assume that mineralization may proceed in situ. Nutrients already available in the contaminated material allow for the mineralization of more than 26% of toluene. Additional amounts of inorganic nitrogen in the forms of NH_4^+ and NO_3^- increased mineralization by a factor of one-fifth, indicating that small additions of N may enhance total contaminant removal. However, N addition may not increase the maximum rate of mineralization, and further addition of moderate amounts of available nitrogen may, in fact, greatly inhibit mineralization.

The modest amount of mineralization observed in the abiotic control may be due to insufficient irradiation, to total oxidation of toluene by cell-free oxygenases, or to contamination of the column following irradiation. It is highly unlikely that the radiation treatment itself was insufficient. Although it is possible for a cell-free system to totally degrade a compound, it is unlikely to occur within 26 days. It is far more likely that the material was inoculated through handling subsequent to radiation treatment.

The apparent suppression of mineralization by the addition of 40 mg-N·kg^{-1} is unexpected but not unprecedented. Swindoll et al. (1988) found that the addition of either 136 ng of $(NH_4)_2SO_4$ or an inorganic nutrient mixture containing 2 ng of $NaNO_3$ inhibited toluene mineralization in slurries containing 1 g of dry aquifer solids. In addition, Morgan and Watkinson (1990) reported nearly 80% inhibition of naphthalene mineralization by an amendment of 5 mg of NH_4NO_3 per g of soil.

Although nitrate is known to be toxic to eucaryotes, including humans, the addition of less than 40 mg NO_3^--N per kg of vadose material should not have a toxic effect. Oligotrophy, with respect to organic carbon, is a well-known phenomenon in microbial ecology. High concentrations of available organic carbon (amounts in excess of 500 mg/L) are known to inhibit certain species. It is possible that bacteria native to subsoil may possess such traits in relationship to available nitrogen levels. It is also possible that the addition of 40 mg nitrogen per kg affected the fate of toluene but not its overall metabolism. Toluene C may have been used primarily for biomass. Results consistent with this hypothesis have been reported by Aelion and Bradley (1991). Clearly further work is needed to resolve this issue.

The results indicate that soil-vapor extraction, in the absence of biodegradation, would be a poor stratagem for contaminant removal from this soil. Significant loss attributable to venting occurred only when the sample had been sterilized. However, less than 4 mg of toluene was added per kg of vadose material. Thus, absorption and adsorption to organic and inorganic material may have prevented efficient partitioning to the vapor phase. This would seem to indicate that biological treatment may be the preferred remediation method when contaminants are present at low concentrations. Although the total amounts of toluene removed from the biotic treatments were low in all cases, more than twice as much parent compound was recovered when the column received 40 mg-N·kg^{-1}, indicating a possible inhibition of the initial step of toluene metabolism.

Bench-scale treatability studies, such as these, offer the advantage of defining remediation design parameters (e.g., nutrient levels and degradation rates) under highly controlled conditions. However, follow-on tests should be planned to augment laboratory data with results under actual field conditions.

REFERENCES

Aelion, C. M., and P. M. Bradley. 1991. "Aerobic Biodegradation Potential of Subsurface Microorganisms from a Jet Fuel-Contaminated Aquifer." *Appl. Environ. Microbiol. 57:* 57-63.

Dupont, R. R., W. Doucette, and R. E. Hinchee. 1991. "Assessment of In Situ Bioremediation Potential and the Application of Bioventing at a Fuel-Contaminated Site." In R. E. Hinchee and R. F. Olfenbuttel (Eds.), *On-Site Bioreclamation: Processes for Xenobiotic and Hydrocarbon Treatment*, pp. 262-282. Butterworth-Heinemann, Boston, MA.

Ely, D. L., and D. A. Heffner. 1988. "Process for In-situ Biodegradation of Hydrocarbon Contaminated Soil." U.S. Patent 4,765,902.

Gibson, D. T., and V. Subramanian. 1984. "Microbial Degradation of Aromatic Hydrocarbons." In D. T. Gibson (Ed.), *Microbial Degradation of Organic Compounds*, pp. 181-252. Marcel Dekker, Inc., New York, NY.

Hosler, K. R., T. L. Bulman, and R. M. Booth. 1992. "The Persistence and Fate of Aromatic Constituents of Heavy Oil Production Waste during Landfarming." In P. T. Kostecki, E. J. Calabrese, and M. Bonazountas (Eds.), *Hydrocarbon Contaminated Soils, Vol. II. Proc. 5th Annual Conference on Hydrocarbon Contaminated Soils*, pp. 591-609. Lewis Publishers, Ann Arbor, MI.

Leahy, J. G., and R. R. Colwell. 1990. "Microbial Degradation of Hydrocarbons in the Environment." *Microbiol. Rev. 54:* 305-315.

Lindstrom, J. E., R. C. Prince, J. C. Clark, M. J. Grossman, T. R. Yeager, J. F. Braddock, and E. J. Brown. 1991. "Microbial Populations and Hydrocarbon Biodegradation Potentials in Fertilized Shoreline Sediments Affected by the *T/V Exxon Valdez* Oil Spill." *Appl. Environ. Microbiol. 57:* 2514-2522.

Madsen, E. L., J. L. Sinclair, and W. C. Ghiorse. 1991. "In Situ Biodegradation: Microbiological Patterns in a Contaminated Aquifer." *Science 252:* 830-833.

Miller, M. E., T. A. Pedersen, C. A. Kaslick, G. E. Hoag, and C. Y. Fan. 1992. "Column Vapor Extraction Experiments on Gasoline Contaminated Soil." In P. T. Kostecki, E. J. Calabrese, and M. Bonazountas (Eds.), *Hydrocarbon Contaminated Soils, Vol II. Proc. 5th Annual Conference on Hydrocarbon Contaminated Soils*, pp. 437-449. Lewis Publishers, Ann Arbor, MI.

Morgan, P., and R. J. Watkinson. 1990. "Assessment of the Potential for *In-situ* Biotreatment of Hydrocarbon-Contaminated Soils." *Water Sci. Technol.* 22(6):63-68.

Russell, E., C. Ohland, and K. Adler. 1992. "In Situ Bioremediation of Petroleum Solvent Contaminated Subsurface Soils." Presented at the 7th Annual Conference on Hydrocarbon Contaminated Soils, held Sept. 21-24, 1992 at the Univ. of Mass., Amherst, MA.

Swindoll, C. M., C. M. Aelion, and F. C. Pfaender. 1988. "Influence of Inorganic and Organic Nutrients on Aerobic Biodegradation and on the Adaptation Response of Subsurface Microbial Communities." *Appl. Environ. Microbiol. 54:* 212-217.

COST EFFECTIVENESS AND FEASIBILITY COMPARISON OF BIOVENTING VS. CONVENTIONAL SOIL VENTING

H. J. Reisinger, E. F. Johnstone, and P. Hubbard, Jr.

ABSTRACT

Over the past 5 years, vacuum-enhanced soil venting has become one of the most efficient and frequently used tools to address fuel hydrocarbons in the vadose zone. The technique is a mass transfer process that removes the volatile compounds from the parent hydrocarbon mixture, vadose soils, and to a lesser extent those dissolved in groundwater, and transfers them to the atmosphere. In some areas, the extracted hydrocarbons may be discharged directly without treatment. In other areas it is necessary to treat the vacuum extraction system discharge prior to release to the atmosphere. Research over the past 5 years has demonstrated that it is efficacious to engineer a vadose zone remediation system that integrates conventional advective soil venting with in situ biodegradation. The advective venting aspect of the integrated system operates in much the same manner as described above. However, the process is modified to introduce oxygen to the vadose zone via either direct air injection or aspiration of air through engineered aspiration points. The product of this combination of technologies is removal of hydrocarbons via volatilization and mineralization of hydrocarbons by an indigenous microbial consortium. In many instances it is possible to minimize or eliminate the need for off-gas treatment prior to discharge to the atmosphere using bioventing techniques. This paper presents the results of a cost effectiveness and feasibility comparison that considers the theoretical and technical justification of the two remediation approaches. The advantages, limitations, and controlling mechanisms are assessed. Cost information for operation of the two systems is presented and compared based on a generalized site configuration. Comparative case histories are then used to support the conclusions of the hypothetical comparison. Results of the comparisons indicate that in some instances conventional venting is the option of choice. However, it is clear that when it is necessary to treat off-gas, bioventing is the more cost-effective approach.

INTRODUCTION

Releases from aboveground and underground petroleum fuel-holding and handling systems have the potential to create environmental impacts that require time and capital-intensive remediation efforts. The U.S. Environmental Protection Agency (U.S. EPA) has estimated that there are approximately 2 million underground storage tank (UST) systems at 700,000 locations across the nation, and that about 25% of them fail tightness tests and may be leaking (U.S. EPA 1988). There is no estimate of the number of releases from aboveground storage and handling systems, but it is safe to assume that such systems do, from time to time, produce releases. Following release of liquid-phase hydrocarbon from a storage or handling system, the hydrocarbon mixture migrates vertically through unsaturated soil under the influence of gravity and capillary forces. As vertical migration through the vadose zone progresses, the liquid mixture interacts with the soil through which it is passing. Some of the liquid is retained in the soil by either occupying pore spaces, adsorbing to the solid matrix, or partitioning into the associated organic carbon, some volatilizes to form a vapor phase, and some dissolves into the vadose water. When the released volume is large enough, liquid-phase hydrocarbon will reach the water table where it will accumulate and float because it is less dense than water. A portion of the liquid will dissolve in the groundwater and migrate with it.

In a hypothetical setting that assumed a 3,785-L release from a gasoline UST, Hinchee et al. (1987) estimated that 49 L of the release would partition into the groundwater as a dissolved phase, 3,642 L would partition into the soil as either pore-filled liquid or adsorbed phase, and 95 L would partition into the soil atmosphere as a vapor phase. It is therefore apparent that the soil compartment serves as a significant secondary source of hydrocarbon that needs to be addressed as a part of any overall site remediation. As such, a number of approaches have been developed to address hydrocarbon in the vadose zone. Among these are soil venting or soil vapor extraction and bioventing. These serve as the focus of the feasibility and cost comparison that follows.

TECHNOLOGY DESCRIPTION

Soil Venting

Soil venting is a vadose zone remediation technique that takes advantage of the volatile character of individual compounds and mixtures of hydrocarbon compounds released to the subsurface. Volatile compounds evaporate from the phase-separated fraction and from the adsorbed fraction to form an equilibrium in the soil atmosphere. The equilibrium established as a result of this process can be described by Raoult's Law as modified to include component interactions:

$$P_i = P_i' * X_i \qquad (1)$$

where P_i = partial pressure of volatile component I over a liquid mixture
P_i' = vapor pressure of pure component i
X_i = mole fraction in parent liquid

Once in the soil atmosphere, the volatilized compounds remain at or near equilibrium until the system is upset. In nature, the equilibrium is upset as a result of molecular diffusion as described by Fick's Law:

$$\delta m/\delta t = -D(\delta^2 C/\delta z^2) \tag{2}$$

where $\delta m/\delta t$ = mass flux through a unit area
D = the gaseous diffusion coefficient
C = constituent concentration
z = distance

Under a diffusion-based scenario, the organic molecules migrate from areas of high concentration along a concentration gradient. As the migration takes place, volatilization continues to maintain the equilibrium. Disruption of this equilibrium then becomes the primary remediation mechanism in soil venting.

In soil venting the equilibrium is disrupted by applying a vacuum to the subsurface. Application of the vacuum disrupts the equilibrium by establishing advective airflow through the vadose zone. This then ideally causes the rate of volatilization to increase to a point that the system becomes diffusion limited. The increase in volatilization is a direct result of the decrease in pressure in the subsurface. This relationship can be shown through an examination of the effect of ambient pressure on the vapor-phase diffusion coefficient. Ehlers et al. (1969) show that vapor-phase diffusion is inversely proportional to ambient pressure according to the following relationship:

$$D = D_s + D_v \ (P_o/P) \tag{3}$$

where D = actual diffusion coefficient in soil
D_s = nonvapor-phase diffusion coefficient
D_v = apparent vapor-phase diffusion coefficient
P_o = reference pressure (standard atmospheric)

Thus, as the pressure in the subsurface is decreased as a result of application of a vacuum, the rate of volatilization increases. The rate of volatilization is further increased as a result of the movement of air through the subsurface, which is also a result of the venting process. As air moves through the contaminated soil, it removes the hydrocarbon that has volatilized thereby reducing the concentration and increasing the rate of volatilization. In this case, the system can be expressed as a source renewal model. The rate of hydrocarbon mass removal can be expressed mathematically per the following expression (Johnson et al. 1990):

$$\frac{\delta M_i}{\delta t} = Q\,C_i - B_i \qquad (4)$$

where M_i = total number of moles of component in soil
Q = volumetric flow rate
C_i = molar concentration of component entering extraction well
B_i = degradation of component

Off-Gas Treatment

Regulatory movement toward increasingly stringent atmospheric discharge standards has made treatment of off-gas from soil venting systems necessary in many parts of the United States. Early soil venting systems often discharged extracted vapors directly to the atmosphere. This, of course, simply transferred the hydrocarbon extracted from the soil compartment to the atmospheric compartment. Direct discharge is, therefore, becoming less common and the off-gas is being treated prior to discharge. Treatment technologies most commonly employed include thermal incineration, catalytic oxidation, granular activated carbon adsorption, oxidation via internal combustion engine, and biofiltration.

Thermal incineration employs heat, generally at a temperature of about 540 to 750°C to oxidize hydrocarbon. In this process, the hydrocarbon is converted to carbon dioxide and water vapor. The extracted hydrocarbon is oxidized either via incineration in a flame or in a combustion chamber. When the hydrocarbon concentration is sufficiently high, the majority of the energy to support combustion is derived directly. When the concentrations are lower, it is necessary to supplement the hydrocarbon with a secondary fuel source, generally propane or natural gas. Thermal incineration is most appropriate for process streams that contain higher hydrocarbon concentrations (Figure 1).

Catalytic oxidation also employs heat to destroy the influent hydrocarbons. These systems generally operate at a temperature range of 300 to 500°C. The off-gas stream is passed through a preheater into a catalyst chamber where the oxidation takes place. The catalyst chamber contains noble metal catalyst (palladium or platinum). The hydrocarbon adsorbs to the catalyst surface where oxidation takes place in the presence of thermal energy. These systems typically are most commonly used for mid-range off-gas streams (Figure 1). It is usually necessary for a supplemental energy source in the form of electricity, propane, or natural gas to be used. Newer systems that combine thermal oxidation and catalytic combustion are also coming into more common use.

Vapor-phase granular activated carbon (GAC) also is used to treat off-gas. In this process, the hydrocarbon is passed through a bed of activated carbon where it is removed from the process stream via adsorption. As such, the process is simply a mass transfer operation in which the hydrocarbon is moved from the subsurface to the carbon. It is then necessary to either regenerate the carbon,

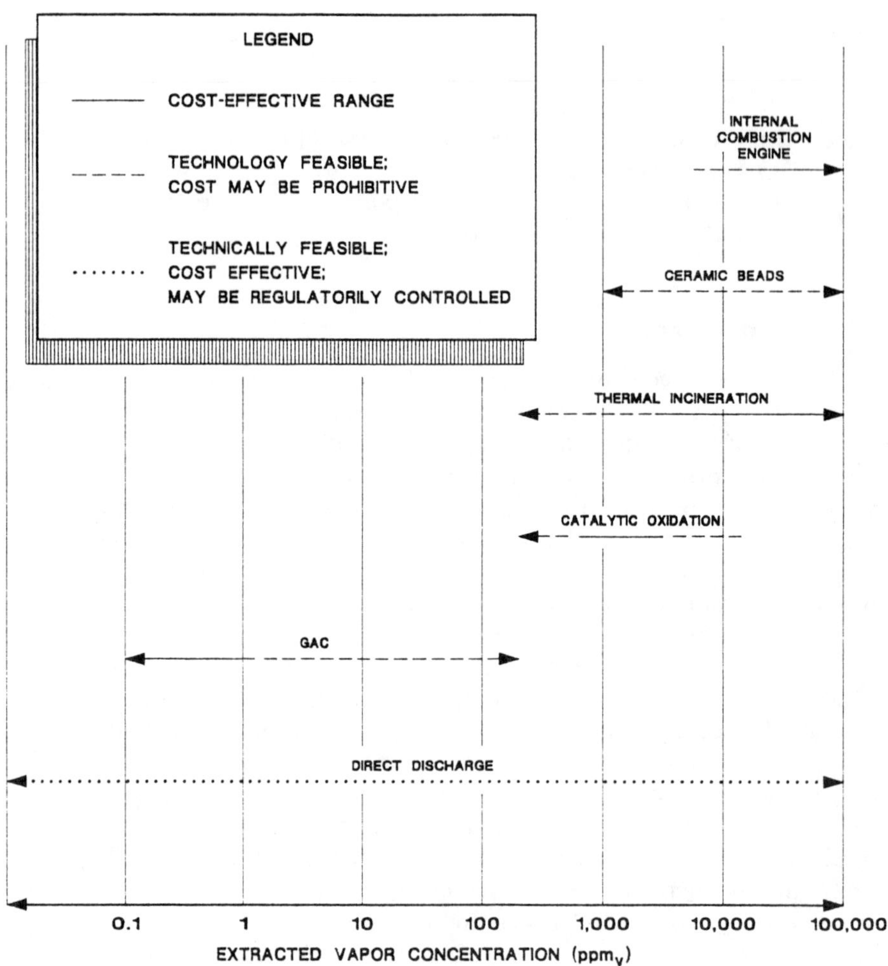

Modified from: Pedersen and Curtis (1991).

FIGURE 1. Extracted vapor concentration (ppm_v).

reactivate it, or dispose of it in a landfill or incinerate it. Vapor-phase granular activated carbon adsorption is most commonly used in lower concentration off-gas streams (Figure 1).

An internal combustion engine also can be used to treat venting system off-gas. In this process, the engine (typically a standard commercially available gasoline engine) is modified to accept fuel in a gaseous rather than a liquid form. The off-gas from the venting system is injected with a supplemental fuel such as propane. The extracted hydrocarbons are then converted to carbon dioxide and water vapor. These units typically are equipped with a catalytic converter on the exhaust to further reduce emissions (Figure 1).

Aboveground biological filters also are used to treat off-gas from soil venting systems. This approach to off-gas treatment is relatively new to the United States. It has, however, been used in Western Europe for several years (Diks & Ottengraf 1991). In the process, the off-gas from the venting system is channelled through an aboveground biofilter. The filter can take a number of configurations and orientations, but generally is packed with medium that is used to support the growth of a biological consortium capable of degrading the compounds of interest in the influent gas. As such, the filter is essentially a fixed-film bioreactor. A variety of packings including soil, sand, compost, and peat have been used. The hydrocarbon compounds introduced into the reactor interact with the packing and the water contained therein, and provided there is adequate retention time, biodegradation occurs. Douglas et al. (1991) established removal rates in excess of 90% for selected petroleum hydrocarbons in such a bioreactor.

Bioventing

Bioventing is an in situ remediation technique that combines conventional advective soil venting with biodegradation. The soil-venting phase of the process takes advantage of the highly volatile nature of the components that comprise the hydrocarbon mixture. Biodegradation is capable of addressing the less volatile high-molecular-weight hydrocarbon compounds, as well as the more volatile organics, removed during conventional venting. Integrating these two processes allows a more diverse range of fuel products to be remediated in situ.

In any soil system, there is a naturally occurring consortium of microbiological organisms comprised of bacteria, fungi, protozoans, and viruses. The distribution of organisms in the overall population is controlled by the nature of the available energy sources and by a variety of other environmental factors (i.e., moisture content, pH, macronutrients, temperature, and terminal electron acceptor). Those organisms with the greatest ability to use the most abundant energy source and to best adapt to the environment will dominate the consortium in terms of both numbers and biomass. This is not to say that the consortium will become a single-species microcosm, instead, the suites of organisms best adapted to the environment and tolerant of each other will dominate. Other species will coexist in the consortium, but at lower numbers and biomass. As the soil environment changes, the species dominance in the microbiological consortium changes.

The numbers and biomass of organisms in a consortium are controlled by the availability of the primary food or energy source, the supplemental macronutrients, (i.e., nitrogen, phosphorus, and potassium) and the presence of a terminal electron acceptor (i.e., oxygen in aerobic systems). The population will continue to grow until an essential nutrient or other factor becomes limiting. The population at this point is then considered to be at its carrying capacity. It will maintain at this point until conditions change, i.e., an increase or decrease in the food source, nutrient, or electron acceptor, or addition of another organism that can compete more efficiently in the new environment.

At a hydrocarbon release site, organisms with the ability to use the released fuel as an energy source generally are present as an element of the indigenous microbial consortium. Prior to the introduction of the hydrocarbon, they generally are present as a relatively minor component of the total population. After the hydrocarbon is introduced, the degrader population grows until another factor becomes limiting. Because the organisms that degrade hydrocarbon most efficiently are aerobes (i.e., use oxygen as the terminal electron acceptor in cellular respiration), oxygen generally becomes the limiting factor. Thus, an increase in the oxygen content in the subsurface increases the population of hydrocarbon degraders and thus, the rate of hydrocarbon removal.

Indigenous hydrocarbon degraders have been shown to have the capacity to mineralize hydrocarbon fuels ranging from gasoline (light refined fuel) through diesel and heavier fuel oils (heavy refined fuel). The following stoichiometric relationships represent the aerobic biodegradation (mineralization) of hydrocarbon fuels.

$$C_6H_6 + 7.5O_2 \rightarrow 6CO_2 + 3H_2O \quad \text{benzene (gasoline)} \tag{5}$$

$$C_6H_{14} + 9.5O_2 \rightarrow 6CO_2 + 7H_2O \quad \text{hexane (JP-4)} \tag{6}$$

$$C_{10}H_{22} + 15.5O_2 \rightarrow 10CO_2 + 11H_2O \quad \text{n-decane (diesel)} \tag{7}$$

It has been shown in numerous studies that in natural systems, macronutrients rarely are limiting (Miller et al. 1991). In situ biodegradation of hydrocarbon in the vadose zone, thus, becomes an efficient and cost-effective remediation tool. Because oxygen generally is the limiting factor, the primary objective in the bioremediation design is to supply oxygen at a rate that shifts the system to a point that hydrocarbon becomes the limiting factor.

For the past two decades, researchers have evaluated a variety of means by which oxygen could be made available to the microbial consortium. Most of these have involved introduction of oxygen in liquid form (i.e., peroxide). Introduction and distribution of oxygen in liquid form is limited by the nature of the solid matrix. The primary controlling factor is hydraulic conductivity. A secondary controlling factor is heterogeneities in the solid matrix. Liquid introduced to the subsurface migrates, preferentially, into the more permeable zones, thereby creating an uneven distribution. When this occurs, oxygen may not be available to organisms in all of the areas that contain hydrocarbon. When water is used as the oxygen-carrying vehicle, the solubility of oxygen in water is the limiting factor for biodegradation. If pure oxygen is used and a concentration of 40 mg/L is achieved, approximately 36,287 kg (35,961 L) of water must be delivered to mineralize 0.45 kg of hydrocarbon. If 500 mg/L hydrogen peroxide is successfully delivered to the hydrocarbon source, 5,896 kg (5,678 L) of water must be delivered to degrade 0.45 kg of hydrocarbon. It is thus apparent that, if oxygen is provided by using an aqueous vehicle, significant volumes of liquid must be delivered. These relationships, of course, assume that all of the oxygen is available for hydrocarbon mineralization. This is not, however, the case. It has been shown in field studies that a significant portion of the oxygen injected in aqueous form is used to satisfy the natural soil oxygen demand, which further increases the volume of water that must be injected. An additional factor that is particularly problematic in vadose remediation is that application of liquid precludes the option of venting as a result of pore flooding. It is obvious that it is impossible to pass air through soil pores that are filled with water. An alternative to introduction of oxygen in liquid form is introduction in gaseous form. Because ambient air is 21% oxygen and is available at virtually no cost, air is an excellent cost-effective oxygen source. Not only does air provide oxygen more efficiently and cost effectively, but it also can be introduced and distributed more efficiently. The ease of air introduction and distribution is a function of its higher diffusivity (up to 4 orders of magnitude) and lower coefficient of friction relative to water.

It is, thus, apparent that both venting and in situ biodegradation are viable means of remediating hydrocarbon-impacted sites. The approach discussed herein integrates conventional venting with in situ bioremediation (bioventing) to address residual hydrocarbon in the vadose zone. In a bioventing system, air is injected into the subsurface to provide oxygen for the indigenous hydrocarbon degrader consortium. It is this process that drives the in situ biodegradation process. Application of a negative pressure results in direct removal of hydrocarbon through volatilization and advective airflow through the vadose. In bioventing, the primary remediation mode is biodegradation and, thus, advective venting is essentially a minor process component. Application of the vacuum does, however, serve to distribute the injected oxygen more evenly. By engineering a system to favor in situ biodegradation, the hydrocarbon atmospheric mass loading can be minimized.

FACTORS AFFECTING TECHNOLOGY APPLICATION (FEASIBILITY)

The feasibility of applying any approach to remediation is governed by a variety of biological, chemical, and physical factors that must be considered as a part of the selection process. The chemical factors for the most part revolve around the composition of the hydrocarbon mixture and the manner in which they interact with the subsurface matrix into which they are introduced. The physicochemical character of the solid matrix also must be considered because it serves to govern the extent to which and the manner in which the introduced hydrocarbon interacts with the solid matrix. The physical character of the solid matrix also must be considered as it controls not only the subsurface movement of the hydrocarbon, but the extent to which amendments added can be distributed.

Soil Venting

Chemical Considerations. The physicochemical character of the contaminant that is the subject of the remediation is of paramount importance in selecting a remediation approach. In the case of soil venting, the factor that drives efficacy to the greatest extent is vapor pressure, which indicates the tendency of a compound to partition into the vapor state. Because venting removes contaminants from the vapor state and increases the rate of volatilization from the soil residual and nonaqueous phases, compounds with higher vapor pressures are better candidates for venting than compounds with lower vapor pressures. Table 1 contains a listing of major gasoline constituents along with important physicochemical parameters. Dragun (1988) states that compounds with vapor pressures less than 1×10^{-7} mm Hg would tend not to volatilize to any great extent. Bennedsen et al. (1985) suggest that compounds with a vapor pressure greater than 0.5 mm Hg are good candidates for vapor extraction. Thus, soil venting could be applied successfully to address the more volatile compounds in gasoline, but would not be an appropriate technology for remediation of the less volatile gasoline fraction, diesel fuel, kerosene, or heavier hydrocarbon mixtures.

Aqueous solubility, a measure of a compound's tendency to partition into the aqueous phase of a subsurface system, also plays an important role in determining the feasibility of soil venting. Compounds that are very soluble in water tend to partition into groundwater and are not good venting candidates. Examples of such compounds in gasoline include methyl tertiary butyl ether (MTBE), the lead scavengers ethylene di-bromide (EDB) and ethylene di-chloride (DCE), and alcohols. The presence of water can effectively reduce a compound's tendency to volatilize. Thus, compounds that are more highly soluble in water relative to their vapor pressures tend to be effectively less volatile and become less well suited for soil venting.

The tendency of a compound to adsorb to the solid matrix into which it has been introduced also has bearing on its suitability for soil venting. Hydrocarbon compounds adsorb to soil in proportion to their octanol/water or organic carbon partitioning coefficients. This can be expressed as the soil partitioning coefficient (Table 1). The tendency to adsorb is roughly proportional to molecular weight, and the higher the molecular weight, the greater the potential to adsorb, and the lower the feasibility for venting. Just as vapor pressure and aqueous solubility interact to control partitioning into the vapor phase, molecular weight and adsorption interact to further control partitioning. Hydrocarbon adsorbed or present in the organic carbon in the soil is less likely to partition into the vapor phase. However, aqueous solubility also plays a role in this interaction. Because in many cases water has a greater affinity for soil, adsorbed hydrocarbon can be desorbed or displaced by water. This then can create a pulse of vapor-phase hydrocarbon as a result of introduction of water into the soil system.

It is, therefore, apparent that those compounds that have higher vapor pressures and a lower tendency to adsorb to the solid matrix, are better candidates for soil venting. Therefore, for most hydrocarbon mixtures, conventional soil venting addresses but a fraction of the total mass, i.e., the more volatile fraction.

Physical Considerations. A number of physical factors must be considered in determining the feasibility of applying soil venting at a site. Among these are factors dealing with the nature of the site proper or with the solid matrix beneath the site. Foremost among the site factors is the depth to groundwater, which determines the thickness of the vadose zone. If the vadose zone is very

TABLE 1. Chemical parameters of environmental interest for the major component in gasoline.

		Component	Oct./Water Part. Coef.	Solubility (mg/L)	Vapor Pressure (mmHg)	Henry's Law Const. (atm-m^3/mole)
*	**n-paraffins**	n-butane	2.86	61.0	1,520[a]	1.90
		n-pentane	3.40	26.0	430	1.05
		n-hexane	3.94	9.5	120	1.44
*	**isoparaffins**	isobutane	2.73	49.0	2356	3.7
		isopentane	2.30	48.0	570	1.1
		2-methylpentane	3.81	13.0	555	1.5
		3-methylpentane	3.81	13.0	160	1.4
		2,3-dimethylbutane	3.68	18.0	198	1.2
		2-methylhexane	4.35	5.6	54	1.3
		3-methylhexane	4.35	5.6	54	1.3
		2,3-dimethylpentane	4.22	5.8	106	2.4
		2,4-dimethylpentane	4.22	5.8	106	2.4
		2,2,4-trimethylpentane	4.76	0.56	41	11.0
		2,3,4-trimethylpentane	4.76	0.56	41	11.0
		2,3,3-trimethylpentane	4.76	0.56	41	11.0
		2,2,3-trimethylpentane	4.76	0.56	41	11.0
		2,2,5-trimethylpentane	4.76	0.56	41	11.0
		2-methyloctane	5.43	0.24	7.6	5.3
		3-methyloctane	5.43	0.24	7.6	5.3
		4-methyloctane	5.43	0.24	7.6	5.3
*	**cycloparaffins**	methylcyclohexane	3.45	14.0	114	1.0
		1,cis,3	3.45	38.0	106	0.36
		1,trans,3	3.45	38.0	106	0.36
		cyclopentane	2.85	161.0	266	0.15

TABLE 1. (continued)

	Component	Oct./Water Part. Coef.	Solubility (mg/L)	Vapor Pressure (mmHg)	Henry's Law Const. (atm-m3/mole)
	methylcyclopentane	2.35	43.0	122	0.32
	propylene	1.31	200.0	7,752	2.1
	2,trans-butene	1.85	78.5	1,520	1.4
	2,cis-butene	1.85	78.5	1,368	1.3
	1-pentene	2.20	202.0	160	0.25
	2,trans-pentene	2.20	203.0	410	0.19
	2,cis-pentene	2.20	203.0	410	0.19
	2-methyl-1-pentene	2.80	78.0	182	0.26
	2-methyl-2-pentene	2.80	78.0	182	0.26
* **aromatics**	benzene	2.13	1,780.0	76	0.0044
	toluene	2.69	515.0	22	0.0052
	ethylbenzene	3.15	152.0	6.8	0.006
	o-xylene	3.12	175.0	5.3	0.004
	m-xylene	3.20	187.0	6.1	0.005
	p-xylene	3.15	198.0	6.8	0.005
	1-methyl,3-ethylbenzene	3.6	57.0	2.3	0.006
	1-methyl,4-ethylbenzene	3.6	57.0	2.3	0.006
	1,2,4-trimethylbenzene	3.65	57.0	2.3	0.006
	naphthalene	3.30	30.0	2.3	0.0013
* **miscellaneous**	ethylene dibromide	1.94	4,310 [b]	10	0.00061
	ethylene dichloride	1.66	8,690.0	61	0.00091
	isopropyl ether	1.98	9,000.0	129	0.0019
	methyl-t-butyl ether	1.20	48,000.0	204	0.00049
	tetraethyllead	------	0.0025	0.15	0.079

NOTES: (a) Estimated based on 25 ° C vapor pressure.
(b) Estimated 20° C solubility equal to 30° solubility.

thin, it may be difficult, if not impossible to apply soil venting to a site as a result of water table upconing when a vacuum is applied. The type of surface at the site also may have some bearing on the feasibility of soil venting. If a site is totally uncovered and the depth to groundwater is very shallow, it may be impossible to apply venting without applying a surface seal to preclude short circuiting. The lack of a competent surface seal also may allow infiltration of precipitation and irrigation water, thereby reducing the effectiveness of soil venting.

Although the physical site factors have bearing on the feasibility of successfully applying soil venting at a site, the character of the vadose zone has the potential to exert an even greater influence. Because soil venting depends on movement of air and vapor-phase hydrocarbon through the impacted porous medium, the ability of the soil matrix to transmit these fluids is of primary importance. The permeability of any material describes this ability. This physical characteristic, to a great extent, is a function of the grain size and sorting of the matrix. Soils comprised predominately of larger particles of uniform size generally have a larger permeability and are, therefore, capable of transmitting larger volumes of fluids. Conversely, soils comprised of smaller particles or poorly sorted soils transmit smaller volumes of fluids due to a reduction in the volume of interconnected pores. Poorly sorted soils or those comprised of smaller particles also have higher tortuosities. That is, the path that a molecule of hydrocarbon needs to take to be transported a unit distance is greater in finer-grained soils than in more coarse-grained soils. Although finer-grained soils generally have lower vapor-phase permeabilities, they sometimes can transport large volumes of fluids through preferential flowpaths. While this means of transport sometimes is viewed as short-circuiting, these flowpaths likely are the same routes that the hydrocarbon has taken as it migrated through the vadose zone. Therefore, it is not always valid to discount venting in a fine-grained soil.

Soil moisture content can have a profound impact on the feasibility of venting at a site. Moisture exerts its influence in a number of ways. It fills soil pores, thereby reducing the permeability. It is not possible to move air and vapors through pores that are filled with water. Water in the vadose also reduces the concentration of the vapor-phase hydrocarbon through solubilization. Chiou and Shoup (1985) reported that hydrocarbons are adsorbed to a greater extent in soils with lower moisture content. Addition of moisture causes the adsorbed hydrocarbon to be displaced by water molecules, thereby creating a pulse in the vapor-phase hydrocarbon concentration. This, of course, is a transient phenomenon and the water soon reduces the effectiveness of venting as a result of a reduction in permeability as discussed above.

Off-Gas Treatment. Because venting is a mass transfer process that moves hydrocarbon from the subsurface to the atmosphere, and a greater environmental focus is on air quality, treatment of off-gas from venting systems is being required more and more frequently. The cost for off-gas treatment can be a significant part of the overall cost of soil venting and can, in some instances, be the most expensive aspect of the process.

Bioventing

Chemical Considerations. Because bioventing has elements of both advective venting and biodegradation, a number of the chemical considerations that impact the feasibility of venting also impact the feasibility of bioventing. Vapor pressure exerts some impact on the efficacy of bioventing. However, whereas soil venting is nearly totally dependent on the volatility of compounds, bioventing addresses all biodegradable compounds in the target hydrocarbon mixture. Thus, bioventing is efficacious for a larger range of compounds than is conventional venting. This results in removal of a larger fraction of the total hydrocarbon mass including the less volatile, higher-molecular-weight compounds. Solubility of hydrocarbons in water in the vadose zone has less impact on the feasibility of bioventing than on conventional venting. In fact, it is necessary for there to be some moisture in the vadose zone to support biodegradation. The organisms that degrade hydrocarbon reside in the vadose water. The tendency for hydrocarbons to adsorb to the solid matrix also has less impact on the feasibility of bioventing.

Physical Considerations. The physical character of the site impacts bioventing in much the same way that it impacts soil venting. A shallow water table makes application of bioventing less feasible in that it is more difficult to move gases and vapors. The absence of a surface seal does not however, adversely affect the feasibility of bioventing to the same extent as its effect on conventional venting. Because the primary objective in bioventing is to introduce oxygen to the subsurface, aspiration of ambient air through the surface is not a problem.

The nature of the solid matrix impacts bioventing in much the same way it impacts conventional venting. Vapor-phase permeability is the primary controlling factor in the process. For bioventing to be applied successfully, it is necessary to distribute oxygen by moving air through the soil. Thus, it is necessary for the solid matrix to have permeabilities that allow for sufficient air movement.

Macronutrient (i.e., nitrogen, phosphorus, and potassium) availability is a factor unique to bioventing. These elements are essential components of cellular respiration and cell growth and must be present in adequate concentrations for biological processes to proceed. In essence, the objective of bioventing is to develop a system in which hydrocarbon is the limiting factor. Therefore, it is desirable for the macronutrients and the terminal electron acceptor to be present in concentrations in excess of those required to support the microbial population. While it is necessary to consider nutrient augmentation as part of a bioventing feasibility assessment, Miller et al. (1991) have found that moisture and nutrient augmentation typically are not limiting. A comparison of feasibility issues for conventional venting and bioventing is given in Table 2.

COMPARISON OF COSTS

The costs for site remediation using the approaches discussed above vary greatly as a function of site configuration, size of contaminated area, contaminant concentration, and the need for off-gas treatment. To objectively compare soil

TABLE 2. **Soil venting vs. bioventing feasibility comparison.**

	SOIL VENTING	BIOVENTING
State of Technology	Proven and in common use	Newer technology; applications in progress
Applicability	Volatile compounds	All biodegradable compounds
Controlling Factors	Permeability; moisture content; depth to groundwater; depth to bedrock; contaminant characteristics; site configuration; off-gas treatment requirements	Permeability; moisture content; depth to groundwater; depth to bedrock; site configuration
Cost Range	Low without off-gas treatment; high with off-gas treatment	Moderate
Advantages	Simple to engineer; low installation and monitoring cost; rapid action	Addresses full range of petroleum hydrocarbons; moderate cost when compared to venting without off-gas treatment, low cost when compared to venting with off-gas treatment; contaminants are totally mineralized - no further treatment is required
Disadvantages	Addresses only volatile compounds; high cost with off-gas treatment; short-circuiting can adversely impact; contaminants are transferred from soil to atmosphere without off-gas treatment	Not fully proven; may require macronutrient and moisture augmentation
Technical Complexity	Low without off-gas treatment; High with off-gas treatment	Low to moderate

venting without off-gas treatment, soil venting with off-gas treatment, and bioventing, a generalized hypothetical site and two case histories have been prepared. The generalized site is assumed to be in an eastern Piedmont setting with soils comprised of sandy silts that are contaminated with weathered gasoline. The contaminated area covers 6,000 m^2, has a volume of 7,200 m^3, holds a hydrocarbon mass of 82,200 kg, and has been shown through pilot testing to have an average in situ biodegradation rate of 17 mg/kg-day. The total flow through a conceptual venting system is 9.12 m^3/min, and the atmospheric off-gas loading is 317 kg/day. Given the parameters outlined above, the cost estimates for soil venting, soil venting with off-gas treatment, and bioventing in Table 3 were prepared. Results of this analysis show that for a one year period of operation soil venting is 62% less expensive than soil venting with off-gas treatment and 12% less expensive than bioventing. Bioventing is 56% less expensive than soil venting with off-gas treatment. The estimated times for cleanup are 400 days for venting and venting with off-gas treatment and 500 days for bioventing. Although bioventing will require 20% longer than venting with off-gas treatment to complete the cleanup, the overall cost for bioventing is considerably lower.

Site No. 1 is in the coastal plain physiographic province and has soils comprised of sands with small amounts of silt that have become contaminated with gasoline and diesel fuel. The contaminated area covers about 2,090 m^2, has a volume of 3,832 m^3, contains a hydrocarbon mass of 10,100 kg, and has been shown through pilot testing to have an average in situ biodegradation rate of 13.6 mg/kg-day. The total flow through a conceptual venting system designed for the site is 4.25 m^3/min, and the atmospheric off-gas loading is 855 kg/day. Given the parameters outlined above, cost estimates were prepared for soil venting, soil venting with off-gas treatment, and bioventing, as shown in Table 3. Results of this analysis show that soil venting is 57% less expensive than soil venting with off-gas treatment and 4% less expensive than bioventing. Bioventing is 55% less expensive than soil venting with off-gas treatment. The estimated times for cleanup are 200 days for venting and venting with off-gas treatment and 273 days for bioventing. The cleanup estimates for the venting portion of this site likely are optimistic, given that a portion of the contamination is diesel fuel which will not be addressed to any great extent by venting alone. It is likely that the ultimate overall costs for both of the venting options will increase given the longer time to complete the cleanup. Even with longer application times, venting will not adequately address the heavier petroleum fractions.

The second site history is a typical service station site situated in the Piedmont physiographic province. The soils at site No. 2 are primarily silts with some minor sand components. Most of the hydrocarbon, which is weathered gasoline, is in the saprolite, which is a sandy gravel. The contaminated area covers 232 m^2, has a volume of 708 m^3, contains a hydrocarbon mass of 917 kg, and has been shown through pilot testing to have an average in situ biodegradation rate of 8 mg/kg-day. The total flow through the conceptual venting system is 1.02 m^3/min, and the atmospheric off-gas loading is 380 kg/day. Given the parameters outlined above, cost estimates were prepared for soil venting, soil venting with off-gas treatment, and bioventing as shown in Table 3. Results of this analysis show

TABLE 3. Soil venting vs. bioventing cost comparison.

	Soil Venting Direct Off-Gas Discharge			Soil Venting With Off-Gas Treatment			Bioventing		
	General Site	Site #1	Site #2	General Site	Site #1	Site #2	General Site	Site #1	Site #2
Pilot Testing	**7,500**	**12,500**	**6,000**	**10,000**	**15,000**	**7,000**	**10,000**	**13,000**	**6,500**
Design/Permitting	**5,000**	**5,000**	**2,500**	**7,500**	**7,000**	**3,000**	**5,000**	**5,000**	**2,500**
System Installation	**14,500**	**21,000**	**17,000**	**19,000**	**27,000**	**21,000**	**16,000**	**20,000**	**17,500**
Labor	7,500	11,000	10,000	9,000	14,000	12,000	9,000	11,000	12,000
Materials	5,000	8,000	5,000	5,000	8,000	5,000	5,000	8,000	5,000
Electrical	2,000	2,000	2,000	5,000	5,000	4,000	2,000	5,000	2,500
Startup	**3,000**	**2,500**	**2,750**	**6,000**	**6,300**	**4,250**	**4,000**	**3,000**	**3,000**
Labor	2,000	1,500	2,000	5,000	5,000	3,500	3,000	2,000	2,500
Analytical	800	500	500	800	500	500	800	500	1,500
Materials	200	500	250	200	800	250	200	500	250
Operation (1Year)	**43,300**	**48,800**	**27,000**	**148,700**	**152,300**	**81,750**	**48,800**	**52,300**	**31,000**
Labor	12,000	20,000	7,500	17,400	25,000	10,250	14,000	20,000	9,500
Analytical	7,500	5,000	2,500	7,500	7,500	3,500	7,500	5,000	2,500
Extraction System Rental	15,000	15,000	12,000	15,000	15,000	12,000	17,500	17,500	13,000
Off-Gas System Rental	NA	NA	NA	72,000	68,000	36,000	NA	NA	NA
Reports (Quarterly)	4,800	4,800	2,000	4,800	4,800	2,000	4,800	4,800	2,000
Electricity	4,000	6,000	3,000	12,000	12,000	6,000	5,000	5,000	4,000
Supplemental Fuel	NA	NA	NA	20,000	20,000	12,000	NA	NA	NA
Total First-Year Cost	**73,300**	**89,800**	**52,250**	**191,200**	**207,600**	**117,000**	**83,800**	**93,300**	**60,500**
Estimated Time (Days)	**400**	**200**	**130**	**400**	**200**	**130**	**500**	**273**	**150**

that soil venting is 55% less expensive than soil venting with off-gas treatment and 14% less expensive than bioventing. Bioventing is 48% less expensive than soil venting with off-gas treatment. The estimated times for cleanup are 130 days for venting and venting with off-gas treatment and 150 days for bioventing. Site No. 2 was used in this analysis to determine the sensitivity of size on the cost comparisons. It is apparent that savings are in the same range for the smaller site as for the larger site.

CONCLUSIONS

Analysis of the data presented above suggests that bioventing and conventional venting with and without off-gas treatment can be applied successfully to a wide variety of site conditions and configurations. Many of the limitations identified for the two remediation approaches are shared by both. Specifically, both require the ability of the solid matrix to transmit gaseous fluids. The estimated ranges of times for cleanup are similar for the two approaches. However, despite the seeming longer time for cleanup using bioventing, it should be noted that the estimates provided assume that biodegradation is the only operator in the bioventing scenarios, and in reality, where permitted, it is possible to simultaneously remove hydrocarbon via venting. The layout of the scenarios was done to illustrate the ability of bioventing to reduce or eliminate the need for off-gas treatment. It therefore is likely that, when venting and in situ bioremediation are combined, the time for cleanup will be less than the time for venting alone. One advantage of bioventing is its ability to address higher-molecular-weight compounds and heavier petroleum fuels that are left totally untouched by venting alone. Thus, bioventing becomes a very valuable tool in addressing sites contaminated with heavier fuels. Bioventing also is a true remediation tool in that it mineralizes the hydrocarbon rather than transferring it to the atmosphere. Bioventing also can be engineered to eliminate or minimize the need for off-gas treatment, and this is a very significant logistical advantage. The analyses performed clearly show that soil venting without off-gas treatment is the most cost effective of the three options. However, when off-gas treatment is required, bioventing definitely is more cost effective. In summary, bioventing is a vadose zone remediation technique that can be readily and cost effectively applied to hydrocarbon-contaminated sites, but as is the case with all remediation approaches, it is not a panacea and needs to be evaluated along with other feasible options.

REFERENCES

Bennedsen, M. B., J. P. Scott, and J. D. Hartley. 1985. "Use of Vapor Extraction Systems for *In-Situ* Removal of Volatile Organic Compounds from Soil." *Proceedings of the 5th National Conference on Hazardous Wastes and Hazardous Materials*, pp. 92-95, Hazardous Materials Control Research Institute, Gaithersburg, MD.

Chiou, C. J., and J. D. Shoup. 1985. "Soil Sorption of Organic Vapors and Effects of Humidity on Sorptive Mechanism and Capacity." *Environmental Science and Technology 19*: 1196-1200.

Diks, R. M. M., and S. P. P. Ottengraf. 1991. "A Biological Treatment System for the Purification of Waste Gases Containing Xenobiotic Compounds." In R. E. Hinchee and R. F. Olfenbuttel (Eds.), *On- Site Bioreclamation: Processes for Xenobiotic and Hydrocarbon Treatment*, pp. 452-463. Butterworth-Heinemann, Boston, MA.

Douglass, R. N., J. M. Armstrong, and W. M. Korreck. 1991. "Design of a Packed Column Bioreactor for On-Site Treatment of Air Stripper Off Gas." In R. E. Hinchee and R. F. Olfenbuttel (Eds.), *On-Site Bioreclamation: Processes for Xenobiotic and Hydrocarbon Treatment*, pp. 209-225. Butterworth-Heinemann, Boston, MA.

Dragun, J. 1988. *The Soil Chemistry of Hazardous Materials*, Hazardous Materials Control Research Institute, Silver Spring, MD.

Ehlers, W. J. Letey, W. F. Spencer, and W. J. Farmer. 1969. "Lindane Diffusion in Soils: I. Theoretical Considerations and Mechanism of Movement." In *Soil Science Society of America Proceedings 33*, 501- 504.

Hinchee, R. E., D. C. Downey, and E. J. Coleman. 1987. "Enhanced Bioreclamation, Soil Venting, and Ground-Water Extraction: A Cost-Effectiveness and Feasibility Comparison." In *Proceedings, Petroleum Hydrocarbon and Organic Chemicals in Ground-Water Prevention, Detection, and Restoration Conference, Houston, Texas*, National Water Well Association, Worthington, OH.

Johnson, P. C., M. W. Kenblowski, and J. D. Colthart. 1990. "Quantitative Analysis for the Cleanup of Hydrocarbon Contaminated Soils by *In Situ* Venting." *Journal of Ground Water 28* (3): 413-429.

Miller, R. N., C. C. Vogel, and R. E. Hinchee. 1991. "A Field-Scale Investigation of Petroleum Hydrocarbon Biodegradation in the Vadose Zone Enhanced by Bioventing at Tyndall Air Force Base, Florida." In R. E. Hinchee and R. F. Olfenbuttel (Eds.), *In-Situ Bioreclamation: Applications and Investigations for Hydrocarbon and Contaminated Site Remediation*, pp. 283-302. Butterworth-Heinemann, Boston, MA.

Pedersen, T. A., and J. T. Curtis. 1991. *Soil Vapor Extraction Technology 316.* Noyes Data Corporation. Park Ridge, NJ.

United States Environmental Protection Agency. 1988. "Underground Storage Tanks: Technical Requirements." *Federal Register 53*, No. 185, 23 September 1988, 37082-37212.

THE EFFECTS OF SOIL TYPE, CRUDE OIL TYPE AND LOADING, OXYGEN, AND COMMERCIAL BACTERIA ON CRUDE OIL BIOREMEDIATION KINETICS AS MEASURED BY SOIL RESPIROMETRY

M. H. Huesemann and K. O. Moore

ABSTRACT

The kinetics of crude oil biodegradation in soil matrices were studied in a soil respirometer system by measuring both oxygen consumption and carbon dioxide production during the bioremediation process. Three soils with varying organic matter content were artificially contaminated with a California crude oil and amended with nutrients and water to stimulate biodegradation. Cumulative oxygen consumption profiles corrected for oxygen consumption in crude oil-deficient control soils varied significantly. Although total oxygen consumption was higher in the topsoil, amounts of total hydrocarbons removed were essentially the same in each soil. Thus, oxygen consumption as measured by soil respirometry cannot be used as a quantitative indicator for hydrocarbon biodegradation kinetics in soils with a high organic matter content. Crude oil additions stimulate biodegradation of soil organic matter that cannot be accounted for by a control. Increased oxygen consumption rates were found to correlate with increased concentrations of hydrocarbon degraders in each soil. The effects of crude oil loading, crude oil type, oxygen, and commercial bacteria addition were all investigated in Texas sand which has a very low (0.4%) total organic carbon (TOC) content and is therefore well suited for soil respirometry. The optimal loading for California crude in Texas sand was found to be 5% (wt), whereas a 10% (wt) loading resulted in a partial inhibition of the biodegradation kinetics. Light (API gravity 39) Michigan crude oil (5% wt) was found to biodegrade faster than heavy (API gravity 21) California crude (5% wt) when added to Texas sand. The presence of pure oxygen-enhanced biodegradation rates up to twofold for both California and Michigan crude oil indicate that crude oil biodegradation can be oxygen-limited under standard atmospheric conditions. Finally, the addition of Hydrobac, a bacterial preparation marketed for the biodegradation of hydrocarbon materials, did not enhance the biodegradation of California crude oil in Texas sand.

INTRODUCTION

Bioremediation has been successfully used to remove petroleum hydrocarbons from contaminated soils. During the bioremediation process, under ideal conditions, the hydrocarbons are converted by indigenous soil microorganisms to carbon dioxide, water, and biomass. Numerous factors are known to affect both the kinetics and the extent of hydrocarbon removal from contaminated soils. These include soil properties such as pH, temperature, moisture, aeration, and nutrient status (e.g., nitrogen, N, and phosphorus, P, fertilizers); contaminant characteristics such as molecular structure and toxicity to microbes; and the ecology of the microbial populations present in the soil (Atlas 1991; Atlas & Bartha 1972, Frankenberger 1992). Several authors have studied the biodegradation of various crude oils and other refined petroleum hydrocarbons under defined laboratory or field conditions (Song et al. 1990; Walker et al. 1975; Walker, Petrakis & Colwell 1976; Westlake et al. 1974).

Few publications report data regarding the effects of environmental factors on the biodegradation kinetics of petroleum hydrocarbons. It has been shown that certain compound classes (e.g., alkanes, saturated cyclics, and aromatic compounds with various numbers of rings) present in South Lousiana crude oil biodegraded at different rates in aqueous, nutrient-amended shaker flask cultures (Walker, Colwell & Petrakis 1976). The effects of compost addition, compost age, compost-to-soil ratio, and optimal moisture content on the biodegradation kinetics of diesel fuel (1% wt) contaminated soil was studied by measuring oxygen consumption rates in several different soil respirometer systems (Stegmann et al. 1991). In another study, an electrolytic respirometer was used to assess the effects of nutrient addition (N and P) and soil tilling on oxygen consumption kinetics during the biodegradation of crude oil added to soil (Harmsen 1991). Finally, the U.S. EPA conducted respirometry experiments in order to screen commercial bacteria preparations for possible enhancements of crude oil biodegradation rates (Venosa et al. 1991).

This paper investigates the effects of soil type, crude oil type and loading, oxygen, and the addition of commercial bacteria on hydrocarbon bioremediation kinetics as measured by soil respirometry.

MATERIAL AND METHODS

Soil Respirometer Setup and Operation

The rate of crude oil biodegradation in a soil matrix was assessed by measuring the rate of carbon dioxide production and oxygen consumption in a simple soil respirometer system consisting of a tightly sealed 2-qt (1.89-L) Mason jar. At the beginning of the experiment approximately 200 g of crude oil contaminated soil (see section on soil treatments below) was transferred to the jar. The jar was flushed with ambient humidified air or pure oxygen for ca. 5 minutes and was tightly closed using Mason jar lids. A rubber septum had been added to the lid to enable sampling of the headspace gases with a syringe.

At the beginning of the experiment and at given time intervals during the operation of the respirometer, a 1-mL gas sample was taken from the jar with the syringe and analyzed for carbon dioxide (CO_2), oxygen (O_2), and nitrogen (N_2) using a Varian gas chromatograph (Model 3700) with an Alltech CTR-I column and a thermal conductivity detector (TCD). The oven temperature was controlled initially at 35°C for 2 min, ramped at a rate of 15°C/min up to 55°C, and maintained at 55°C for 10 min. The injection port and TCD temperatures were set at 200°C.

In cases where the headspace gas composition reached less than 5% (vol) oxygen as a result of aerobic biodegradation activities in the soil, the jar was opened and flushed with humidified air or pure oxygen (see section on soil treatments below) to replenish the oxygen content in the jar. The jar was closed and headspace gas analyses commenced as outlined above.

Soil Treatments

A number of different soil treatment experiments were designed to study the effects of soil type, crude oil loading, molecular oxygen, crude oil type, and the addition of commercial bacteria on crude oil biodegradation kinetics. Three different soils were used in this study. These were Texas sand (TXS), Baccto topsoil (BAC), and Hyponex topsoil (HYP). All soils were purchased at local nurseries. The characteristics and composition of each soil are summarized in Table 1.

All soils were sieved prior to use in the respirometer experiments to minimize soil sampling errors. After the removal of rocks and bulky organic materials (sticks, leaves, roots, etc.), each soil was thoroughly mixed and homogenized. Distilled water was added to each soil in order to adjust the moisture levels required for optimal biodegradation activity. The moisture contents for TXS, BAC, and HYP were adjusted to approximately 9% (wt), 60% (wt), and 42% (wt), respectively. The variation in moisture content for these three soils is the result of differences in water-holding capacity for each soil due to specific soil characteristics such as density, organic matter content, and the particle-size distribution.

Soils were amended with varying concentrations of nitrogen and phosphorus fertilizer based on the amount of crude oil added to the soil. For each gram of crude oil, 107 mg NH_4NO_3 and 42 mg K_2HPO_4 were added to provide the nitrogen and phosphorus required for stimulating microbial growth of hydrocarbon degraders. This is equivalent to a C:N:P ratio (wt) of 100:5:1, assuming that 75% (wt) of crude oil consists of carbon (C). The pH of the soil was adjusted to around 7.0 by adding diluted NaOH solution where necessary.

Two fresh crude oils of different API gravity were used to contaminate the clean soils. The first oil, California crude (CA-CR), was a heavy crude (API gravity 21) from Huntington Beach, California, and the second crude oil, Michigan crude (MI-CR), was obtained from a production facility in Michigan. This oil was relatively light as indicated by its comparatively higher API gravity of 39. The level of crude oil loading is expressed in weight percent based on moist, clean,

TABLE 1. Properties of Texas sand, Baccto topsoil, and Hyponex topsoil.

Parameter[(a)]	Texas Sand	Baccto Topsoil	Hyponex Topsoil
pH	5.8	5.1	5.8
Hydrometer Texture			
Sand, %	92.3	63.2	66.2
Silt, %	0.5	15.6	17.6
Clay, %	7.3	21.1	16.2
TOC[(b)], %	0.4	21.5	14.0
NO_3-N, mg/kg	2.3	79.1	24.1
NH_4-N, mg/kg	7.0	53.2	12.3
EDTA PO_4-P, mg/kg	4.6	67.2	169.0

(a) All soil analyses were performed by Soil Analytical Services, College Station, Texas.
(b) Total organic carbon was determined using the Walkley Black titration method.

fertilizer-amended soil. For instance, a 5% (wt) loading indicates that 5 g fresh crude oil is added to 100 g soil prior to initiating the respirometer experiments. The following treatments were performed:

Treatment A: 2% (wt) CA-CR in TXS with air.
Treatment B: 5% (wt) CA-CR in TXS with air.
Treatment C: 10% (wt) CA-CR in TXS with air.
Treatment D: 5% (wt) CA-CR in BAC with air.
Treatment E: 5% (wt) CA-CR in HYP with air.
Treatment F: 5% (wt) CA-CR in TXS with 100% oxygen.
Treatment G: 5% (wt) MI-CR in TXS with air.
Treatment H: 5% (wt) MI-CR in TXS with 100% oxygen.

Control respirometer experiments were run for each soil treatment to account for carbon dioxide production or oxygen consumption from background respiration activity in the soil. The compositions of the controls were similar to the corresponding treatments with the exception that the crude oil was omitted. All treatment and control experiments were incubated at room temperature (20°C).

Sampling and Analyses

Sampling. Soil samples were taken at specified time intervals from the jars for determining either the remaining crude oil concentration levels (O&G, TPH) or the magnitude of bacterial counts (hydrocarbon degraders) in the soil. Prior to sampling, the soil in the jar was mixed to minimize sampling errors. After the removal of 10 to 20 g soil, the jar was flushed with humidified air (or 100% oxygen). The jar was closed and the GC headspace gas analysis was performed as outlined above. A record of the total amount (grams) of soil present in the jar was kept in order to compute the volume (L) of carbon dioxide production or oxygen uptake normalized on a per kg soil basis.

O&G/TPH. Gravimetric O&G and TPH analyses were performed according to Standard Methods 5520 E & F (1989). The soil is Soxhlet extracted in Freon™, and the resulting O&G extract is subjected to silica gel treatment for the removal of polar compounds. Whereas the concentration of O&G is indicative of the total Freon extractable compounds in the waste, the magnitude of TPH reflects the concentrations of nonpolar compounds in the soil such as saturated and aromatic hydrocarbons.

Soil Moisture. The moisture content (or inversely the total solids content) of the soil was determined according to Standard Method 2540 B (1989).

Soil pH. Approximately 10 ml distilled water was added to 10 g soil, and the pH of the soil slurry mixture was measured with a standard, calibrated pH probe.

Hydrocarbon Degrader Counts. To perform the counts, 10 g of soil were placed into a 250-mL of sterile serum vial and slurried with 90 mL of sterile Bushnell-Haas medium. Slurries were shaken in a wrist action shaker for 60 min and then serially diluted (1:10) up to 10^{-12} in 9-mL sterile Bushnell-Haas mineral medium. Approximately 0.3 mL California crude oil was added and vortexed into each tube serving as carbon and energy source for bacterial metabolism. In order to obtain a more statistically accurate estimate of the hydrocarbon degrader concentration in the soil, the most probable number (MPN) counting technique was used employing five replicate dilution series. All cultures were incubated for 7 days at 30°C, and the highest dilutions showing turbid growth were selected to compute the approximate number of bacteria/gram of wet soil. For example, if the 10^{-5} dilution tube was the highest dilution showing turbid growth, the number of bacteria present were approximately 10^5/g wet wt soil.

RESULTS AND DISCUSSION

Cumulative Oxygen Consumption and Carbon Dioxide Production Profiles

Figure 1 depicts cumulative oxygen (O_2) consumption and carbon dioxide (CO_2) production profiles for treatment D and the respective control for a 489-day time period. For both treatment and control, cumulative O_2 consumption volumes were approximately 1.5 times the cumulative CO_2 production volumes. The observed O_2/CO_2 ratio of ca. 1.5 is in close agreement with the theoretical

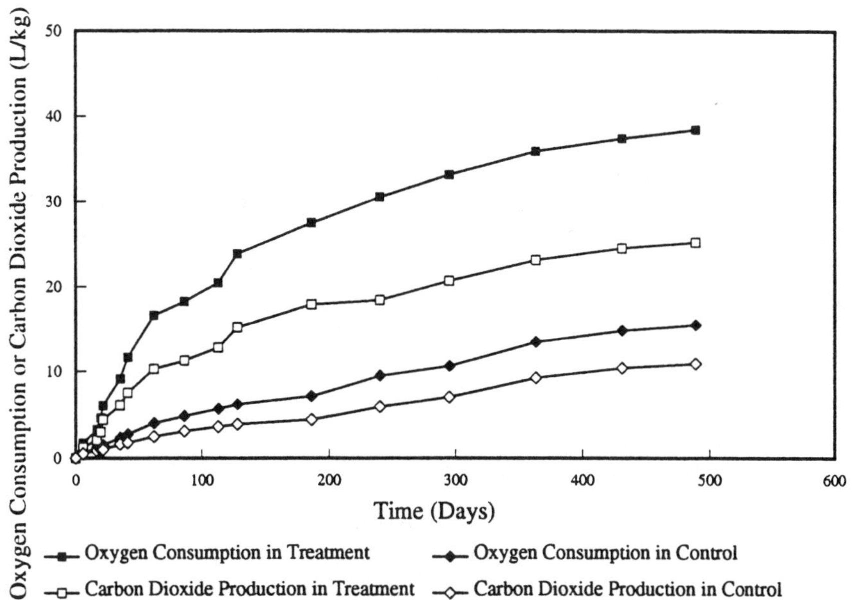

FIGURE 1. Cumulative oxygen consumption and carbon dioxide production in treatment D (5% wt California crude in Baccto topsoil with air) and the respective control.

stoichiometry for the complete oxidation (mineralization) of a hydrocarbon (CH_2) compound to carbon dioxide and water: $CH_2 + 1.5\ O_2 \rightarrow CO_2 + H_2O$.

The relatively high O_2 consumption and CO_2 production rates in the control for treatment D are indicative of biodegradation of soil organic matter present in the Baccto topsoil. During the 489-day experiments, the cumulative oxygen consumption totals in controls using Texas sand, Baccto topsoil, and Hyponex topsoil were 1.0, 15.6, and 22.2 L/kg, respectively. The relatively high respiratory activity in Baccto and Hyponex topsoils when compared to the low activity in Texas sand is due to differences in soil total organic carbon (TOC) content for each soil as shown in Table 1. In an effort to simplify the graphical presentation of biodegradation kinetics, all subsequent plots for crude oil biodegradation treatments depict only cumulative oxygen consumption curves that have been corrected (via subtraction) for the background respiratory activity in the respective control soils.

The reproducibility of O_2 consumption and CO_2 production curves was verified by performing treatment B in triplicate, which resulted in almost identical curves. The addition of 2% (wt) of respiratory toxin sodium azide to treatment B completely inhibited CO_2 production, whereas up to 0.8 L/kg O_2 consumption was measured during a 435-day incubation period. The observed low levels of oxygen consumption may be the result of chemical oxidation processes in the soil (Texas sand).

Effects of Crude Oil Loading

Soil respirometry has been used to optimize the biodegradation kinetics in a given soil matrix while varying parameters such as temperature, moisture, nutrient addition, or waste loading. Figure 2 presents cumulative oxygen consumption (g O_2/kg moist soil) profiles for Texas sand contaminated with 2% (wt) (treatment A), 5% (wt) (treatment B), and 10% (wt) (treatment C) California crude oil and amended with fertilizer and water. Despite the presence of more biodegradable hydrocarbon substrate in treatment C (10% wt), cumulative oxygen consumption is less than in treatment B (5% wt). This may be indicative of inhibitory effects of high petroleum hydrocarbon levels on soil microbial biodegradation activity. In addition, treatment C (10%) exhibited a 2-week longer lag or acclimation period than did treatments A and B.

Effects of Crude Oil Type

Figure 3 depicts the cumulative oxygen consumption profiles for 5% California crude oil (treatment B) and 5% (wt) Michigan crude oil (treatment G) in Texas sand amended with fertilizer and water. Both the rate and the extent of oxygen consumption are significantly higher for the Michigan crude oil when compared to California crude oil. This observation can be explained by the fact that the high API gravity crude oil contains more low-molecular-weight (lighter) hydrocarbons that are more readily and rapidly biodegraded than higher-molecular-weight,

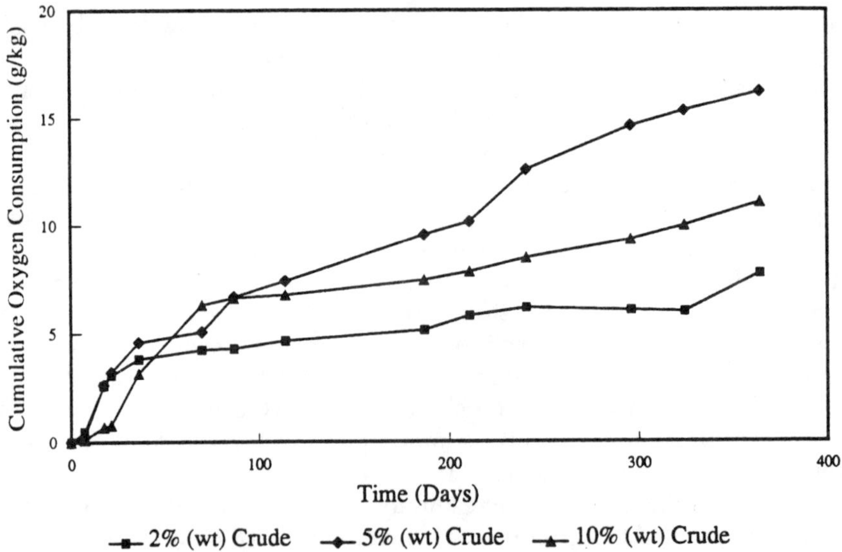

FIGURE 2. Cumulative oxygen consumption during biodegradation of California crude (2%, 5%, 10% wt) in Texas sand.

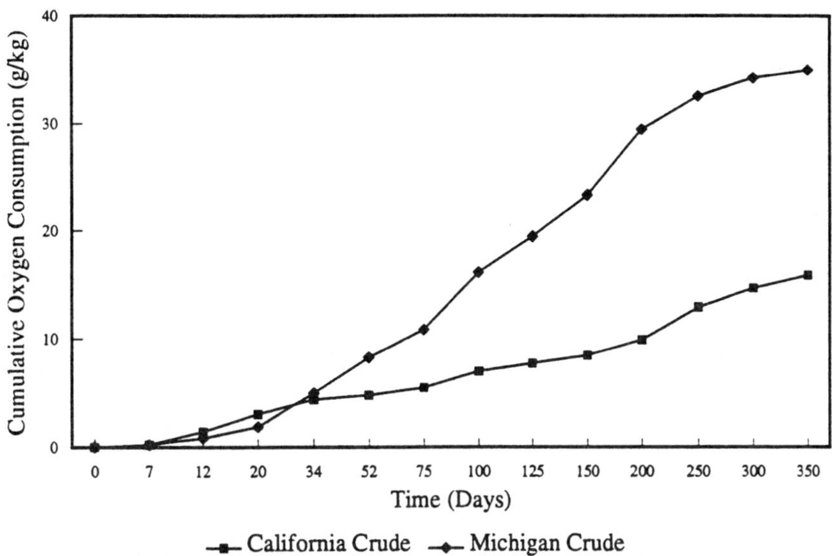

FIGURE 3. Cumulative oxygen consumption during biodegradation of California and Michigan crude (5% wt) in Texas sand.

polynuclear aromatics or polar compounds in the heavier California crude oil (Walker et al. 1975; Walker, Colwell & Petrakis 1976; Walker, Petrakis & Colwell 1976; Westlake et al. 1974). Finally, it is noteworthy that none of the oxygen consumption profiles leveled off after 1 year of incubation. This indicates that complete crude oil biodegradation may take longer than 1 year under the test conditions.

Effects of Pure Oxygen versus Air

Texas sand contaminated with 5% (wt) California or Michigan crude oil were incubated separately in the presence of either air or 100% oxygen. The resulting cumulative oxygen consumption profiles for each treatment (e.g., treatments B, F, G, and H) are shown in Figure 4. As expected, the oxygen consumption rates are up to two times higher in the presence of pure oxygen when compared to air, indicating that crude oil biodegradation kinetics are limited by oxygen availability. Because improvement of biodegradation rates due to the presence of pure oxygen is only marginal (e.g., a factor of two), it is questionable whether it will be economical to use pure oxygen in practical bioremediation applications. However, it must be pointed out that the oxygen consumption profiles in Figure 4 were measured in the presence of variable oxygen concentrations, because oxygen was used up during the biodegradation process. For instance, headspace oxygen concentrations decreased from the initial 100% to approximately 50% in both treatments F and H during the first 34 days of incubation.

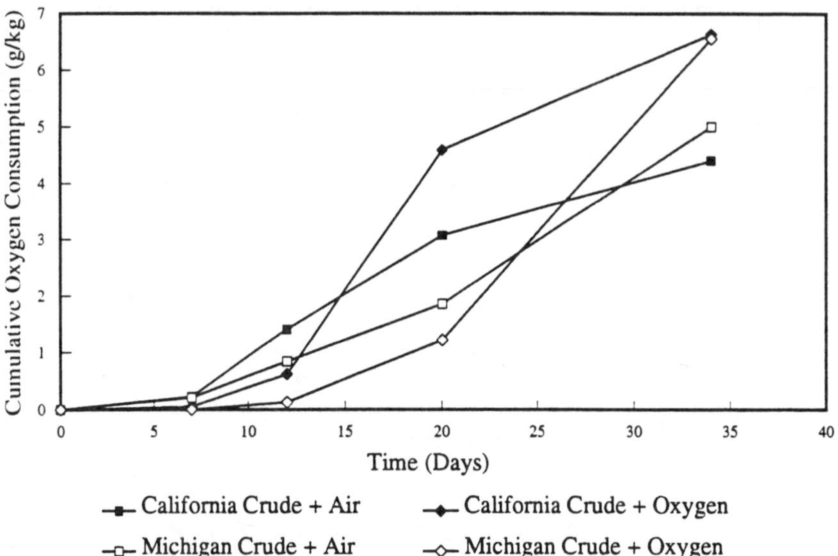

FIGURE 4. Cumulative oxygen consumption during biodegradation of California and Michigan crude (5% wt) in Texas sand with air or oxygen.

Relationship between Cumulative Oxygen Consumption and O&G/TPH Removal

Despite the widespread application of respirometry (Harmsen 1991, Stegmann et al. 1991, Venosa et al. 1991) for measuring biodegradation kinetics, this technique is useful only if one can correlate contaminant removal with oxygen consumption or carbon dioxide production. Figure 5 presents the relationship between the amounts of hydrocarbons removed (either O&G or TPH corrected for uncontaminated controls) and the cumulative oxygen consumption in treatment E. Despite the analytical uncertainties with O&G and TPH analyses in soil matrices, it can be seen that generally higher oxygen consumption is correlated with increased O&G or TPH removal. The fact that hardly any hydrocarbon removal was observed during the consumption of the initial 12 g/kg oxygen is indicative of the biodegradation of volatile hydrocarbons that are not measurable with the gravimetric O&G and TPH methods. It is not entirely clear why the O&G and TPH curves reach a plateau above 45 g/kg oxygen consumption. It is possible that soil organic matter instead of petroleum hydrocarbons is biodegraded. This topic will be discussed in more detail below when considering the O&G removal and cumulative oxygen consumption in three different soil types.

Finally, the ratio of oxygen consumption to hydrocarbon (O&G) removal in the "linear" range between 10 and 45 g/kg cumulative oxygen consumption can be calculated as 3.5 g O_2 consumed per g O&G removed. This observed ratio is in close agreement with the theoretical ratio of 3.4, which can be derived from

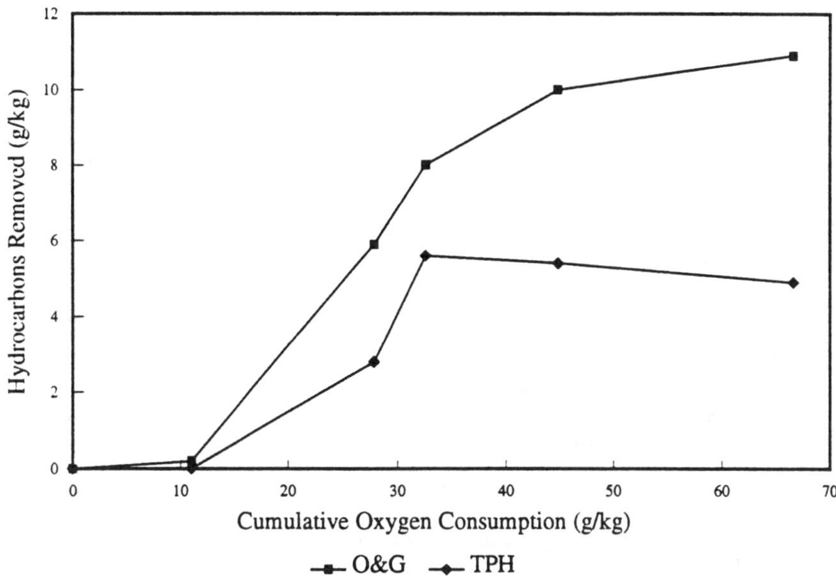

FIGURE 5. O&G and TPH removal vs. cumulative oxygen consumption during biodegradation of 5% wt California crude in Hyponex topsoil.

the complete mineralization stoichiometry relationship (O_2/CH_2 = 1.5 moles/ 1 mole = 48 g/14 g = 3.4).

Effects of Soil Type

In order to study the effect of soil type on oxygen consumption kinetics, 5% (wt) of California crude oil was added to three different soil matrices, namely Texas sand (treatment B), Baccto topsoil (treatment D), and Hyponex topsoil (treatment E). All soils were amended with the same amounts of fertilizers and were incubated in the presence of air. As shown in Figure 6, the soil type has a significant effect on oxygen consumption kinetics as well as on the extent of total oxygen consumption, both of which were highest in Hyponex topsoil, less in Baccto topsoil, and least in Texas sand. Because all oxygen consumption profiles are corrected for oxygen consumption in controls, it appears that the observed significant differences in cumulative oxygen consumption must be related to differences in the extent of crude oil biodegradation in these soils. It was shown in Figure 5 that, at least for treatment E (Hyponex topsoil + California crude), O&G removal appears to correlate with cumulative oxygen consumption. At day 435, the cumulative oxygen consumption in Hyponex topsoil was 67 g/kg, in Baccto topsoil 32 g/kg, and in Texas sand only 20 g/kg. During the same period, however, O&G removal from each soil was almost *exactly* the same, namely 10 to 11.5 g/kg. No significant O&G removal was observed in uncontaminated

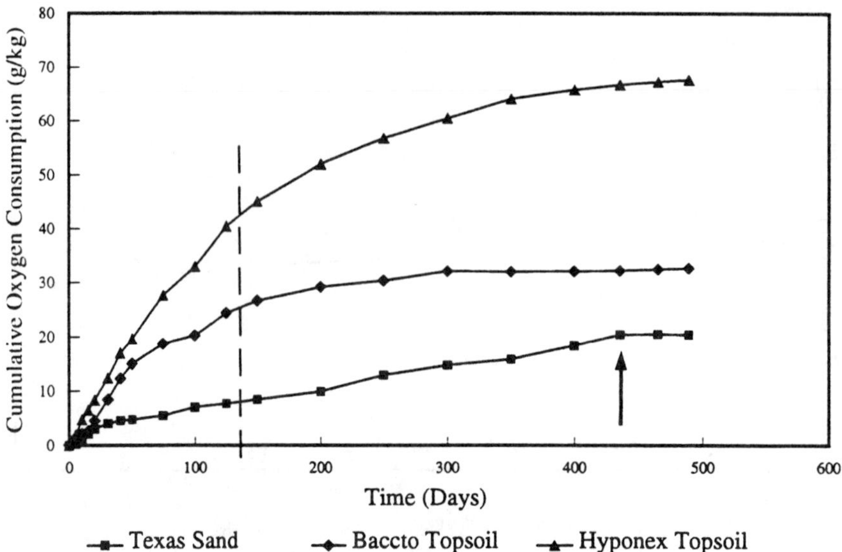

FIGURE 6. Cumulative oxygen consumption during biodegradation of California crude (5% wt) in three different soils.

controls (O&G concentrations in uncontaminated soils were approximately 1 g/kg, 3 g/kg, and 2 g/kg for Texas sand, Baccto topsoil, and Hyponex topsoil, respectively).

These observations clearly indicate that differences in cumulative oxygen consumption in the three soils are *not* related to differences in the extent of O&G biodegradation. One hypothesis to explain this observation is that the addition of crude oil to soil stimulates the biodegradation of the soil organic matter resulting in additional oxygen consumption not accounted for in the control. The soil in the control culture received only fertilizer and water. The addition of crude oil could cause both an increase of bacterial numbers and a change in bacterial ecology resulting in enhanced biodegradation of the inherent soil organic matter compared to the crude oil-deficient control. The fact that the addition of biodegradable materials such as straw stimulates the decomposition of native humus in soil has been reported earlier (Stevenson 1982) and is referred to as priming action by soil scientists.

These findings are important, because they indicate that it is not possible to run a proper control experiment in respirometry. This is particularly the case for soils with a high organic matter content. The addition of substrates such as petroleum hydrocarbons stimulates the biodegradation of soil organic matter which cannot be accounted for in a control. Consequently, the determination of biodegradation kinetics by soil respirometry is likely to result in biased results (see Figure 6) if soils with high organic matter contents are used.

In order to determine the relationship between the microorganism concentration and oxygen consumption rates, the most probable number (MPN) technique was used to assess the number of hydrocarbon degraders in each soil at day 140. The concentrations (MPN) of hydrocarbon degraders were $2.3*10^5/g$, $4.6*10^6/g$, and $1.2*10^8/g$ in Texas sand, Baccto topsoil, and Hyponex topsoil, respectively. At the same time, the cumulative oxygen consumption rates in these three soils were 0.029, 0.090, and 0.182 g/kg day, respectively. Consequently, higher oxygen consumption rates are correlated with higher bacterial concentrations in the soil. It appears that soils with high soil organic matter contents are able to support a larger and possibly more diverse microbial population after crude oil addition than soils deficient in natural organics. An increase in population size and diversity seems to be associated with enhanced biodegradation of the natural soil organic matter as hypothesized above.

Effects of Commercial Bacteria Addition

Numerous microbial preparations are on the market that are purported to reduce acclimation times, enhance biodegradation rates, and even increase the total extent of contamination removed. At day 435, 2% (wt) of Hydrobac, a bacterial preparation marketed by Polybac Corporation specifically for petroleum hydrocarbon materials, was added to treatment B (5% California crude in Texas sand) *as well as* to the respective control (see arrow in Figure 6). Immediately after the addition of the microbial product, the oxygen consumption rate in treatment B increased dramatically by a factor of approximately 3.7. However, because a strong increase of oxygen consumption also was observed in the control, the cumulative oxygen consumption profile (see Figure 6) corrected for the background control actually flattened out after the addition of Hydrobac. This indicates that the addition of this microbial preparation does not enhance oxygen consumption due to petroleum hydrocarbon biodegradation.

The drastic increases in oxygen consumption rates in both treatment B and its respective control are most likely attributable to the biodegradation of the nutrient support for the microbes. Upon the addition of Hydrobac microbes, petroleum hydrocarbon biodegradation appears to slow down in favor of the more readily biodegradable compounds in the nutrient support material. As such, this bacterial preparation appears to have the opposite effect as expected: it actually slows down the crude oil biodegradation kinetics instead of enhancing it. The ineffectiveness of microbial products for enhancing hydrocarbon biodegradation has been reported earlier (Venosa et al. 1991). Reasons for the possible failure of inoculation to enhance biodegradation may include toxins or predators in the environment, the use of other organic compounds in preference to the pollutant, or the inability of the microbes to move through the soil to sites containing the contaminant (Atlas 1991, Goldstein et al. 1985).

CONCLUSIONS

1. Soil respirometry appears to provide accurate estimates of contaminant biodegradation kinetics mainly in soils with low organic matter content.

The addition of petroleum hydrocarbons to high organic carbon soils may result in an enhanced biodegradation of the soil organic matter that cannot be accounted for in a control experiment. Consequently, the use of oxygen consumption profiles from soil respirometer experiments is likely to result in biased biodegradation kinetics for soils with a high organic matter content.

2. A light, high API gravity crude oil was more easily biodegradable than a heavy, low API gravity crude oil.
3. The biodegradation of crude oil in soil is limited by oxygen availability. The presence of pure oxygen enhanced biodegradation rates up to twofold compared to air.
4. The optimal loading of California crude oil in Texas sand was around 5% (wt). At a 10% (wt) loading, biodegradation kinetics were partially inhibited.
5. The addition of Hydrobac, a commercial bacterial preparation, did not enhance oxygen consumption rates due to crude oil biodegradation.

ACKNOWLEDGMENTS

We would like to thank Drs. Joseph Salanitro and Gary Drinkard for reviewing this paper and for providing valuable insights during our technical discussions.

REFERENCES

Atlas, R. M. 1991. "Bioremediation of Fossil Fuel Contaminated Soils." In R. E. Hinchee and R. F. Olfenbuttel (Eds.), *On Site Bioreclamation: Applications and Investigations for Hydrocarbon and Contaminated Site Remediation*, pp. 14-32. Butterworth-Heinemann, Stoneham, MA.

Atlas R. M., and R. Bartha. 1972. "Degradation and Mineralization of Petroleum in Sea Water: Limitation by Nitrogen and Phosphorus." *Biotechnology and Bioengineering* 14(3): 319-330.

Frankenberger, W. T. 1992. "The Need for a Laboratory Feasibility Study in Bioremediation of Petroleum Hydrocarbons." In E. J. Calabrese and P. T. Kostecki (Eds.), *Hydrocarbon Contaminated Soils and Groundwater*, Vol. 2, pp. 237-293. Lewis Publishers, Ann Arbor, MI.

Goldstein, R. M., L. M. Mallory, and M. Alexander. 1985. "Reasons for Possible Failure of Inoculation to Enhance Biodegradation." *Applied and Environmental Microbiology 50*(4): 977-983.

Harmsen, J. 1991. "Possibilities and Limitations of Landfarming for Cleaning Contaminated Soils." In R. E. Hinchee and R. F. Olfenbuttel (Eds.), *On Site Bioreclamation: Processes for Xenobiotic and Hydrocarbon Treatment*, pp. 255-272. Butterworth-Heinemann, Stoneham, MA.

Method 2540 B. 1989. "Total Solids Dried at 103 - 105°C." In *Standard Methods for the Examination of Water and Wastewater*, 17th ed., pp. 2-72 to 2-73.

Method 5520 E. 1989. "Extraction Method for Sludge Samples." In *Standard Methods for the Examination of Water and Wastewater*, 17th ed., pp. 5-46 to 5-47.

Method 5520 F. 1989. "Hydrocarbons." In *Standard Methods for the Examination of Water and Wastewater*, 17th ed., pp. 5-47 to 5-48.

Song, H.-G., X. Wang, and R. Bartha. 1990. "Bioremediation Potential of Terrestrial Fuel Spills." *Appl. Environ. Microbiol. 56*(3): 652-656.

Standard Method. *See* Method.

Stegmann, R., S. Lotter, and J. Heerenklage. 1991. "Biological Treatment of Oil Contaminated Soils in Bioreactors." In R. E. Hinchee and R. F. Olfenbuttel (Eds.), *On Site Bioreclamation: Processes for Xenobiotic and Hydrocarbon Treatment*, pp. 188-208. Butterworth-Heinemann, Stoneham, MA.

Stevenson, F. J. 1982. *Humus Chemistry*, pp. 13-14. John Wiley, New York, NY.

Venosa, A. D., J. R. Haines, W. Nisamaneepong, R. Govind, S. Pradhan, and B. Siddique. 1991. "Screening of Commercial Bioproducts for Enhancement of Oil Biodegradation in Closed Microcosms." In *Remedial Action, Treatment, and Disposal of Hazardous Waste*, pp. 388-403. Proceedings of the 17th Annual Hazardous Waste Research Symposium, Cincinnati, OH.

Walker, J. D., R. R. Colwell, and L. Petrakis. 1975. "Microbial Petroleum Degradation: Application of Computerized Mass Spectrometry." *Can. J. Microbiol. 21*: 1760-1767.

Walker, J. D., R. R. Colwell, and L. Petrakis. 1976. "Biodegradation Rates of Components of Petroleum." *Can. J. Microbiol. 22*: 1209-1213.

Walker, J. D., L. Petrakis, and R. R. Colwell. 1976. "Comparison of the Biodegradability of Crude and Fuel Oils." *Can. J. Microbiol. 22*: 598-602.

Westlake, D. W. S., A. Jobson, R. Phillippe, and F. D. Cook. 1974. "Biodegradability and Crude Oil Composition." *Can. J. Microbiol. 20*: 915-928.

BIOSTIMULATION VERSUS BIOAUGMENTATION—THREE CASE STUDIES

M. E. Leavitt and K. L. Brown

ABSTRACT

An ongoing debate has existed since the beginning of the bioremediation industry. Is it really necessary to add bacteria to bioremediation systems, or can bacteria that are indigenous to the matrix get the job done? Some professionals indicate that adding bacteria shortens the treatment period, thus offsetting the cost of the culture. Direct comparisons of the performance of bioaugmentation versus biostimulation are not abundant in the literature. IT Corporation (IT) has had several opportunities to evaluate the benefits of adding commercially available bacteria to bioremediation systems. To understand the benefit while considering the cost, side-by-side comparisons of treatments with and without specialized bacteria were made. Three pilot-scale studies are presented, and the resulting contaminant removal rates are compared. One study focused on using bioreactors to treat tank bottoms generated during crude oil storage and compared indigenous organisms to known petroleum degraders. A second study tested treatment of acetone and *bis*-2-chloroethyl ether in a wastewater with elevated dissolved solids; a commercial culture was compared to activated sludge. The third study demonstrated land treatment of weathered crude oil in drilling mud; one plot examined indigenous organisms, the other a commercial culture with a recommended nutrient blend. At the completion of these comparisons, we concluded that for some bioremediation applications biostimulation of indigenous organisms is the best choice considering cost and performance.

INTRODUCTION

Bioremediation has gained considerable recognition in recent years as an innovative remedial technology for the reduction of petroleum hydrocarbons in soil and groundwater. One of the major concerns raised when comparing bioremediation to other remedial technologies is the time required to achieve the treatment target level. Significant strides have been made in optimizing systems to maximize the degradation rate. Examples of supplements include the use of hydrogen peroxide as an enriched oxygen sources for in situ groundwater bioremediation, surfactants to increase the solubility of otherwise insoluble organics, and the use of a specialized bacterial culture to increase the rate of contaminant loss.

Many of these enhancements require significantly more effort and money to implement. Therefore, most bioremediation users choose to carefully determine the advantages to such supplements relative to their cost prior to including them in full-scale remedial designs. Examples of such determinations have been published. Leavitt et al. (1992) demonstrated enhanced biodegradation of polycyclic aromatic hydrocarbons (PAH) in a column study with surfactant enhancement. Graves and Leavitt (1991) concluded that different surfactants would not benefit the land treatment of petroleum contamination in soil. Berkey et al. (1991) described the undertaking to evaluate bioremediation products for the *Exxon Valdez* oil spill. Hardaway et al. (1991) described their approach to determining the potential for utilizing hydrogen peroxide as an enhancement and approaches to avoiding contaminant toxicity effects in a subsurface system.

Relative to the bioaugmentation approach, Roy (1992) describes the site evaluation process and some successful applications of bioaugmentation; however, there were no biostimulation controls included. Similarly, Williams and Lieberman (1992) describe a successful in situ aquifer remediation system utilizing acclimated bacteria for both the surface bioreactor and the subsurface system, but no data were provided on the efficiency of bacteria transport through the subsurface or the activity attributable to indigenous bacteria. Jobson et al. (1974) compared bioaugmentation and biostimulation in field plots. There was a small but apparently significant benefit in the bioaugmented plots. However, the crude oil was added to pristine soil simultaneously with the fertilizer and bacteria; therefore, there was no acclimation period that indigenous bacteria typically have in a contaminant release scenario. Compeau et al. (1991) compared two different commercially available cultures to uninoculated and sterilized inocula for petroleum degradation in soil. Compared to indigenous microorganisms, neither of the cultures led to an increase in petroleum hydrocarbon degradation even though the microbial density was higher.

This paper presents the results from three demonstrations designed to evaluate the benefit of bioaugmentation using specialized bacterial cultures. Each of these demonstrations was conducted in pilot-scale systems to provide a more realistic performance for full-scale estimations.

DEMONSTRATIONS OF BIOSTIMULATION VS. BIOAUGMENTATION

Case Study No. 1 — Biodegradation of Crude Oil Sludges

A 2-month pilot study was conducted in late 1991 at the United States Environmental Protection Agency (U.S. EPA) Test and Evaluation Facility in Cincinnati, Ohio. The objective of the pilot study was to demonstrate the biodegradation of crude oil sludges generated during storage. The sludge, collected directly from the storage tank, was determined to be 100% total petroleum hydrocarbon (TPH), and Toxicity Characteristic Leaching Procedure (TCLP) volatile and semivolatile analyses indicated the presence of benzene and *o*-cresol. The flashpoint of the sludge as measured by the Pensky Marten Closed Cup method was 58°F.

The project utilized two 64-L, stainless steel EIMCO Biolift® slurry reactors operated in batch mode. Reactor 1 (R1) was augmented with nutrients and the pH was adjusted to 7.0. Reactor 2 (R2) was bioaugmented with a naturally isolated, petroleum-degrading bacteria culture (J. Bonner, Texas A&M). The culture, a mixed consortium isolated from a metals plating wastewater treatment lagoon, had demonstrated significant TPH removal during bench-scale fermentation experiments. Prior to inoculation, the culture was maintained in a nutrient broth supplemented with diesel fuel. Fertilization and pH adjustment were conducted similarly to R1. The microbial density of hydrocarbon degraders, as well as heterotrophs, was monitored throughout the study. The organic compounds in the different phases of the bioreactors were also monitored regularly throughout the study.

Microbiological Data. The bacterial analyses included plate counts for heterotrophs and hydrocarbon degraders and gene probe analysis for naphthalene degraders (King et al. 1990). The results are presented in Table 1. The initial heterotrophic population in R1 was 10^5 colony-forming units (CFU/g) solids. The population reached 10^7 CFU/g solids after 3 weeks. This density was maintained for the duration of the treatment. Hydrocarbon degraders were equal in concentration to the heterotrophs throughout the first 6 weeks of the study. Hydrocarbon-degrading populations decreased to 10^6 CFU/g solids during the remaining 2 weeks of the investigation. The naphthalene-degrading population was initially 10^4 CFU/g solids, increasing to 10^5 after 3 weeks, and dropped to below 10^4 CFU/g solids after 5 weeks.

R2 was inoculated at the initiation of the project, and again after 5 weeks of operation with 2 L of a 10^8 CFU/mL bacterial culture. The R2 population

TABLE 1. Crude oil sludge biodegradation study microbial enumerations.

	Total Heterotrophs		Specific Degraders		nah A Probe[(1)]	
Week	R1	R2	R1	R2	R1	R2
0	1.6×10^5	3.3×10^6	1.6×10^5	1.5×10^6	4.5×10^4	2.6×10^5
3	8.5×10^5	4.3×10^7	5.4×10^5	8.3×10^6	3.5×10^5	5.0×10^6
4	3.7×10^7	8.3×10^7	1.7×10^7	1.6×10^7	1.2×10^5	2.0×10^6
6	3.3×10^7	1.4×10^7	5.8×10^6	4.6×10^6	2.0×10^5	2.0×10^5
8	1.2×10^7	1.2×10^7	6.9×10^6	4.1×10^5	8.0×10^4	2.0×10^4

Note: All bacterial densities are reported as CFU/g solids.
(1) nah A gene probe for naphthalene degraders.

densities were initially one order of magnitude higher than R1. However, the R2 hydrocarbon-degrading population dropped from a peak of 10^7 CFU/g solids to 10^5 CFU/g solids after 8 weeks. Similarly, the naphthalene-degrading population reached a peak of 10^6 CFU/g solids and dropped to less than 10^3 CFU/g solids over the same period of time.

Contaminant Data. Contaminants were analyzed in the mixed liquor and headspace of each reactor weekly over a 2-month period; no analytical replicates were performed. Through the combined mechanisms of volatilization and biodegradation, 100% of the compounds having 8 or fewer carbon atoms were removed from both reactors. R1 and R2 demonstrated 98 and 97% removal of compounds containing 9 to 12 carbon atoms. R1 demonstrated 60, 53, and 30% removal of compounds containing 13 to 14 carbon atoms, 15 to 18 carbon atoms, and greater than 18 carbon atoms, respectively. R2 illustrated reduced performance, with only 14% removal of compounds containing 12 to 14 carbon atoms and no reduction of higher-molecular-weight compounds. The poor performance in R2 prompted the conclusion that bioaugmentation did not benefit this bioreactor system.

Case Study No. 2 — Industrial Wastewater Treatment

During the feasibility study segment of the Superfund process, technologies appropriate for the remediation of a high salinity industrial wastestream combined with contaminated groundwater were demonstrated. Analyses of the existing wastewater treatment aeration lagoon effluent and contaminated groundwater confirmed the presence of organic priority pollutants and metals, as well as high salt concentrations. Pretreatment of the wastestream prior to activated sludge treatment included metals removal, metals sludge treatment, and air stripping.

A pilot-scale system was operated to demonstrate removal efficiencies of 98, 82, and 88% for 5-day biological oxygen demand (BOD_5), total organic carbon (TOC), and chemical oxygen demand (COD), respectively. Once steady-state operation had been achieved, a commercially available bacterial culture was added to the system for 1 month to improve flocculation and COD removal. The Munox culture supplied by Microlife Technics (Sarasota, Florida) was added daily per vendor specifications. Monthly averages of carbon removal obtained during the addition of Munox were compared to those obtained one month following the cessation of culture addition. Due to the extended treatment period examined, analytical replicates were not performed.

It was determined that the culture did not significantly affect either the removal of the influent carbon or the system clarification (Figure 1). Additionally, Figure 1 illustrates that the system acclimation period did not decrease as system performance continued to improve following the cessation of culture addition. Therefore, the biokinetic constants obtained with the nonaugmented reactor were used to aid in the design of the full-scale system.

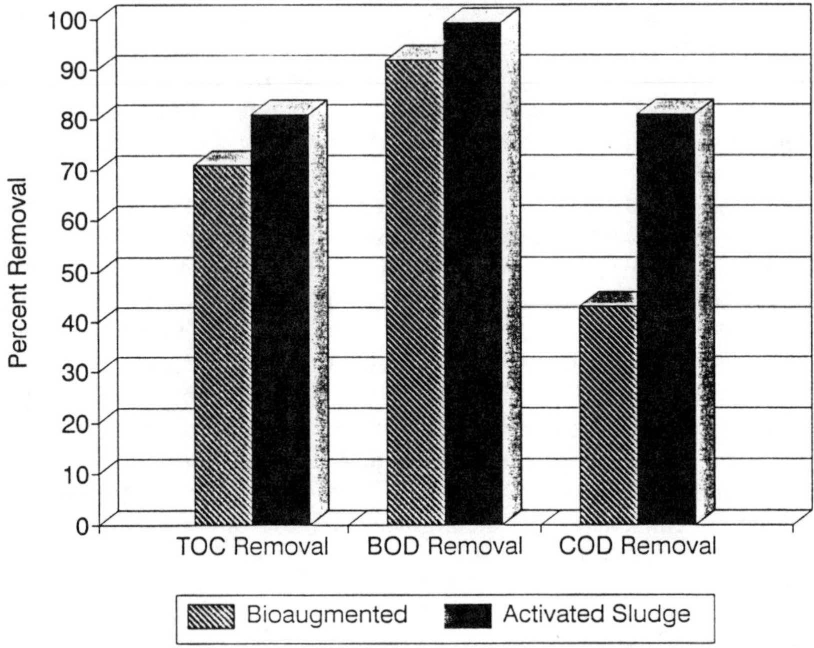

FIGURE 1. Case Study No. 2 — Performance measurements.

Case Study No. 3 — Land Treatment of Weathered Crude Oil in Drilling Mud

The goal of this project was to demonstrate accelerated biodegradation of weathered crude oil in drilling mud using pilot-scale soil volume and equipment. Additionally, the demonstration was intended to compare the extent of biodegradation with conventional treatment, with additional bulking agents, and with bioaugmentation. Three treatment plots were constructed at the site in Southern California. Each plot contained approximately 500 yd^3 (382 m^3) of waste soil material (soil). Plot 1 received fertilizer, mixing, and irrigation as needed. Plot 2 received the same treatment, but initially was mixed with 7% straw (by volume). Plot 3 was augmented with a proprietary culture and nutrient blend (Tesoro Petroleum Environmental Services Inc., San Antonio, Texas), and was treated as recommended by Tesoro Environmental Products Company. Mixing and irrigation in Plot 3 were the same as for Plots 1 and 2. During operation of the plots, pH adjustment became necessary, and each plot was treated with lime according to its average pH value. The demonstration was maintained for 6 months, from November 1991 until May 1992.

Prior to the initiation of this study, it was indicated that a single dose of the bacterial culture and nutrients would be sufficient for the duration of the demonstration. The recommendation included adding five 55-gal drums of PES-31 bioculture, representing an inoculation rate of 1 gal for every 2 yards (1.83 m) of material. After observing low levels of nitrogen and phosphorus, an additional 200 lb (75 kg) of nutrient (15:30:15, N:P:K) was applied after 1.5 months of treatment. Three 55-gal drums of PES-31 and 400 lb (149 kg) of nutrient blend were added to Plot 3 after 5 months of treatment.

Plots 1 and 2 received nutrients in two doses during the study. At the initiation of the study, 250 lb (93 kg) of ammonium sulfate, 50 lb (19 kg) of potassium phosphate, and 1.5 tons (1,500 kg) of urea were added to each plot. After 1.5 months, an additional 300 lb (112 kg) of urea was added to both Plots 1 and 2. The aggressive fertilization rate was recommended due to the high level of contamination and the extremely low bacterial density observed in the baseline plot samples.

An extensive soil compositing procedure was applied to minimize the variability in analytical results. Two methods were used. Samples were collected from ten locations within a grid for each plot. This composite was mixed and divided into five subunits. One subunit was then further mixed with an equal weight of water. After settling for several days, the solids from this slurry were sampled and submitted for analysis. Some analyses were also conducted on samples collected just prior to slurrying with water. All biomonitoring analyses were conducted on samples prior to slurrying.

Microbiological Data. Figure 2 illustrates the bacterial growth trend during the presentation. All but 2 of the initial 15 samples collected had contaminant degraders below the 10^3 CFU/g soil detection limit. Within the first month of treatment, the population density increased in all plots to between 10^5 and 10^7 CFU/g. The populations peaked at approximately 10^8 CFU/g after 2 months of

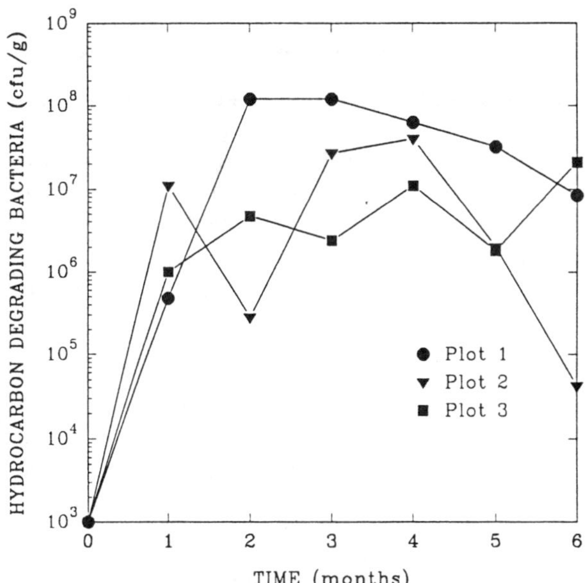

FIGURE 2. Case Study No. 3 — Microbial density before and during treatment.

treatment, and steadily declined for the remainder of the treatment. Plot 2 exhibited a more extreme decline than Plots 1 and 3, presumably due to a more drastic pH shift during treatment (data not shown).

Contaminant Data. A summary of the organic analyses is provided in Table 2. Oil and grease, as well as total recoverable petroleum hydrocarbon (TRPH) analyses were conducted on samples after slurrying. TRPH is also reported for samples collected just prior to slurrying. Considering oil and grease results, Plot 3 was the best performer, with a loss of 86% at the end of the study. Plot 2 exhibited an 82% loss in oil and grease, whereas Plot 1 exhibited only 32% loss. TRPH analyses in slurried samples proved Plot 2 to be the best performer (55% loss), followed by Plot 1 (45%) and then Plot 3 (27%). TRPH results in samples that were not slurried had a 62% loss in Plot 3, 59% in Plot 2, and 59% in Plot 3. When all analyses were averaged for each plot, Plot 2 showed the highest percent loss at 65%. Plot 3 averaged a 58% loss, and Plot 1 a 45% loss.

The recommendations drawn from this study would apply to a full-scale remediation system for more than 70,000 yd^3 (53,519 m^3) of soil. Considering the scale, the cost of nutrients was a significant factor in determining these recommendations. The nutrient cost for Plots 1 and 2 was $0.35/lb. At the time of the study, the cost of PES-31 was $1,700/drum. The proprietary nutrient blend was $3/lb. The recommended ratio of 1 drum/2 yd^3 (1.5 m^3) would result in more than $1 million cost for bacteria alone. Considering the cost and the marginal benefit in reduction, bioaugmentation was not recommended for the full-scale treatment. The bulking agent was recommended because it did render the soils more workable and was relatively inexpensive.

TABLE 2. Case Study No. 3 — Bioremediation performance by analysis.

		Oil and Grease (mg/kg)	TRPH (mg/kg) Slurried Samples	TRPH (mg/kg) Non-slurried Samples	Overall Average Reduction (percent)
Plot 1	Before	82,444	51,111	130,000	
Nutrients Only	After	55,711	28,222	53,000	
	Percent Reduction	32	45	59	45
Plot 2	Before	85,125	54,375	140,000	
Nutrients,	After	15,636	24,667	58,000	
Bulking Agent	Percent Reduction	82	55	59	65
Plot 3	Before	113,000	77,556	150,000	
Bulking Agent,	After	16,278	56,544	57,000	
Bacteria,	Percent Reduction	86	27	62	58
Nutrient Blend					

SUMMARY

Although many successes have been reported using bioaugmentation, those who fund and implement remediation technologies must be aware of the overall benefit of such an approach. Bioaugmentation can cause public and regulatory alarm, resulting in unnecessary time delays and cost escalation. While there is a niche for bioaugmentation, the results of the case studies described suggest that some conventional applications may not require bioaugmentation.

REFERENCES

Berkey, E., J. M. Cogen, V. J. Kelmeckis, L. T. McGeeham, and A. T. Merski. 1991. "Evaluation Process for the Selection of Bioremediation Technologies for Exxon Valdez Oil Spill." In G. S. Sayler, R. Fox, and J. W. Blackburn (Eds.), *Environmental Biotechnology for Waste Treatment*, pp. 85-90. Plenum Press, New York, NY.

Compeau, G. C., W. D. Mahaffey, and L. Patras. 1991. "Full-Scale Bioremediation of a Contaminated Soil and Water." In G. S. Sayler, R. Fox, and J. W. Blackburn (Eds.), *Environmental Biotechnology for Waste Treatment*, pp. 91-110. Plenum Press, New York, NY.

Graves, D. A., and M. E. Leavitt. 1991. "Petroleum Biodegradation in Soil: The Effect of Direct Application of Surfactant." *Remediation*, Spring Edition: 147-166.

Hardaway, K. L., M. S. Katterjohn, C. A. Lang, and M. E. Leavitt. 1991. "Feasibility and Other Considerations for Use of Bioremediation in Subsurface Areas." In G. S. Sayler, R. Fox, and J. W. Blackburn (Eds.), *Environmental Biotechnology for Waste Treatment*, pp. 111-126. Plenum Press, New York, NY.

Jobson, A., M. McLaughlin, F. D. Cook, and D. W. S. Westlake. 1974. "Effects of Amendments on the Microbial Utilization of Oil Applied to Soil." *Applied Microbiology 27*: 166-171.

King, J. H. M., P. M. DiGrazia, B. Applegate, R. Burlage, J. Sanseverino, P. Dunbar, F. Larimer, and G. S. Sayler. 1990. "Rapid and Sensitive Bioluminescent Reporter Technology for Naphthalene Exposure and Biodegradation." *Science 249*: 778-781.

Leavitt, M. E., C. A. Lang, J. Sanseverino, K. Hague, D. Dameron, J. Rightmyer, and D. A. Graves. 1992. "Implications of Surfactant Augmentation for In Situ Bioremediation Systems." Proceedings from the Air and Waste Management Association, Pittsburgh, PA, pp. 109-121.

Roy, K. A. 1992. "Petroleum Company Heals Itself — and Others." *Hazmat World*, May: 75-80.

Williams, C. M., and M. T. Lieberman. 1992. "Bioremediation of Chlorinated and Aromatic Organic Solvent Waste in the Subsurface." *The National Environmental Journal*, Nov/Dec: 40-44.

NITRATE-BASED BIOREMEDIATION OF PETROLEUM-CONTAMINATED AQUIFER AT PARK CITY, KANSAS: SITE CHARACTERIZATION AND TREATABILITY STUDY

S. R. Hutchins and J. T. Wilson

ABSTRACT

In the late 1970s, leakage from an underground pipeline at Park City, Kansas, had resulted in approximately 2.4 ha of a shallow water-table aquifer being contaminated with mixed oily-phase material. Aerobic in situ bioremediation was initiated but was unsuccessful due to plugging of the injection wells by gas and iron precipitates. Laboratory and field studies are therefore under way to evaluate the use of nitrate as an alternative electron acceptor for anaerobic bioremediation of the aquifer. Twelve continuous cores were obtained and analyzed for total petroleum hydrocarbon (TPH) and benzene, toluene, ethylbenzene, and xylenes (BTEX) to delineate the vertical extent of contamination and to obtain aquifer material for a treatability study. Based on the zero-order rate constants for alkylbenzene and nitrate removal, the feed nitrate concentration, and the design application rate, the time required for remediation is estimated to be 200 days. The rate of remediation will be governed by the permissible influent nitrate concentrations rather than by the intrinsic rate of the microbial processes. Six cells in the field site will receive different experimental treatment regimes: (1) methanogenic, (2) denitrifying, and (3) mixed denitrifying/microaerophilic conditions, with no flow being delivered to the other three cells to serve as controls.

INTRODUCTION

Aquifers contaminated with gasoline and other fuels often exhibit BTEX levels of in excess of the regulatory limits mandated by the U.S. Environmental Protection Agency (Cotruvo & Regelski 1989). In many cases, bioremediation is required to bring the aquifers and associated groundwater into compliance. This is most commonly done through the addition of nutrients and either oxygen or hydrogen peroxide (Thomas & Ward 1989). However, the high oxygen demand of severely contaminated aquifers limits the usefulness of this approach (Wilson et al. 1986). The problem is exacerbated when the aquifers are initially anaerobic, because high concentrations of reduced metal ions also can exhibit oxygen demand. Oxygen addition in these circumstances can have a further deleterious effect by causing precipitation of metal complexes, leading to plugging of the aquifer formation (Aggarwal et al. 1991).

For these reasons, several investigators are now evaluating in situ bioremediation under anaerobic conditions, specifically with nitrate as an alternative electron acceptor. The capability of subsurface bacteria to degrade reduced monoaromatic hydrocarbons under denitrifying conditions has now been firmly established in microcosms (Hutchins et al. 1991a, Major et al. 1988), enrichments (Jorgensen et al. 1991), and pure cultures (Evans et al. 1991, Schocher et al. 1991). These processes also have been evaluated in the field at pilot scale (Hilton et al. 1992, Hutchins et al. 1991b), but there are few documented full-scale projects that demonstrate the applicability of nitrate-based bioremediation.

An aquifer contaminated with petroleum hydrocarbons at Park City, Kansas, offers such an opportunity and is being considered for nitrate-based bioremediation. This report describes the site characterization and treatability study used to assess the current design of the remediation system and to estimate the time required to reduce BTEX levels to below drinking water standards.

SITE DESCRIPTION

The site is located in the Little Arkansas River Valley northwest of Wichita, Kansas, and overlies the Equus Beds aquifer. The aquifer is approximately 14 m thick in the study area and is underlain by gray fissile shale comprising the Hennessy formation. The aquifer comprises sand and gravel interlaced with silt and clay lenses and capped by clay overburden (Figure 1a). The depth to water is 5 to 6 m, and the hydraulic conductivity ranges from 0.12 to 0.36 cm/sec. Additional site information has been published by Kennedy & Hutchins (1992).

The aquifer became contaminated during the late 1970s when a mixture of petroleum hydrocarbons leaked slowly from a pipeline carrying a variety of refined products, including gasoline and diesel (Figure 1b). The leak was discovered in 1980 when BTEX was detected in a municipal well (City Well #6, Figure 1b). Following closure of the well, two trenches were installed on either side of a flood control channel at the site to accumulate free product. Several thousand gallons of hydrocarbon were pumped from these trenches, after which the remaining seepage was ignited and allowed to burn for more than 1 year. The trenches were backfilled in 1984, and groundwater pumping was resumed in 1986 in an attempt to remove contaminated water and control the contaminant plume.

In 1989, a subsurface injection system was completed to provide an evaluation of aerobic in situ bioremediation. The system consisted of 1,500 m of infiltration gallery piping located in the bottom of the flood control channel and 530 injection wells spaced 6 m on center located east and west of the flood control channel (Figure 2a). Each well was approximately 5.5 m in depth and screened over a 1-m

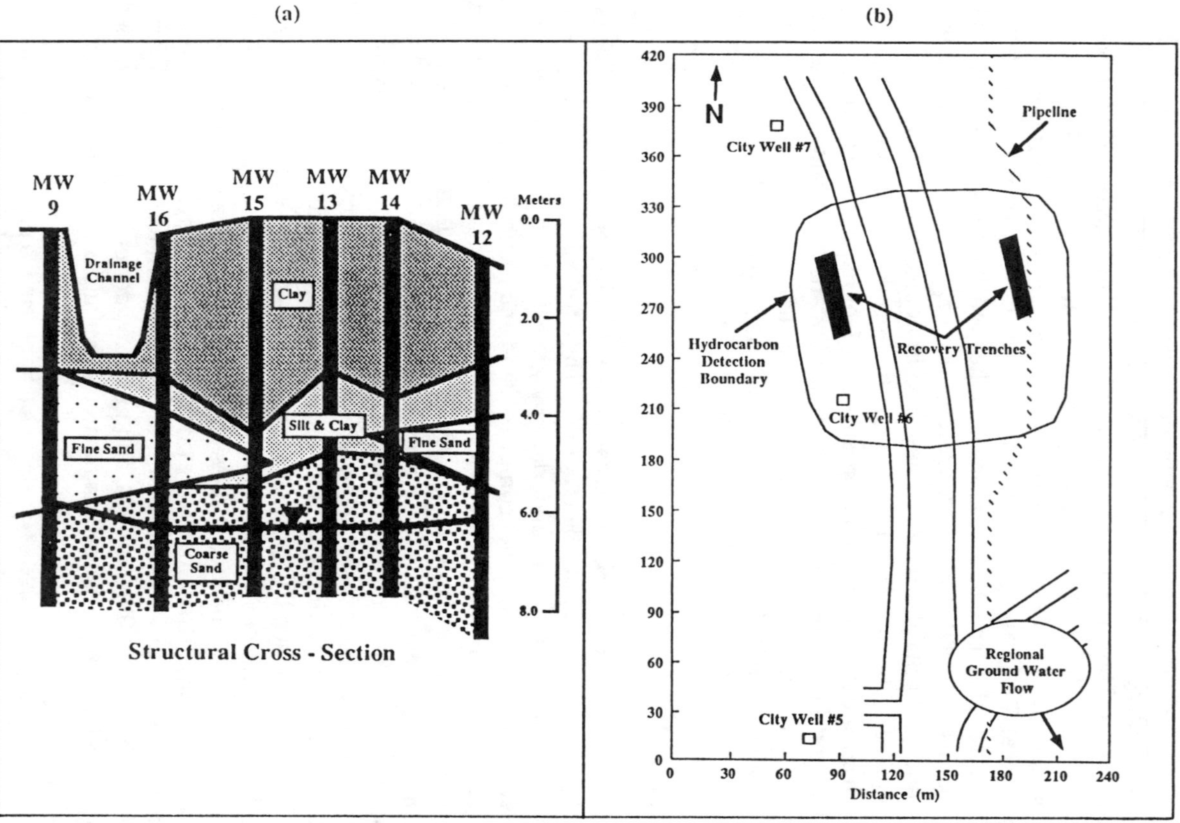

FIGURE 1. Park City field site, showing (a) cross section of site geology, and (b) location of original pipeline and recovery trenches.

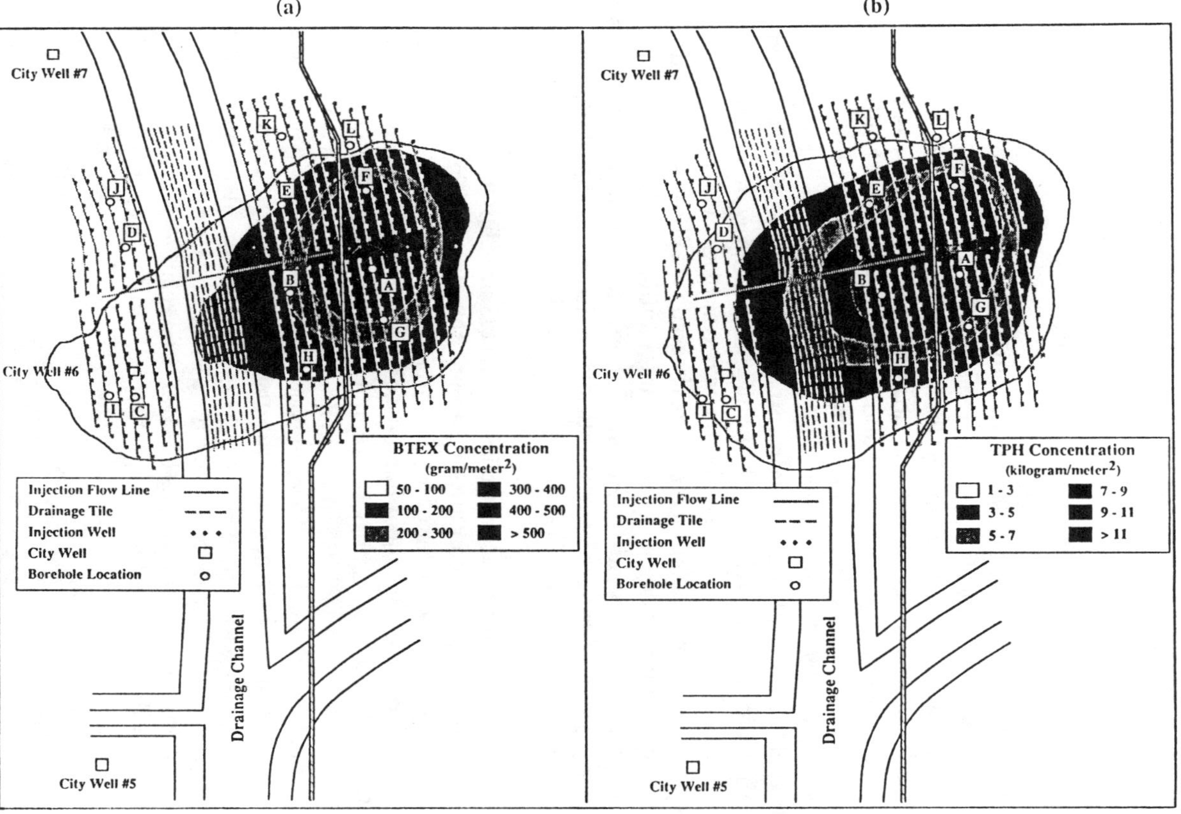

FIGURE 2. Park City field site, showing (a) distribution of BTEX based on soil cores, and (b) distribution of TPH based on soil cores.

interval in the sandy vadose zone below the clay layer but just above the water table. During a 1-month test period, oxygen-saturated water amended with nutrients was injected above the formation. However, a continued loss in injection capability was observed, due primarily to air entrainment and secondarily to iron precipitation (L. Kennedy, personal communication). Based on these results, a decision was made to evaluate nitrate-based bioremediation for anaerobic treatment of the aquifer.

Contaminant Distribution

Discrete 15-cm cores were obtained over continuous lengths from 3 to 10 m below land surface for 12 locations within the study area to define the vertical extent of contamination, using methods described previously (Hutchins et al. 1991b). These cores were extracted and analyzed for BTEX by gas chromatography/mass spectroscopy (GC/MS) and for total petroleum hydrocarbon (TPH) by GC using JP-4 jet fuel as a reference standard. In addition, selected extracts were subjected to an extensive GC/MS search to better define the composition of the residual volatile hydrocarbons, using a Restek XTI-5 30 m × 0.25 mm capillary column with 0.5-μ film thickness. Data were obtained in a scan mode (m/z = 34 to 450), and peak spectra were compared with library spectra to provide tentative identifications. These were checked visually and then grouped into selected compound categories, and a semiquantitative analysis was provided estimating the relative distribution of the selected compound classes, assuming that response factors were identical and that mass distribution was proportional to the cumulative area distributions in the reconstructed ion chromatogram.

Contours for the distribution of BTEX and TPH are shown in Figures 2a and 2b, respectively. These contours represent the cumulative mass of the two parameters per square meter of surface area, based on the integrated core analyses. The focus of the contamination occurs near the original pipeline, although there is some discrepancy between the distribution of BTEX and that of TPH. Based on subsequent information, it is believed that active weathering (natural attenuation) of the BTEX fraction is occurring west of the original spill location (Wilson et al. 1992). Analysis of reconstructed ion chromatograms from sample extracts reveals that the residual hydrocarbon at the site contains more higher-molecular-weight components than does JP-4 jet fuel, primarily as alkanes and polycyclic aromatic hydrocarbons (PAHs). The relative distribution of the various compound classes in four selected sample extracts ranges from 66 to 71% alkanes and alkenes, 12 to 19% alkylbenzenes, 5.6 to 7.3% cycloalkanes and cycloalkenes, 5.2 to 6.9% naphthalenes, 1.9 to 3.9% other PAHs, and 0.3 to 1.1% unidentified compounds, based on the tentative identification of approximately 250 compounds. From other research (Al-Bashir et al. 1990, Hutchins et al. 1989, Mihelcic & Luthy 1988), it is expected that the alkylbenzenes, naphthalenes, and selected PAHs will be the most likely candidates for biodegradation under denitrifying conditions.

TREATABILITY STUDY

Microcosm tests were conducted to determine whether BTEX could be degraded under denitrifying conditions by the indigenous subsurface microorganisms at Park City, and to determine the rate and extent of nitrate consumption and alkylbenzene biodegradation. These rates, in conjunction with detailed core analyses from the site, were then used to estimate nitrate demand and assess the current design of the remediation system.

Details of microcosm preparation, sampling, and analyses have been described previously (Hutchins 1991). In brief, coarse sand aquifer material was obtained for microcosm preparation from both contaminated and uncontaminated intervals at the site using aseptic sampling techniques under anaerobic conditions. Microcosms were prepared in an anaerobic glovebox using 13-mL headspace vials with 10-g aquifer solids and sterile diluted spring water, and amended with ammonium and phosphorus nutrients. Sets prepared with both contaminated and uncontaminated aquifer material received either nitrate alone, a mixed BTEX spike alone, nitrate plus BTEX spike, or nitrate plus BTEX spike with biocides. The BTEX spike contained benzene, toluene, ethylbenzene, *m*-xylene, *p*-xylene, *o*-xylene, 1,2,3-trimethylbenzene, 1,2,4-trimethylbenzene, and 1,3,5-trimethylbenzene. Final concentrations were 20 mg/L mixed BTEX, 40 mg/L NO_3-N, and/or 500 mg/L sodium azide and 250 mg/L mercuric chloride as biocides. Microcosms were sealed without headspace using Teflon™-lined septa and incubated in an anaerobic glovebox at 20°C. Periodically, three replicates were sacrificed for each set and analyzed for aqueous BTEX, nitrate, nitrite, and nutrients.

Toluene, ethylbenzene, *m*-xylene, *p*-xylene, 1,3,5-trimethylbenzene, and 1,2,4-trimethylbenzene were degraded to less than 5 µg/L within 20 days in the clean-aquifer microcosms amended with nitrate and BTEX. About half of the *o*-xylene was removed, whereas benzene and 1,2,3-trimethylbenzene were recalcitrant. The data (mean with standard error, three replicates) are summarized for benzene and the alkylbenzenes (TEX) in Figures 3a and 3b, respectively. There was no degradation of the compounds in the microcosms without nitrate addition during the same time period, relative to controls. However, nitrate consumption occurred without BTEX addition, indicating a background nitrate demand of 15 mg/L in the uncontaminated aquifer material (Figure 3c). Alkylbenzene biodegradation also occurred in the contaminated-aquifer microcosms, but sorption and leaching effects mediated by the residual petroleum hydrocarbons precluded valid quantitation of the rates of removal. Although the measured nitrate consumption rate was equivalent to that observed in the clean-aquifer microcosms, the actual nitrate demand was at least four times as high in the contaminated-aquifer microcosms (Figure 3d).

ESTIMATION OF FIELD PERFORMANCE

Zero-order rate constants were obtained for removal of the alkylbenzenes (TEX) and nitrate, and expressed on a dry-weight basis (Table 1). Although

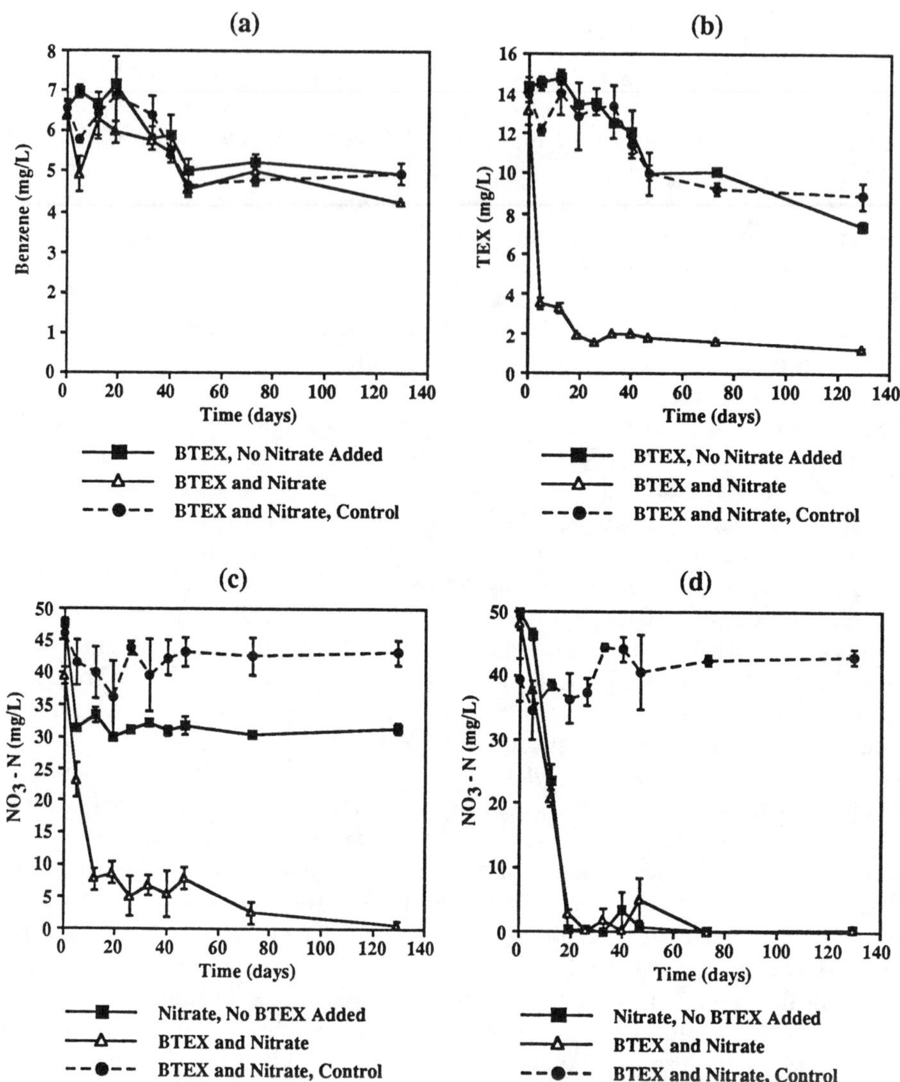

FIGURE 3. Treatability study data, showing (a) removal of benzene in clean-aquifer microcosms, (b) removal of TEX in clean-aquifer microcosms, (c) removal of nitrate in clean-aquifer microcosms, and (d) removal of nitrate in contaminated-aquifer microcosms.

benzene was recalcitrant in the microcosms, previous work had shown benzene to be degraded at field scale, presumably due to the presence of residual oxygen in the system (Hutchins et al. 1991b). It is expected to be degraded at Park City as well, and hence these rates were used to calculate the distance of nitrate travel (penetration zone) and to estimate the time required for remediation of the BTEX

TABLE 1. Zero-order rate constants for TEX and nitrate-nitrogen removal in microcosms prepared with uncontaminated and contaminated Park City aquifer material (all units in mg/kg dry weight).

Parameter	Treatment	Uncontaminated	Contaminated
TEX	Nitrate only	—	0.248[(a)]
	Nitrate plus BTEX	1.65	0.275[(a)]
NO_3-N	Nitrate only	2.80	2.92
	Nitrate plus BTEX	2.80	3.06

(a) These rates are suspect due to interferences described in the text, and were not used to estimate the time required for remediation.

fraction. All calculations were made for Cell #2, a 3,400-m^2 area in the center of the well field that is to receive groundwater recharge containing 10 mg/L NO_3-N (the maximum allowable nitrate concentration) at an application rate of 680 m^3/d. These calculations are shown in more detail in the Appendix. In brief, continuous core data obtained from the two locations within this cell, summarized in Table 2, indicate that Cell #2 contained approximately 1,300 kg BTEX and 58,000 kg TPH. The nitrate demand of the contaminated zone can then be calculated to be 1,400 kg NO_3-N using stoichiometric relationships discussed elsewhere (Hutchins 1991), and the time required for remediation can be estimated based on the projected application rate. These estimates were compared to those obtained using the measured microbial rates of TEX degradation and nitrate removal, considering the average BTEX values for either the entire depth interval or the most contaminated interval within the aquifer profile.

TABLE 2. Summary core data for locations 60B and 60H in Cell #2.

Parameter	Units	60B	60H
Depth Interval	m BLS[(a)]	2.4-8.5	2.7-8.9
Contaminated Zone	m BLS	2.4-6.9	6.0-7.2
Uncontaminated Zone(s)	m BLS	6.9-8.5	2.7-6.0
		7.2-8.9	
Most Contaminated Interval[(b)]	m BLS	6.4-6.6	6.7-6.8
Mean BTEX Concentration	mg/kg dry wt	42.6	24.3
Mean TPH Concentration	mg/kg dry wt	2,360	686
Peak BTEX Concentration	mg/kg dry wt	353	412
Peak TPH Concentration	mg/kg dry wt	10,500	13,400

(a) Meters below land surface.
(b) For BTEX, not necessarily TPH.

Based on projected nitrate demand alone, 210 days would be required to supply enough nitrate to remediate the aquifer (Appendix A-1). This assumes that BTEX constitutes the bulk of the nitrate demand, and that desorption and reaction rates are instantaneous relative to the application rate. The reaction rates, at least, should be fast. Only 13 days would be required to satisfy the average theoretical nitrate demand if nitrate were continuously present, based on the microbial reaction rate for nitrate removal observed in the microcosm tests (Appendix A-2). Similarly, only 21 days would be required on the average to degrade all of the BTEX, based on the observed microbial reaction rate for BTEX biodegradation (Appendix A-3). This illustrates that the main factor affecting remediation time will be the rate of application of nitrate.

This can also be shown by calculating the penetration zone of nitrate once the infiltrate reaches the water table. Based on the application rate, the design nitrate concentration, and the microbial rate of nitrate removal, the penetration zones would be 5.8 to 6.1 m for the Core 60B location and 6.3 to 6.6 m for the Core 60H location (Appendix A-4). As shown in Figure 4a, the actual contaminated interval extends to below these zones in both locations, and hence nitrate breakthrough is not expected until substantial remediation has occurred.

Although overall remediation will be controlled by the nitrate application rate, microbial reaction rates could become the limiting process parameters, if the extent of remediation is assessed solely within the most contaminated interval which contains the highest BTEX concentrations. For example, it would take 310 days to remediate the most contaminated interval in Cell #2, even given the fast rates of TEX removal by the microorganisms (Appendix A-5). Of course, if this interval is small relative to the entire volume being treated, it may not contribute significantly to the final average BTEX concentration in the groundwater after remediation.

DESIGN OF REMEDIATION SYSTEM FOR FIELD DEMONSTRATION

To better evaluate the efficacy and optimize the use of nitrate for anaerobic bioremediation, the field site has been divided into six experimental treatment cells (Figure 4b). The cells are not isolated physically, but the boundaries can be established because the existing distribution network can be segmented so that different cells receive different types of recharge.

The treatment regimes will be delineated by selected amendment and application of the recharge water. The treatment regimes are (1) methanogenic for Cell #4, (2) denitrifying for Cell #2, and (3) mixed microaerophilic/denitrifying for Cell #3. The working hypothesis is that addition of electron acceptors (Cells #2, #3) will result in more extensive BTEX biodegradation than that observed with nutrient addition alone (Cell #1), and that addition of small amounts of oxygen (Cell #3) will facilitate biodegradation of benzene. No flow is delivered to Cells #1, #5, and #6, which serve as corresponding controls. The infiltration recharge for each of the active treatment cells will be amended with nutrients to yield 8 mg/L ammonia-nitrogen, but the Cell #2 recharge will receive an additional nitrate

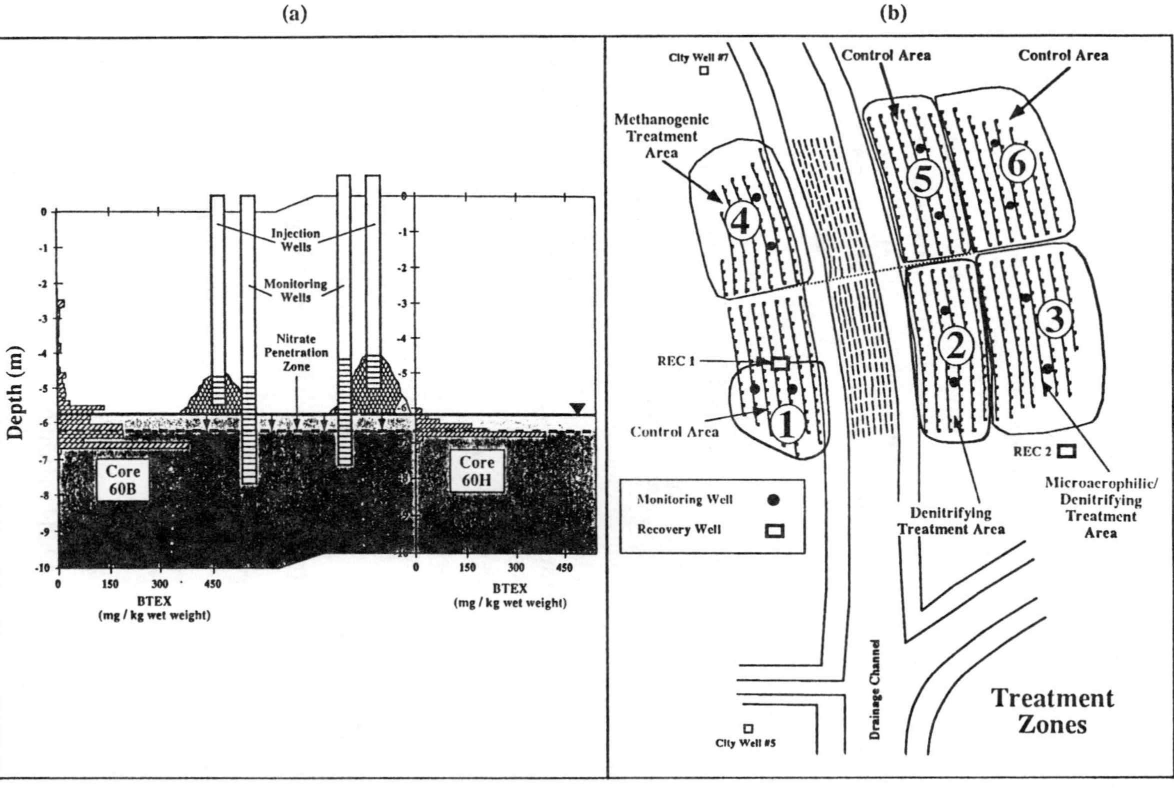

FIGURE 4. (a) Cross section of BTEX distribution in Cell #2, showing location of injection and monitoring wells, and calculated depth of nitrate penetration zone, and (b) treatment design schematic, showing location of treatment cells and recovery wells.

supplement to yield approximately 8 mg/L nitrate-nitrogen, and the Cell #3 recharge will receive the nitrate amendment as well as oxygen to yield an additional 1.5 mg/L oxygen. Each active treatment cell will receive 654 m^3/d recharge. Recharge passing through the contaminated zone will be captured by recovery wells REC #1 and REC #2 (Figure 4b). REC #1 will extract 1,960 m^3/d that will be amended and recirculated back through the injection wells, and REC #2 will extract 1,090 m^3/d that will be treated and disposed. Although this will result in some blending of the water from each of the active treatment cells, concentrations of the added electron acceptors remaining after passage through the contaminated interval should be insignificant, and the recharge will once again be selectively amended prior to redistribution.

Based on flow modeling, the estimated time required to recirculate 1 pore volume is 14 days (Kennedy & Hutchins 1992). The field demonstration is scheduled to operate for 1 year, with routine monitoring of water quality and periodic analyses of core samples to determine the rate and extent of bioremediation.

CONCLUSIONS

The microcosm tests have shown that many of the alkylbenzenes can be degraded under denitrifying conditions in this aquifer, although benzene may be recalcitrant unless some oxygen is available. These data can then be used to predict the limiting factors for remediation and indicate that regulatory constraints on the permissible nitrate loading will govern the length of time required for remediation, because the in situ microbial processes are sufficiently rapid to not be a limitation. The system is scheduled to begin operation in April 1993, and should provide a quantitative assessment of the efficacy and economics of nitrate-based bioremediation.

APPENDIX A

A-1. Time required for remediation of Cell #2, based on nitrate demand and nitrate application rate:

Cell #2 dimensions	34 m x 99 m x 6.1 m = 2.1×10^4 m^3
Bulk density	1.8 g/cm^3 = 1800 kg/m^3
Cell #2 mass	3.8×10^7 kg
Average BTEX concentration	33.5 mg BTEX/kg dry wt
Total mass BTEX	1.3×10^3 kg BTEX
Stoichiometry	$C_{61}H_{67} + 62.2H^+ + 62.2NO_3^- \rightarrow 31.1N_2 + 61CO_2 + 64.6H_2O$
Nitrate demand ratio	1.07 kg NO_3-N/kg BTEX
Nitrate demand	1.4×10^3 kg NO_3-N required
Application rate	470 L/min
Design nitrate concentration	10 mg/L NO_3-N
Time required	210 days

A-2. Time required for remediation of Cell #2, based on nitrate demand and microbial rate of nitrate removal measured in microcosms:

Cell #2 mass	3.8×10^7 kg dry wt
Nitrate demand	1.4×10^3 kg NO_3-N required
Measured rate of nitrate removal	2.8 mg NO_3-N/kg dry wt/d
Time required	13 days

A-3. Time required for remediation of Cell #2, based on BTEX mass and microbial rate of BTEX removal measured in microcosms:

Cell #2 mass	3.8×10^7 kg dry wt
BTEX mass	1.3×10^3 kg BTEX
Measured rate of BTEX removal	1.65 mg BTEX/kg dry wt/d
Time required	21 days

A-4. Calculation of nitrate penetration zone, based on nitrate application rate and microbial rate of nitrate removal measured in microcosms:

Cell #2 areal dimensions	3.4×10^3 m^2
Application rate	0.20 m/d
Porosity	0.30
Seepage velocity	0.67 m/d
Measured rate of nitrate removal	2.8 mg NO_3-N/kg dry wt/d
Water/solid ratio	0.17 L H_2O/kg aquifer
Nitrate removal rate	17.5 mg NO_3-N/L/d
Design nitrate concentration	10 mg/L NO_3-N
Time required	0.57 days
Penetration zone	0.40 m

A-5. Time required for remediation of Cell #2, based on BTEX concentration in the most contaminated interval and microbial rate of BTEX removal:

BTEX concentration in most contaminated interval	515 mg BTEX/kg dry wt
Measured rate of BTEX removal	1.65 mg BTEX/kg dry wt/d
Time required	310 days

REFERENCES

Aggarwal, P. K., J. L. Means, and R. E. Hinchee. 1991. "Formulation of Nutrient Solutions for In Situ Biodegradation." In R. E. Hinchee and R. F. Olfenbuttel (Eds.), *In Situ Bioreclamation: Applications and Investigations for Hydrocarbon and Contaminated Site Remediation*, pp. 51-66. Butterworth-Heinemann, Stoneham, MA.

Al-Bashir, B., T. Cseh, R. Leduc, and R. Samson. 1990. "Effect of Soil/Contaminant Interactions on the Biodegradation of Naphthalene in Flooded Soil under Denitrifying Conditions." *Appl. Microbiol. Biotechnol. 34*: 414-419.

Cotruvo, J. A., and M. Regelski. 1989. "National Primary Drinking Water Regulations for Volatile Organic Chemicals." In E. J. Calabrese, C. E. Gilbert, and H. Pastides (Eds.), *Safe Drinking Water Act: Amendments, Regulations, and Standards*, pp. 29-34. Lewis Publishers, Chelsea, MI.

Evans, P. J., D. T. Mang, K. S. Kim, and L. Y. Young. 1991. "Anaerobic Degradation of Toluene by a Denitrifying Bacterium." *Appl. Environ. Microbiol. 57*: 1139-1145.

Hilton, J., B. Marley, T. Ryther, and J. Forbes. 1992. "Pilot Test of Nitrate-Enhanced Bioremediation in a Moderate- to Low-Permeability Aquifer." In NGWA/API (Eds.), *Proceedings, Petroleum Hydrocarbons and Organic Chemicals in Ground Water: Prevention, Detection, and Restoration*, pp. 527-540. Water Well Journal Publishing, Dublin, OH.

Hutchins, S. R., J. T. Wilson, R. H. Douglass, and D. J. Hendrix. 1989. "Field and Laboratory Evaluation of the Use of Nitrate to Remove BTX from a Fuel Spill." In I. P. Murarka and S. Cordle (Eds.), *Proceedings, Environmental Research Conference on Groundwater Quality and Waste Disposal*, Chpt 29. Electric Power Research Institute, Palo Alto, CA.

Hutchins, S. R. 1991. "Optimizing BTEX Biodegradation under Denitrifying Conditions." *Environ. Toxicol. Chem. 10*: 1437-1448.

Hutchins, S. R., G. W. Sewell, D. A. Kovacs, and G. A. Smith. 1991a. "Biodegradation of Aromatic Hydrocarbons by Aquifer Microorganisms Under Denitrifying Conditions." *Environ. Sci. Technol. 25*: 68-76.

Hutchins, S. R., W. C. Downs, J. T. Wilson, G. B. Smith, D. A. Kovacs, D. D. Fine, R. H. Douglass, and D. J. Hendrix. 1991b. "Effect of Nitrate Addition on Biorestoration of Fuel-Contaminated Aquifer: Field Demonstration." *Ground Water 29*: 571-580.

Jorgensen, C., E. Mortensen, B. K. Jensen, and E. Arvin. 1991. "Biodegradation of Toluene by a Denitrifying Enrichment Culture." In R. E. Hinchee and R. F. Olfenbuttel (Eds.), *On-Site Bioreclamation: Processes for Xenobiotic and Hydrocarbon Treatment*, pp. 480-486. Butterworth-Heinemann, Stoneham, MA.

Kennedy, L., and S.R. Hutchins. 1992. "The Geologic, Microbiological, and Engineering Constraints of In-Situ Bioremediation as Applied to BTEX Cleanup in Park City, Kansas." *Remediation J.* (in press).

Kopp, J. F., and G. D. McKee. 1979. "Manual — Methods for Chemical Analysis of Water and Wastes." EPA-600/4-79-020. U.S. Environmental Protection Agency, Washington, DC.

Major, D. W., C. I. Mayfield, and J. F. Barker. 1988. "Biotransformation of Benzene by Denitrification in Aquifer Sand." *Ground Water 26*: 8-14.

Mihelcic, J. R., and R. G. Luthy. 1988. "Microbial Degradation of Acenaphthene and Naphthalene Under Denitrification Conditions in Soil-Water Systems." *Appl. Environ. Microbiol. 54*: 1188-1198.

Schocher, P. J., B. Seyfried, F. Vazquez, and J. Zeyer. 1991. "Anaerobic Degradation of Toluene by Pure Cultures of Denitrifying Bacteria." *Arch. Microbiol. 157*: 7-12.

Thomas, J. M., and C. H. Ward. 1989. "In Situ Biorestoration of Organic Contaminants in the Subsurface." *Environ. Sci. Technol. 23*: 760-766.

Wilson, J. T., L. E. Leach, M. Henson, and J. N. Jones. 1986. "In Situ Biorestoration as a Ground Water Remediation Technique." *Ground Water Monitor. Rev. 6*: 56-64.

Wilson, J. T., D. H. Kampbell, S. R. Hutchins, B. H. Wilson, and L. G. Kennedy. 1992. "Geochemical Indicators of Anaerobic Biodegradation of BTEX." Presented at the Air and Waste Management 85th Annual Meeting and Exhibition, Kansas City, MO, June 21-26.

APPLICATION OF NITRATE AS ELECTRON ACCEPTOR AT AN IN SITU BIOREMEDIATION OF AN ABANDONED REFINERY SITE: PILOT STUDY AND LARGE-SCALE OPERATION

G. Battermann, R. Fried, M. Meier-Löhr, and P. Werner

ABSTRACT

The groundwater under an abandoned refinery site is highly polluted with benzene, toluene, ethylbenzene, and xylene (BTEX) compounds (10 to 100 mg/L). The light nonaqueous-phase liquid (LNAPL) was removed, but the immobile residual saturations still exist and are endangering the groundwater quality. Laboratory tests proved the biodegradability of the hydrocarbons under both aerobic and denitrifying conditions. A pilot bioremediation was conducted to test nitrate and oxygen (through disintegration of hydrogen peroxide) as additional electron acceptors. Based on the results of the pilot study, a large-scale operation with about 500 m^3/h circulating water was installed and operated.

INTRODUCTION

Remediation of a hydrocarbon spill usually is considered successful if the contaminants in free phase (LNAPL) at the surface of the groundwater have been removed (Arbeitskreis "Wasser und Mineralöl" 1970). Due to the remaining hydrocarbons bound to the soil (immobile residual saturations), groundwater contamination after such measures can still be high. Pollutants perpetually can dissolve into the groundwater flow. Today the authorities demand more effective requirements and higher standards for remediation to reduce the contaminant risk. This paper presents the strategy applied at an abandoned refinery site, the preinvestigation necessary to perform large-scale remediation, and the preliminary experiences in applying the measures at the site.

HYDROGEOLOGICAL CONDITIONS AND GROUNDWATER CONTAMINATION

The site subsurface consists of flood-originated loam of 05 to 1 m thickness. A mixture of sand, gravelly sand, or sandy gravel with a thickness of about 20 m is located underneath. This layer is intersected at different depths with patchy silt layers of small areal extent. The permeability ranges from 7×10^{-4} to 2×10^{-3} m/s.

The discharged mixture of hydrocarbons was an intermediate of gasoline production consisting mainly of BTEX aromates. Its solubility in water is relatively high, in the range of about 250 mg/L. Upon recognition of the spill in 1974, hydraulic cleanup measures were applied immediately to prevent spreading of the LNAPL pollution. The free oil at the groundwater surface had nearly been completely removed in 1986, with the recovery of about 2,000 metric tonnes (MT) of oil.

Only the immobile residual saturations of oil in the pore volume remained in the subsurface. At a depth between 6 to 10 m in the saturated zone, the contamination covered an area of only about 7 hectares. With average concentrations of about 1 g/kg, the total amount of oil is calculated to be about 500 MT in 500,000 MT of polluted soil. Concentrations of hydrocarbons in the groundwater varied from 10 to 100 mg/L, spread over an area of about 20 hectares.

REMEDIATION STRATEGY

To reduce the costs of the permanent discharge and water treatment operations, a cleanup method with an acceptable time period had to be developed and applied before the area could be reused for another purpose. Conventional remediation techniques could be excluded because of the widespread contamination located deep beneath the surface. The most efficient and most economical solution seemed to be in situ bioremediation. Previous experiences in enhanced bioremediation of a hydrocarbon-polluted area of a large-scale case could be applied (Battermann 1986, Battermann & Werner 1984, DVWK 1991). Additional experiences using nitrate were available (Hutchins & Wilson 1991). Laboratory studies showed clearly that hydrocarbons are readily biodegradable by indigenous microflora (DVGW-Forschungsstelle 1986) under both aerobic and denitrifying conditions.

The cleanup strategy we developed consisted of the following action steps. First, intensive investigations using groundwater models were undertaken to achieve complete and effective groundwater transport in all the contaminated areas. About 520 m^3/h of groundwater are discharged by 16 wells; 400 m^3/h are recharged via infiltration ditches; the average residence time of the circulating water in the remediation area ranges from about 40 to 60 days. To protect against spreading pollution into surrounding areas, a net discharge of 100 m^3/h (Figure 1) is required.

Second, enhancement of the bioremediation in the groundwater requires an addition of electron acceptors and nutrients (N and P compounds). Due to the large amount of contaminants at the site, the dissolved oxygen in the reinjected water was not sufficient, so nitrate had to be added as an electron acceptor.

Third, as a complementary measure to increase the kinetics of biodegradation, the groundwater temperature was raised to 20°C.

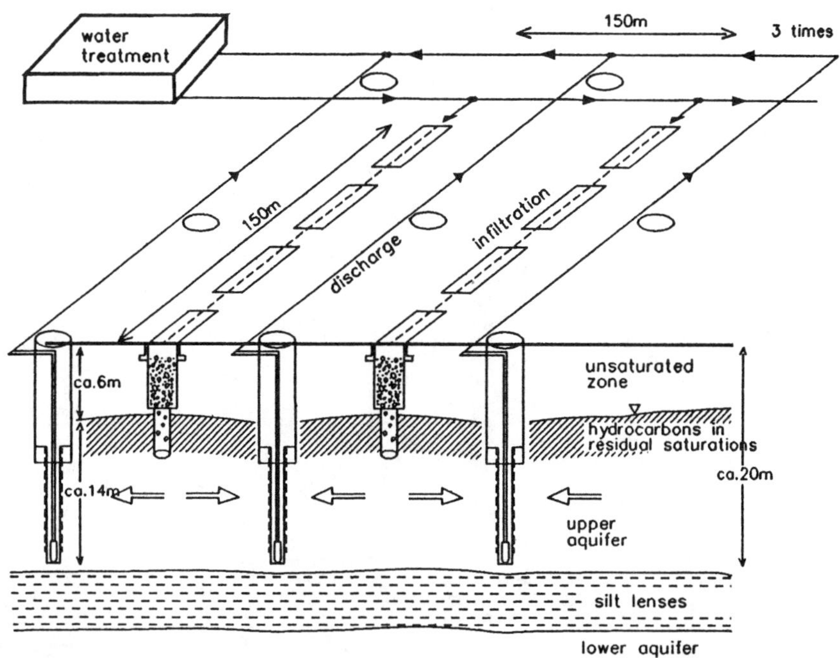

FIGURE 1. Schematic cross section of installations for large-scale remediation.

Fourth, the discharged water had to be treated before reinfiltration to remove hydrocarbons, iron, and manganese. The concentrations of the above-mentioned compounds are so high that they would cause a damaging reduction of recharge capacity in the infiltration ditches due to precipitation and biomass production.

Fifth, a pilot-scale field experiment had to be performed to verify and evaluate the results from the laboratory study.

PILOT-SCALE FIELD TEST

The pilot plant was constructed to enable a parallel but separate test for two different electron acceptors (test field I: nitrate; test field II: oxygen from disintegration of hydrogen peroxide). At a recharge rate of 5 m^3/h the mean residence time of the water in the subsurface ranged from 3 to 5 days (see Figure 2). The pilot plant operated for 27 weeks divided into 6 experimental stages focusing on different investigation targets. No loss in hydraulic conductivity was observed.

The overall result of the field experiment with respect to the applicability of nitrate and hydrogen peroxide for in situ bioremediation of the contaminants at this specific site showed that only nitrate is efficient enough and can be distributed over all contaminated areas. In contrast, hydrogen peroxide is efficient only in the vicinity of the infiltration due to its rapid catalytic disintegration (Lawes 1991, Spain et al. 1988) and the limitations in transport of dissolved oxygen. The following results therefore focus only on the results obtained with nitrate.

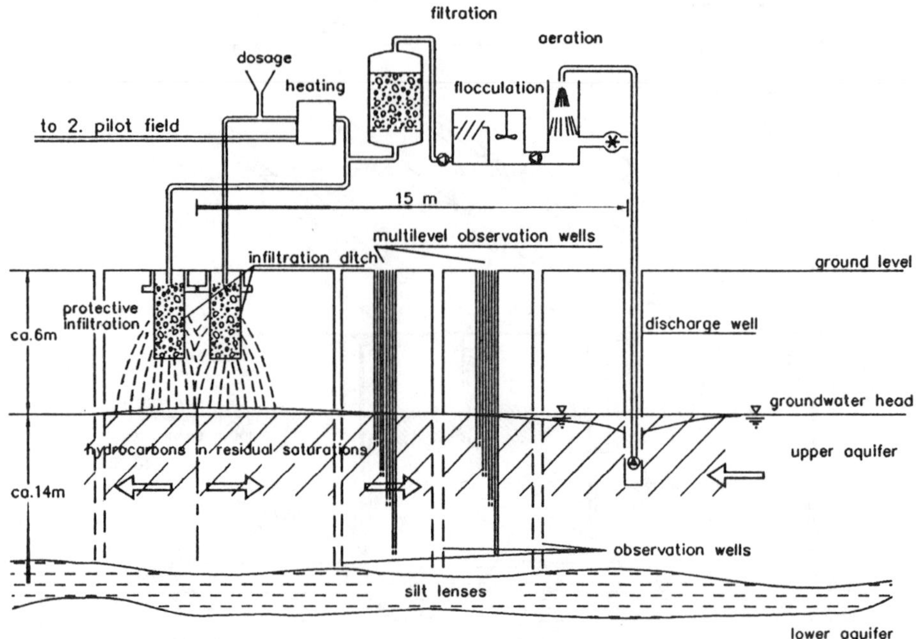

FIGURE 2. Schematic cross section of pilot plant.

Figure 3 shows average values of relevant parameters depending on the experimental stages in the direction of flow within the test field (bottom line). For better presentation of the results it was necessary to select the sequence of the different experimental stages backward or forward. When starting the test runs (experimental stage 2, tracer experiment), the environmental conditions were reducing due to the high content of dissolved aromatic hydrocarbons resulting in high concentrations of iron and manganese and a very low redox potential.

The beginning of experimental stage 3 is defined with the addition of nitrate, which immediately changed the environmental conditions. Dissolved iron was no longer detectable The concentration of manganese was reduced to half. The circulating water was almost free of hydrocarbons, indicating the degradation rate is higher than the dissolution rate of the immobilized residuals causing up to 20 mg/L hydrocarbons before the measure.

The intermediate formation of nitrite caused average concentrations of 1 mg/L during experimental stages 3 and 4. Even high nitrate concentrations (up to 200 mg/L in experimental stage 6) caused only a slight increase of nitrite up to 4 mg/L. Nitrite formation under field conditions is much less than in laboratory studies (DVGW-Forschungsstelle 1986).

According to the mass balance based on nitrate consumption, a degradation rate of 1 to 2 g of hydrocarbons per cubic meter of soil per day was calculated. This rate could be doubled by raising the groundwater temperature from 12°C to 20°C.

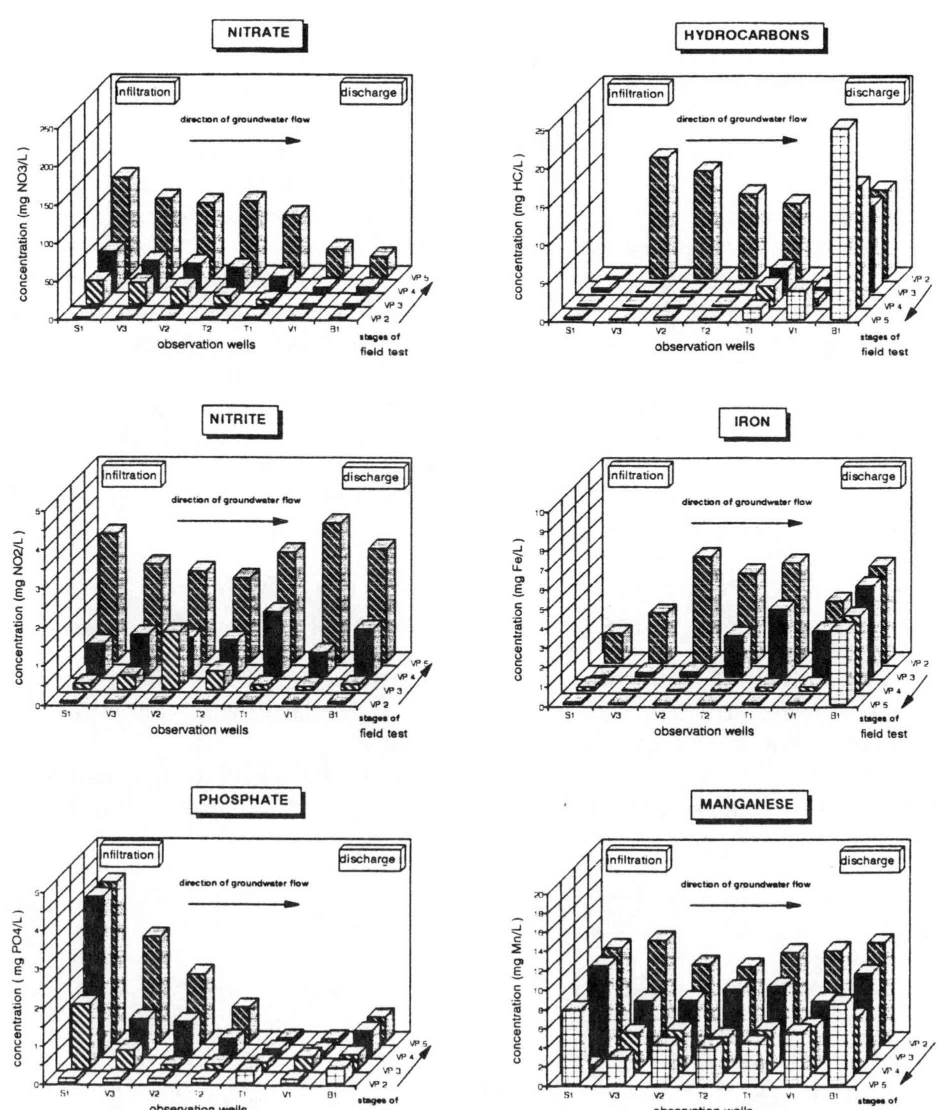

FIGURE 3. Average values of chemical parameters in groundwater during the pilot-scale field test.

A rough calculation based on the data provided from the test field shows that, for degradation of all bioavailable hydrocarbons at the site, about 5 years of operation time is necessary. A completed project (Battermann 1986) proved that the remaining hydrocarbons generally are less soluble and cause only negligible concentrations in the groundwater.

LARGE-SCALE REMEDIATION

The large-scale plant has been operating since 1991 with a capacity of 520 m^3/h. The hydraulic conditions calculated with a groundwater flow model are verified by a tracer test over the entire area. The next step in 1992 was to increase the nitrate concentration and inorganic growth factors in the recharge water to enhance bioactivity in the subsurface. Furthermore, the hydraulic system was optimized and recharge water was heated to 25°C.

A nitrate consumption of 70 to 100 mg/L during an average residence time of 50 days could be observed at an average groundwater temperature of 18°C. The degradation rate of about 1 to 15 g hydrocarbons per cubic meter of soil per day is less then the rate observed in the pilot test.

Figure 4 shows a mass balance of the overall degradation processes in the subsurface based on the measured parameters of oxygen, nitrate, carbon dioxide, and bicarbonate. During the year 1992, mineralization of about 70 MT of hydrocarbons by consuming 26 MT of oxygen and 270 MT of nitrate was calculated, using the stoichiometric balance equation

$$\begin{aligned} C_7 + 65\ NO_{3-\,\geq\,015\ C5}H_7O_2N\ + 624\ HCO_{3-} \\ + 026\ OH^- + 317\ N_2 + 02\ 2H_2O \end{aligned} \quad (1)$$

Mineralization of hydrocarbons under denitrifying conditions can be proved mainly by the formation of bicarbonate. Observed and calculated concentrations of bicarbonate correspond within the limits of 10%.

Next to the biodegradation, the stripping of volatile dissolved hydrocarbons contributes to the cleanup with 30 MT/year. The concentration of hydrocarbons

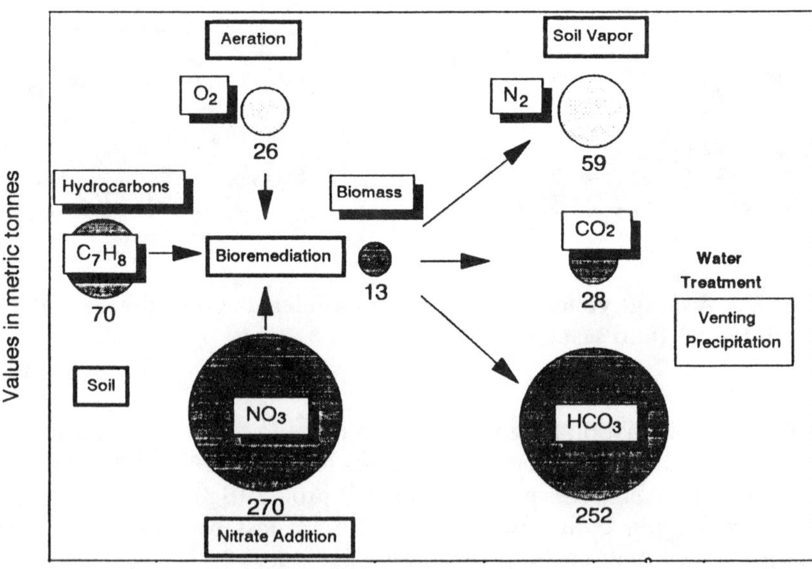

FIGURE 4. Mass balance of bioremediation (1992).

increases slightly in some wells and, on average, about 8 mg/L dissolved hydrocarbons can be detected. The explanation for this unexpected increase is a mobilizing effect probably caused by the formation of biosurfactants. Laboratory investigations have shown that biosurfactants are produced under growth-limiting conditions (Francy et al. 1991).

The rise in redox potential in the groundwater caused, as expected, a decrease of iron and manganese. Concentrations of 1 mg/L for iron and 2 mg/L for manganese currently exist in the influent of the treatment plant.

The operation of the large-scale remediation so far confirms that the experiences of the pilot plant cannot be transferred completely due to multifactorial interference in the system. Nevertheless the pilot study has provided the necessary and valuable background as a basis for a large-scale remediation.

The removal of the hydrocarbons endangering the groundwater at the site is expected to be finished in about 5 years of operation, as scheduled. Optimizing the system to shorten the operation time and reduce costs is the target.

REFERENCES

Arbeitskreis "Wasser und Mineralöl." 1970. *Beurteilung und Behandlung von Mineralölunfällen auf dem Lande im Hinblick auf den Gewässerschutz*, 2nd ed. Bundesministerium des Innern, Bonn, Germany.

Battermann, G., and P. Werner. 1984. "Beseitigung einer Untergrundkontamination mit Kohlenwasserstoffen durch mikrobiellen Abbau." *GWF-Wasser/Abwasser 125*(8):366-373.

Battermann, G. 1986. "Decontamination of Polluted Aquifers by Biodegradation." In *Contaminated Soil*, pp. 711-722. Martinus Nijhoff Publishers, Dordrecht, Netherlands.

DVGW-Forschungsstelle am Engler-Bunte-Institut der Universität Karlsruhe. 1986. *Chemische und mikrobiologische Untersuchungen von Boden- und Wasserproben.* Karlsruhe, Germany.

DVWK-Schriften 98. "Sanierungsverfahren für Grundwasserschadensfälle und Altlasten — Anwendbarkeit und Beurteilung" (especially chapter 43). Verlag Paul Parey, Hamburg and Berlin, Germany.

Francy D. S., J. M. Thomas, R. L. Raymond, and C. H. Ward. 1991. "Emulsification of hydrocarbons by subsurface bacteria." *J. Ind. Microbiol. 8*:237-246.

Hutchins S. R., and J. T. Wilson. 1991. "Laboratory and Field Studies on BTEX Biodegradation in a Fuel-Contaminated Aquifer under Denitrifying Conditions." In R. H. Hinchee and R. F. Olfenbuttel (Eds.), *In Situ Bioreclamation*, pp. 157-172. Butterworth-Heinemann, Boston, MA.

Lawes, B. C. 1991. "Soil-Induced Decomposition of Hydrogen Peroxide." In R. H. Hinchee and R. F. Olfenbuttel (Eds.), *In Situ Bioreclamation*, pp. 143-156. Butterworth-Heinemann, Boston, MA.

Spain, J. C., J. D. Milligan, D. C. Downey, and J. K. Slaughter. 1988. "Enhanced Biodegradation: Rapid Decomposition of H_2O_2 by Bacteria." *Ground Water 27*: 163-167.

APPLYING FIELD-EXPEDIENT BIOREACTORS AND LANDFARMING IN ALASKAN CLIMATES

C. M. Reynolds, M. D. Travis, W. A. Braley, and R. J. Scholze

ABSTRACT

Many contaminated soil sites in cold regions are isolated and remote. We have evaluated the feasibility of on-site treatment by using landfarming and field-expedient bioreactors, such as recirculating leachbeds. Landfarming can be used to treat the less-contaminated soil which often comprises the bulk of the contaminated soil volume. Highly contaminated soils can be readily contained and treated on site using recirculating leachbeds. In field evaluations, the spatial average of total petroleum hydrocarbon concentration in a diesel-contaminated soil decreased from 6,200 mg/kg dry soil to 280 mg/kg in approximately 7 weeks. We used geostatistical techniques to delineate the spatial variability in total petroleum hydrocarbon concentration in the landfarm soil. Spatial variability decreased with time and provided information that may provide guidance for cost-effective sampling. At another site, a recirculating leachbed was used to decrease TPH concentration in diesel-contaminated soil from between 300 mg/kg and 47,000 mg/kg to between 240 and 570 mg/kg in a 5-week period. The complementary use of these two technologies provides a cost-effective treatment option.

INTRODUCTION

The extreme climate, short operating season, inaccessibility of sites, and limited alternatives present unique challenges to the successful bioremediation of contaminated soils in the arctic and subarctic regions of Alaska. These constraints increase the importance of both optimizing biodegradation rates and using relatively rugged, uncomplicated designs. We are currently evaluating the techniques of landfarming and recirculating leachbeds for both individual and complementary use. Landfarming can be used to treat large quantities of soil. Recirculating leachbeds are readily adapted to specific sites and provide for reuse of mechanical and electrical systems. Our objectives include application and demonstration of these technologies in conjunction with identifying and reducing restrictions to optimum biodegradation rates.

Landfarming is a frequently chosen treatment for contaminated soil because of containment, relatively low cost, and high potential for success (Harmsen 1991). Lining and leachate recovery systems, frequently used to assure that soluble fractions or high concentrations do not leach, essentially double the construction costs (Bell et al. 1989, Lynch 1991). Two possibilities for maximizing cost effectiveness of landfarms are using centrally located, lined landfarms to treat multiple batches of contaminated soil where the volume of contaminated soil is large, and developing the ability to appropriately ease liner needs at sites where the volume of contaminated soil does not require multiple batches.

Many contaminated-soil sites in cold regions are isolated and remote, thereby limiting the feasibility of using centrally located treatment facilities. Depending on soil volume, highly contaminated soils can be readily contained and treated on site using recirculating leachbeds, resulting in two benefits. First, at remote locations, on-site treatment is possible, and second, at other areas, "hot-spots" that contain higher concentrations or more mobile fractions than the general area can be locally treated. The latter use, coupled with additional information on the less-contaminated soil at a site, could provide a mechanism for landfarming the bulk of the soil without liners and without increasing the leaching potential.

MATERIALS AND METHODS

Landfarm

The landfarm is located at Fairbanks International Airport in Fairbanks, Alaska. Specifics have been documented previously (Braley 1991, Braley 1993, Reynolds 1993, Walker & Travis 1990). In brief, soil contaminated with #1 fuel oil from crash-fire training and a release of between 6,000 and 10,000 gal. (23,000 and 38,000 L) of a mixture containing diesel #2 fuel and unknown petroleum waste products was placed in an on-site landfarm that covered approximately 1 acre (0.405 ha). Nutrient amendments were 270 lb (122 kg) nitrogen as ammonium nitrate, 34 lb (15.4 kg) of phosphorus as triple super phosphate, and 26 lb (11.8 kg) of potassium as muriate of potash. Nutrients were added on June 22 and July 22 and the soil was periodically tilled to an approximate 8-in. (20 cm) depth. The volume of soil being actively treated was approximately 1200 cu. yd. (917 m^3).

Soil was placed in the landfarm in August 1991. This report focuses on the soil remediation that occurred during the 1992 summer, which was the first full summer of treatment after the soil had over-wintered. Twenty-five soil samples, each a composite of samples, were taken on June 4, June 24, and August 18, 1992, from a uniformly spaced grid. Sampling procedures were described earlier (Reynolds 1993). Extraction was done using methylene chloride and sonification. Values reported herein are gravimetric values representing total methylene chloride extractable hydrocarbons. Volatile losses of hydrocarbons from the soil were not measured. To evaluate spatial and temporal variability of total petroleum hydrocarbon (TPH) values within the landfarm, contour maps of TPH concentrations were generated using Kriging, the geostatistical technique that provides the best estimate of TPH contours from the available data. The disadvantage of Kriging is that such figures can be misinterpreted to represent degradation processes rather than the combined effects of degradation processes and spatial variability. The advantages are increased information concerning the highly variable spatial distribution of TPH in the landfarm and knowledge of the temporal changes in spatial variability. The spatially averaged TPH values for each sampling date were calculated by integrating the volume under the TPH contour maps.

Recirculating Leachbeds

Recirculating leachbeds are similar to slurry reactors. The concept is to develop a lined containment area to serve as a bioreactor (Figure 1). Either a pit

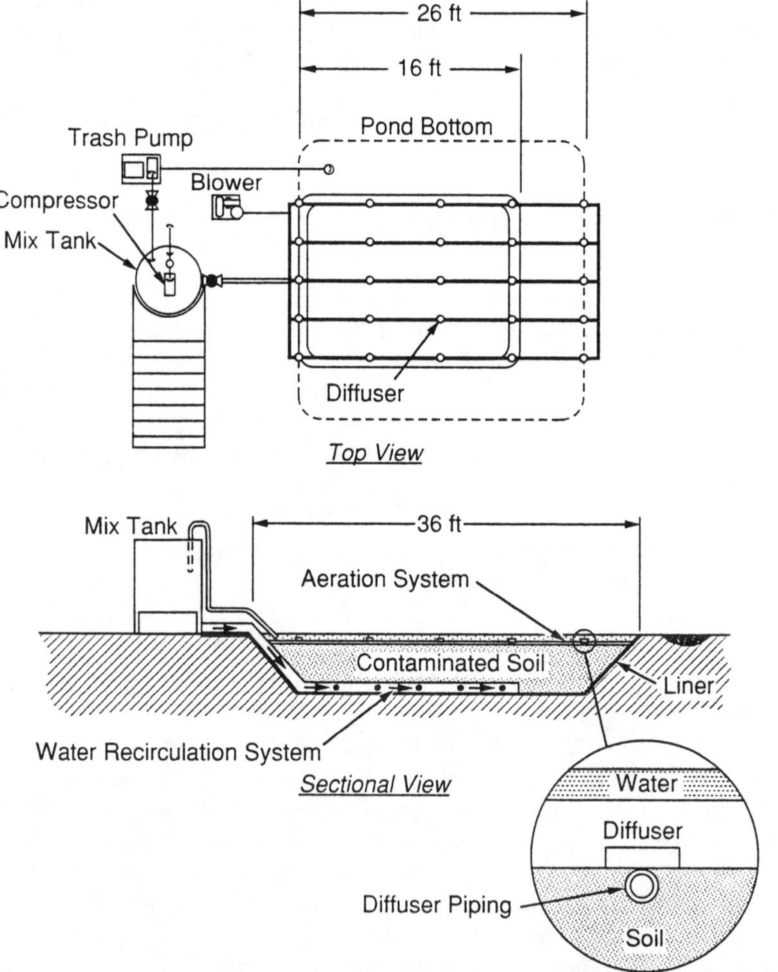

FIGURE 1. Typical design for a recirculating leachbed (adapted from Braley 1993).

(generally resulting from the excavation), a bermed perimeter, or a combination can be used depending on available materials. Contaminated soil is placed into the bioreactor and, through an inexpensive PVC distribution system, aerated and nutrient-amended water is recirculated into the bottom of the bioreactor, upwards through the contaminated soil, and then through overlying ponded and aerated water. Skid-mounted mechanical systems include a mixing tank and circulation pumps for water and air.

RESULTS AND DISCUSSION

Landfarm

The TPH, estimated from spatially averaged values at 25 nodes in the landfarm in June 1992, was 6,200 mg/kg dry soil (Figure 2). The corresponding spatial average decreased to 280 mg/kg by August 1992. Kriging provided an estimate of the spatial distribution of TPH within the landfarm (Figure 3). Time-series maps of kriged TPH values provided an indication of the temporal variability that was observed, notwithstanding extensive sampling precautions (Reynolds 1993). Temporal changes in the arithmetic TPH means and kriged TPH means were similar, although kriged values gave a substantially higher initial TPH estimate on June 4 (Figure 2). Air and soil temperature increased from approximately 15°C to between 20 and 25°C during the latter part of June (Figure 4). On the August 18 sampling, the relative uniformity of the TPH map and the smaller range of values suggested that, as remediation processes proceeded

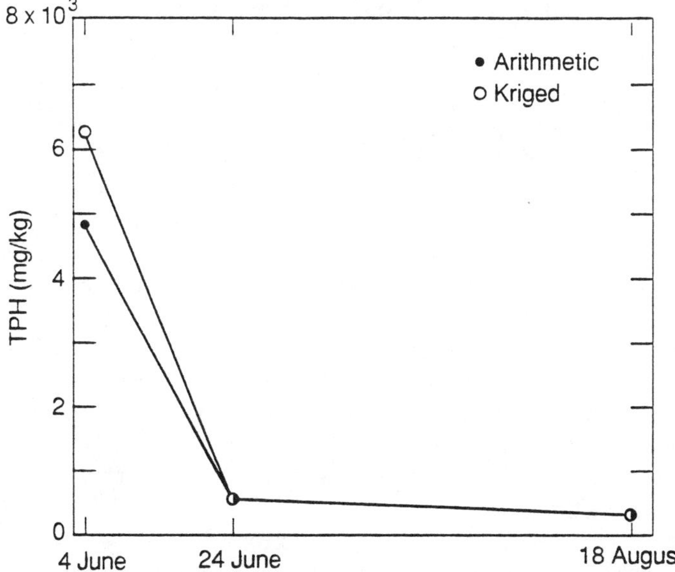

FIGURE 2. Arithmetic mean TPH and kriged mean TPH values for the landfarmed soil.

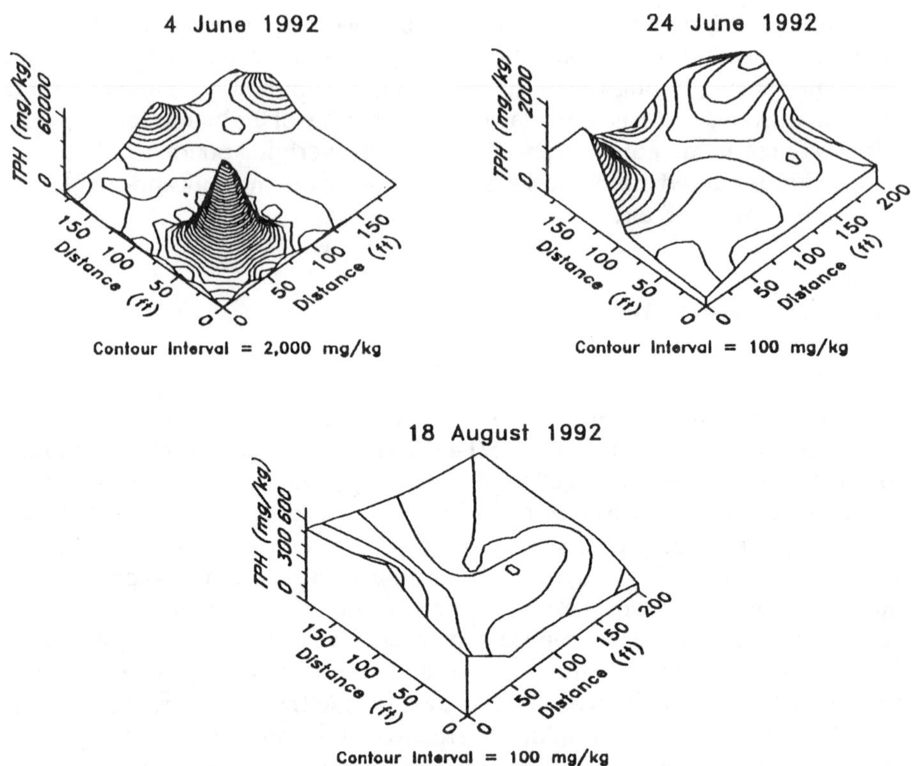

FIGURE 3. TPH distribution in the landfarm at three sample times. Note the differences in the TPH axis and contour intervals.

through a season, fewer samples would be needed to provide similar levels of certainty in estimating residual TPH levels.

Recirculating Leachbed

Due to the more dynamic nature of the treatment, recirculating leachbeds appeared to provide more uniform treatment. TPH levels in a diesel and waste-oil-contaminated soil decreased from between 300 and 47,000 mg/kg to between 240 and 570 mg/kg in a 5-week period at Anatuvuk Pass, a northern Alaskan site. Corresponding values for petroleum-hydrocarbon-degrading microorganisms, as determined by the sheen screen technique (Brown & Braddock 1990), increased from 1.8 E4/g to 4.5 E6/g. Final diesel-range organics after 8 weeks of treatment were less than 200 mg/kg.

The more rapid remediation of the recirculating leachbed can be used alone or in conjunction with landfarming and could provide an expedient means to

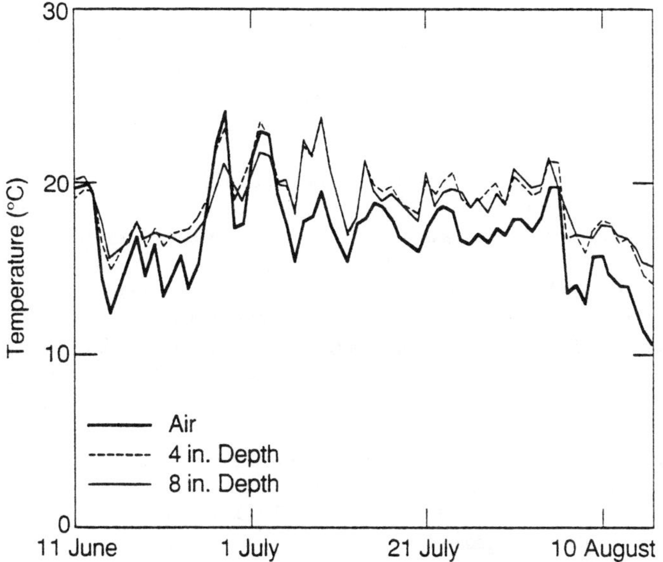

FIGURE 4. Air and soil temperatures at the landfarm.

treat highly contaminated soil, thereby increasing the potential for landfarming the remaining soil without liner requirements.

ACKNOWLEDGMENTS

Support for this work was provided by joint funding from the U.S. Army Corps of Engineers Construction Productivity Advancement Research (CPAR) program, the Alaska Department of Transportation and Public Facilities (AKDOT&PF), the Cold Regions Research and Engineering Laboratory (CRREL) Work Unit Number AF25-RT-05 Low Temperature Biotreatment and the Construction Engineering Research Laboratory (CERL). D. C. Leggett (CRREL) provided useful discussions during the project. L. B. Perry and P. W. Schumacher (CRREL) conducted the laboratory analysis.

REFERENCES

Bell, C. E., P. T. Kostecki, and E. J. Calabrese. 1989. "State of Research and Regulatory Approach of State Agencies for Cleanup of Contaminated Soils." In E. J. Calabrese and P. T. Kostecki (Eds.), *Petroleum Contaminated Soils*, Vol. 2, pp. 73-96. Lewis Publishers, Chelsea, MI.

Braley, W. A. 1991. "The Fairbanks International Airport Experimental Bioremediation Project." Presented at the BP Exploration Alaska Soil Remediation Workshop, Anchorage, AK, Nov. 19-20.

Braley, W. A. 1993. "Fairbanks International Airport Bioremediation Project Preliminary Report." Alaska Department of Transportation and Public Facilities, Fairbanks International Airport, Fairbanks, AK.

Brown, E. J., and J. F. Braddock. 1990. Sheen Screen, A Miniaturized Most-Probable Number Method for Enumeration of Oil-Degrading Microorganisms. *Appl. Environ. Microbiol.* *56*:3895-3896.

Harmsen, J. 1991. "Possibilities and Limitations of Landfarming for Cleaning Contaminated Soils." In R. E. Hinchee and R. F. Olfenbuttel (Eds.), *On-Site Bioreclamation, Processes for Xenobiotic and Hydrocarbon Treatment*, pp. 255-272. Butterworth-Heinemann, Stoneham, MA.

Lynch, J. E. 1991. "Application of Bioremediation to Contaminated Soils." Presented at Bioremediation Workshop at Sixth Annual Conference on Hydrocarbon Contaminated Soils, University of Massachusetts, Amherst, MA, Sept. 24.

Reynolds, C. M. 1993. "Field Measured Bioremediation Rates in a Cold Region Landfarm: Spatial Variability." In P. T. Kostecki and E. J. Calabrese (Eds.), *Proceedings of the Seventh Annual East Coast Conference on Hydrocarbon Contaminated Soils.* Lewis Publishers, Ann Arbor, MI (in review).

Walker, G. D., and M. D. Travis. 1990. *Electromagnetic Induction Survey of the Fairbanks International Airport Crash, Fire, and Rescue Burn Site for Hydrocarbon Contamination Plumes in Frozen Soil.* Report NO. AK-RD-90-11, Alaska Department of Transportation and Public Facilities, Statewide Research, Fairbanks, AK.

BIOREMEDIATION OF THE *EXXON VALDEZ* OIL SPILL: MONITORING SAFETY AND EFFICACY

R. C. Prince, J. R. Clark, J. E. Lindstrom, E. L. Butler, E. J. Brown, G. Winter, M. J. Grossman, P. R. Parrish, R. E. Bare, J. F. Braddock, W. G. Steinhauer, G. S. Douglas, J. M. Kennedy, P. J. Barter, J. R. Bragg, E. J. Harner, and R. M. Atlas

ABSTRACT

The application of fertilizers to stimulate oil biodegradation by indigenous microorganisms was a major part of the cleanup activities following the *Exxon Valdez* oil spill in Prince William Sound, Alaska. Here we describe a substantial field-monitoring program designed to assess the efficacy and safety of the technique on a time scale of relevance to the decision-making process directing the cleanup. We show that fertilizer application at the study sites was generally successful in delivering nutrients throughout the oiled sediment, that it stimulated microbial hydrocarbon degradation activity, and that it increased the degradation rate of the spilled oil severalfold. We also show that no adverse ecological effects were observed following fertilizer application. Fertilizer application was thus a safe and effective means to enhance the natural cleansing of the oil spill, and may well have applicability in other marine spills where oil reaches similar shorelines.

INTRODUCTION

The grounding of the *Exxon Valdez* on Bligh Reef on 24 March, 1989, released almost 11 million U.S. gallons (37,000 metric tons) of North Slope crude oil into the waters of Prince William Sound, Alaska. A major storm a few days later spread the oil, which by this time had lost perhaps 30% of its bulk by evaporation and dissolution, onto the shorelines of the numerous islands in the western part of the Sound and out into the Gulf of Alaska. Subsequent cleanup efforts involved physically washing the oil off the shoreline and collecting it with skimmers, followed by the application of carefully chosen fertilizers to stimulate the biodegradation of the remaining oil by the indigenous microbial populations. More than 74 miles (120 km) of shoreline were treated in this way in 1989, and approximately a third of this shoreline was treated again in 1990 (Harrison 1991,

Nauman 1991, Galt et al. 1991, Pritchard & Costa 1991, Bragg et al. 1992). Much smaller bioremediation programs continued in 1991 and 1992 (Bragg et al. 1992).

Two fertilizers were used in the bioremediation program. Customblen™ 28-8-0 (Grace-Sierra Chemicals, Milpitas, California) is a slow-release formulation of soluble nutrients encased in a polymerized vegetable oil; it contains ammonium nitrate, calcium phosphate, and ammonium phosphates with a nitrogen to phosphorus ratio of 28:3.5. Inipol EAP22™ (CECA S.A. 92062 Paris La Defense, France), an oleophilic fertilizer designed to adhere to oil (Ladousse & Tramier 1991), is a microemulsion of a saturated solution of urea in oleic acid, containing tri(laureth-4)-phosphate and 2-butoxyethanol. It was applied only where there was surface oil. The application rates were approximately 360 g/m^2 Inipol EAP22™ plus 17 g/m^2 of Customblen™ to areas that had surface oil, or 95 g/m^2 Customblen™ to areas that were clean on the surface but had subsurface oil.

The monitoring program described here was designed at the request of the State of Alaska to assess the efficacy and environmental safety of the fertilizer application program. Methods and measurements were selected to provide data within a 6-week time period to assist in decision-making on the continued use of bioremediation in 1990. To assess its efficacy, the program focused on the presence of fertilizer nutrients in the beach interstitial water, on the stimulation of microbial populations and biodegradation potential at the surface and in subsurface sediments, and on changes in the amount and composition of oil in the sediments. To assess the environmental safety of the application protocols, the program examined the potential toxicity of the fertilizers to aquatic biota, nutrient loading in the water that might stimulate algal growth, and planktonic chlorophyll in nearshore water. It also assessed the amount of dissolved petroleum hydrocarbons in nearshore water to address the potential that enhanced microbial activity on the shorelines might release petroleum into the water column.

The monitoring program continued beyond the initially mandated six weeks to characterize the efficacy and safety more completely. A "Final Report" including all the data and the conclusions to date was given to the Federal On-Scene Coordinator in December 1990 (Prince et al. 1990). This confirmed and extended the preliminary reports provided during the summer which concluded that bioremediation was effective and safe. At that time effectiveness was assessed by combining information gained from the oil chemistry, which indicated that the natural rate of degradation had been substantial, with microbial analyses that showed that hydrocarbon degradation activity was stimulated severalfold following fertilizer application. Subsequent multivariate statistical analyses have confirmed and extended these conclusions by showing that changes in the oil chemistry during the summer can be definitively attributed to enhanced degradation following fertilizer application (Bragg et al. 1992, 1993).

Three separate shoreline sections were monitored (Figure 1), all at the northern end of Knight Island (Lethcoe 1990, Teal 1991). Key selection criteria were the size of the section and the presence of two oiled areas within the section that appeared to be similar so that one could be treated with fertilizer while the other could be left as a reference. One site (KN-211E) was on a high-energy beach facing the Sound; the other two sites (KN-132B and KN-135B) were low-energy beaches

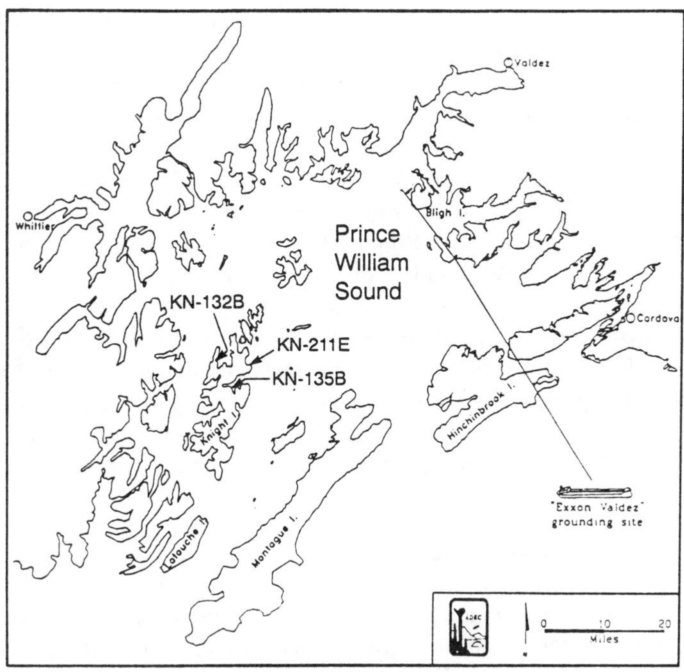

FIGURE 1. Prince William Sound, Alaska, showing the site of the grounding of the *Exxon Valdez*, and the three sites monitored in the experiments reported here.

in protected bays. Site KN-132B was in Herring Bay at the mouth of an anadromous stream, and site KN-135B was in the Bay of Isles. KN-211E principally had subsurface oil, KN-132B had only surface oil, and KN-135B had both. Although more heavily oiled than the majority of shorelines receiving bioremediation treatment in 1990, these sites provided the opportunity to assess bioremediation of surface and subsurface oil, alone and in combination. They were chosen because each had appropriate sediment (small gravel) beneath the surface armor, the appearance of reasonably uniform oiling throughout the oiled zone, no obvious input of surface water from the supratidal zone, and an area large enough to be subdivided into treated and reference areas. Both portions of each site received similar manual treatment (removal of oiled debris and tarmats, rolling of large rocks to expose oiled surfaces, etc.) before fertilizer application.

The site in the Bay of Isles, KN-135B, had both surface and subsurface oil, and had not received bioremediation treatment in 1989; we discuss this beach in this paper. (Data from the other sites are presented in Prince et al. 1990, with a full discussion of the microbial data in Lindstrom et al. 1991, and of the interpretation of the oil chemistry data in Bragg et al. 1992, 1993.) Typically the fine sediment was overlain by 10 cm of mixed pebble and cobble, and there was substantial surface and subsurface oiling. Although the extent of oiling and depth of penetration were quite variable within the segment, the areas chosen for sampling had the appearance of heavy oiling to a depth of about 40 cm. An area of approximately 2,025 m^2 received 103 g/m^2 of Customblen™ and 361 g/m^2 of Inipol EAP22™ (55 g total N/m^2) on May 21, 1990. A second application, of 17 g/m^2 of Customblen™ and 303 g/m^2 of Inipol EAP22™ (27 g total N/m^2), was made on July 13. The entire segment was treated with Customblen™ at an application rate of 91 g/m^2 (25 g total N/m^2) on August 1, and a final application of Customblen™ at 17 g/m^2 and Inipol at 361 g/m^2 (31 g total N/m^2) was applied to the entire segment on September 5, 1990.

These applications were intended to follow the application guidelines being used for the 1990 cleanup program (26 g total N/m^2), but in fact the initial application of Customblen™ was six-fold higher than planned. The site thus received approximately twice the recommended amount of available nitrogen on the first application. Subsequent applications conformed to the application guidelines.

SAMPLING STRATEGY

Three undisturbed, visually similar sampling areas were selected on the areas destined to be the fertilized and unfertilized portions of each shoreline, and a perforated steel pipe (5 cm diameter, 70 cm long) was driven into the beach material at each sampling location; the pipes enabled the gathering of interstitial water and served as the centers of each sediment sampling area. An additional well, coated with silicone sealant so that only the bottom 10 to 15 cm remained permeable, was installed on the fertilizer side of the beach. Samples of interstitial water were collected before and at 2, 4, 8, 15, 32, 57, 70 and 78 days after the first fertilizer application, and analyzed for the presence of dissolved nitrogen and phosphorus nutrients (Gilbert & Loder, 1977).

Measurements initially were made on samples from both the top and the bottom of each well, but because there were no differences, only samples from the bottoms of the wells were analyzed after day 32. There also were no significant differences between samples collected from the open and from the sealed wells in the fertilized area. Dissolved oxygen, salinity, temperature, and pH also were measured. With the same frequency, triplicate surface and subsurface samples of sediment (approximately 200 g) were taken within a 1.5-m radius of each well, for analysis of hydrocarbon degradation potentials and microbial populations (Brown & Braddock 1990, Brown et al. 1992, Lindstrom et al. 1991). For sediment sampling, the surface was defined as the uppermost stratum of the fine-grained gravel sediment; surface samples were taken after mixing the top 2 to 5 cm of this material. Subsurface samples, again of small gravels, were taken 30 cm deeper. Additional samples (500 g), taken before and at 32, 70, and 109 days after the first application, were analyzed for oil loading by solvent extraction, and the chemical composition of this oil was quantified with gas chromatographic techniques (Butler et al. 1991, Douglas et al. 1992, 1994). Each sample came from previously undisturbed sediment.

Toxicity issues were addressed in the first few tides after application, when the concentrations of added nutrients were greatest in nearshore waters. Samples were collected in the fertilized area of the site and at a reference site (control) uninfluenced by the fertilizer applications (approximately 0.8 km distant). The strategy was designed to obtain worst-case representations of fertilizer entering the nearshore environment by sampling at a place along the shoreline where there was minimal dilution, and at a time during the tide that allowed the maximum opportunity for fertilizer release into overlying water. Water samples were collected at 0.5 m depth when water depth was 1 m. Samples were collected 1 hour after fertilizer application and then at the midpoint of the first, second, third, fifth and seventh outgoing tides. Samples from the reference site were collected at the same time as those from the fertilized area before application, and on the second and fifth subsequent tides. Samples were kept on ice and shipped via air express to MEC Analytical Systems Inc. (Tiburon, California). A shrimp-like crustacean, *Mysidopsis bahia*, was used as a surrogate for indigenous species, and 96-hour acute toxicity tests were conducted following accepted effluent testing guidelines (Peltier & Weber 1985). Testing began the day samples were received by the laboratory, i.e., 1, 2, or 3 days after field collection. Because no toxicity was detected at any of the three sites after the first application of fertilizer, even with the over-application of fertilizer described above, toxicity testing was not continued for subsequent fertilizer applications.

Nearshore water samples for ammonia and nitrate analyses were collected concurrently with the toxicity samples, using the same collection protocol and schedule, but with replicate samples. As with the toxicology assessment, this strategy was designed to characterize the exposures of nearshore biota to toxic components of the fertilizer nutrients under presumed worst-case conditions. Samples were analyzed using standard oceanographic methods for nutrient analyses (Gilbert & Loder 1977, United States Environmental Protection Agency 1988a).

Water samples for chlorophyll and total petroleum hydrocarbon (TPH) analyses were collected in the same manner as those for toxicity testing, but on the schedule used for monitoring interstitial water. The three samples were taken offshore of the fertilized area, the unfertilized area, and from a distant reference site. Chlorophyll analyses followed oceanographic fluorometric techniques, and water samples were analyzed by infrared spectrometry for total petroleum hydrocarbons (Parsons et al. 1984, United States Environmental Protection Agency 1988b). Because no significant levels of petroleum hydrocarbons were detected, this analysis was discontinued after the second application of fertilizer. Chlorophyll analyses were continued to monitor for potential stimulation of algal growth.

RESULTS

Interstitial Water

Figure 2A shows the total nitrogenous nutrients measured in the interstitial water collected from sampling wells on the fertilized and unfertilized portions

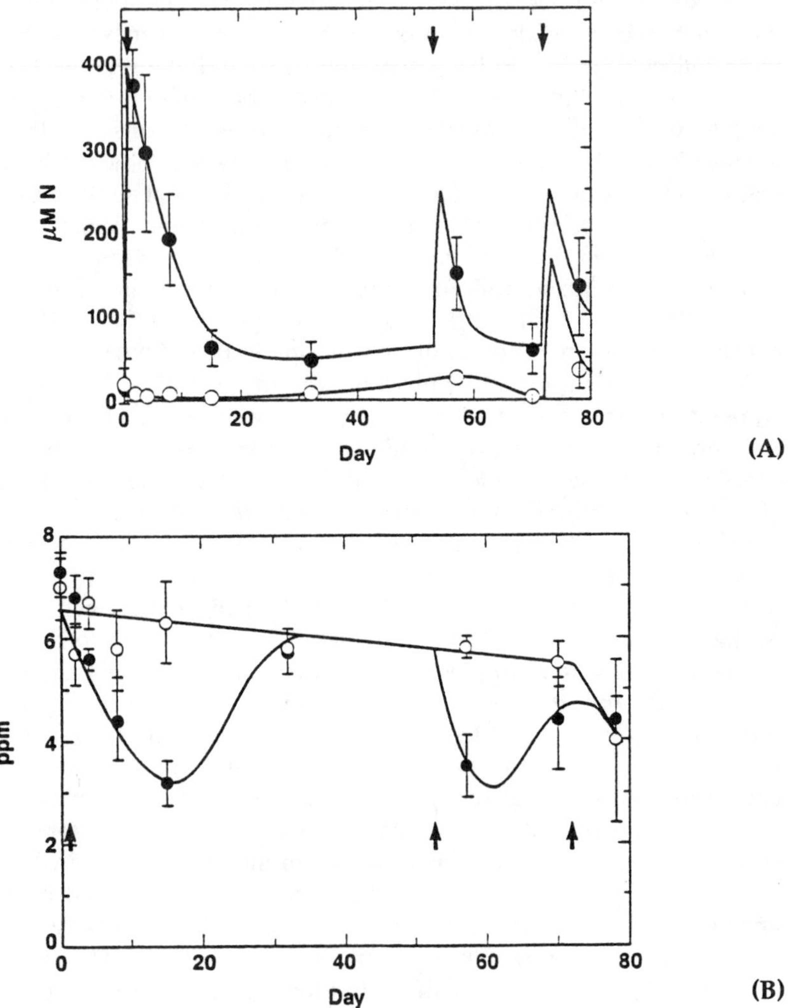

FIGURE 2. (A) Soluble nitrogenous nutrients in interstitial water from (•) the fertilized and (o) the unfertilized portions of the test site. The data represent the sum of nitrate, nitrite, ammonium, and organic nitrogen. All samples filtered through 45-μm filters on the support vessel; samples for Kjeldahl analysis acidified; those for inorganic analyses frozen before shipment. No significant differences between samples from the top/bottom of each well. Means (±1 s.d.) of all samples from each side of the beach. Arrows indicate application of fertilizers (third to both portions of the beach). Line through the points following second,third applications indicates predicted kinetics if they followed those measured in more detail after the first application. Day 0 was May 21, 1990. (B) Dissolved oxygen in the interstitial water, measured in the field with CHEMetrics K-7512 ampules. Symbols, arrows, and line as above.

of the site. With the exception of the measurements on day 2, when there was approximately 90 μM soluble organic nitrogen which we attribute to the urea in the Inipol EAP22™, the total was made up of essentially equimolar concentrations of ammonium and nitrate, as expected from the composition of the Customblen™ fertilizer. Clearly, the application of fertilizer to the surface of the beach was an effective means of getting nitrogenous nutrients into the oiled zone of the subsurface for at least 30 days. Analyses of Customblen™ pellets collected 52 days after application indicated that they had lost approximately 80% of their contents at this time.

The fertilizers also contained phosphate, but no substantial increase in this nutrient was detected in the interstitial water; it ranged from 0.1 to 0.5 μM during the monitoring period on both portions of the beach. Analyses of Customblen™ pellets collected 52 days after application indicated that phosphate had left the pellets with the nitrogenous nutrients, so we assume that the phosphate was either taken up by indigenous organisms or precipitated, perhaps as calcium phosphate, upon leaving the pellet.

There was clear evidence of increased oxygen consumption in the interstitial water following the application of fertilizer (Figure 2B). The temperature of the interstitial water varied from 9 to 17.5°C during the 3 months of the program, but typically was around 13°C. The salinity varied from 0.5 to 2.8%, reflecting substantial freshwater input from the backshore on occasion, but typically was around 1.5%. The pH varied from 7.0 to 7.4.

Microbial Analyses

The rationale for the fertilizer applications was that they would stimulate the metabolism of the indigenous oil-degrading microorganisms. This would be reflected by increases in the hydrocarbon biodegradation potential, and perhaps by increased numbers of microbes in fertilized portions of each beach. Because there is some evidence that the biodegradation of paraffins and aromatic molecules may involve different microbial populations (Foght et al. 1990 [although see for example Gauthier et al. 1992]), the mineralization potentials for both ^{14}C-*n*-hexadecane and ^{14}C-phenanthrene to $^{14}CO_2$ were assessed for shoreline populations (Lindstrom et al. 1991). Figure 3 shows the assayed mineralization potentials, presented as the ratio of potentials in fertilized to potentials in unfertilized samples. As expected for environmental samples, the range of individual values is quite high, but there was a statistically significant increase in the mineralization potential following fertilizer application (Table 1).

Microbial populations in the sediments were enumerated by standard Most Probable Number (MPN) techniques (Lindstrom et al. 1991); there were usually significantly more bacteria on the fertilized area, especially after the second application of fertilizer (Table 1).

The facts that the relative differences in mineralization potentials (Figure 3) are not directly proportional to the relative levels of nutrients in the interstitial water (Figure 2A), and that the microbial populations increased following fertilizer

TABLE 1. Statistical treatment of the data from the microbial analyses[(a)].

Day	Heterotroph MPN	Oil Degrader MPN	Hexadecane Mineralization	Phenanthrene Mineralization
Surface Sediments				
0	86.9*	0.0	28.1	72.9*
2	78.5*	80.6	99.9	60.5*
4	77.0*	75.0	69.2	66.3*
8	66.8	86.6*	99.9	27.4*
15	27.4	40.4	99.9	99.8
32	71.1	N.A.	99.9	99.9
52	62.1	40.4*	99.9	99.9
56	99.9	99.4	99.9	98.5
70	72.9	97.1	99.9	99.5
78	N.A.	62.1	99.7	62.1
Subsurface Sediments				
0	16.6*	46.5	58.8	54.1*
2	81.3	84.1	82.0	6.4
4	14.3*	99.0	99.9	83.8
8	99.9	62.1	99.9	95.8
15	70.1	88.8	99.9	70.6
32	11.1	N.A.	99.9	99.9
52	83.8	20.5	99.9	99.9
56	98.9	99.9	99.9	99.9
70	99.0	96.0	99.9	98.2
78	N.A.	4.8	99.9	95.0

(a) Results of two-sample Mann-Whitney U tests (Zar 1984) of sample differences. Units are percent probability that the sampled fertilized and unfertilized data populations are different at each sampling day. The number of samples for each treatment ranged from 14 to 18 for mineralization data, and from 7 to 9 for MPN data. All values represent the probability that the fertilized data are greater than the unfertilized ones, with the exception of those marked *, where they are greater in unfertilized samples. N.A. = not available. See Lindstrom et al. (1991) for further details.

addition, indicate that the radiorespirometric assays are not merely a laboratory bioassay for nutrients. The assayed mineralization potentials and the in situ microbial population data, taken together, provide convincing evidence that bioremediation markedly stimulated microbial activity in the oiled sediment compared to the unfertilized area. This stimulation was some three to five-fold higher after one application of fertilizer at the surface, and as high as ten-fold in the subsurface. Relative hexadecane mineralization potential increased still

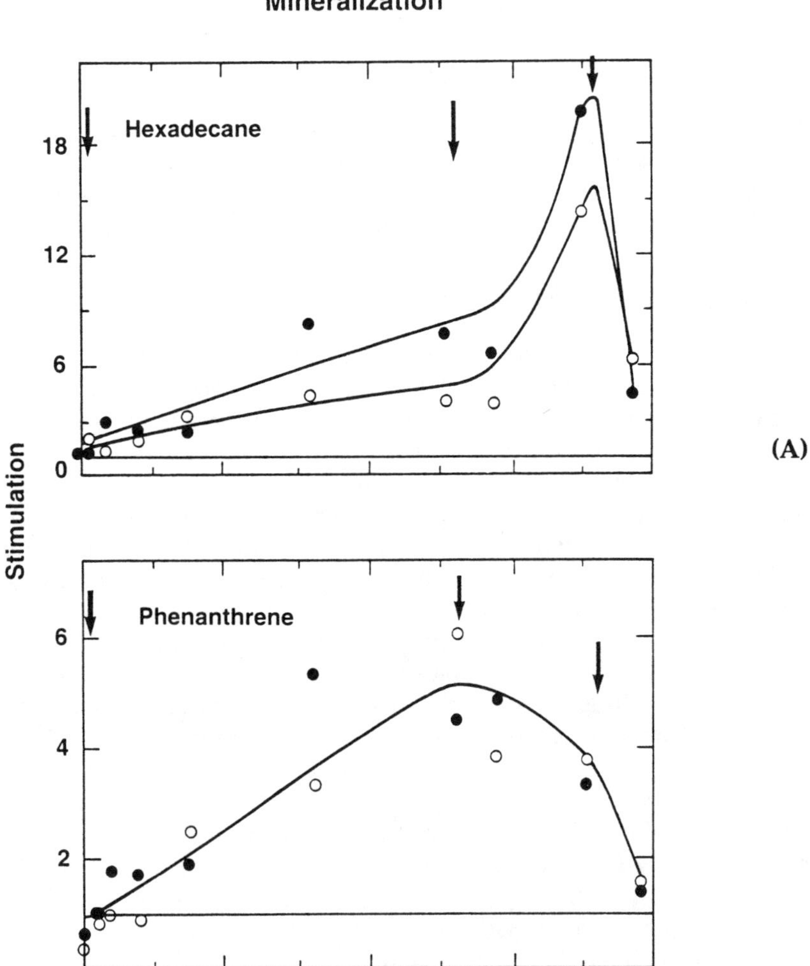

FIGURE 3. Enhancement of hexadecane (A) and phenanthrene (B) mineralization potential following fertilizer application. Samples were processed according to the University of Alaska, Fairbanks, standard respirometric protocol (Brown et al. 1991). Data are plotted as ratios of median $\%CO_2$ produced in this assay, Fertilized:Unfertilized (9 samples from each section of beach at each sampling, run in duplicate). (o) surface sediments, (•) subsurface sediments. The lines merely aid the eye, and the arrows indicate the application of fertilizer. The third application of fertilizer was to both portions of the beach.

further after the second application of fertilizer, although relative phenanthrene mineralization showed a slight decline during this period.

Fertilizer was applied to both portions of the beach on day 72, and, as expected, the difference between the two areas was markedly diminished by day 78. Unfortunately, the absolute rates of hydrocarbon degradation are difficult, and perhaps even impossible, to estimate from these measurements. As we discuss below, estimates based on changes in oil composition have now yielded statistically significant estimates of rate enhancement in absolute units.

Oil Analyses

The most relevant determination of enhanced biodegradation would be a stimulated rate of disappearance of oil from the treated shoreline. Unfortunately, the distribution of oil in the affected sediments precluded this simple analysis. Despite considerable effort to find sites with apparently uniform oiling, care to take samples of similar sediment size, and care to filter extracts to remove micron-sized mineral particles that otherwise contributed substantially to estimates of the oil loading, the oil loadings of what were thought to be similar sites did not follow a normal distribution. Analysis using D'Agostino's D parameter (Zar 1984) suggests that the distribution is more nearly lognormal, allowing us to calculate 95% confidence limits about the means of our measurements of total extractable hydrocarbon (Table 2). Such broad ranges for estimates of the average oil loading on a beach preclude detecting a statistically significant effect of bioremediation on the oil loading alone on the time scale of this program.

We therefore chose to examine the loss of oil based on changes in its composition. Previous studies of biodegradation in the field have often relied on ratios of biodegradable to less biodegradable compounds to estimate the importance of biodegradation; for example, the ratio of n-alkanes to isoprenoids has been widely used (Gundlach et al. 1983, Lee & Levy 1987). We found, however, that isoprenoids were readily degraded in Prince William Sound, so changes in such ratios substantially underestimate the extent of biodegradation (Butler et al. 1991). We have used C_{30}-17α(H),21β(H)-hopane as the relatively nondegradable compound in our analyses (Butler et al. 1991). Hopanes are biodegradable (Peters & Moldowan 1991), but the process is thought to be very slow (Requejo & Halpern 1989). Hopane thus becomes enriched within the residual oil as the oil weathers by evaporation and biodegradation.

Two potential pitfalls in using C_{30}-17α(H),21β(H)-hopane as a conserved marker were addressed in laboratory experiments. On the one hand, it was important to assess whether the molecule was likely to be degraded on time scales of relevance to the cleanup, since if it were, this would lead to an underestimate of the true rate of degradation. On the other hand, hopanes are molecular fossils that were originally derived from the biomass that gave rise to the crude oil (Prince 1987, Ourisson & Albrecht 1992), and although the stereochemistry at the 17 and 21 positions in current biomass is different from that found in fossil fuels, it was important to test whether any 17α(H),21β(H)-hopane was being produced

TABLE 2. Total extractable hydrocarbon in oiled sediments. The units are mg total extractable hydrocarbon per kg sediment dry weight[(a)].

	Surface		Subsurface	
	Unfertilized	Fertilized	Unfertilized	Fertilized
Day 0				
Mean	5,836	3,870	6,346	3,674
95% Confidence	2,409-14,139	1,596-9,384	2,712-14,850	1,399-9,644
Day 32				
Mean	9,223	5,023	2,724	2,862
95% Confidence	4,649-18,298	1,496-16,861	406-4,502	1,105-7,412
Day 70				
Mean	9,089	8,945	2,128	1,912
95% Confidence	4,115-20,077	1,752-8,884	453-9,995	701-5,214
Day 109				
Mean	7,478	4,831	3,551	752
95% Confidence	3,381-16,540	1,704-13,695	1,429-8,525	317-1,786

(a) 30 g samples of gravel sediment, spiked with appropriate surrogate standards (o-terphenyl, $^2H_8$-naphthalene, $^2H_{10}$-fluorene, and $^2H_{12}$-chrysene) were serially extracted with 1:1 methylene chloride/acetone, dried with Na_2SO_4, filtered through a 3 μm filter, evaporated to dryness, and weighed. Each mean value is the geometric mean of nine samples, with the 95% confidence limits calculated assuming a lognormal distribution. Analysis assuming a normal distribution gave substantial probabilities of negative oil in the samples. It should be emphasized that the sediment analyzed here, small gravel, has a much greater surface area per unit weight than the typical sediment on the beach, which is principally large cobbles; these estimates of oil loading should not be extrapolated to the entire beach matrix.

from the bacterial biomass intimately associated with the oil, for this would lead to an overestimate of biodegradation.

Laboratory experiments to test these possibilities (Prince et al. 1993) indicated that 17α(H),21β(H)-hopane was neither generated nor degraded on time scales of relevance to the cleanup. Furthermore, m/e 191 mass chromatograms of all the field samples exhibited hopane and homohopane distributions characteristic of North Slope crude oil, and monitoring the principal hopane biodegradation products (25-nor-hopane m/e 171, Peters & Moldowan 1991) did not indicate any selective biodegradation of C_{30}-17α(H),21β(H)-hopane. No chemical species that we examined, including the paraffins and a range of polyaromatic hydrocarbons (see Douglas et al. 1992, 1994), were consistently enriched with respect to hopane, indicating that hopane is the least biodegradable compound we resolved.

To use hopane as an internal standard, we developed an accurate assessment of hopane concentration in the oil (Butler et al. 1991). The standard deviation of 37 measurements of the reference oil was 12%, but this variation was evident primarily in differences between different runs, and was substantially smaller, typically 2 to 3%, in replicate samples before and after each batch of sediment samples. Using the hopane analysis, we calculate that the depletion of an artificially weathered oil, with a known depletion of 30.3% by weight with respect to fresh Alaska North Slope crude oil, was 30.1%.

To assess biodegradation, we have examined the changing ratios of fractions of the oil to the conserved hopane (Bragg et al. 1993). First, however, we must describe the data available for analysis. Many individual components of the oil were determined using gas chromatography (GC) coupled with either flame ionization detection (FID) or mass spectrometry (MS) (Douglas et al. 1992). Polar components of the oil are not analyzable by such techniques, so they are removed from the oil, and quantified, by a column chromatography step prior to GC. The change in polar compounds and other unresolved hydrocarbons was determined by measuring their weight fraction in the oil relative to the hopane concentration on a total oil basis. We have focused on key hydrocarbon groups for monitoring effectiveness. The first is the total oil detectable by gas chromatography, which we have termed the total GC-detectable hydrocarbon (TGCDHC); this represents more than 50% of the oil components. A second group is the total resolved hydrocarbons (TRHC) detected by GC/FID, which includes the C_{10} to C_{32} linear alkanes plus pristane and phytane, representing some 4% of the oil. A third group is the sum of selected 2- to 5-ring parent and alkylated polycyclic aromatic hydrocarbons (TPAH) (Douglas et al. 1992, 1994), which represent <1% of the oil. As biodegradation proceeds, the ratio of each of these classes of compounds to hopane will decline.

Biologically reasonable models for microbial degradation of the spilled oil under the nitrogen-limited conditions in Prince William Sound predict that the rate of biodegradation, measured as change of the ratio of an oil component to hopane (e.g., TGCDHC/hopane) will depend on the amount of oil present, the prior weathering that the oil has undergone, and the amount of nitrogen delivered to the oil-degrading microbes per unit oil. The prior weathering can be estimated from the polar fraction of the oil, because this fraction increases as degradation proceeds.

Regression models based on these factors indicate that the amount of nitrogen delivered per unit oil is the controlling factor. In analytical form the best-fitting model may be expressed as:

$$C_h(t) = \alpha\,[1-p(t)]^{\gamma}\, e^{\delta r(t)+\omega t}\,\varepsilon$$

where $C_h(t)$ is the ratio of, for example TGCDHC/hopane, at time t, p is the polar fraction of the oil, r is the average nitrogen concentration:individual oil load ratio, and ε is the assumed multiplicative error term. The average nitrogen concentration was obtained by integrating the nitrogen profile of Figure 2A and dividing by the time interval in days since application. This model fits all the data collected in the Monitoring Program very well (Bragg et al. 1993). The fit for KN-135B is shown in Table 3 and Figure 4. Note that the fits are indeed biologically reasonable; the ratio of TGCDHC/hopane is smaller in samples that have had

TABLE 3. Estimates and standard errors for the parameters of the regression model[(a)].

	R^2	$\gamma \pm$ s.e. 1-polar fraction	$\omega \pm$ s.e. days (×1000)	$\delta \pm$ s.e. ave N conc/oil load (µM days/mg/kg sed.)
Surface	0.57	0.483±0.380	-2.760±1.070	-16.44±6.76
Subsurface	0.88	0.770±0.456	-2.020±0.835	-7.67±1.23
Combined	0.71	1.263±0.226	-2.867±0.750	-5.59±1.16

(a) The "polar fraction" of the oil is that fraction adsorbed to the alumina column during work-up (Douglas et al. 1992); "days" is the number of days since the beginning of the monitoring program began; "ave N conc/oil load" is the average nitrogen concentration, in µM/days (Figure 2A), divided by the oil load of the sample under consideration. Statistical analyses were performed using JMP 2.0 (SAS Institute, Cary, North Carolina). Data from surface and subsurface samples are treated separately and combined. Note that the polar fraction is only a statistically significant part of the model when surface and subsurface samples are treated in the same model, but that all three analyses provide statistically significant and biologically reasonable values showing the importance of time and the nitrogen:carbon ratio in stimulating biodegradation.

more nitrogen per unit oil, and that have had more time for bacterial activity (ω and δ are negative). The derivative of the model can be used to predict the rate of degradation, and the stimulation of that rate upon the addition of fertilizer (Bragg et al. 1993). For example, the rate of background degradation in subsurface sediments on KN-135B can be expressed as 0.2% of the TGCDHC/hopane per day on the unfertilized portion of the beach, and about 1% per day on the fertilized side; a rate enhancement of approximately five-fold. Very similar estimates of the stimulation are obtained if TRHC or TPAH are considered instead of TGCDHC. These are in excellent agreement with the estimates based on microbial mineralization potential (Figure 3 and Lindstrom et al. 1991), with the added dimension that they predict rates in terms of the amount of oil being consumed.

Our results also suggest that fertilizer application could have been even more successful in stimulating degradation if more nutrients had been added, because there is no evidence in our data that we came close to saturating the stimulatory effect of increased available nitrogen.

Water Quality of Nearshore Water

As described above, water samples were collected for the first several tides after fertilizer application, shipped to a marine toxicity testing facility, and tested for acute toxicity to marine life using *Mysidopsis bahia* as a surrogate for indigenous species. Including the pretreatment samples, a total of 30 four-day tests from the three beaches were run, in triplicate (Peltier & Weber 1985). Survival in all

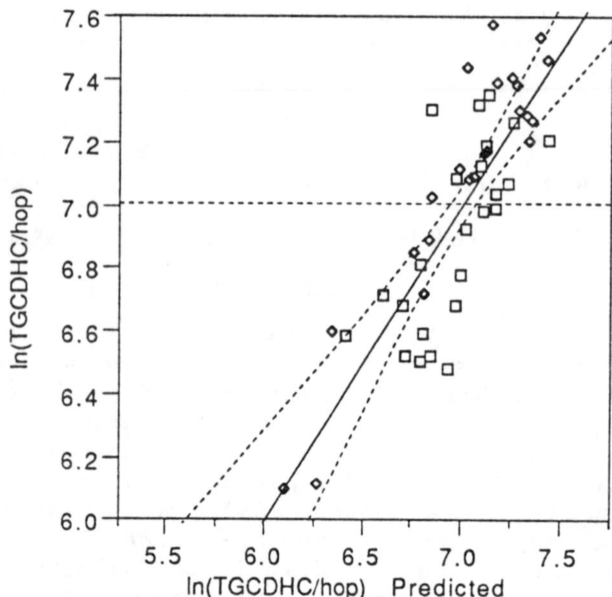

FIGURE 4. Multiple regression fit of the oil chemistry data. The data are the total GC-detectable hydrocarbon/hopane (TGCDH/hop) for the samples collected at the surface (□) and subsurface (◇). The solid line is the best fit to the equation, using the parameters listed in Table 3, and the dashed lines indicate the 95% confidence limits.

undiluted field samples collected after fertilizer application ranged from 90 to 100%, indicating no acute toxicity caused by fertilizer application.

The concentrations of ammonium and nitrate ion also were determined in nearshore water samples. Background concentrations were approximately 1 μM each, and following fertilizer application peaked at up to 35 μM between 7 and 57 hours post-application at the three sites, and returned to baseline concentrations within four days of the fertilizer application. These values are much less than those published for short-term acute toxicity to most marine biota (576 μM at pH 8; United States Environmental Protection Agency 1988a).

The potential that algal growth might be stimulated in the nearshore water by fertilizer run-off was assessed by measuring the amount of chlorophyll in the water by fluorescence. All readings were less than 67 μg/L, and there was no evidence of fertilizer stimulating algal productivity at any of the three monitoring sites.

The potential that bioremediation might stimulate the flushing of oil from the sediment into the nearshore water was assessed by infrared (IR) spectroscopy. Only 16 of 174 samples from the three sites had detectable concentrations of oil; none was higher than 41 mg/L, and this was from a remote reference site. Although the technique has a rather high detection limit (0.2 mg/L), it was adequate to demonstrate that there was no substantial release of hydrocarbons into the nearshore water following fertilizer application. The releases detected seem to be an occasional consequence of the physical removal of oil from impacted shorelines.

DISCUSSION

Microbial degradation is a major, perhaps the major, clearing mechanism for the removal of oil from the marine environment (National Research Council 1985), and the process has been the subject of considerable study (e.g., Atlas 1981, 1984; Oudot 1984; Bertrand et al. 1989; Leahy & Colwell 1990; Prince 1992). In most marine environments, the growth of oil-degrading microorganisms is limited by the supply of hydrocarbons. An oil spill provides a massive input of carbon, but only negligible amounts of nitrogen and phosphorus, so degradation is soon limited by these nutrients. There have been many proposals for adding nutrients to overcome these limitations (e.g., Townsley 1975, Bartha & Atlas 1976, Atlas 1977, Tellier et al. 1984, Ladousse & Tramier 1991), but the technique had not been used extensively to aid the cleanup of oil spills until the spill from the *Exxon Valdez*. As such there has been little appreciation for the scope and limitations of the concept, or for the difficulties of monitoring its effectiveness in the environment.

It is easy to characterize oil biodegradation in the laboratory (e.g., Atlas 1981, 1984; Oudot 1984; Bertrand et al. 1989; Leahy & Colwell 1990), and most workers have assumed that this success would be readily extrapolated to the field. Nevertheless, there have been few attempts to measure the process in the field. The data presented here demonstrate that biodegradation is an important process contributing to the removal of oil from the shorelines of Prince William Sound. While we have focused here on only one of our study sites, generally similar results were obtained at the other two beaches (Prince et al. 1990, Lindstrom et al. 1991, Bragg et al. 1993). Notable differences were that the first application of fertilizer on KN-211 was unsuccessful at delivering nutrients to the interstitial water, although subsequent applications were effective, and that the oil on KN-132 was so biodegraded at the beginning of the test that, although there was a statistically significant increase in microbial degradation potential (Lindstrom et al. 1991), we could not resolve a statistically significant effect of fertilizer on stimulating the degradation of the oil (Bragg et al. 1993).

All the samples collected during this program show substantial evidence of biodegradation, and this includes not only the alkane fraction of the oil, but also PAHs with alkyl substituents (Wang et al. 1990; Bragg et al. 1992, 1993; Elmendorf et al. 1994). By the end of the 1990 cleanup season, the majority of surface samples were already more than 50% depleted in alkanes (nC_{10} to nC_{32}) and more than 20% depleted in C_2-chrysenes, whereas subsurface samples indicated that the subsurface depletion typically was one-half to one-third of that at the surface. The most degraded samples, collected from relatively lightly oiled samples, were already approximately 80% depleted in total hydrocarbons (detected by GC or gravimetric weight) in the summer of 1990, indicating that the beached oil had the potential for at least that much biodegradation. Similar biodegradation potentials have been found in continuous culture experiments (Bertrand et al. 1983).

CONCLUSION

Assessing the safety and efficacy of bioremediation in the field on a time scale of relevance to making decisions about a cleanup is not a trivial task. Nevertheless, the joint Exxon/USEPA/Alaska Department of Environmental Conservation Monitoring Program succeeded in demonstrating that bioremediation was safe and effective. Bioremediation, although not a panacea, is nevertheless a very useful part of an overall plan for spill cleanup. Bioremediation was clearly successful in stimulating the biodegradation of the more bioavailable components of the oil, and thus can accelerate ecological recovery by diminishing the duration of oil impact. As an example, extrapolating the data from the subsurface of KN-135B reported here, it would have taken approximately 500 days for the unfertilized area to achieve the same depletion achieved on the fertilized side in 100 days. Even in the worst case, where the degradation of the nonanalyzable components of the oil was not stimulated at all by the application of fertilizer, this translates to a stimulation of two- to four-fold.

Our results show that fertilizer application was a successful and safe strategy for stimulating the natural biodegradation of spilled oil in Prince William Sound. It seems reasonable to extrapolate to other environments with similar nutrient limitations, and to suggest that bioremediation will play an important role in cleanup activities following any future spills in similar environments.

REFERENCES

Atlas, R. M. 1977. "Stimulated petroleum degradation". *CRC Crit. Rev. Microbiol. 6*:371-386.

Atlas, R. M. 1981. "Microbial degradation of petroleum hydrocarbons: An environmental perspective." *Microbiol. Rev. 45*:180-209.

Atlas, R. M. (Ed.) 1984. *Petroleum Microbiology*. Macmillan, New York, NY.

Bartha, R., and R. M. Atlas. 1976. "Biodegradation of oil on water surfaces." U.S. Patent 3,959,127.

Bertrand, J. C., E. Rambeloarisa, J. F. Rontani, G. Giusti, and G. Mattei. 1983. "Microbial degradation of crude oil in sea water in continuous culture." *Biotechnol. Letts. 5*:567-572.

Bertrand, J. C., P. Caumette, G. Mille, M. Gilewcz, and M. Denis. 1989. "Anaerobic biodegradation of hydrocarbons." *Sci. Prog., Oxf.* 73:333-350.

Bragg, J. R., R. C. Prince, J. B. Wilkinson, and R. M. Atlas. 1992. *Bioremediation for Shoreline Cleanup Following the 1989 Alaskan Oil Spill*. Exxon Co. USA, Houston, TX.

Bragg, J. R., R. C. Prince, E. J. Harner, and R. M. Atlas. 1993. "Bioremediation effectiveness following the *Exxon Valdez* spill." *Proceedings of the 1993 International Oil Spill Conference*, American Petroleum Institute, Washington DC. In press.

Brown, E. J., and J. F. Braddock. 1990. "Sheen-screen; a miniaturized most-probable-number method for enumeration of oil-degrading microorganisms." *Appl. Environ. Microbiol. 56*:3895-3896.

Brown, E. J., S. M. Resnick, C. Rebstock, H. V. Luong, and J. E. Lindstrom. 1991. "UAF radiorespirometric protocol for assessing hydrocarbon mineralization potential in environmental samples." *Biodegradation* 2:121-127.

Butler, E. L., G. S. Douglas, W. G. Steinhauer, R. C. Prince, T. Aczel, C. S. Hsu, M. T. Bronson, J. R. Clark, and J. E. Lindstrom. 1991. "Hopane, a new chemical tool for measuring oil biodegradation." In R. E. Hinchee and R. F. Olfenbuttel, (Eds.), *On Site Bioreclamation: Processes for Xenobiotic Hydrocarbon Treatment*, pp. 515-521. Butterworth-Heineman, Stoneham, MA.

Douglas, G. S., K. J. McCarthy, D. T. Dahlen, J. A. Seavey, W. G. Steinhauer, R. C. Prince, and D. L. Elmendorf. 1992. "The use of hydrocarbon analyses for environmental assessment and remediation." In P. T. Kostecki and E. J. Calabrese (Eds.), *Contaminated Soils; Diesel Fuel Contamination*, pp. 1-21. Lewis Publishers, Ann Arbor, MI.

Douglas, G. S., R. C. Prince, E. L. Butler, and W. G. Steinhauer. 1994. "The use of internal chemical indicators in petroleum and refined products to evaluate the extent of biodegradation". In R. E. Hinchee, B. C. Alleman, R. E. Hoeppel, and R. N. Miller (Eds.), *Hydrocarbon Bioremediation.* Lewis Publishers, Ann Arbor, MI.

Elmendorf, D. L., C. E. Haith, G. S. Douglas, and R. C. Prince. 1994. "Relative rates of biodegradation of substituted polyaromatic hydrocarbons." In R. E. Hinchee, B. C. Alleman, R. E. Hoeppel, and R. N. Miller (Eds.), *Hydrocarbon Bioremediation.* Lewis Publishers, Ann Arbor, MI.

Foght, J. M., P. M. Fedorak, and D. S. Westlake. 1990. "Mineralization of [^{14}C]hexadecane and [^{14}C]phenanthrene in crude oil; specificity among bacterial isolates." *Can. J. Microbiol. 36*:169-175.

Galt, J. A., W. J. Lehr, and D. L. Payton. 1991. "Fate and tranport of the *Exxon Valdez* oil spill." *Environ. Sci. Tech. 25*:202-209.

Gauthier, M. J., B. Lafay, R. Christen, L. Fernandez, M. Acquaviva, P. Bonin, and J.-C. Bertrand. 1992. "*Marinobacter hydrocarbonoclasticus* gen. nov., sp. nov., a new, extremely halotolerant hydrocarbon-degrading marine bacterium." *Int. J. Syst. Bacteriol. 42*:568-576.

Gilbert, P. M., and T. C. Loder. 1977. "Automated analysis of nutrients in sea water; a manual of techniques." *WHOI Technical Report 47*, 46 pp. Woods Hole, MA.

Gundlach, E. R., P. D. Boehm, M. Marchand, R. M. Atlas, D. M. Ward, and D. A. Wolfe. 1983. "The fate of *Amoco Cadiz* oil." *Science 221*:122-129.

Harrison, O. R. 1991. "An overview of the *Exxon Valdez* oil spill." *Proceedings of the 1991 International Oil Spill Conference*, pp. 313-319. American Petroleum Institute, Washington DC.

Kennicutt, M. C. 1988. "The effect of biodegradation on crude oil bulk and molecular composition." *Oil. Chem. Pollut.* 4:89-112.

Ladousse, A., and B. Tramier. 1991. "Results of twelve years of research in spilled oil bioremediation; Inipol EAP22™." *Proceedings of the 1991 International Oil Spill Conference*, pp. 577-582. American Petroleum Institute, Washington DC.

Leahy, J. G., and R. R. Colwell. 1990. "Microbial degradation of hydrocarbons in the environment." *Microbiol. Rev. 54*:305-315.

Lee, K., and E. M. Levy. 1987. "Enhanced biodegradation of a light crude oil in sandy beaches." *Proceedings of the 1987 International Oil Spill Conference*, pp. 411-416. American Petroleum Institute, Washington DC.

Lethcoe, J. R. 1990. *Geology of Prince William Sound, Alaska.* Prince William Sound Books, Valdez, AK.

Lindstrom, J. E., R. C. Prince, J. R. Clark, M. J. Grossman, T. R. Yeager, J. F. Braddock, and E. J. Brown. 1991. "Microbial populations and hydrocarbon biodegradation potentials in fertilized shoreline sediments affected by the t/v *Exxon Valdez* oil spill." *Appl. Environ. Microbiol. 57*:2514-2522.

National Research Council. 1985. *Oil in the Sea*, National Academy Press, Washington, DC.

Nauman, S. A. 1991. "Shoreline cleanup; equipment and operations." *Proceedings of the 1991 International Oil Spill Conference*, pp. 141-148. American Petroleum Institute, Washington DC.

Oudot, J. 1984. "Rates of microbial degradation of petroleum components as determined by computerized capillary gas chromatography and computerized mass spectrometry." *Mar. Environ. Res. 13*:277-302.

Ourisson, G., and P. Albrecht. 1992. "Hopanoids. 1. Geohopanoids: The most abundant natural products on earth?" *Acc. Chem. Res. 25*:398-402.

Parsons, T. R., Y. Maita, and C. M. Lalli. 1984. *A Manual of Chemical and Biological Methods for Seawater Analyses.* Pergamon Press, Oxford, U.K.

Peltier, W. H., and C. I. Weber (Eds.). 1985. *Methods for measuring the acute toxicity of effluents to freshwater and marine organisms*, 3rd ed. EPA/600/4-85/013. United States Environmental Protection Agency, Environmental Monitoring and Support Laboratory, Cincinnati, OH.

Peters, K. E., and J. M. Moldowan. 1991. "Effects of source, thermal maturity, and biodegradation on the distribution and isomerization of homohopanes in petroleum." *Org. Geochem. 17*:47-61.

Prince, R. C. 1987. "Hopanoids; the world's most abundant biomolecules." *Trends Biochem. Sciences 12*:334-335.

Prince, R. C. 1992. "Bioremediation of oil spills, with particular reference to the spill from the *Exxon Valdez*." In J. C. Fry, G. M. Gadd, R. A. Herbert, C. W. Jones, and I. A. Watson-Craik (Eds.), *Microbial Control of Pollution*, pp. 19-34. Cambridge University Press, Cambridge, UK.

Prince, R. C., J. R. Clark, and J. E. Lindstrom. 1990. *Bioremediation Monitoring report to the U.S. Coast Guard.* Alaska Department of Environmental Conservation, Anchorage AK. 85 pp.

Prince, R. C., S. M. Hinton, J. R. Bragg, D. L. Elmendorf, J. R. Lute, M. J. Grossman, W. K. Robbins, C. S. Hsu, G. S. Douglas, R. E. Bare, C. E. Haith, J. D. Senius, V. Minak-Bernero, S. J. McMillen, J. C. Roffall, and R. R. Chianelli. 1993. "Laboratory studies of oil spill bioremediation; toward understanding field behavior." ACS meeting, in press.

Pritchard, P. H., and C. F. Costa. 1991. "EPA's Alaska oil spill bioremediation project." *Environ. Sci. Tech. 25*:372-379.

Requejo, A. G., and H. I. Halpern. 1989. "An unusual hopane biodegradation sequence in tar sands from the Pt. Arena (Monterey) formation." *Nature 342*: 670-673.

Teal, A. R. 1991. "Shoreline cleanup; reconnaissance, evaluation and planning following the *Valdez* oil spill." *Proceedings of the 1991 International Oil Spill Conference*, pp. 149-152. American Petroleum Institute, Washington DC.

Tellier, J., A. Sirvins, J.-C. Gautier, and B. Tramier. 1984. "Microemulsion of nutrient substances," U.S. Patent 4,460,692.

Townsley, P. M. 1975. "Material for biological degradation of petroleum." U.S. Patent 3,883,397.

U.S. Environmental Protection Agency. 1988a. *Ambient water quality criteria for ammonia (saltwater).* EPA 440/5-88-004. NTIS, Springfield, VA.

U.S. Environmental Protection Agency. 1988b. *EPA 600 Method 418.1.* NTIS, Springfield, VA.

Wang, X., X. Yu, and R. Bartha. 1990. "Effect of bioremediation on polycyclic aromatic hydrocarbon residues in soil." *Environ. Sci. Technol. 24*:1086-1089.

Zar, J. H. 1984. *Biostatistical Analysis.* Prentice-Hall, Englewood Cliffs, NJ.

Zobell, C. E. 1946. "Action of microorganisms on hydrocarbons." *Bact. Rev. 10*: 1-49.

IN SITU BIOREMEDIATION: AN INTEGRATED SYSTEM APPROACH

C. H. Nelson, R. J. Hicks, and S. D. Andrews

ABSTRACT

In situ bioremediation represents a technologically feasible and cost-effective method for the remediation of soils and groundwaters contaminated with petroleum hydrocarbons. The ultimate goal of bioremediation is to convert organic wastes in biomass and harmless by-products of metabolism such as carbon dioxide, water, and inorganic salts. This paper presents a case study describing a successful application of in situ bioremediation at a nonvolatile hydrocarbon-impacted site. At a public utility facility in Colorado, soil and water sampling confirmed the presence of petroleum hydrocarbons, and laboratory studies showed that the chemical, microbiological, and hydrogeological characteristics of the site were conducive to the implementation of a bioremediation system. Vadose zone contamination was addressed using a bioremediation system coupled with a vapor extraction system and a groundwater recovery well. The bioremediation system began operating in July 1989 and is now in the postclosure monitoring phase, which is scheduled for completion in April 1993. Approximately 36,000 pounds of hydrocarbons were removed, with 94% of the contaminant mass degraded biologically. Approximately 9 million gallons of groundwater were recovered, amended with nutrients, and reinfiltrated. The site has been included in U.S. EPA's Bioremediation Field Initiative program.

INTRODUCTION

For almost 30 years, nonvolatile petroleum hydrocarbons from used motor oil, diesel, gasoline, and other automotive fluids had been released to a used oil sump at a truck maintenance facility in Denver, Colorado. The site had both vadose- and saturated-zone contamination. The facility's owner, Public Service Company of Colorado, hired Groundwater Technology, Inc., of Englewood, Colorado, to complete a site assessment and to initiate and design an on-site remediation system, one that would not disrupt ongoing maintenance and building expansion.

After reviewing several remedial operations, in situ bioremediation was selected. The conceptual design included stimulating indigenous bacterial populations through the introduction of oxygen and inorganic nutrients. For most in situ bioremediation systems, groundwater is recovered, amended with nutrients, and reinfiltrated. Oxygen is added by either injecting hydrogen peroxide (which decomposes to oxygen and water) or installing an air-sparging system. For vadose zone contamination, nutrients are added through percolation or injection. Oxygen is normally supplied through a negative pressure vapor extraction system.

In situ bioremediation is very site specific, and its successful implementation depends on a thorough understanding of the physicochemical, hydrogeological, and microbiological factors controlling not only the biodegradation of the contaminant at the site but also the mass transport of nutrients and oxygen. For this reason, an extensive site assessment and treatability study was conducted that included the installation of nine monitoring wells, about 7.6 m deep, using 102-mm-diameter polyvinyl chloride casings and a 3-m screen interval (Figure 1).

LOCAL GEOLOGY

The site is underlain at a depth of about 11 m by interbedded claystone and sandstone bedrock of the Denver formation. Overlying the bedrock are recent alluvial sands and gravels deposited in the South Platte River floodplain. These floodplain deposits are dense and interbedded, and average about 9 m thick. A layer of stiff, very sandy and silty clay ranging from 1.2 to 3 m thick rests on the sand and gravel. The principal aquifer impacted at this site lies within the interbedded alluvial sands and gravels. From the surface topography, drainage, and water levels encountered during initial drilling, it was determined that the groundwater (found about 4 m below ground surface) flows north to northeast, roughly parallel to the flow of the South Platte River and 300 m east of the site.

SITE ASSESSMENT

Groundwater samples collected from monitoring well 8 (MW 8), located near the former oil sump, were the most contaminated. This monitoring well contained the highest levels of benzene, toluene, ethylbenzene, and xylenes (BTEX) and of total petroleum hydrocarbons (TPH) with relatively low but detectable levels of chlorinated organics. Total oil and grease (TOG) measurements ranged from phase-separated product in MW 1 and MW 8 to concentrations below detection levels in perimeter wells. Soil samples taken during the installation of MW 8 showed relatively low concentrations of BTEX, but relatively high concentrations of TPH and TOG.

Groundwater samples were analyzed for ammonium, nitrate, nitrite, phosphate, potassium, pH, and background bacterial levels. The results indicated detectable levels of potassium only and pH levels near neutral. Background bacterial population densities were relatively high.

Initial mass balance estimates indicated that about 272 kg of hydrocarbons existed in the zone of groundwater fluctuation, and about 11.7 tons existed in

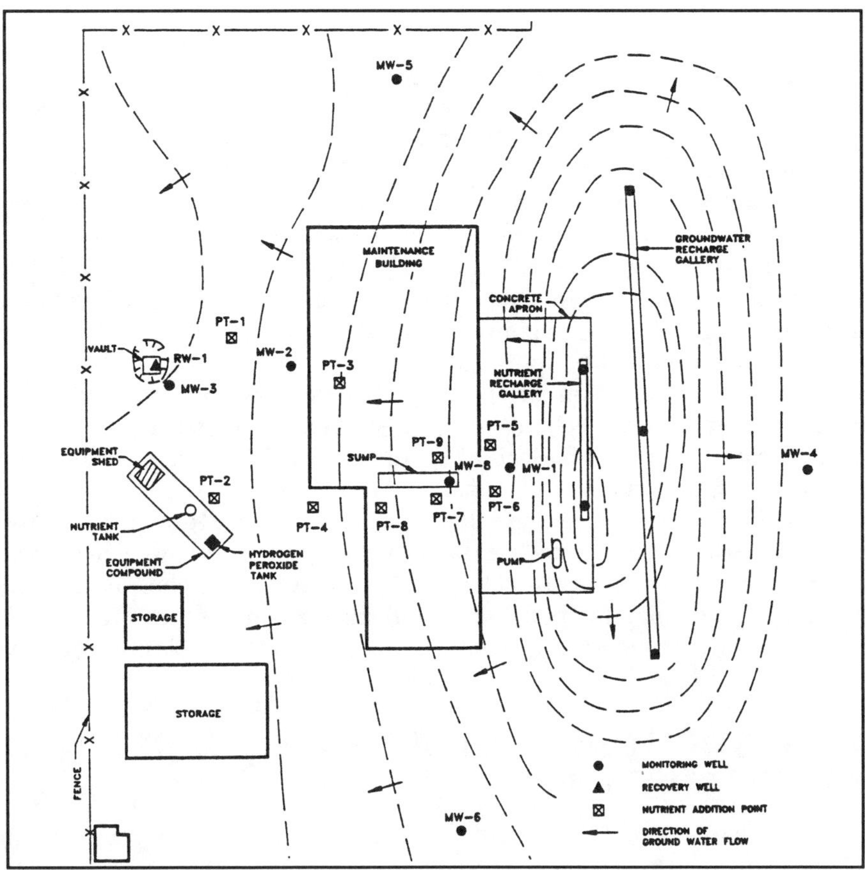

FIGURE 1. Site map.

unsaturated and saturated sediments located beneath the former used oil sump. Mass balance estimates for groundwater showed that significantly less hydrocarbon mass (136 g) existed in the dissolved phase. We directed remedial efforts toward the partially saturated and unsaturated sediments, the source zones for groundwater contamination. In summary, the site assessment indicated that:

1. The primary site contaminate was waste oil located in the saturated and unsaturated sediments beneath the former used oil sump.
2. A relatively large population of hydrocarbon using bacterial existed within the zone of contamination. Their growth appeared to be restricted by limiting nutrient and oxygen conditions.
3. Groundwater occurred at a shallow depth (4 m) in sediments that appeared to be relatively permeable.
4. Groundwater samples showed high levels of BTEX, TPH, and TOG, with localized but detectable levels of chlorinated organics.

FEASIBILITY STUDY

Soils and sediment samples were collected and shipped to our treatability laboratory in Concord, California to determine the biodegradability of the contaminants under various nutrient loads and aerobic conditions. Aerobic testing was conducted to simulate optimal conditions for the bioremediation of hydrocarbons.

Column studies were also conducted to determine how nutrient and hydrogen peroxide loading would affect the hydraulic conductivity of sediments in the subsurface above the groundwater table. Hydrogen peroxide can provide large quantities of dissolved oxygen to the aqueous phase as it breaks down into oxygen and water. Due to hydrogen peroxide's reactive nature, it is important to test its stability in native soil samples.

Tests were conducted on each primary soil type in the immediate subsurface under consideration for injection. Test results showed that in situ bioremediation would be feasible, although the loading of nutrients and hydrogen peroxide would be critical to the project's success and should be minimized in the silty sand zone due to its high reactivity. Loading in the coarse sand zone and gravel zone caused only moderate hydrogen peroxide reactivity and no change in hydraulic conductivity. A nutrient adsorption test of soils indicated relatively high phosphate adsorption and low ammonia adsorption. Based on estimated nutrient requirements, we determined that ammonia and phosphate loading would be feasible in all soil zones.

REMEDIATION SYSTEM DESIGN AND INSTALLATION

The primary targets for the in situ bioremediation system were sorbed hydrocarbons in the unsaturated and saturated sediments located beneath the area of the former used oil sump and, to a lesser extent, dissolved hydrocarbons in groundwater located beneath and surrounding the maintenance building. Small amounts of phase-separated product also existed in the groundwater. The primary functions of the remediation system included groundwater recovery, treatment, and reinjection; vapor extraction and discharge; simulation of in situ bioremediation by subsurface inorganic nutrient and oxygen additions; and phase-separated hydrocarbon recovery (Figure 2).

A vapor extraction test showed that a soil vapor extraction point, screened from 1.2 m to 4.6 m below ground surface in the unsaturated zone, could achieve 15 m of radial influence at 0.5 m of water vacuum. Vacuum measurements did not vary with depth, indicating a fairly uniform area of permeability in the unsaturated soil. We assumed a 3.7-m radius of influence at the water-table interface to determine percolation rates and spacing of injection points, and to predict lateral distribution of dilute inorganic and hydrogen peroxide solutions. A 72-hour pump test generated aquifer data for groundwater modeling. The modeling showed that a recovery well located near RW 1 and pumping at a rate of 10 gpm would induce a sufficient hydraulic gradient to capture injected nutrients and dissolved hydrocarbons.

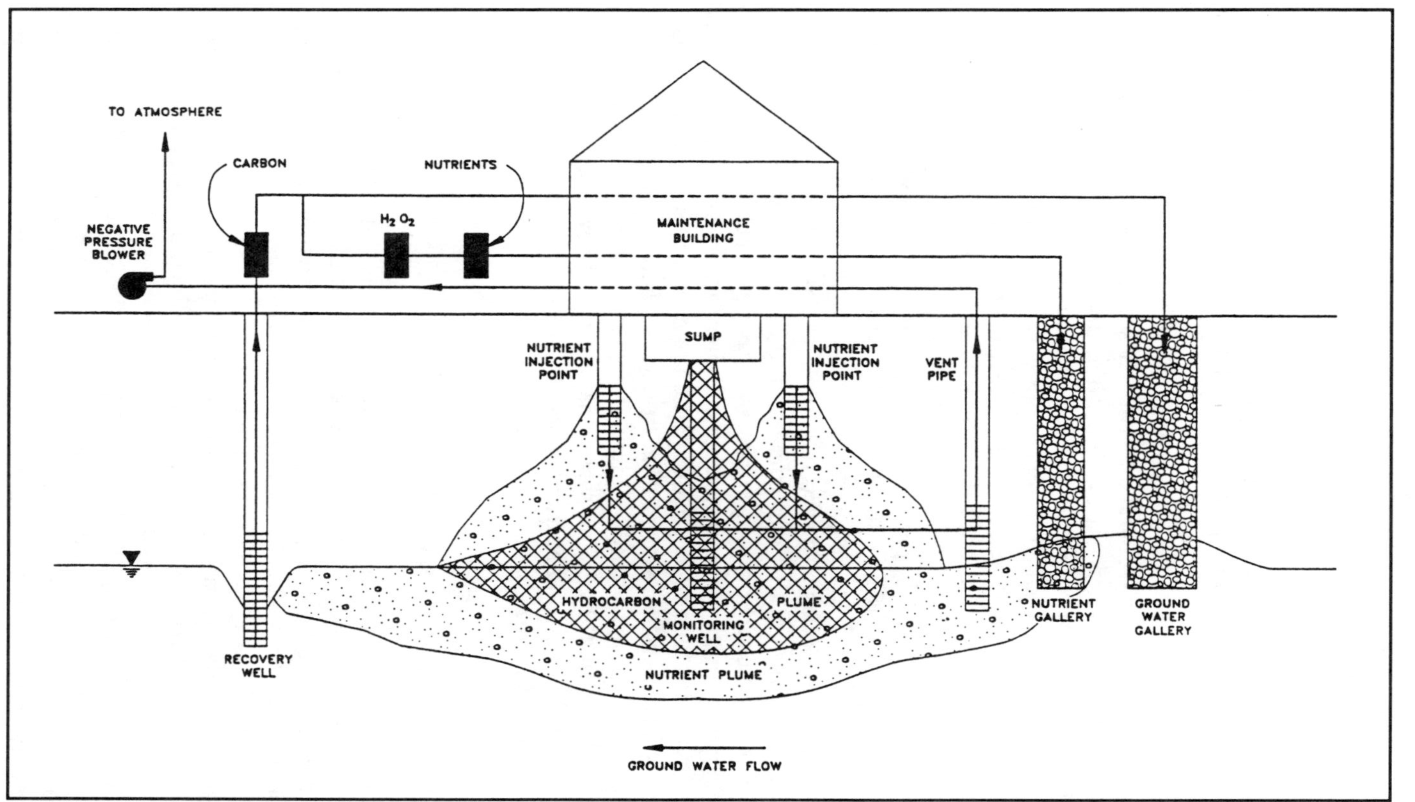

FIGURE 2. Remediation system schematic.

Recovery well 1 (RW 1) captured groundwater and pumped it through activated carbon. At this point, the flow was split. A portion of the water was pumped to the nutrient gallery after being amended with inorganic nutrients and hydrogen peroxide. The remainder went directly into the groundwater gallery, thereby creating a closed-loop system. A negative pressure blower connected to MW 1 induced airflow through the nutrient injection points. We confirmed vacuum influence in injection points PT 5 through PT 8 and a 15-m radius of influence.

Laboratory tests showed that hydrogen peroxide and nutrient loading worked best in sediments from the coarse sand interval. These sediments also contained high levels of hydrocarbons near the groundwater/unsaturated zone interface. So to stimulate bioremediation in this highly contaminated area, groundwater reinjection galleries were installed to this depth to facilitate the loading of the hydrogen peroxide and nutrients. The water flow from the recovery well was split into two galleries: the nutrient gallery and the groundwater gallery. This ensured high concentrations of hydrogen peroxide and nutrients in the nutrient gallery would be loaded just upgradient from the used oil sump area.

Crews drilled injection points (PT 1 through PT 9) into unsaturated sediments for the direct injection of hydrogen peroxide and nutrients at depths of 2.7 to 3.7 m. These injection points also served as preferential airflow pathways for the vapor extraction system that was connected to MW 1. Nutrient injections and the addition of atmospheric oxygen from the vapor extraction system stimulated bioremediation in the unsaturated zone and enhanced the desorption of adsorbed hydrocarbons for recovery in the monitoring wells. Several components of the bioremediation system served dual functions. For example, the vapor extraction system removed volatile hydrocarbons and provided atmospheric oxygen to the unsaturated zone. This system had designed flexibility for adjustment of hydrogen peroxide and nutrient concentrations in the subsurface while controlling the hydraulic gradient across the site.

OPERATION AND MONITORING

Once the bioremediation system was installed, it was inspected weekly to adjust and maintain the water table depression pump, the hydrogen peroxide and nutrient injection equipment, and the soil vapor extraction system. Crews conducted field analyte tests, sampled groundwater, and gauged monitoring wells. The system operated from July 1989 to March 1992.

Hydrogen peroxide migrating from the nutrient gallery and the injection points toward the recovery well decomposed to dissolved oxygen and water. Dissolved oxygen measurements showed a consistent trend with high levels (greater than or equal to 10 mg/L near the nutrient gallery and decreasing but measurable values as groundwater migrated toward the recovery well. Low concentrations of dissolved oxygen, especially near the zones of highest contaminant concentrations (MWs 1 and 8), were probably due to high rates of bacterial consumption associated with hydrocarbon metabolism.

High levels (greater than or equal to 10 mg/L) of ammonia nitrogen and phosphate, required for a successful bioremediation program, existed across the area of groundwater contamination after about a year of system operation. Analytical data from MW 2A (taken from July 1989 through March 1992) showed significant increases in ammonia, phosphate, and dissolved oxygen concentrations starting in the summer of 1990. Average quarterly results showed that benzene, total BTEX, and TPH decreased significantly after 2 years of system operation. TPH decreased from phase-separated product to below detection limits in MW 1 and to approximately 15 ppm in MW 8. Benzene concentrations decreased from 220 mg/L to < 1 µg/L in MW 1 and from 180 µg/L to 3 µg/L in MW 8. Groundwater samples from surrounding monitoring wells had BTEX concentrations ranging from 3 µg/L to < 1 µg/L after the system had operated for about 2 years. Bacterial analyses indicated that relatively high concentrations of background heterotrophs and hydrocarbon-utilizing bacteria existed in groundwater throughout the site averaging 10^6 colony forming units/mL.

Throughout the project, crews took soil borings and analyzed them for BTEX, TPH, and TOG. The primary source of soil contamination occurred near MWs 1 and 8 at depths between 4 m and 5 m. BTEX in this zone ranged from an initial 420 mg/kg to below detection limits as the project progressed. The TPH level decreased from 5,200 mg/kg to 55 mg/kg. TOG decreased from 12,000 mg/kg to 1,900 mg/kg.

Analyses for polycyclic aromatic hydrocarbons (PAHs) and total metals from groundwater samples showed concentrations below Colorado Department of Health and EPA guidelines. Only indeno [1,2,3-*cd*] pyrene was detected at a level of 0.49 µg/L. Nitrate analyses of groundwater from all monitoring wells ranged from below detection limits to less than 50 mg/L, with an average range of 10 to 20 mg/L maintained during the project. Nitrate concentrations quickly decreased when nutrient injection ceased, suggesting that nitrate may have acted as a terminal electron acceptor during the remediation process.

RESULTS

In 2½ years of operation, a total of 16.3 tons of contaminant mass was removed from the site. Phase-separated hydrocarbon recovery removed 680 kg (4% of the contamination) from MWs 1 and 8. The vapor-extraction system volatilized 354 kg (2%).

Bioremediation removed 15.1 tons (94%) of contamination. Approximately 33,400 m^3 of groundwater was extracted, treated, and reinjected resulting in more than 15 pore volumes of groundwater that were circulated though the zone of contamination. The estimated contaminant mass removed compared favorably to the initial contaminant mass estimate of 11.8 tons and provided further evidence of site remediation.

Bacteria responsible for aqueous-phase bioremediation may have used dissolved oxygen, produced by the disassociation of hydrogen peroxide and/or nitrate from the oxidation of ammonia compounds in the nutrient solution, as terminal electron acceptors. Several researchers have shown that nitrate is a very effective electron acceptor in the bioremediation of alkylbenzene compounds in groundwater. Nitrate has a relatively high electron accepting capacity, compared to oxygen and hydrogen peroxide, and tends to be more stable during groundwater migration, thereby having a greater potential to affect downgradient plume areas.

EPA conducted additional soil sampling and groundwater modeling under the Bioremediation Field Initiative Program to further investigate the results of the remedial effort. The site is currently undergoing closure with the Colorado Department of Health. Additional information on this project can be found in *Bioremediation: Field Experience*, edited by Jurgen H. Exner, Douglas E. Jerger, and Paul E. Flathman, currently being published by Lewis Publishers, Ann Arbor, Michigan.

IN SITU BIOREMEDIATION OF A GASOLINE AND DIESEL FUEL CONTAMINATED SITE WITH INTEGRATED LABORATORY SIMULATION EXPERIMENTS

A. Robertiello, G. Lucchese, C. Di Leo, R. Boni, and P. Carrera

ABSTRACT

Following an accidental spill of gasoline and diesel fuel, a large area of sandy soil was contaminated to a depth of up to around 6 m. A 120-m^2 area was subjected to an in situ bioremediation experiment. Within the treatment area two different well-point series were introduced aimed at extracting contaminated water from the subsoil and with reinjecting it after microbial degradation treatment with indigenous bacteria. The bioremediation process lasted for 3 months (August to October). Both water and soil were monitored on the basis of chemical/physical and microbiological criteria. The average pollutant content of the treated soil decreased notably during the first 25 days of continuous operations, from 1.0 to 0.57%. Due to unfavorable weather conditions, the activity was later downscaled to the laboratory level by using suitable simulation apparatus. The laboratory experiment was performed to (1) continue the in-field experience by reproducing and monitoring the same operative conditions used in the bioremediation process, and (2) study in more detail the influence of the major biological and chemical/physical parameters.

INTRODUCTION

Over recent years, the need to reclaim polluted soil and groundwater has stimulated the initiation of a number of research projects aimed at developing new technologies. Among these, those based on the use of microorganisms are interesting both for their versatility and for their economical benefits (EPRI-EEI 1989, Miljoplan 1987).

As with other forms of contamination, microbiological treatment (Thayer 1991, Werner 1991) of hydrocarbon-contaminated soils may be conducted both in situ (Bartha et al. 1990, Mathewson et al. 1988) and on site (API 1980, CONCAWE 1980). Although in situ biodegradation involves the use of complex technologies, it usually consists of an apparatus to collect and extract water from the subsoil. This water is then directed into an aerated collection vessel, where nutrients are added and the pH level is adjusted, and is finally returned to the soil (EPRI-EEI 1989).

This paper focuses on the experimental activities conducted both in field and in laboratory to remediate a gasoline and diesel fuel-contaminated site. The study was conceived to (1) investigate the bioavailability of the pollutants to the naturally selected hydrocarbon-degrading strains, and (2) analyze the degradation kinetics stemming from the use of an in situ treatment plant.

IN SITU ACTIVITY

Following an accidental spill of gasoline and diesel fuel, a large area of sandy soil was contaminated to a depth of up to around 6 m. A 120-m^2 area was subjected to an in situ bioremediation experiment. Within the area, two different series of well-points were introduced (Figure 1).

In proximity of this site, a water treatment plant was set up consisting of tank No. 1 which directly received water extracted from the groundwater and served to volatilize the lighter components of the pollutants; tank No. 2 used as a fermenter for the growth of the microorganisms (the tank was initially inoculated with a starter prepared in laboratory); tank No.3 used for further oxygen enrichment of the culture broth and as a settler for the recovery (via skimming) of residual hydrocarbons. Ahead of and behind the three tanks, two rotary pumps were fitted to ensure both the extraction and reintroduction of the water via the

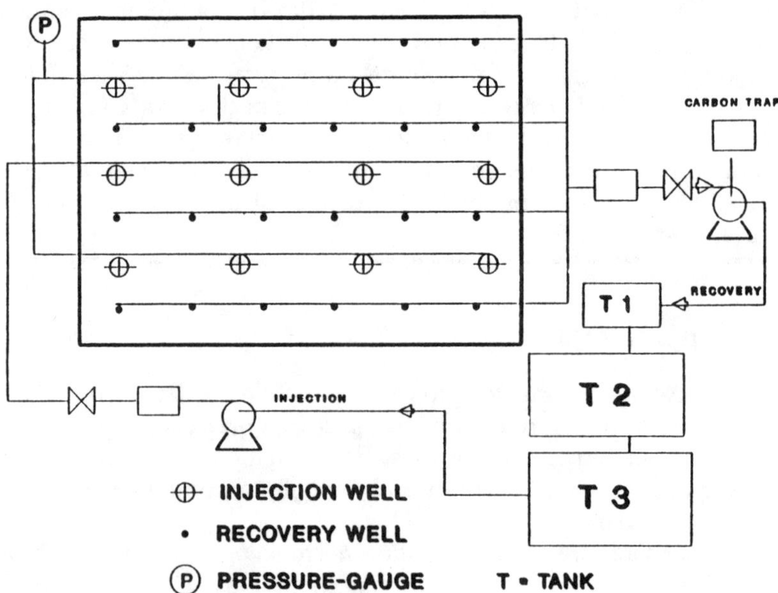

FIGURE 1. In situ bioremediation plant.

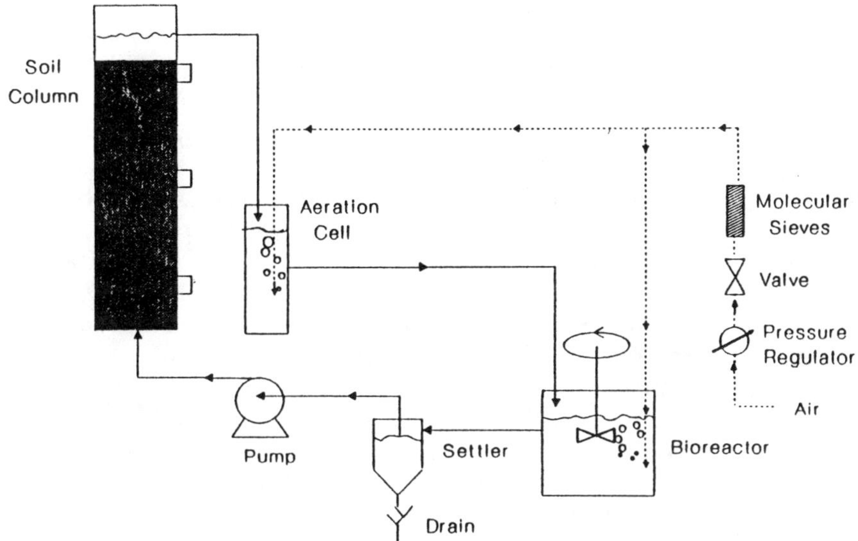

FIGURE 2. Laboratory simulation apparatus.

well-points established in the soil being treated. The culture broth was introduced via well-points established in the soil to a depth of approximately 4 m, and was recovered through other well-points sunk to a depth of approximately 2 m (groundwater level).

The apparatus was operated intermittently for 7 weeks so that all the process parameters could be checked and optimized. After that period, the plant operated continuously for approximately 45 days.

Throughout the experiment, nitrogen and phosphorus (urea and biammonium phosphate) were added in sufficient quantities so as to roughly maintain a C:N:P ratio of 100:5:1 (Bartha et al. 1990).

LABORATORY SIMULATION

The apparatus used for the experiment (Figure 2) consisted of a column filled with soil taken from the site with an average pollutant content of 1.6%. Process water was passed upwards through the soil (at a flowrate of 0.09 L per liter of sample per hour, as had been done in the field) and than sent to an aerator to remove the most volatile components. From the aerator, the process water passed to the fermenter for biodegradation and than returned to the column. Both at the beginning and during the experiment subsequent additions of nitrogen and phosphorus nutrients were made, as had been done during the in situ experiment. An identical apparatus to act as a control was setup to monitor reclamation phenomena of the soil through non biological mechanisms through the addition of mercuric chloride to the process water.

RESULTS AND DISCUSSION

Figure 3 summarizes the most significant parameters monitored during the in-field experiment and indicates that the number of microorganisms was constant (10^7 to 10^8 CFU/mL) over the first 50 days of in situ activity and then fell constantly over the following 45 days (continuous operation), at which point an inversion of this trend occurred. The hydrocarbon concentration at the point where the water entered the tank system was lower than expected. (A visual inspection of the waters showed a thick layer of hydrocarbons at the surface sign of saturation.) This could be explained by the fact that a significant part of the lighter fraction was stripped by the depression of the pump and captured by an activated carbon trap (Figure 1).

The hydrocarbon concentration in the water at the output of the tank system was at all times lower than at the input, which would indicate that biodegradation was occurring within the external water treatment system (Figure 4).

The average pollutant content of the treated soil show in Figure 3 (expressed as the average value of the 5 different sample points taken from the 3-m depth within the area under study), decreased notably during the first 25 days of continuous operations, from 1.0 to 0.57%. This fall was analogous to the fall in the number of hydrocarbon-oxidizing microorganisms. This situation changed around the 75th day when the hydrocarbon content of the soil treated rose sharply along with a sudden drop and rise in the groundwater level (Figure 3). This phenomenon probably allowed entry of pollutants from the area beyond the barriers. The increase in quantity of hydrocarbons and number of microorganisms in the process water should be interpreted as the natural reaction of the system to a new increase in pollutants. It is interesting to note that during the experiment

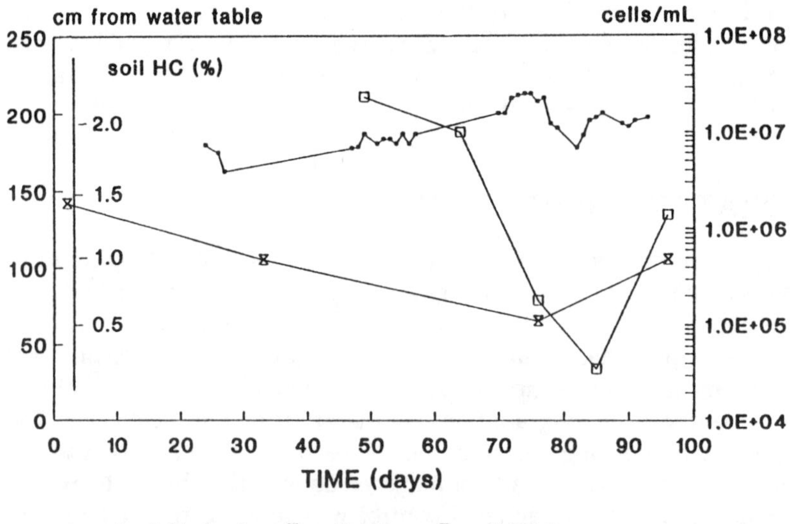

FIGURE 3. Parameters monitored during the in-field experiment.

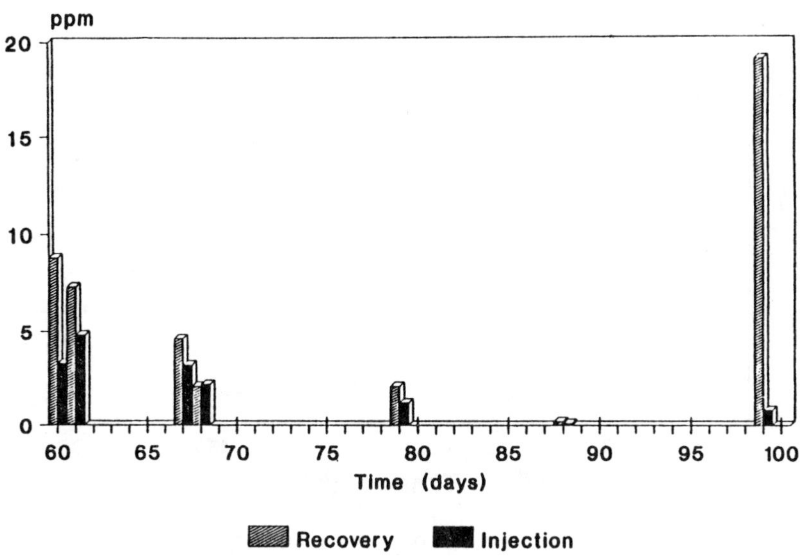

FIGURE 4. Hydrocarbons in process water.

the oxidizing hydrocarbon microbial flora in the process water rapidly altered to the point that they comprised *Alcaligenes faecalis* and *Acinetobacter calcoaceticus*. The latter is recognized for its capacity to produce an extremely active biosurfactant known as Emulsan (Kosaric et al. 1987). After transfer to the laboratory for a period of 203 days, during the first 50 days the pH level was maintained to 7.5 by automatic addition of KOH. This tendency in the pH decrease is in agreement with what observed in the process waters of the in situ activity. From approximately the 30th day, nitrates formed in the soil and water (data not shown).

From the beginning of the continuous operation to the 85th day, the total population of heterotrophs and hydrocarbon-oxidizing microorganisms in the fermenter slowly decreased from 10^7 to 10^5 CFU/mL (Figure 3), parallel with the progressive decrease in carbon sources. The major part of the microorganisms present showed hydrocarbon-oxidizing activity. Furthermore, specific microbiological tests demonstrated the presence of denitrifying flora in the soil samples. There was a reduction in the pollutant content of the soil in the order of 75%. (The experimental values produce a curve described by the function $y = A x^n$, Figure 5) (Spain 1982).

An approximately 45% reduction in the concentration of hydrocarbons in the soil occurred in the control apparatus due mainly to the elution of the hydrocarbon fraction, which is more soluble in water, and the later removal of the lighter fractions through the action of aeration.

Figure 6 shows the experimental values for the hydrocarbon concentration in the soil against time for both the in situ and laboratory experiments. Comparison of the two curves demonstrates substantially similar trends.

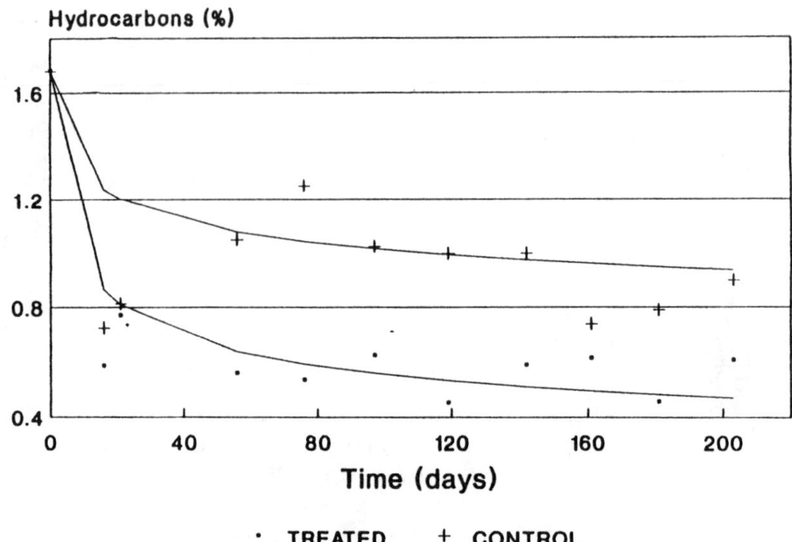

FIGURE 5. Biodegradation trend in the lab.

CONCLUSIONS

Despite the fact that the duration of the in situ activity was insufficient, the experimental data collected both in field and in laboratory give rise to the several conclusions. The efficacy of the in situ bioremediation technique was confirmed by the fact that it was demonstrated that the quantity and quality of hydrocarbon-degrading microorganisms present in the system were closely linked to (and proved to be in equilibrium with) the quantity of the contaminants to be broken down.

The treatment of the soil using a culture broth containing hydrocarbon-degrading microorganisms and the biosurfactants they produce allowed for the separation of hydrocarbons from the contaminated soil, partly by being stripped and captured by a trapping system, and partly from breakdown in the external oxidization tanks.

The main factor limiting the biodegradation kinetics appeared to be the availability of oxygen in the soil. This fact is confirmed by the increasing number of facultative anaerobic and denitrifying bacteria detected in the soil under investigation (data not shown).

Inocula consisting of different strains from those which naturally tend to select in hydrocarbon-polluted soils would appear of little use. In both the field and laboratory, after a few weeks of treatment, a naturally selected microbial flora tended to prevail over those initially introduced.

On the basis of the results, one may reasonably conclude that, left to operate for two consecutive seasons, the system would produce a hydrocarbon content in line with the Dutch guidelines (Moen 1988).

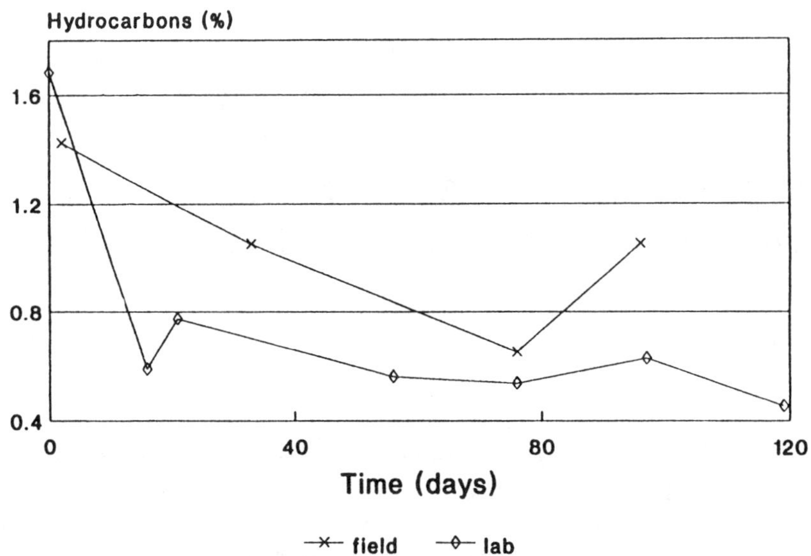

FIGURE 6. Biodegradation trend: field vs. laboratory.

ACKNOWLEDGMENTS

Thanks are particularly due to Aquater S.p.A. and Messrs. Alberto Bittoni and Adalberto Massetti for their technical assistance.

REFERENCES

API. 1980. *Landfarming: An Effective and Safe Way to Treat/Dispose of Oily Refinery Wastes.* American Petroleum Institute, Solid Wastes Management Committee.

Bartha R., H. G. Song, and X. Wang. 1990. "Bioremediation potential of terrestrial fuel spills." *Appl. Environ. Microbiol. 56*(3): 652-656.

CONCAWE. 1980. *Sludge Farming: A Technique for the Disposal of Oily Refinery Wastes.* Report No. 3/80.

EPRI-EEI. 1989. *Remedial Technologies for Leaching Underground Storage Tanks.* Lewis Publishers Chelsea, MI.

Kosaric, N., W. L. Cairns, and N. C. C. Gray. 1987. *Biosurfactants and Biotechnology*, pp. 214-217. Marcel Dekker, Inc.

Mathewson, J. R., R. B. Grubbs, and B. A. Molnaa. 1988. "Innovative techniques for the bioremediation of contaminated soils." In *California Water Pollution Control Association* (Oakland, CA, 7-8 June 1988).

Miljoplan, A. H. 1987. "Soil decontamination." In *Meeting of the Working Party on Environmental Protection* (Genova, 26-27 Nov. 1987).

Moen, J.E.T. 1988. *Soil Protection in the Netherlands*, pp. 289-298. Kluwer Academic Publ.

Spain, J. B. 1982. *Basic Microcomputer Models in Biology*, pp.43-62. Addison-Wesley Publishing Co.

Thayer, A. M. 1991. "Bioremediation: Innovative technology for cleaning up hazardous waste." *C&EN*, August 26:23-43.

Werner, P. 1991. "Trattamento microbiologico dei terreni contaminati." *Rifiuti Solidi* 2:121-126.

EFFECTS OF HYDROGEN PEROXIDE ON THE IN SITU BIODEGRADATION OF ORGANIC CHEMICALS IN A SIMULATED GROUNDWATER SYSTEM

C. J. Lu

ABSTRACT

Effects of the addition of hydrogen peroxide on the biodegradation of aromatic compounds and volatile fatty acids were studied with a series of sand column reactors. Dissolved oxygen was a major limiting factor for the in situ biodegradation of benzene and *n*-butyric acid in the simulated groundwater system. Hydrogen peroxide was used as an alternative oxygen source. Experimental results showed that the addition of hydrogen peroxide enhanced the biodegradation of benzene, propionic acid, and *n*-butyric acid. However, the ratio of organics biodegraded to the amount of hydrogen peroxide added decreased with the increase of influent hydrogen peroxide concentration, indicating that hydrogen peroxide was not efficiently utilized when its concentration was high. The requirement of hydrogen peroxide for the complete biodegradation of benzene was twice that theoretically calculated from the stoichiometric equation.

INTRODUCTION

Contamination of groundwaters by organic chemicals is widespread and is increasingly a major concern. Current research on the remediation of groundwater contamination has focused on in situ biodegradation, which is considered a cost-effective and environmentally sound remediation method (Lee et al. 1988, Thomas et al. 1991, Wilson & Brown 1989). However, one of the major limiting factors in bioremediation of organic contaminated groundwaters is the amount of dissolved oxygen available for microorganisms. Hydrogen peroxide can be used as an alternative oxygen source to supply more oxygen to enhance microbial activity during the bioremediation of contaminated soils and groundwaters (Flathman et al. 1991, Lee et al. 1988, Pardieck et al. 1992). Although the benefit of adding hydrogen peroxide as the oxygen source might be counteracted by its high decomposition rate (Aggarwal et al. 1991, Hinchee et al. 1991), the use of hydrogen peroxide still could overcome the limitation of oxygen solubility during conventional aeration (Schlegel 1977).

Hydrogen peroxide is an oxygen concentrate; more than 100 L O_2 would be produced from 1 L of commercially available H_2O_2 (30%) (Schlegel 1977). The oxygen produced from the decomposition of hydrogen peroxide could be used directly by microorganisms to enhance microbial activity and thus increase the biodegradation rate (Lu & Hwang 1992, Pardieck et al. 1992). Therefore, this study focused on the effects of the addition of hydrogen peroxide on the biodegradation of benzene, propionic acid, and *n*-butyric acid in a simulated groundwater system.

MATERIALS AND METHODS

Chemicals

The chemicals used in this study were benzene, propionic acid, and *n*-butyric acid (HBu). These organic chemicals are commonly detected in the groundwaters contaminated by leachate from domestic and industrial waste landfill sites.

In addition to the organic chemicals, the artificial influent solution also contained nutrients to support microbial growth. The compositions of the nutrients are as follows: K_2HPO_4, 100 mg/L; KH_2PO_4, 60 mg/L; $MgSO_4$, 8.6 mg/L; $CaCl_2$, 23 mg/L; and NH_4Cl, 3.5 mg/L (or higher, if necessary).

Simulated Groundwater System

The experiments were conducted with a series of sand column reactors. The schematic diagram of the column reactor is shown in Figure 1. One set of column reactors contained two or three columns running in series. Generally, two sets of columns were operated in parallel to study the effects of the presence

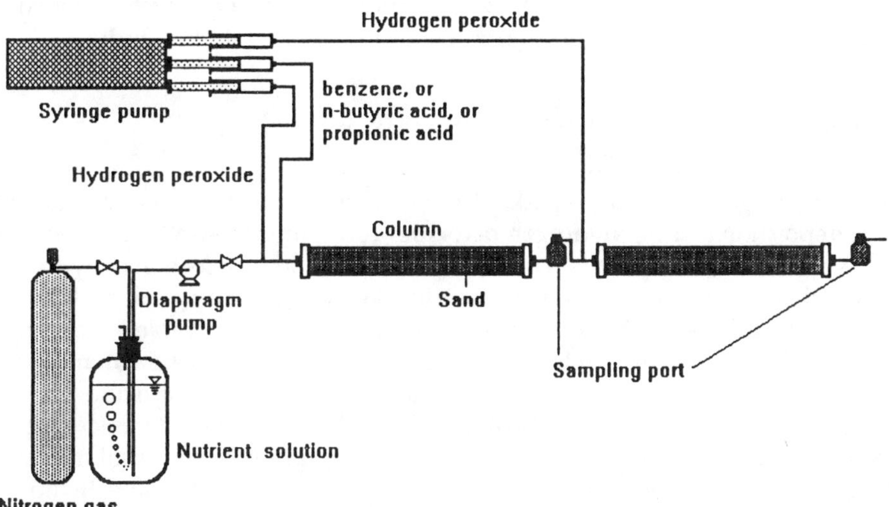

FIGURE 1. Schematic diagram of a sand column reactor.

of hydrogen peroxide on the biodegradation of organic chemicals. Sand with average diameter of 0.5 mm was used as the medium to simulate the aquifer. The length of each column was 50 (or 60) cm. Artificial simulated groundwater was provided with a diaphragm pump. The passing velocity was controlled at 1 m/day (or less). Hydrogen peroxide, supplied with a syringe pump at 20 mL/day, was added at the influent side of the column reactor or between the column reactors. To simulate the low dissolved oxygen concentration in the aquifer, the simulated groundwater was continuously purged with nitrogen gas to keep the oxygen concentration at less than 0.4 mg/L. Benzene, propionic acid, and *n*-butyric acid were injected with a syringe pump to avoid loss through volatilization. Samples also were collected with a syringe and analyzed by high-performance liquid chromatography (HPLC).

RESULTS AND DISCUSSION

Chemical Oxidation and Adsorption

Hydrogen peroxide was used as the alternative oxygen source for this study. Hydrogen peroxide also is a strong oxidant and may chemically oxidize the organic chemicals of interest, resulting in an overestimation of the biodegradation within the column reactors. The results of the batch study indicated that benzene and *n*-butyric acid were kept intact, even when the concentration of hydrogen peroxide reached 2,000 mg/L. The results suggested that neither benzene nor *n*-butyric acid was chemically oxidized by hydrogen peroxide.

The results of the adsorption study showed that the organic chemicals used in this study were almost not adsorbed by sand. The K values (adsorption capacity) for the Freundlich adsorption equation were very small. The K values for benzene and *n*-butyric acid were 8.4×10^{-19} and 3.5×10^{-11}, respectively. Therefore, the adsorption effect was insignificant and was ignored in this study.

Oxygen Produced from Hydrogen Peroxide

Table 1 shows oxygen concentrations at the end of a sterilized sand column, when hydrogen peroxide was added at the influent side of the column. Ideally, decomposition of 1 mg hydrogen peroxide would produce 0.47 mg of oxygen. However, if the rate of oxygen production from the decomposition of hydrogen peroxide exceeds the rate of oxygen utilization, oxygen may form air bubbles and escape from water because of its low solubility in water. Table 1 shows that dissolved oxygen concentrations at the end of the sterilized sand column were not proportional to the influent concentrations of hydrogen peroxide. In this test, oxygen decomposed from hydrogen peroxide was not used by microorganisms. The composition of the influent contained a phosphate buffer solution that might decrease the decomposition rate of hydrogen peroxide. Therefore, hydrogen peroxide still could be detected at the end of the sand column, even if it took 12 hours to travel through the column.

TABLE 1. Dissolved oxygen (DO) concentrations at the end of a sterilized sand column.

Conc of H_2O_2 (mg/L)	DO (mg/L)
0	1.6
10	6.0
20	10.3
40	16.7
60	22.3
80	25.5
100	28.0
150	30.3

Biodegradation of Volatile Fatty Acids

Volatile fatty acids are the major components in the leachate from domestic waste landfill sites. Oxygen concentration is also one of the major limiting factors for the bioremediation of organic acid-contaminated groundwaters. Figure 2 shows the effects of the addition of hydrogen peroxide on the biodegradation of

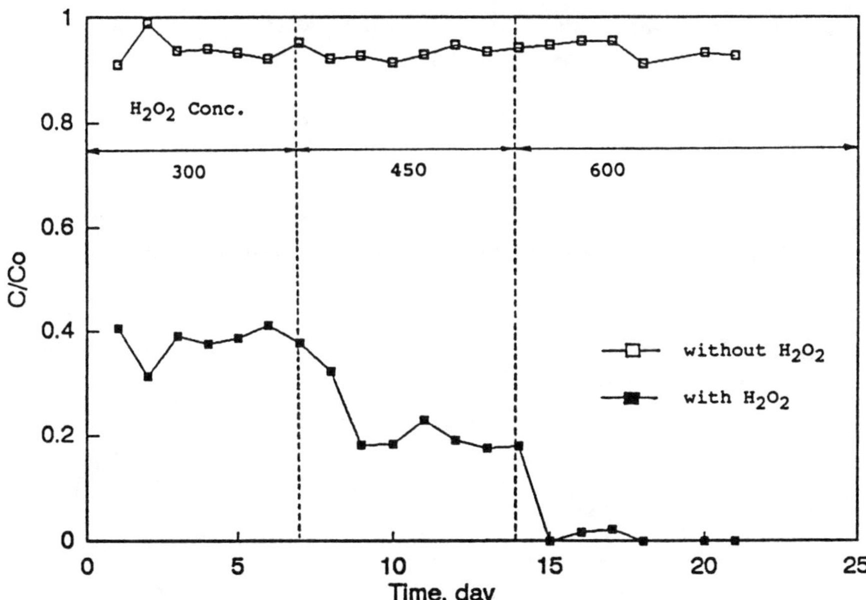

FIGURE 2. Effects of hydrogen peroxide on the biodegradation of *n*-butyric acid in the sand columns (concentration of *n*-butyric acid: 306 mg/L; H_2O_2 concentration: day 0-7, 300 mg/L; day 7-14, 450 mg/L; day 14-20, 600 mg/L).

n-butyric acid at the concentration of 306 mg/L in a 60-cm-long sand column. This figure indicates that only about 10% of *n*-butyric acid was removed from the column without the presence of hydrogen peroxide. With the addition of hydrogen peroxide at concentrations of 300 mg/L and 450 mg/L, the removal of *n*-butyric acid increased to 60% and 80%, respectively. When the concentration of hydrogen peroxide was increased to 600 mg/L, *n*-butyric acid was almost completely biodegraded. The results indicated that the addition of hydrogen peroxide increased the biodegradation of *n*-butyric acid in the simulated groundwater system.

Table 2 summarizes the removal of *n*-butyric acid with hydrogen peroxide added in the sand column reactors. This table suggests that the removal of *n*-butyric acid in the sand column was not proportional to the concentration of hydrogen peroxide. On the contrary, an increase in the concentration of hydrogen peroxide decreased the ratio of the amount of *n*-butyric acid removed to each milligram of hydrogen peroxide added. When the concentration of hydrogen peroxide was increased from 300 mg/L to 600 mg/L, the removal efficiency of *n*-butyric acid decreased from 0.57 to 0.467 mg HBu removed/mg H_2O_2 added. The results showed that oxygen was not used efficiently in the simulated groundwater system when the concentration of hydrogen peroxide was high. Table 1 also indicates that oxygen produced from the decomposition of hydrogen peroxide was not complete in the soluble form, when the concentration of hydrogen peroxide was high. The result in Table 2 suggests that hydrogen peroxide added at relatively high concentration in one injection point was not effectively utilized by microorganisms. Hydrogen peroxide may be utilized more efficiently if added at a relatively lower concentration. The study on the biodegradation of propionic acid in the sand column also showed that the presence of hydrogen peroxide enhanced the removal of propionic acid.

Biodegradation of Benzene

Benzene is one of the major pollutants in petroleum waste-contaminated groundwaters. Generally, aromatic compounds are biodegraded more effectively within an aerobic environment. Figure 3 shows the effects of adding hydrogen

TABLE 2. Removal of *n*-butyric acid (*n*-HBu) at various concentrations of hydrogen peroxide.

Conc of H_2O_2 (mg/L)	*n*-HBu removed (mg/L)	Net *n*-HBu removed (mg/L)	$\frac{n\text{-HBu removed}}{H_2O_2\text{ added}}$ (mg *n*-HBu/mg H_2O_2)
0	20.4	—	—
300	191.5	171.1	0.570
450	256.9	236.5	0.526
600	300.8	280.4	0.467

peroxide on the biodegradation of benzene in the sand column reactor. The influent concentration of benzene for both columns was 60 mg/L. The dissolved oxygen concentration was 8 mg/L. Figure 3 shows that the removal of benzene was about 47% for the column without the addition of hydrogen peroxide. However, the addition of hydrogen peroxide increased the biodegradation of benzene. The results indicated that 10 mg/L of hydrogen peroxide was still too low to significantly enhance the biodegradation of benzene. Therefore, the concentration of hydrogen peroxide was stepwise increased from 10 mg/L to 50, 100, 150, and 200 mg/L. The results indicated that an increase in the concentration of hydrogen peroxide increased the removal of benzene. At the steady state, an increment of 50 mg/L of hydrogen peroxide resulted in the enhanced benzene removal of 7.5 mg/L. According to the stoichiometry of benzene biotransformation (McCarty 1971, 1988), an increment of 50 mg/L of hydrogen peroxide would theoretically result in enhanced benzene removal of 15.3 mg/L.

Figure 3 shows that 200 mg/L of hydrogen peroxide was required to completely biodegrade benzene in the sand column. However, the theoretical requirement of hydrogen peroxide was only 104 mg/L. Again, the hydrogen peroxide required to completely biodegrade the remaining benzene was almost doubled compared to that calculated from the theoretical stoichiometric equation (McCarty 1988). The results suggested that the theoretical requirement of hydrogen peroxide for cleanup of contaminated groundwater might be underestimated.

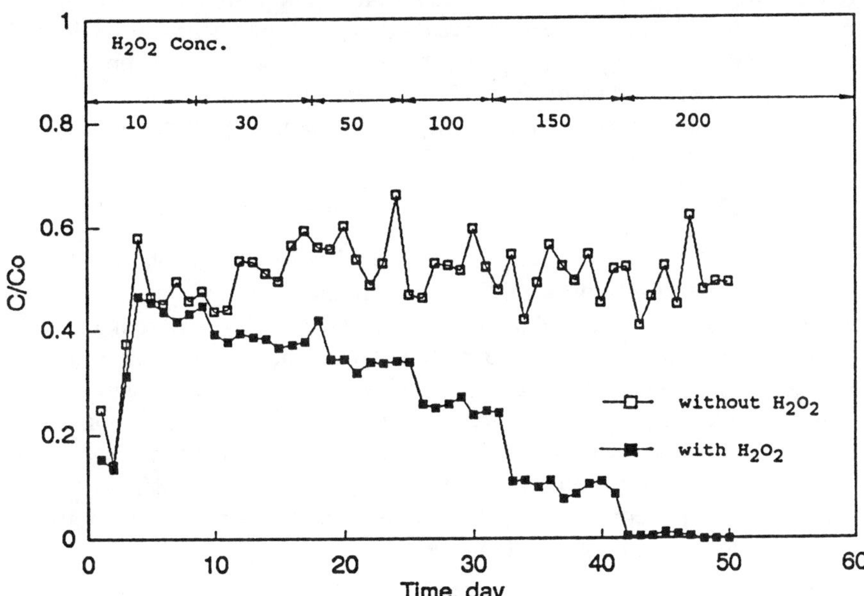

FIGURE 3. Effects of hydrogen peroxide on the biodegradation of benzene in the sand columns (concentration of benzene: 60 mg/L; H_2O_2 concentration: day 0-9, 10 mg/L; day 9-18, 30 mg/L; day 18-25, 50 mg/L; day 25-32, 100 mg/L; day 32-42, 150 mg/L; day 42-52, 200 mg/L).

TABLE 3. Removal of benzene at various concentrations of hydrogen peroxide.

Conc of H_2O_2 (mg/L)	Net benzene removed (mg/L)	Benzene removed / H_2O_2 added (mg benzene/mg H_2O_2)
30	7.96	0.265
50	12.2	0.244
100	16.6	0.166
150	24.4	0.161
200	29.9	0.146

Table 3 summarizes the amount of benzene removed at the various concentrations of hydrogen peroxide in the sand columns. Table 3 suggests that the removal of benzene was not proportional to the concentration of hydrogen peroxide. The ratio of benzene biodegraded to the amount of hydrogen peroxide added decreased with an increase in the concentration of hydrogen peroxide. For instance, when the concentration of hydrogen peroxide was 30 mg/L, the amount of benzene removed per unit of hydrogen peroxide added was 0.265 mg benzene/mg H_2O_2. However, it was only 0.146 mg benzene/mg H_2O_2 when the concentration of hydrogen peroxide was 200 mg/L. The results showed that the utilization efficiency of hydrogen peroxide decreased about 45% when its influent concentration increased from 30 mg/L to 200 mg/L. Again, the utilization efficiency of hydrogen peroxide was better when its influent concentration was lower.

SUMMARY

Biodegradation of an aromatic compound (benzene) and volatile fatty acids (propionic acid and *n*-butyric acid) in the simulated groundwater system was limited by the concentration of dissolved oxygen. Hydrogen peroxide could be used as an alternative oxygen source to enhance the biodegradation of these compounds. Oxygen produced from the decomposition of hydrogen peroxide was not completely utilized by microorganisms in the simulated groundwater system. The utilization efficiency of hydrogen peroxide decreased with an increase in its influent concentration. The removals of organic chemicals in the sand column were not proportional to the concentrations of hydrogen peroxide added because of the low utilization efficiency.

REFERENCES

Aggarwal, P. K., J. L. Means, D. C. Downey, and R. E. Hinchee. 1991. "Use of Hydrogen Peroxide as an Oxygen Source for In Situ Biodegradation, Part II. Laboratory Studies." *J. Hazardous Materials* 27:301.

Flathman, P. E., J. H. Carson, Jr., S. J. Whitehead, K. A. Khan, D. M. Barnes, and J. S. Evans. 1991. "Laboratory Evaluation of the Utilization of Hydrogen Peroxide for Enhanced Biological Treatment of Petroleum Hydrocarbon Contaminants in Soils." In R. E. Hinchee and R. F. Olfenbuttel (Eds.), *In Situ Bioreclamation: Applications and Investigations for Hydrocarbon and Contaminated Site Remediation*, pp. 125-142. Butterworth-Heinemann, Stoneham, MA.

Hinchee, R. E., D. C. Downey, and P. K. Aggarwal. 1991. "Use of Hydrogen Peroxide as an Oxygen Source for In Situ Biodegradation, Part I. Field Studies." *J. Hazardous Materials 27*:287.

Lee, M. D., J. M. Thomas, R. C. Borden, P. B. Bedient, and C. H. Ward. 1988. "Biorestoration of Aquifers Contaminated with Organic Compounds." *CRC Crit. Rev. Environ. Control 18*:29.

Lu, C. J., and M. C. Hwang. 1992. "Effects of Hydrogen Peroxide on the In Situ Biodegradation of Chlorinated Phenols in Groundwater." Paper presented at the *Wat. Environ. Federation 65th Annual Conference*, Sept. 20-24. New Orleans, LA.

McCarty, P. L. 1971. "Energetics and Bacterial Growth." In S. D. Faust and J. V. Hunter (Eds.), *Organic Compounds in Aquifer Environments*, pp. 495-531. Marcel Dekker, Inc., New York, NY.

McCarty, P. L. 1988. "Bioengineering Issues Related to In Situ Remediation of Contaminated Soils and Groundwater." In G. S. Omenn (Ed.), *Environmental Biotechnology: Reducing Risks From Environmental Chemicals Through Biotechnology*, pp. 143-162. Plenum Press, New York, NY.

Pardieck, D. L., E. J. Bouwer, and A. T. Stone. 1992. "Hydrogen Peroxide Use to Increase Oxidant Capacity for In Situ Bioremediation of Contaminated Soils and Aquifers: A Review." *J. Contaminant Hydrology 9*:221.

Schlegel, H. G. 1977. "Aeration without Air: Oxygen Supply by Hydrogen Peroxide." *Biotechnol. Bioeng. 19*:413.

Thomas, J. M., V. R. Gordy, S. Fiorenza, and C. H. Ward. 1991. "Biodegradation of BTEX in Subsurface Materials Contaminated with Gasoline: Granger, Indiana." *Wat. Sci. Tech. 22*:53.

Wilson, S. B., and R. A. Brown." 1989. "In Situ Bioreclamation: A Cost-Effective Technology to Remediate Subsurface Organic Contamination." *GWMR, Winter*:173.

THE EVOLUTION OF A TECHNOLOGY: HYDROGEN PEROXIDE IN IN SITU BIOREMEDIATION

R. A. Brown and R. D. Norris

ABSTRACT

In situ bioremediation was one of the first technologies with the potential to address both dissolved- and adsorbed-phase organic contamination with minimal disturbance to the site. Pioneering work by Richard L. Raymond beginning in 1972 demonstrated commercial potential. In the early 1980s it was recognized that the lack of an efficient oxygen supply limited implementation of the technology. Early systems used diffusers to saturate injected water with air. These systems, however, introduced limited amounts of oxygen and were prone to fouling. The innovation of using hydrogen peroxide (H_2O_2) provided oxygen at a rate up to two orders of magnitude faster than did the existing technology. Although H_2O_2 was used successfully at a number of sites, problems including too rapid decomposition, gas blockage, and inefficient use were encountered at other sites. Subsequently, alternatives such as the use of nitrate as an electron acceptor, bioventing, and air sparging have been evaluated and implemented based on cost and/or technical advantages. This paper discusses H_2O_2 use from the promise of a significant breakthrough to its current status as one of several electron acceptor choices and examines the driving forces behind the use of H_2O_2, examples of successful application, problems associated with its use, and the use of alternative oxygen sources that will compete with H_2O_2 for use in in situ bioremediation.

INTRODUCTION

Bioremediation is one of the oldest in situ remedial processes, having first been practiced commercially in 1972 to treat a gasoline pipeline spill in Ambler, Pennsylvania (Raymond 1976). Since that time, bioremediation has evolved from a novel process to an important remedial technology. Two key factors have fostered the evolution of bioremediation: microbiology and engineering. The microbiological aspects have focused on basic metabolic processes and how to manipulate them.

Prior to 1972, basic research had identified many biooxidation mechanisms, reaction products, and reaction conditions that would serve as the basis for simple aerobic biodegradation processes. For example, studies by Tausson (Tausson 1927) isolated bacterial strains capable of oxidizing naphthalene, anthracene, and phenanthrene. Subsequent studies (Sisler & ZoBell 1947) demonstrated that marine bacteria could rapidly oxidize benzo[a]anthracene to CO_2. Senez (Senez 1956) observed that normal alkanes were enzymatically attacked at the C-1 position. In 1959, Leadbetter and Foster made one of the most important discoveries in biodegradation. They were the first to observe, define, and report on the cooxidation of hydrocarbons previously considered resistant to oxidation and assimilation (Leadbetter & Foster 1959). Their finding stimulated the current research in the cometabolism of chlorinated ethenes.

Although the understanding of microbial processes has been a cornerstone in the development of bioremediation, there has been an often lengthy delay between successful laboratory studies and commercial, field-scale application. Often lacking are the systems necessary to apply and scale-up laboratory discoveries to field situations. Thus the second aspect of the evolution of bioremediation has been the development of engineering systems to provide the conditions necessary to support the metabolic processes identified in the laboratory, i.e., supplying the proper oxygen tension, moisture, nutrients, pH, temperature, etc. These engineering systems have also evolved from rudimentary systems to effective processes.

The first level of engineering applied to commercial bioremediation was the attempt to stimulate indigenous bacteria by enriching the subsurface environment with oxygen, nitrogen, phosphorus, and trace minerals. This concept of introducing nutrient- and oxygen-amended water to promote biodegradation for site remediation was first attempted by Raymond in 1972 at the Ambler, Pennsylvania Sun Oil pipeline spill (Raymond 1976), and was subsequently patented in 1974. The Ambler site used the injection of nutrient solutions and simple, in-well aeration (Figure 1).

From 1975 to 1983 Raymond and coworkers conducted several demonstration projects with the support of the American Petroleum Institute (API) to demonstrate the feasibility of bioremediation using simple infiltration of nutrient-amended water and in-well aeration. Although in general the objective of many of these tests was not complete site remediation, the resultant reduction in soil and groundwater contamination was sufficiently encouraging to stimulate a widespread interest in the technology. A key finding of this early work was the importance of the rate of oxygen delivery (Floodgate 1973, ZoBell 1973). Where bioremediation did not work effectively, the cause of failure generally was found to be a lack of sufficient oxygen.

Oxygen supply was thus identified as the central issue to be resolved if the technology was going to have a general applicability. This focus on oxygen supply lead to the first major innovation in bioremediation, the use of H_2O_2 as an oxygen "carrier." H_2O_2 was viewed as a chemical oxygen source because it decomposes to generate oxygen. H_2O_2 held promise because it is miscible in water, and each milligram per liter of H_2O_2 added could supply ~0.5 mg/L of oxygen. Its danger

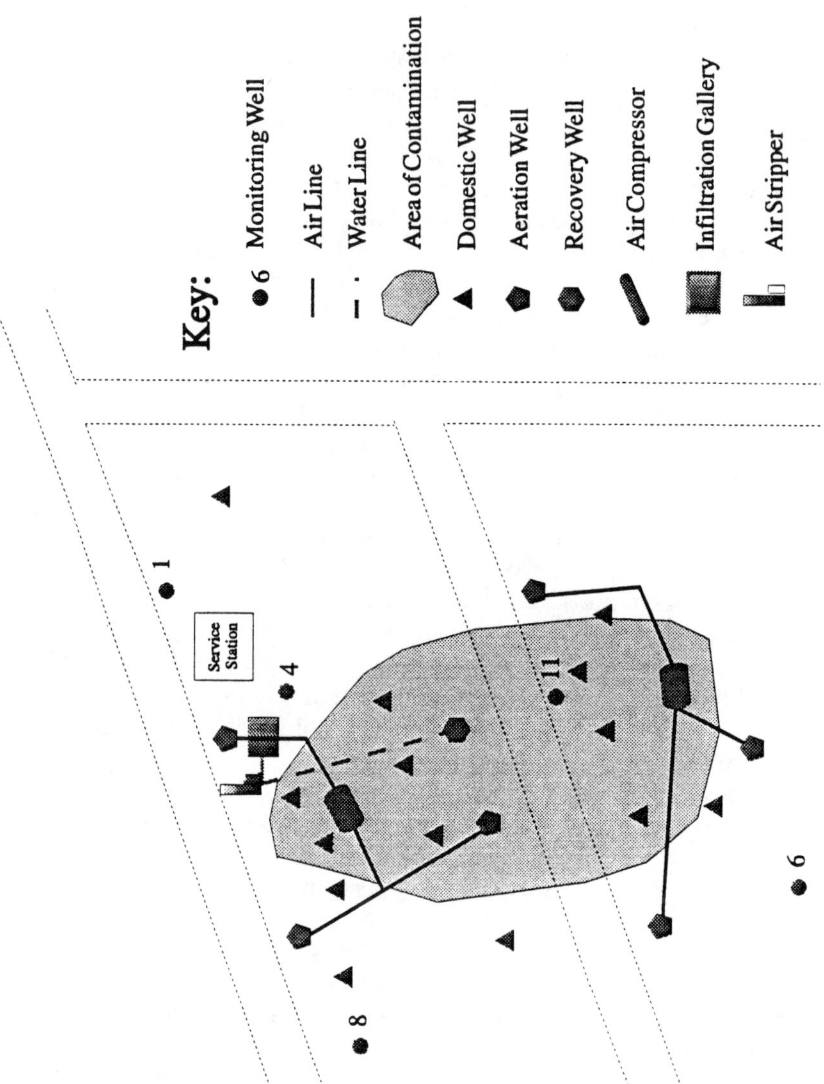

FIGURE 1. Schematic of bioreclamation system with in-well aeration.

was, however, its potential biotoxicity. Nevertheless, because of its potential for increasing oxygen availability beyond the 8 to 10 mg/L dissolved oxygen (DO) limit of in-well aeration, H_2O_2 held promise for accelerating remediation to make bioremediation a viable commercial process. However, H_2O_2 never became as generally applicable, efficient, and cost effective as was first anticipated.

THE DEVELOPMENT OF HYDROGEN PEROXIDE BIOREMEDIATION

Laboratory tests conducted at the Texas Research Institute (TRI) in the early 1980s (TRI 1982) demonstrated that under the conditions of their tests, H_2O_2 could be used by petroleum hydrocarbon-degrading bacteria as a source of oxygen and that the bacteria could tolerate H_2O_2 concentrations up to 1,000 mg/L, the point at which H_2O_2 could potentially increase oxygen availability by ~50 fold over that attained by saturating water with air.

Parallel to the TRI work, Raymond, Brown, and Norris conducted a series of column and field studies to further demonstrate the feasibility of using H_2O_2 for in situ bioremediation (Brown et al. 1984). The laboratory studies demonstrated that the growth of aerobic bacteria in general and hydrocarbon-degrading bacteria in particular, as well as the removal of gasoline, were significantly enhanced by the H_2O_2 use. The field studies were the first to demonstrate that the injection of H_2O_2 solutions to groundwater could increase the DO content of groundwater at distances of 6 to 15 m from the injection point (Brown et al. 1986). A secondary finding was that short, spike concentrations (>2,000 mg/L H_2O_2) would remove biofilm growth from injection wells and thus maintain injectivity. The loss of injectivity had been a problem with the earlier in-well aeration systems. Based on this research (Brown et al. 1986), a U.S. patent claiming the continuous injection of H_2O_2 solutions with the use of periodic spikes to maintain injectivity was issued.

Subsequently, a joint API/FMC Corporation field test conducted in Granger, Indiana (API 1987) demonstrated the successful use of H_2O_2 in a full-scale in situ bioremediation system. The treatment system used a series of injection galleries and recovery wells to circulate a nutrient-enriched H_2O_2 solution. After approximately 160 days of treatment at a flow rate of 37 to 75 lpm and an H_2O_2 concentration of ~500 mg/L, approximately 4,400 kg of gasoline hydrocarbons were removed by the system. The conclusion of this study was that H_2O_2 could be used on a commercial scale.

Commercialization – Phase One

The commercial use of H_2O_2 has gone through two phases. In the first phase, H_2O_2 was used as the sole oxygen source, treating both saturated and unsaturated regimes. An estimated 6 to 10 projects used this approach. In the second phase, H_2O_2 was used to treat only saturated regimes.

The first phase, lasting from 1984 to ~ 1986, several commercial in situ bioremediation projects using H_2O_2 as the sole oxygen source were implemented and in some cases led to a reduction in hydrocarbons (Frankenberger et al. 1989) and benzene, toluene, ethylbenzene, and xylene (BTEX) (Norris & Dowd 1993) to below the detection limit. Peroxide- and nutrient-amended water typically was injected into the vadose zone through galleries and drawn through the saturated zone by recovery wells (Figure 2).

Several of these early applications used retrofits of existing aeration or injection systems; others were specifically designed peroxide injection systems as summarized in Table 1. In the first case, nutrients and aerated water were introduced into the aquifer through one gallery and several wells (Brown & Crosbie 1989). The reduction in hydrocarbon levels appeared to reach a plateau after a 90% reduction in BTEX levels. Inclusion of H_2O_2 led to a more rapid decline in the residual BTEX levels. In a second case, H_2O_2 and nutrients were introduced through a series of injection wells and galleries leading to an overall 70% reduction in BTEX with as much as 99% reduction in several areas of the site (Smallbeck & Leland 1991). In a third example H_2O_2 and nutrients were added through recharge galleries and an interception trench over a 36-month period leading to closure by the Ministry of the Environment of Quebec (Brown & Tribe 1990).

Commercialization – Phase Two

Although H_2O_2 provided a significant improvement in oxygen supply compared to simple in-well aeration, it too had significant limitations, especially for the treatment of vadose zone (unsaturated) soils. Injection of dilute H_2O_2 solutions (100 to 1,000 mg/L, with levels as high as 2,000 mg/L), can result in two potential problems. First, in low permeable soils the rate of percolation may be sufficiently slow that H_2O_2 may decompose and release oxygen before reaching the contaminated soils. Second, increased populations of catalase-secreting bacteria can lead to too rapid decomposition and thus loss of oxygen. Because of these limitations and concerns with cost, H_2O_2 use in the vadose zone was quickly supplanted by the development and use of soil vapor extraction (SVE) technology. As with the replacement of aeration with H_2O_2, some of the early uses of SVE were to retrofit existing sites (Brown & Crosbie 1989). Although the focus of SVE has been primarily to remove volatile organic compounds, it was observed early in SVE development (Thorton & Wooten 1982), that the SVE process resulted in increased biodegradation rates because oxygen was supplied and used by soil microbes. This approach is now commonly referred to as bioventing.

While improving the effectiveness of bioremediation in unsaturated soils, SVE did not address contaminants below the water table and H_2O_2 remained the system of choice for saturated matrixes. Thus bioremediation systems became hybrid, integrated systems combining SVE with H_2O_2 solution injection (Figure 3). This second phase of commercialization began in approximately 1986 and lasted until approximately 1990.

The examples summarized in Table 2 illustrate this phase, including both retrofitted and designed systems. In the first example, sandy soils contaminated

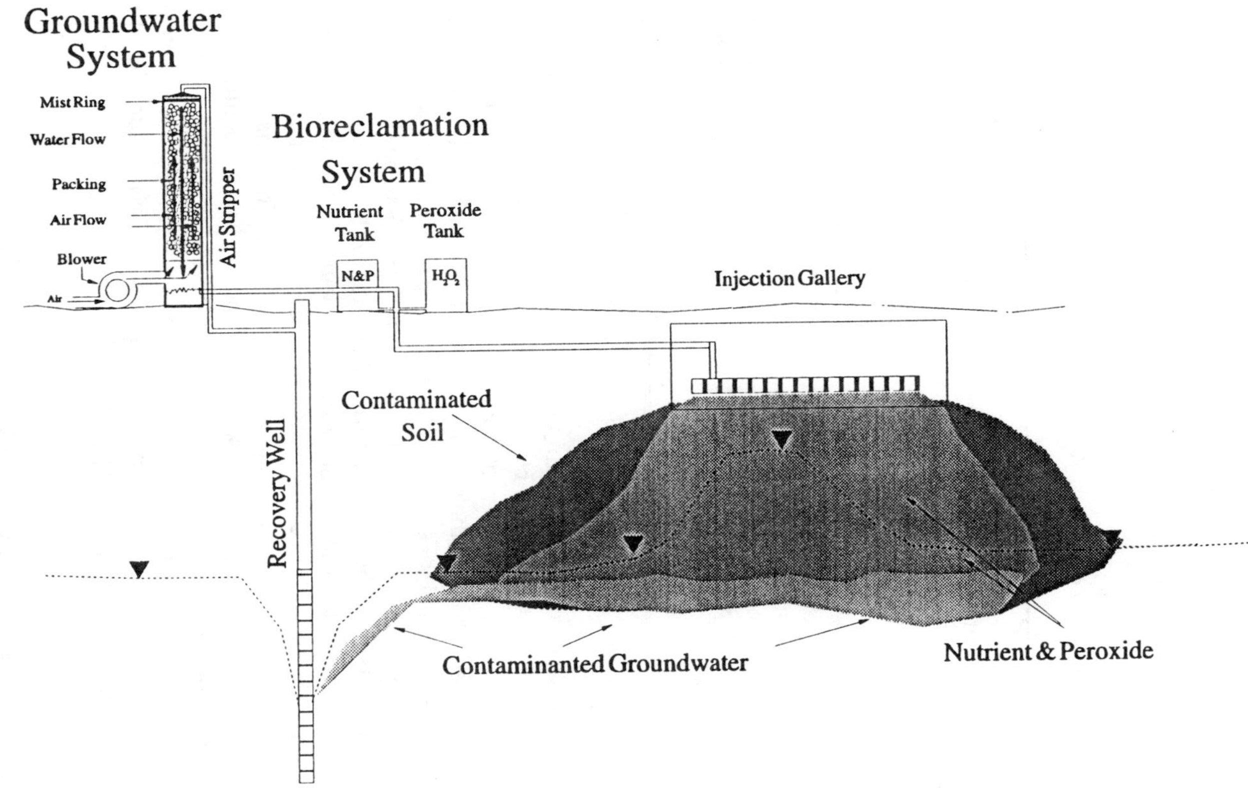

FIGURE 2. Conceptual view of H_2O_2 injection system.

TABLE 1. Case histories – Phase I.

Site/Location Type	System Design	Length of Operation	Results	Comments
Gasoline station, Pennsylvania Fractured bedrock	Retrofit of in-well aeration system. 6 aeration wells. H_2O_2 through injection gallery.	9 months H_2O_2 injection	BTEX reduced ~1 – 1.5 ppm to <250 ppb.	Further reduction limited by fractured bedrock geology.
Gasoline station, Oakland, CA Silty sand with clay	14 injection wells. 10 infiltration basins. 22 recovery wells.	15 months 4 pore volumes	Soil concentrations reduced 70% from start of 100 ppm. Some areas 99%.	Some hot spots removed by excavation.
Gasoline station, Montreal Limestone bedrock	2 injection galleries with recovery trench. 2,000 ppm H_2O_2.	36 months operation at 30-40 lpm	BTEX reduced in groundwater by 85-95%. Asymptote reached.	Residual levels due to bedrock contamination.

from gasoline losses from a Long Island service station were addressed through first free phase recovery followed by in situ bioremediation using H_2O_2 (Lee & Raymond 1991). After 36 months of operation, soil vapor extraction was used to accelerate remediation. In the second case, soils contaminated with gasoline, unsaturated and saturated soils were treated using soil vapor extraction and in situ bioremediation which utilized H_2O_2 at a concentration of 500 mg/l, respectively (Norris & Dowd 1993). The saturated soils were treated to below the detection limit in ten months. After recovery of hydrocarbon vapors reached an asymptote, remediation of the unsaturated soils was completed using addition of nutrients and H_2O_2 through two horizontal vapor extraction wells. In a final example, in situ bioremediation using H_2O_2 was used in conjunction with bioventing/vapor recovery, and bioremediation of excavated soils in an above ground cell (Brown et al. 1991). Off-gases from all systems including an air stripper were treated with a catalytic converter. Air sparging was also used at this site which illustrates a high degree of integration of compatible technologies.

LIMITATIONS OF HYDROGEN PEROXIDE USE

These early uses of H_2O_2 demonstrated its effectiveness but did not allay the primary concerns with its use.

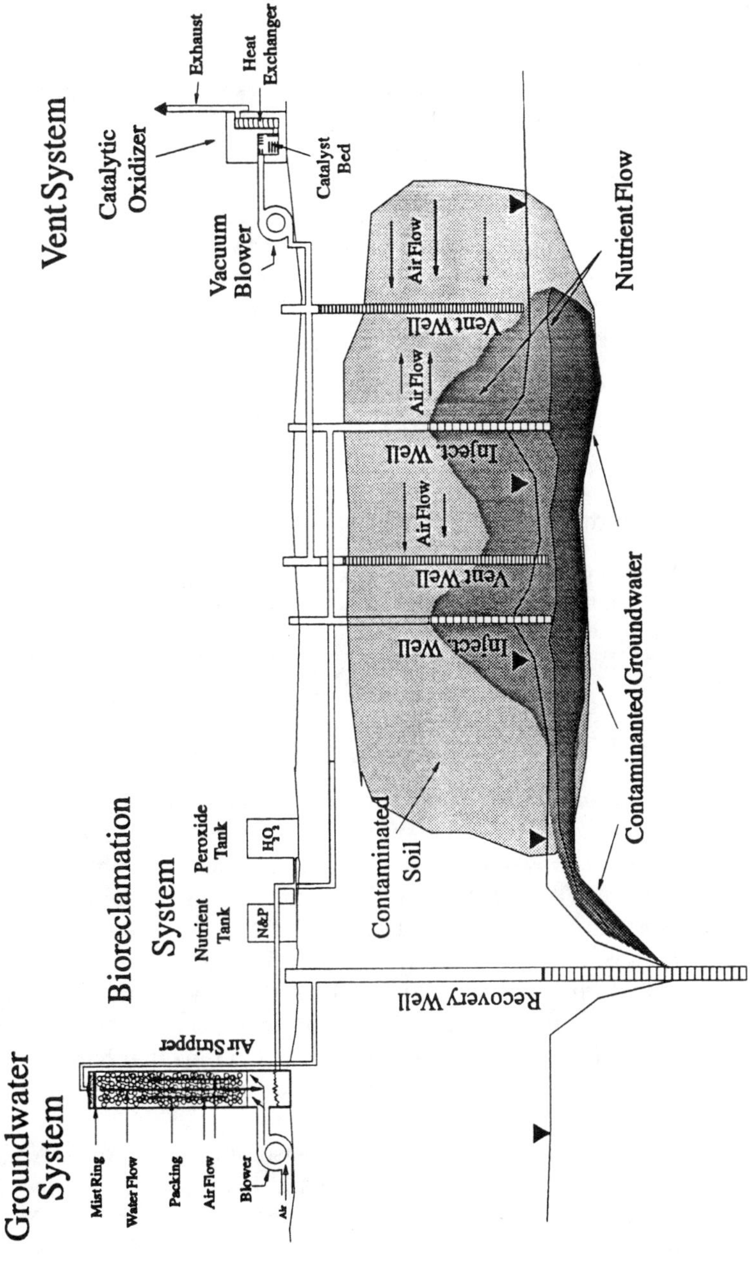

FIGURE 3. Conceptual view of integrated remedial system.

TABLE 2. Case histories – Phase II.

Site/Location Type	System Design	Length of Operation	Results	Comments
Gasoline station, Long Island Medium sands	Continuous H_2O_2 through gallery. Batch fed wells. Vent system with nutrient injection under roadway.	36 months H_2O_2 injection 12 months venting	25,000 kg gas removed. Vent system 1,000 ppm to <10 ppm.	Vent system was retrofit of Phase I system
Gasoline station, Los Angeles Vadose – sands Saturated – silts	4-8 lpm H_2O_2 at 1,000 ppm. 7 wells. Vent with vertical and horizontal wells. H_2O_2 polish after vent.	10 months H_2O_2 injection 32 months total system	110 kg saturated 4,400 kg vent Groundwater BDL Soil above regs with vent alone Soils BDL with Bio	Soil structure limited vent performance
Fuel terminal, New Jersey Medium sands	5 injection wells. 2 recovery wells. 4 vent wells. Batch nutrient galleries in vadose.	18 months (till new spill)	15,500 kg removed 90% reduction in BTEX levels	Persistent contamination due to periodic leak

Concerns with the Stability of Hydrogen Peroxide

The use of H_2O_2 is predicated on its catalyzed conversion to oxygen and water. The rate of this conversion is critical to the success of using H_2O_2. Because a basic issue with H_2O_2 has always been its cost, ~\$2.81 (large use bulk) to \$4.63 (drums)/kg O_2 supplied (1993 \$), uncontrolled decomposition or the loss of oxygen equivalents is a serious concern. With the typical additions of 100 to 1,000 mg/L (50 to 500 mg/L O_2 equivalents), too rapid a decomposition can result in water being supersaturated with oxygen (>50 mg/L) and the subsequent loss of oxygen. Such conditions may result in degassing, which can cause gas blockage and the reduction of permeability around injection points. On the other hand, too slow H_2O_2 decomposition would reduce the oxygen availability in the area being treated.

The key to H_2O_2 use is controlling its decomposition. The conversion of H_2O_2 to O_2 can be catalyzed by metals such as iron and manganese, by the enzyme catalase which is secreted by many microorganisms, and by surfaces. Many of the early researchers felt that the most problematic decomposition mechanism was metal-catalyzed decomposition, whose reactions are sensitive to pH and sometimes are modified by the presence of complexing species i.e., phosphates used primarily as nutrient sources.

A detailed study on the behavior of H_2O_2 in a large number of soils failed to determine any correlation between quantifiable soil characteristics and degradation rates except possibly manganese levels (Lawes 1991). Degradation rates of separate soil samples collected from the same site or even the same split spoon sample gave distinctly different results. Furthermore decomposition rate behavior appeared to decrease over time.

This issue of decomposition was brought to a head in a study sponsored by the U.S. Air Force. Laboratory tests conducted during the design of a bioremediation system at Eglin Air Force Base, Florida, suggested that the rate of conversion of H_2O_2 to O_2 would be satisfactory (Hinchee & Downey 1988). Nutrient- and H_2O_2-amended groundwater was introduced through infiltration galleries located above the water table, but after a period of operation the rate of O_2 conversion was much too fast and the O_2 was lost largely to the unsaturated zone. Thus the O_2 was not available for hydrocarbon degradation in the saturated zone because the H_2O_2 converted to O_2 before the water reached the aquifer. Further study gave evidence that the substantial increase over time in the conversion rate was a result of an increase in the population of catalase-secreting bacteria. Thus H_2O_2 use was complicated by the very entity that it was intended to aid, the bacteria.

Concerns With Toxicity

The potential toxicity of H_2O_2 to microorganisms has always been a concern with its use in bioremediation. Its use as a household "disinfectant", however, is at a level significantly above the level typically used in in situ bioremediation: 3% (~30,000 mg/L) versus 100 to 1,000 mg/L. A continuing question, therefore, has been the concentration at which peroxide is toxic. The upper boundary of H_2O_2 concentration is the level at which peroxide will begin to oxidize and "dissolve" cell material, an effect which is used commercially for slime and bulking control (FMC undated).

The issue of toxicity is complex. First peroxide is produced as a result of normal aerobic respiration. Because of this, many microorganism produce enzymes (peroxidase and catalase) which decompose hydrogen peroxide and thus prevent its buildup. Thus, tolerance to peroxide is a function of the level and activity of these enzymes. A second complicating factor is that some microorganism lack these enzymes. In such cases, even low concentrations may be toxic. A third factor is that microbial population adapt to the presence of H_2O_2 and gradually tolerate higher levels but may lose tolerance as their food supply is depleted. Because it is not possible to predetermine the concentration at which H_2O_2 toxicity becomes significant, most applications H_2O_2 have used "safe" levels, generally less than 2,000 mg/L which still provide oxygen at a much faster rate than simple, in-well aeration.

Concerns With Iron Precipitation

As with the addition of all oxygen sources, the introduction of H_2O_2 into contaminated aquifers can lead to precipitation of iron which is frequently elevated within hydrocarbon plumes. Iron precipitation can plug the formation within the vicinity of the injection well requiring frequent maintenance. The potential for iron precipitation can be lessened by using tripolyphosphate as the phosphorous source and/or by flushing the aquifer with clean low metal content water such as might be obtained from a deeper aquifer.

THE SEARCH FOR ALTERNATIVES

Because the issues of toxicity, stability, and cost have never been addressed conclusively, the search for alternatives to H_2O_2 use have continued in two directions: alternative means of supplying oxygen and an alternative (non-oxygen) electron acceptor.

Alternative Oxygen Sources

Oxygen can be supplied by means other than aeration of the injection water and amendment with H_2O_2. The most obvious is to sparge the injection water with oxygen instead of air. Potentially this can result in a 5 fold increase in the oxygen concentration in the water. Either liquid oxygen (Sloan et al. 1992) or pressurized gas can be used. Alternatively, oxygen can be generated on site (Prosen et al. 1992).

Oxygen microbubbles or colloidal gas aphrons, created by mixing oxygen under pressure with surfactant-amended water (Michaelsen & Lofti 1990), have been used in place of pure oxygen. Laboratory data indicate that oxygen can be delivered through soils and used for biodegradation with the aphrons remaining intact for up to 3 months. This approach is being evaluated within the Superfund Site Program (Anonymous 1992).

Most recently, air sparging has been used to provide oxygen below the water table (Brown & Jasiulewicz 1992). This method involves injecting air below the water table to transfer volatile components to the unsaturated zone for capture by a vapor recovery system. Air sparging can distribute oxygen uniformly across the entire site rather than requiring oxygen to move across the aquifer with the injected water and it is relatively inexpensive to implement and operate. Air sparging provides the same benefits to saturated zone treatment that SVE has for vadose zone treatment. Several operating bioremediation systems consist of air sparging and SVE combined with nutrient infiltration (Figure 4).

Alternative Electron Acceptors

Other electron acceptors include nitrate, carbon dioxide, iron, and sulfate. Of these, nitrate has received the most attention as a substitute for oxygen (Hutchins et al. 1991) because it is relatively inexpensive, is very soluble in water, is not adsorbed to soil matrices, and does not decompose. Thus it can be readily distributed within an aquifer. Its disadvantages are that nitrate concentrations in groundwater typically are limited by regulatory standards to 46 mg/L as NO_3, and that it is effective for fewer classes of compounds than is oxygen. For instance,

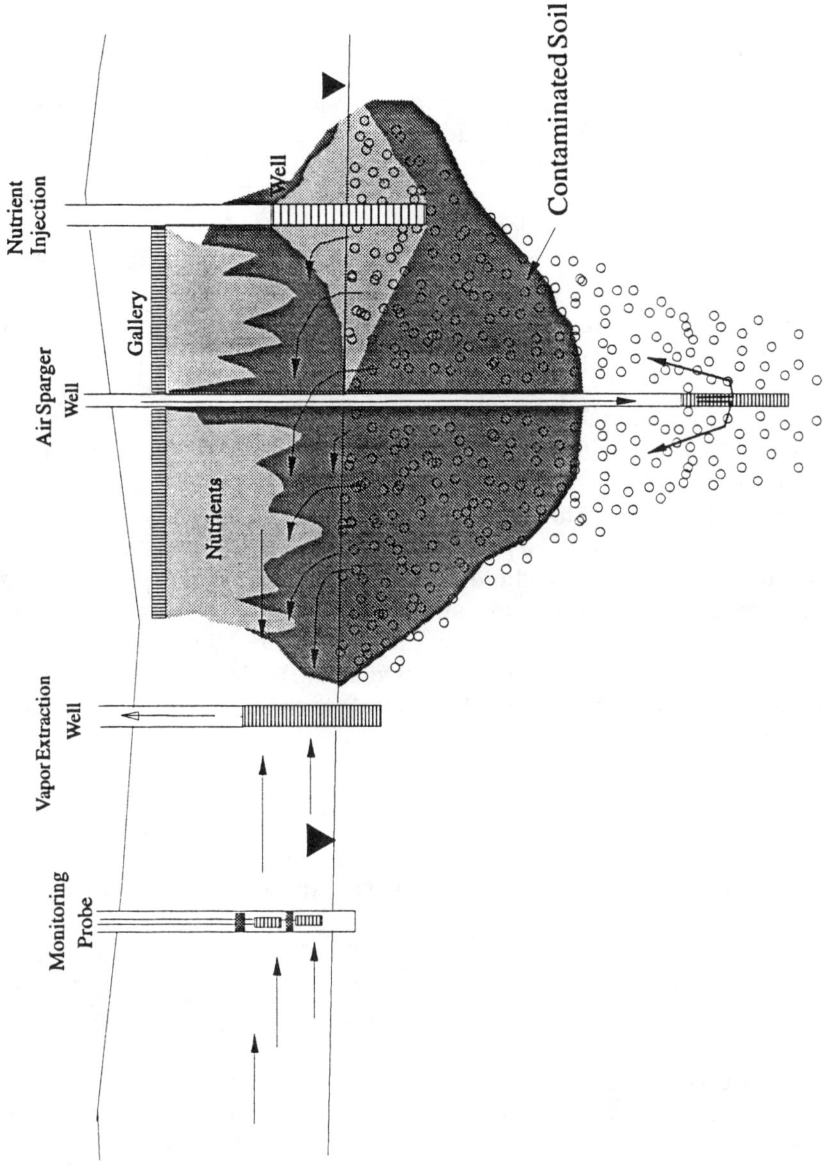

FIGURE 4. Conceptual view of air sparging system.

nitrate-utilizing bacteria do not degrade aliphatic compounds, which are a major component of most commercial hydrocarbon blends. Furthermore, benzene is not readily degraded under denitrifying conditions.

THE FUTURE OF HYDROGEN PEROXIDE

At most sites the selection of an electron acceptor delivery system will involve a number of trade-offs, which may include regulatory, political, and technical concerns. Resolution of the technical issues requires an understanding of the benefits and limitations of each system, the contaminant's physical, chemical, and biological properties, and the site's geological characteristics and infrastructure.

Cost will be the primary deciding factor where more than one electron acceptor delivery system may work quite well as was demonstrated in the USEPA study conducted at the Traverse City, Michigan Coast Guard Station (Wilson 1993). In that case air sparging, hydrogen peroxide, and nitrate all worked reasonably well for the destruction of BTEX components. Hydrogen peroxide was the most costly alternative while not providing any clear benefit over the alternatives. Thus many sites where H_2O_2 may be technically feasible and even reasonably cost effective, other choices may prevail.

However, H_2O_2 is likely to remain the preferred electron acceptor in specific situations, provided the potential problems discussed earlier do not occur or can be managed. These situations include sites where sparging is not applicable (e.g. bedrock, narrow saturated intervals, and sites where SVE systems can not be installed), when electron acceptor requirements are moderate to low but where aeration of the injected water will not be cost effective because of the longer time of remediation, and where oxygen but not nitrate is effective for biodegradation of all of the components of interest.

CONCLUSION

The use of H_2O_2 as an oxygen source for in situ bioremediation was first viewed by many as widely applicable. Experience and research has identified potential problems and limitations of its use. Furthermore, other approaches such as air sparging for aquifer remediation and vapor recovery for unsaturated soils promise to be more cost effective even at sites where H_2O_2 can be used with minimal difficulty. There remain, however, specific conditions for which H_2O_2 is likely to be the electron acceptor source of choice. In designing a bioremediation system it is important to be aware of the benefits and limitations of all of the available electron acceptor sources so that the most cost-effective system can be implemented. As with most technologies, continued experience with implementation and testing of bioremediation will add to our understanding of when and how to use the various electron acceptor sources.

REFERENCES

Anonymous. 1992. *SITE Emerging Technology Awards* 2 (9).

API. 1987. *Field Study of Enhanced Subsurface Biodegradation of Hydrocarbons Using Hydrogen Peroxide as an Oxygen Source.* American Petroleum Institute Publication 4448.

Brown, R. A., and J. Crosbie. 1989. *Oxygen Sources for In Situ Bioremediation.* Hazardous Materials Control Research Institute, Baltimore, MD, June.

Brown, R. A., J. C. Dey, and W. E. McFarland. 1991. "Integrated Site Remediation Combining Groundwater Treatment, Soil Vapor Extraction, and Bioremediation." In R. E. Hinchee and R. F. Olfenbuttel (Eds.), *In Situ Bioreclamation: Application and Investigation for Hydrocarbons and Contaminated Site Remediation*, pp. 444-449. Butterworth-Heinemann, Stoneham, MA.

Brown, R. A., and F. Jasiulewicz. 1992. "Air Sparging Used to Cut Remediation Costs." *Pollution Engineering*, pp. 52-57, July.

Brown, R. A., R. D. Norris, and R. L. Raymond. 1984. "Oxygen Transport in Contaminated Aquifers." *Proceedings of Petroleum Hydrocarbons and Organic Chemicals in Groundwater: Prevention, Detection, and Restoration* (November 5-7, Houston, TX). Water Well Journal Publishing Company, Dublin, OH.

Brown, R. A., R. D. Norris, and R. L. Raymond. 1986. "Stimulation of Bio-oxidation Process in Subterranean Formations," May 13, 1986. U.S. Patent 4,588,506.

Brown, R. A., and R. Tribe. 1990. "In Situ Physical and Biological Treatment of Volatile Organic Contamination: A Case Study Through Closure." Presented at USEPA Second Forum on Innovative Hazardous Waste Treatment Technologies: Domestic and International. Philadelphia, PA, May.

Floodgate, G. D. 1973. "The Microbial Degradation of Oil Pollutants." In D. G. Ahearn and S. P. Meyers (Eds.), Publ. No. LSU-SG-73-01. Center for Wetland Resources, Louisiana State University, Baton Rouge, LA.

FMC Marketing Literature. Undated. "Bulking Control with Hydrogen Peroxide."

Frankenberger Jr., W. T., K. D. Emerson, and D. W. Turner. 1989. "In Situ Bioremediation of an Underground Diesel Fuel Spill: A Case History." *Environ. Manager 13*(3): 325-332.

Hinchee, R. E., and D. C. Downey. 1988. "The Role of Hydrogen Peroxide in Enhanced Bioreclamation." *National Water Well Association*, Vol. 2, pp. 715-721.

Hutchins, S. R., W. C. Downs, G. B. Smith, J. T. Wilson, et al. 1991. *Nitrate for Biorestoration of an Aquifer Contaminated with Jet Fuel.* EPA 600/2-91/009. Robert S. Kerr Environmental Research Laboratory, Ada, OK.

Lawes, B. C. 1991. "Soil-Induced Decomposition of Hydrogen Peroxide." In R. E. Hinchee and R. F. Olfenbuttel (Eds.), *In Situ Bioreclamation: Application and Investigation for Hydrocarbons and Contaminated Site Remediation*, pp. 143-156. Butterworth-Heinemann, Stoneham, MA.

Leadbetter, E. R., and J. W. Foster. 1959. *Arch. Biochem. Biophys. 82*: 491-492.

Lee, M. D., and R. L. Raymond, Sr. 1991. "Case History of the Application of Hydrogen Peroxide as an Oxygen Source for In Situ Bioreclamation." In R. E. Hinchee and R. F. Olfenbuttel (Eds.), *In Situ Bioreclamation: Application and Investigation for Hydrocarbons and Contaminated Site Remediation*, pp. 429-436. Butterworth-Heinemann, Stoneham, MA.

Michaelsen, D. L., and M. Lofti. 1990. "Oxygen Microbubbles Injection for In-Situ Bioremediation: Possible Field Scenarios." *Innovative Hazardous Waste Systems.* Technotric Publishing Co., New York, NY.

Norris, R. D., and K. Dowd. 1993. "Successful In Situ Bioremediation in a Low Permeability Aquifer." In P. E. Flathman and J. Exner (Eds.), *Bioremediation: Field Experiences.* Lewis Publishers, Ann Arbor, MI.

Prosen, B. J., W. M. Korreck, and J. M. Armstrong. 1992. "Design and Preliminary Results of a Full-Scale Bioremediation Utilizing an On-Site Oxygen Generator System." In R. E. Hinchee and R. F. Olfenbuttel (Eds.), *In Situ Bioreclamation: Application and Investigation for Hydrocarbons and Contaminated Site Remediation*, pp. 523-526. Butterworth-Heinemann, Stoneham, MA.

Raymond, R. L. 1976. "Beneficial Stimulation of Bacterial Activity in Groundwater Containing Petroleum Hydrocarbons." *AICHE Symposium Series 73*: 390-404.

Senez, J. C. 1956. *Compt. Rend. Acad. Sci. 24*: 2873-2895.

Sisler, F. D., and C. E. ZoBell. 1947. *Science 106*: 521-522.

Sloan, R., R. D. Norris, and M. J. Dey. 1992. Unpublished results.

Smallbeck, D. R., and D. F. Leland. 1991. "Enhanced In Situ Biodegradation of Petroleum Hydrocarbons in Soil and Groundwater." In *Proceedings of the Petroleum Hydrocarbon and Organic Chemicals in Groundwater: Prevention, Detection, and Restoration*, pp. 393-408.

Tausson, W. C. 1927. *Planta 4*: 214-256.

Texas Research Institute, Inc. 1982. *Enhancing the Microbial Degradation of Underground Gasoline by Increasing Available Oxygen*. Final Report, American Petroleum Institute, Washington, DC.

Thorton, J. C., and W. L. Wooten. 1982. "Venting for the Removal of Hydrocarbon Vapors From Gasoline Contaminated Soil." *J. Environ Sci Health AI7*(1): 31-44.

U.S. Navy, Naval Civil Engineering Laboratory. 1990. *Bioreclamation Studies of Subsurface Hydrocarbon Contamination*. Contract # N62474-87-C-3058. Naval Air Station, Patuxent River, MD, December.

Wilson, J. 1993. "Bioremediation of Hydrocarbon Spills in the Subsurface Environment." Keynote address presented at In Situ and On-Site Bioreclamation Conference, San Diego, CA, April.

ZoBell, C. E. 1973. "The Microbial Degradation of Oil Pollutants." In D. G. Ahearn and S. P. Meyers (Eds.), Publ. No. LSU-SG-73-01. Center for Wetland Resources, Louisiana State University, Baton Rouge, LA.

BIOREMEDIATION OF OIL-CONTAMINATED SHORELINES: THE ROLE OF CARBON IN FERTILIZERS

P. Sveum, L. G. Faksness, and S. Ramstad

ABSTRACT

This paper presents the results from a series of experiments designed to evaluate the effect of different carbon and nitrogen/phosphorus sources on the activity of oil-degrading bacteria in shoreline systems and their effect on the biodegradation of oil. The experiments were designed based on observations from previous field experiments. The experiments have been conducted with mesophilic mixed cultures of bacteria, either in batch cultures or in experimental mesoscale beaches in basins with simulated tide and continuous exchange of seawater. The effect of different nutrient combinations (C, N, P) was investigated by monitoring the total number of bacteria (TNB), the metabolically active bacteria (MAB), and oil degradation (indicated by the decrease in nC_{17}/pristane). The results demonstrated that nutrients or carbon immobilized in biomass increased when easily available C:N:P compounds were added to batch cultures with crude oil. A similar effect was observed in the basin experiments. In both types of experiments, the increased nutrients or carbon immobilized in biomass appeared to result in an increased reduction of the nC_{17}/pristane ratio, compared to non-treated crude oil or crude oil treated with Inipol EAP22.

INTRODUCTION

Fertilizer treatment has proven to be a possible way to enhance biodegradation of spilled oil on shorelines, as well as in other habitats (e.g., Atlas 1991). As shorelines are inundated with seawater during high tide, water-soluble fertilizers will have a restricted retention time. To increase the nutrient retention, and thus oil degradation, specially formulated oil spill fertilizers (e.g., Inipol EAP22) have been developed. Many of these products contain substantial amounts of carbon in addition to nitrogen, phosphorous, and other limiting compounds (e.g., Atlas & Bushdosh 1976, Bergstein & Vestal 1978, Olivieri et al. 1978, Olivieri et al. 1976).

In previous Norwegian studies, oleophilic fertilizers proved to be more effective than water-soluble fertilizers when the spilled oil resided in the intertidal zone. In the supralittoral zone, the rates of biodegradation were similar for both water-soluble and oleophilic fertilizers (Halmø et al. 1985; Halmø & Sveum 1987; Sendstad 1980; Sendstad et al.; 1982 Sendstad et al. 1984). This would indicate that if water transport is limited, there is probably no need for oleophilic fertilizers, although they are essential in environments with high water exchange.

If the applied fertilizer is water-soluble or transformed to water-soluble compounds, the fertilizer components will be transported out of the treated area when exposed to water before they can be assimilated by the indigenous microbial populations. The influence of C:N ratios on nitrogen assimilation and regeneration has been studied by Goldmann et al. (1987) and Tezuka (1990), who found that for marine and freshwater bacteria, increasing the C:N substrate ratios resulted in increased biomass assimilation and decreased ammonia regeneration. Therefore, assimilation of nutrients will depend not only on the availability of the nitrogen and phosphorus as such, but also on the availability of carbon sources.

The mechanism by which fertilizers aid oil biodegradation is not completely clear. Efficiency of nitrogen assimilation and subsequent immobilization will depend on the availability of carbon sources offered along with nitrogen. It was assumed, based on preliminary experiments carried out on arctic beaches and in laboratory systems (Sveum 1986, 1987a, 1987b; Sveum & Ladousse 1989), that the availability of carbon regulates the assimilation of nutrient. Figure 1 conceptualizes possible interactions between carbon availability and nitrogen assimilation, with emphasis on the biodegradation of hydrocarbons.

This paper summarizes some experiments that were done to study the importance of carbon combined with nitrogen and phosphorus to enhance biodegradation of oil deposited on shorelines. The experiments were done both in batch cultures and on shorelines in continuous-flow seawater exchange basins with simulated tides. The results from the batch culture experiments were used for mechanistic studies, whereas the basin studies were excellent simulations of natural open systems.

MATERIAL AND METHODS

Statfjord crude oil was used for all experiments. Table 1 lists the additives used, with their relevant components.

Batch culture experiments were performed in Erlenmeyer flasks (500 mL), containing 300 mL filtered seawater (Millipore 0.2 μm), 0.1% crude oil, and 0.01% additive. The pH was adjusted to 8.1 (TRIS buffer, 20 mM). The growth medium was sterilized (120°C, 20 min) and inoculated with a mixed bacterial culture to give an initial bacterial density of approximately 10^5 cells/mL. The cultures where incubated in darkness at 18°C. Microbial and chemical analyses were done three times during 10 days of incubation. One sample was taken with sterile disposable pipettes from each batch culture at each sampling time, after homogenization of the culture.

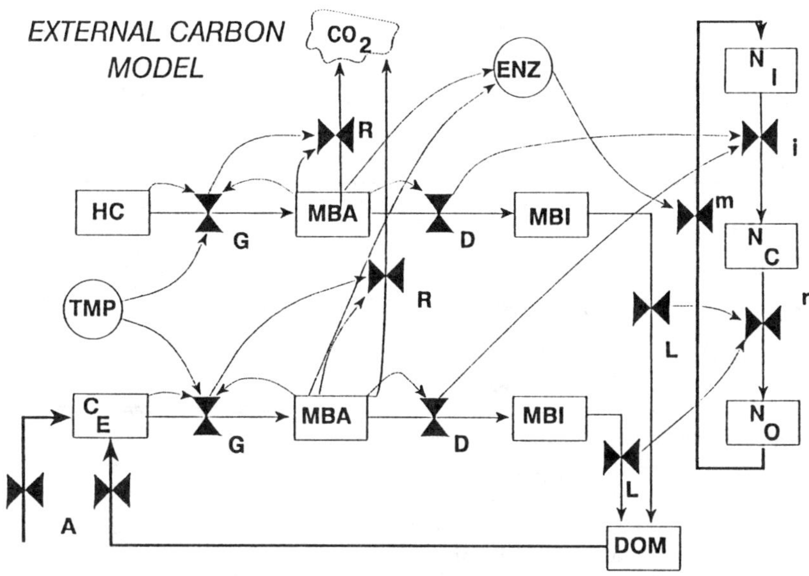

FIGURE 1. Overview of the hypothesized influence of carbon additives on the biodegradation of oil-contaminated habitats. Continuous lines show flow of matter. Dotted lines show flow of information (i.e., influence). HC: hydrocarbons; C_E: external carbon; MBA: metabolically active biomass; MBI: metabolically inactive biomass (necromass); N_I: inorganic nitrogen (or other nutrients); N_C: nitrogen (or nutrients) in cells; N_O: organic nitrogen (or nutrients), not in cells; DOM: dead organic matter; ENZ: enzymes (e.g., urease); TMP: temperature; G: growth; D: death; L: lysis; A: assimilation; m: mineralization; i: immobilization; r: release of nitrogen (or nutrients) due to cell lysis.

Mesoscale experiments (Table 2) were performed in continuous-flow seawater exchange basins (CFB) measuring h × w × l = 1 × 2 × 4 m. The shorelines (h × w × l = 0.9 × 2 × 2.2 m, i.e., the slope was 0.4) consisted of sandy beach material. Basin seawater levels varied according to a normalized tidal cycling. The seawater exchange rate was 1,200 L/h, and the average seawater temperature was approximately 10°C. Crude oil was applied at high tide and was allowed to contaminate the sediment during the falling tide. Additives were applied at the first low tide. Only bacteria indigenous to the seawater and the sediment where used in these basin experiments. Samples were taken from the upper 3 cm of the shoreline sediment during low tide, according to a randomized and stratified system. The sampling intensity is reflected in the figures.

Microbial analyses included direct counts, i.e., with DAPI (4,6-diamidino-2-phenylindole dihydrochloride hydrate) giving the TNB (Porter & Feig 1980) and with FDA (fluorescein diacetate) giving the number of MAB (Lundgren 1981).

TABLE 1. Nitrogen and phosphorus sources used in continuous-flow seawater basins.

Name	Description	N (%)	P (%)	C (%)	Other comp.
Fish meal	Organic protein	10.5[(a)]	2.0[(b)]	46.8[(a)]	40[(c)]
Soybean oil	Fatty acid			~75[(d)]	
Stickwater	Organic protein	11.3[(a)]	1.4[(b)]	41.8[(a)]	45[(e)]
Inipol EAP22	Commercially available fertilizer[(f)]				

(a) Analyzed on a Carlo Elmer Element Analyzer.
(b) Data given from the suppliers.
(c) Fat (12%), water (10%), minerals and vitamins (11%).
(d) Content of 70 to 90% C_{18} fatty acids (20 to 30% oleic acid).
(e) Water (30%), minerals and vitamins.
(f) Content of oleic acid (26.2%), lauryl phosphate (23.7%), 2-butoxy-1-ethanol (10.8%), urea (15.7%), and water (23.6%).

Total heterotrophic respiration in the sediment was measured as the total amount of CO_2 produced, using a Siemens Ultramat 10 infrared gas analyzer modified for septum injection. CO_2 was measured three times during 30 hours of incubation in sealed biological oxygen demand (BOD) bottles.

Samples from the batch culture experiments used for gas chromatographic (GC) analysis were prepared as follows: a 20-mL sample was mixed and filtered (black band filter). The oil was extracted with n-hexane (5 mL). Hydrocarbons in sediment samples were extracted with n-hexane using a Soxtec System HT6 (Tecator AB, Sweden). The resulting oil-hexane solution was filtered through a 0.22-μm filter before analysis on a Hewlett Packard 5890 Series II GC equipped with flame ionization detector (FID) and a splitless injector. Oil biodegradation was evaluated from the nC_{17}/pristane ratio.

TABLE 2. Description of experiments performed in continuous-flow basins.

	Crude oil	Additives
EXP-1	6 L/m^2	None
EXP-2	6 L/m^2	Fish meal (0.6 kg/m^2)
EXP-3	6 L/m^2	Fish meal (0.6 kg/m^2) and soybean oil (0.6 kg/m^2)
EXP-4	4 L/m^2	Inipol EAP22 (0.4 kg/m^2)
EXP-5	4 L/m^2	Stickwater (0.4 kg/m^2)
EXP-6	4 L/m^2	Fish meal (0.4 kg/m^2)

RESULTS AND DISCUSSION

In the batch culture experiments, the TNB and the MAB were counted. In the first experiments (Figure 2), cultures incubated with fish meal and stickwater resulted in the highest TNB, whereas Inipol and the inorganic nutrients gave lower TNB. The cultures incubated with fatty acids alone (i.e., soybean oil and oleic acids) resulted in only 0.1% of the TNB observed with the organic nutrient (Figure 2). In another batch culture experiment, the TNB/MAB ratio was higher for fish meal and stickwater than for fatty acids and crude oil without additives (Figure 3). The higher ratios indicate that organic N-source additions result in an enhanced immobilization of bacterial cells, and that the nitrogen and phosphorus added to the system are stored as necromass, rather than as readily available compounds. MAB measured in cultures incubated with crude oil only, or with fatty acids, were on the order of only 0.1% of those measured in cultures containing organic N/P-sources, or soybean oil and fish meal combined.

The decrease in nC_{17}/pristane ratios during the first experiment discussed above is given in Figure 4. These results are in agreement with the microbial results presented in Figure 2. In the cultures containing stickwater, the decrease in the nC_{17}/pristane ratio was twice that found in cultures treated with Inipol EAP22, inorganic nutrients, and fish meal. Both fatty acid amendments resulted in only limited decrease in the nC_{17}/pristane compared to the reference additives.

Batch culture experiments do not simulate biodegradation in an open system. However, the results do give an indication of the effect of the individual elements in the system when there is optimal retention of nutrients. In order to determine

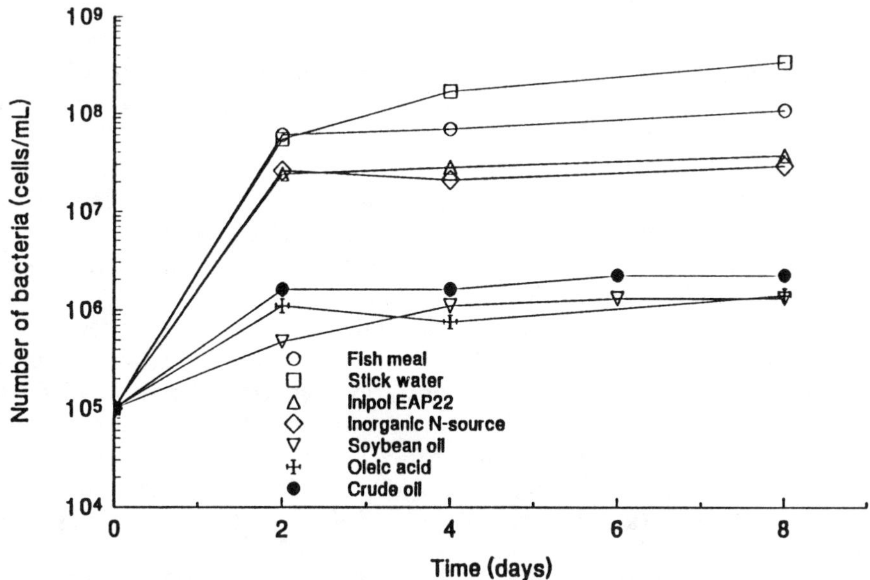

FIGURE 2. Total number of bacteria in batch culture experiments. Statfjord crude oil either alone or in combination with additives.

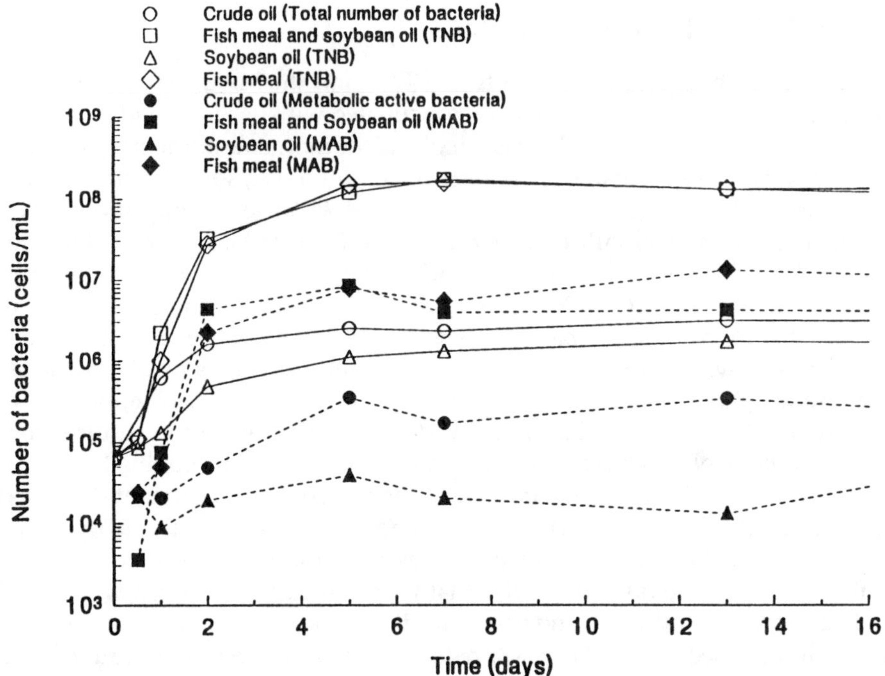

FIGURE 3. Total number (TNB) and the number of metabolically active bacteria (MAB) in batch culture experiments. Statfjord crude oil either alone or in combination with additives.

whether the batch culture results were comparable to those encountered in an open system, a series of experiments were conducted in continuous-flow seawater exchange basins (Table 2).

Figure 5 gives the bacterial numbers from experiments with no additions (Exp. 1), fish meal alone (Exp. 2), and fish meal and soybean oil (Exp. 3). The experiment with only crude oil (Exp. 1) served as a reference. An increase in both TNB and MAB was found in all three experiments during the first 6 days. The increase in MAB and TNB was more rapid in the two treated basins than in the basin with only crude oil. During the first 30 days, the bacterial numbers were higher in the experiment with fish meal than in the experiment with fish meal/soybean oil, indicating that this fatty acid inhibited bacterial growth during the initial period. After 30 days, there was virtually no difference between the two types of additives used. TNB was much lower in the nontreated experiment than in the treated experiment, and remained more or less constant throughout the experimental period. The increase in MAB after approximately 22 days in the reference basin can be attributed to decreased crude oil toxicity.

Figure 6 gives the bacterial number from the second series of basin experiments. The first experiment was treated with Inipol EAP22 (Exp. 4). The

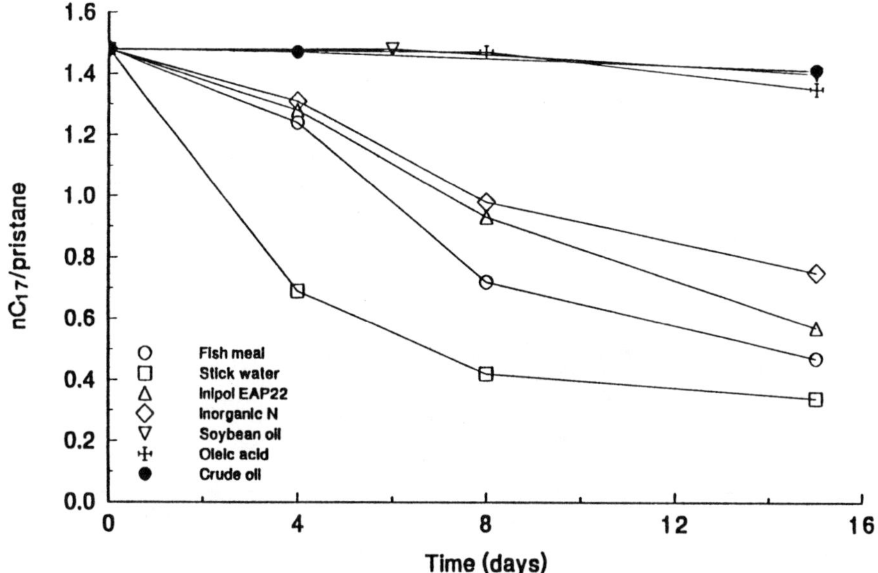

FIGURE 4. Change in the nC_{17}/pristane ratio in the batch culture experiments.

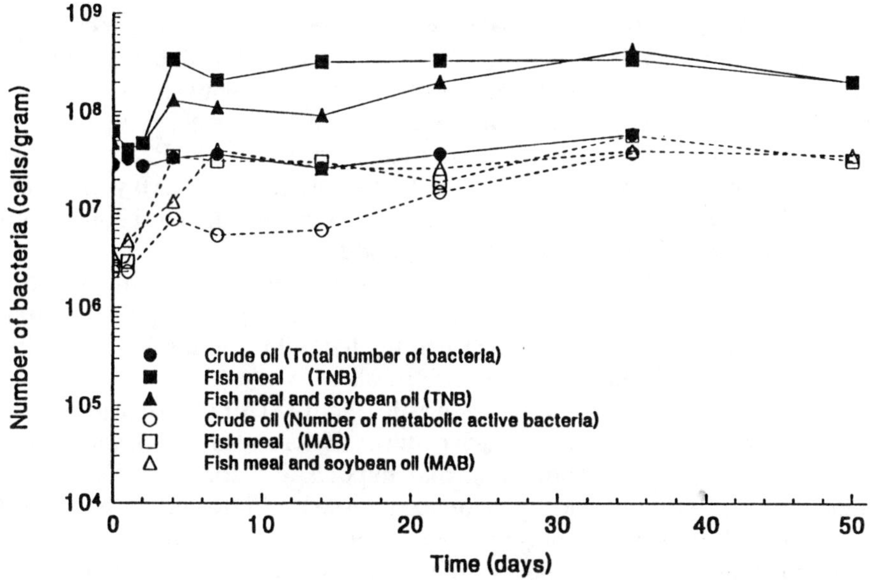

FIGURE 5. Total number (TNB) and the number of metabolically active bacteria (MAB) in the first series of basin experiments. Statfjord crude oil either alone or treated with fish meal or with fish meal and soybean oil combined.

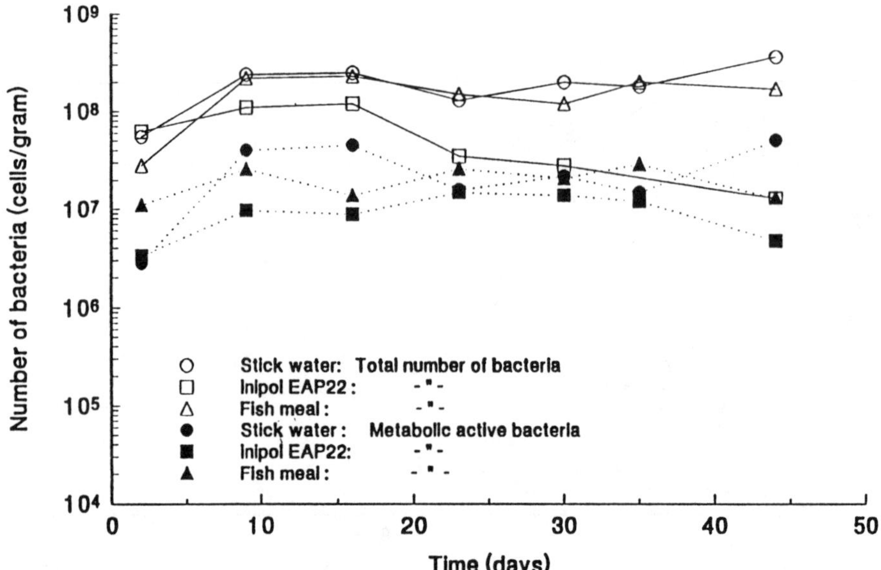

FIGURE 6. Total number and the number of metabolically active bacteria in the second series of basin experiments. Statfjord crude oil either alone or treated with fish meal, with stickwater, or with Inipol EAP22.

other two experiments were treated with stickwater (Exp. 5) and fish meal (Exp. 6). An initial increase in both TNB and MAB was observed in all basins during the first 10 days, but was most pronounced in those treated with stickwater and fish meal. Following this period, TNB remained at a high level in the basins with stickwater and fish meal. In the basin treated with Inipol EAP22, a decrease in TNB was observed after 20 days. The highest number of MAB was observed with stickwater. Inipol EAP22 resulted in a lower MAB than did both the other additives throughout the experimental period.

The heterotrophic activity in the sediment measured as CO_2 emission during the second series of experiments (Figure 7) agrees well with the bacterial numbers. A lower activity was found in the basin treated with Inipol EAP22 than in the basins treated with stickwater and fish meal.

The decrease in the nC_{17}/pristane ratio (Figure 8) for the basin experiments agrees with the results from the batch cultures. Fish meal and stickwater addition resulted in a more rapid degradation than Inipol EAP22 addition. The nC_{17}/pristane ratio decrease was most pronounced in the basin treated with stickwater.

Figure 9 shows the nC_{17}/pristane ratios measured during the second experimental series in relation to the sample location on the shoreline. The nC_{17}/pristane ratios for the Inipol EAP22-treated basin are independent of the location on the beach, whereas in the basins treated with the fish meal and stickwater, the nC_{17}/pristane ratio decreases less in the lower sections of the experimental beach than in the upper sections. In the lower sections, the water

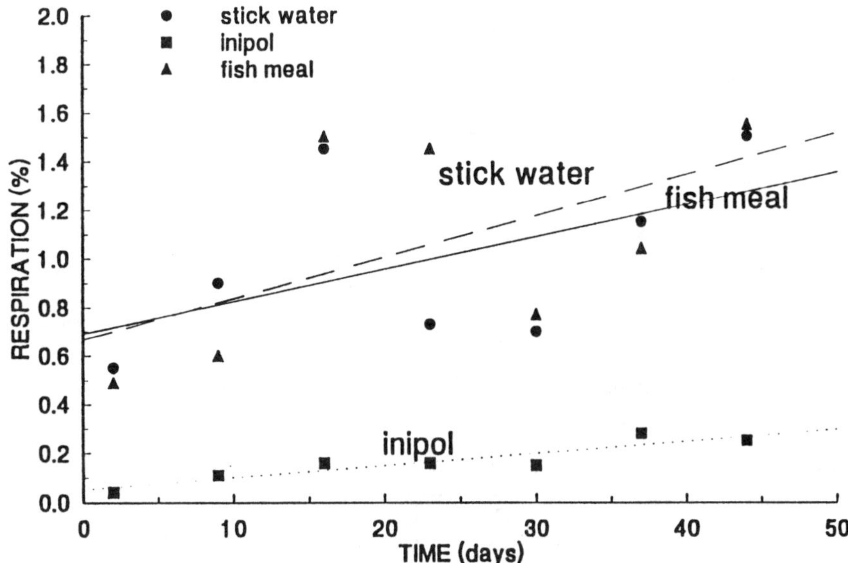

FIGURE 7. Total heterotrophic respiration in the second series of basin experiments given as the percentage of CO_2 in samples taken from the sealed incubation bottles.

content is higher, and gas transport can be a limiting factor. The differences found between Inipol EAP22 and the other two additives may reflect the lower metabolic activity of the former. There was no input of wave energy to these beaches, and indications of anaerobic activity were observed during parts of the experimental period. Had more energy been added, degradation in the basins treated with fish meal and stickwater may have been even more pronounced.

The above results illustrate the effect of offering an easily available carbon source, containing nitrogen and phosphorus, as a fertilizer additive to the indigenous microbial population on an oil-contaminated shoreline. The addition of readily available carbon gives an increased immobilization of nutrients in microbial biomass. This necromass constitutes an excellent storage area for nitrogen and phosphorous. Results from both the batch culture and the basin experiments show agreement between necromass and the biodegradation measured. This would indicate that immobilized nutrients stored in necromass are released slowly, and are made available at concentrations sufficient to maintain the activity of oil-degrading microbes, and thus enhance biodegradation.

ACKNOWLEDGMENTS

This study is part of the Esso SINTEF Coastal Oil Treatment Program (ESCOST), which is supported by Esso Norge A/S. This is report 007 from the ESCOST Program.

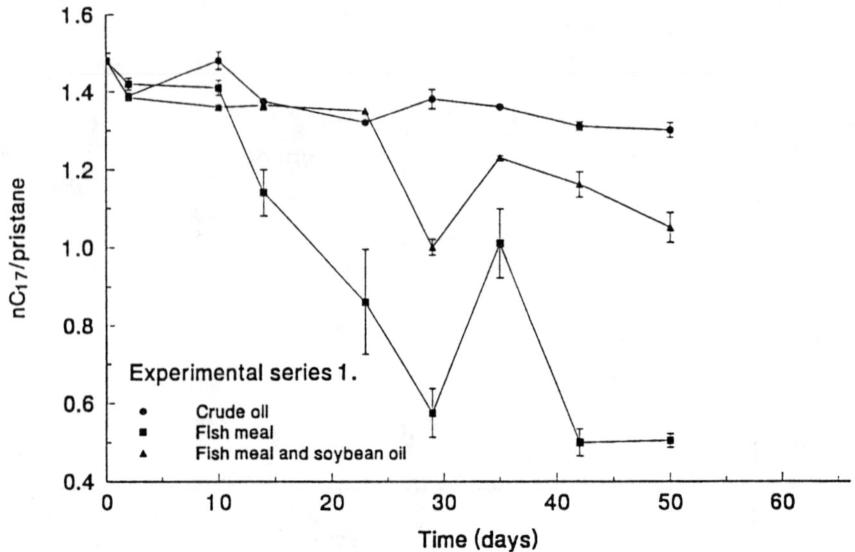

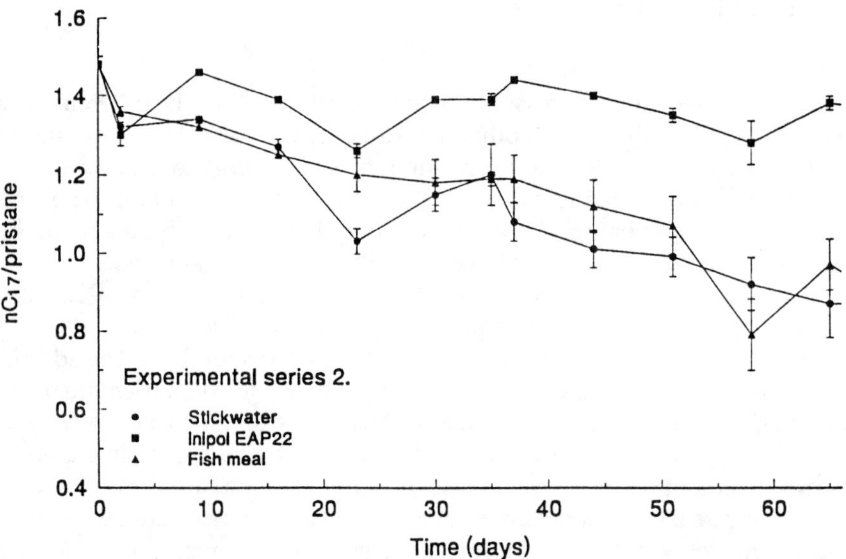

FIGURE 8. Change in the nC_{17}/pristane ratio in the basin experiments throughout the experimental period. In experiments 1, 2, and 3, the results are given as the average of two samples with the range indicated. In experiments 4, 5, and 6, the results are given as the mean of 5 samples, with standard deviation.

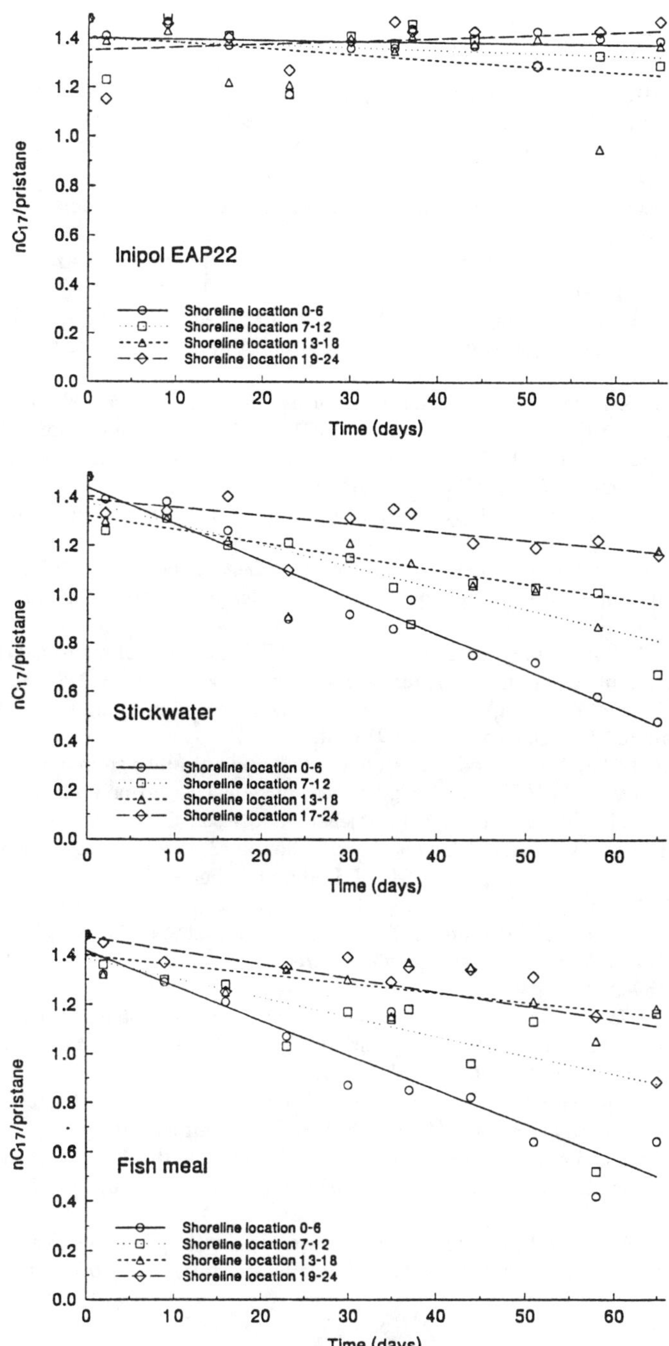

FIGURE 9. Change in the nC_{17}/pristane ratio in the basin experiments, as a function of the shoreline location. (The experimental beaches are divided into four sections: 0-6 is the upper section and 17-24 is the lowest section of the shoreline.)

REFERENCES

Atlas, R. 1991. "Bioremediation of fossil fuel contaminated soils." In R. E. Hinchee and R. F. Olfenbuttel (Eds.), *In Situ Bioreclamation: Applications and Investigations for Hydrocarbon and Contaminated Site Remediation*, pp. 14-32. Butterworth-Heinemann, Boston, MA.

Atlas, R. M., and M. Bushdosh. 1976. "Microbial degradation of petroleum in the arctic." In J. M. Sharpley and A. M. Kaplan (Eds.), *Proc. 3rd Internat. Biodegradation Symp.*, pp. 79-85. Applied Science, London, UK.

Bergstein, P. E., and J. R. Vestal. 1978. "Crude oil biodegradation in Arctic tundra ponds." *Arctic 31*: 159-169.

Goldmann, J. K, D. A. Caron, and M. R. Dennet. 1987. "Regulation of gross growth efficiency and ammonium regeneration in bacteria by substrate C:N ratio." *Limnol. Oceanogr.* 32: 1239-1252.

Halmø, G., E. Sendstad, P. Sveum, A. Danielsen, and T. Hoddø. 1985. *Enhanced Biodegradation Through Fertilization.* SINTEF Report STF21 F85019. Trondheim, Norway.

Halmø, G., and P. Sveum. 1987. *Biodegradation and Photooxidation of Crude Oil in Arctic Conditions.* SINTEF Report STF21 F87007. Trondheim, Norway.

Lundgren, B. 1981. "Fluorescein diacetate as a stain of metabolically active bacteria in soil." *Oikos 36*: 17-22.

Olivieri, R., P. Bacchin, A. Robertiello, N. Oddo, L. Degen, and A. Tonolo. 1976. "Microbial degradation of oil spills enhanced by a slow-release fertilizer." *Appl. Environ. Microbiol.* 31: 629-634.

Olivieri, R., A. Robertiello, and L. Degen. 1978. "Enhancement of microbial degradation of oil pollutants using lipophilic fertilizers." *Marine Pollut. Bull.* 9: 217-220.

Porter, K. G., and Y. S. Feig. 1980. "The use of DAPI for identifying and counting aquatic microflora." *Limnol. Oceanogr.* 25: 943-948.

Sendstad, E. 1980. "Accelerated biodegradation of crude on arctic shorelines." *Proc. 3rd Arctic and Marine Oil Spill Program Tech. Seminar*, pp. 402-416. Edmonton, Alberta.

Sendstad, E., T. Hoddø, P. Sveum, K. Eimhjellen, K. Josefsen, O. Nilsen, and T. Sommer. 1982. "Enhanced oil biodegradation on an arctic shoreline." *Proc. 5th Arctic and Marine Oil Spill Program Tech. Seminar*, pp. 331-340. Edmonton, Alberta.

Sendstad, E., G. Halmø, P. Sveum, A. Danielsen, and T. Hoddø. 1984. *Enhanced Oil Biodegradation in Cold Regions.* SINTEF Report STF21 F84032. Trondheim, Norway.

Sveum, P. 1986. *Marine Gas Oil in Arctic Shoreline Sediments.* SINTEF Report STF21 A86105. Trondheim, Norway.

Sveum, P. 1987a. "Accidentally spilled gas oil in a shoreline sediment on Spitsbergen: Natural fate and enhancement of biodegradation." *Proc. 10th Arctic and Marine Oil Spill Program Tech. Seminar*, pp. 177-192. Edmonton, Alberta.

Sveum, P. 1987b. *Oleophilic Fertilizer Treatment of Spilled Oil on Arctic Beaches— Mud Flats and Silt Beaches.* SINTEF Report STF21 F87113. Trondheim, Norway.

Sveum, P., and A. Ladousse. 1989. "Biodegradation of oil in the Arctic: Enhancement by oil-soluble fertilizer application." *Proc. 1989 Oil Spill Conference*, pp. 439-446. San Antonio, Texas.

Tezuka, Y. 1990. "Bacterial regeneration of ammonium and phosphate as affected by the carbon:nitrogen:phosphorous ratio of organic substrates." *Microb. Ecol.* 19: 227-238.

LIMITATIONS ON THE BIODEGRADATION RATE OF DISSOLVED BTEX IN A NATURAL, UNSATURATED, SANDY SOIL: EVIDENCE FROM FIELD AND LABORATORY EXPERIMENTS

R. M. Allen-King, K. E. O'Leary, R. W. Gillham, and J. F. Barker

ABSTRACT

The potential for remediation of gasoline-contaminated groundwater by infiltration through natural sandy soil was investigated with laboratory and field column experiments, batch microcosms, and a prototype application. Biodegradation was the primary removal mechanism in all systems tested. Biodegradation in the A horizon was rapid in all experimental systems, demonstrating that surface application may be a viable method for remediation of water contaminated by benzene, toluene, ethylbenzene, and xylenes (BTEX). In the A horizon, biodegradation was apparently controlled only by availability of the BTEX compounds and could be modeled by substrate-limited kinetics. Where the oxygen concentration in the laboratory column experiments was very low, no BTEX biotransformation was evident. In the presence of sufficient oxygen, the transformation rates observed in the B and C horizons were much lower than those observed in the A horizon. Addition of nitrogen (as an ammonium salt) resulted in increased rates in these circumstances. These results suggest that in natural soil, the apparent kinetic formulation and biodegradation rate may change with nutrient availability in the soil. Therefore, although correct kinetic formulas are important in forecasting behavior in the subsurface, the availability of mineral nutrients or electron acceptors may be more significant in determining BTEX fate.

INTRODUCTION

Groundwater contaminated by gasoline typically is pumped from extraction wells and treated on site by air stripping, filtration through activated carbon, or a combination of these options. In many cases, treatment of the contaminated water is an expensive aspect of the site remediation. A potentially cost-effective alternative treatment method is to apply the contaminated groundwater to soil such that the gasoline contaminants are biodegraded during infiltration to the water table.

It has long been recognized that petroleum hydrocarbons, such as benzene, toluene, ethylbenzene, and the xylene isomers (BTEX) are degraded by natural soil microorganisms (Claus & Walker 1964, McGill et al. 1981). An understanding of the kinetics and the rate limiting processes of the transformation reactions is important in predicting the behavior of these compounds in the subsurface and in enhancing our ability to exploit degradation reactions to effectively remediate contaminated sites. Although there is significant literature on degradation of oily waste in soil, the kinetics of biodegradation of dissolved aromatic hydrocarbons in soil is not well understood. Often, first-order kinetics are assumed for dissolved groundwater contaminants (Angley et al. 1992, for example). However, published rate constants for the transformation process in soil and groundwater vary over several orders of magnitude (see Alvarez et al. 1991, and references therein).

In the current study, the fate and transport of dissolved BTEX during steady infiltration through unsaturated soil was investigated as an alternative method for remediation of gasoline-contaminated groundwater. The objectives of the study were to evaluate surface application as a method for remediation of groundwater containing dissolved BTEX, to determine the appropriate kinetic model and rate parameters, and to identify conditions that limit the biodegradation rate in unsaturated soil. Laboratory column and batch experiments, a field column experiment, and a prototype field application (field plot) were used to achieve these objectives.

METHODS AND MATERIALS

The experiment methods are summarized below. For more complete descriptions, the reader is referred to Allen (1991) and Allen-King et al. (1993a, 1993b) for the laboratory experiments, Allen et al. (1987; 1989) for the field column experiments, and to O'Leary (1991) for the prototype field application.

Laboratory Experiments

Soil for the laboratory experiments was collected from an uncontaminated location at the field site. In most of the laboratory experiments, toluene was used as a model aromatic hydrocarbon representing BTEX. Gasoline-contaminated groundwater was applied for one of the experiments.

Batch Microcosms. The batch microcosms were used to determine the kinetics of the degradation reaction in a simple, closed system. Toluene was added repeatedly to the microcosms to simulate the effects of continuous exposure to toluene. Soil for these experiments was taken from the A, B, and C horizons in the layered laboratory column experiment.

The microcosms were 1 L glass bottles containing 90 g of soil and 40 to 60 mL groundwater to make a slurry. Microcosms for each soil horizon were prepared in triplicate. The remaining volume of headspace in the microcosms maintained aerobic conditions throughout the experiments. The microcosms could be amended or sampled repeatedly without volatile loss. Control microcosms showed no significant leakage or abiotic removal of toluene. The toluene added created solution concentrations of 24-27 mg/L (96 µg/g) in the C-horizon microcosms (C microcosms) and 41-42 mg/L (154 µg/g) in the A microcosms. Toluene concentration in the gas-phase was monitored over time at 1 to 24 hr intervals, depending on the rate of change of concentration. Oxygen and carbon dioxide were monitored intermittently.

Unsaturated Columns. A schematic of the column apparatus is shown in Figure 1. The soil columns were 10.8 cm diameter and 165 to 167 cm in length. For one experiment, the natural soil layering observed in the field was re-created in the laboratory column (layered column experiment). In other cases, soil from a particular depth interval was homogenized and packed to the full length of the column (homogeneous column experiments). Toluene- and chloroform-amended groundwater was applied continuously to the unsaturated soil column a drip applicator, located 11 cm above the soil surface. Chloroform was added as a

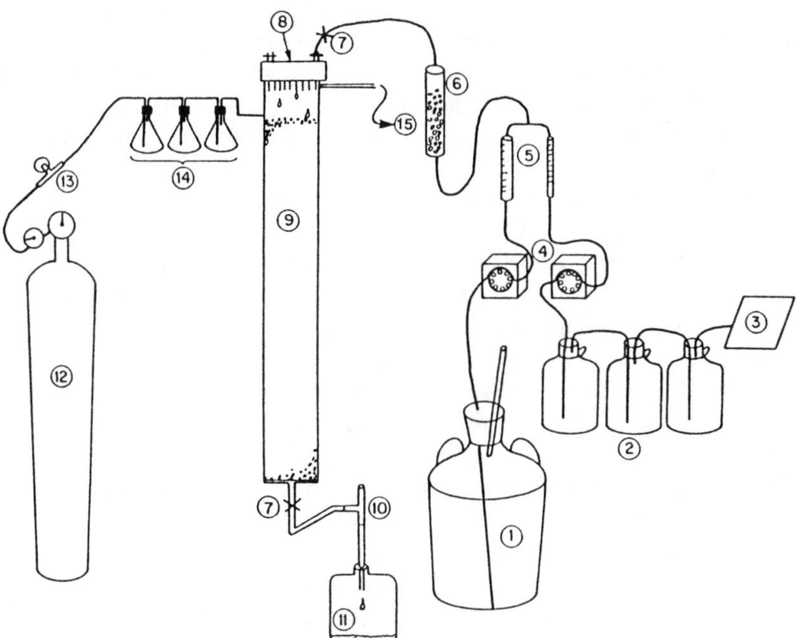

FIGURE 1. Schematic of laboratory column experiment: (1) pristine groundwater; (2) contaminated source water; (3) Teflon™ bag; (4) peristaltic pumps; (5) flow meters; (6) mixing chamber; (7) aqueous samples; (8) applicator; (9) soil column; (10) water table elevation; (11) effluent water; (12) compressed air; (13) needle valve; (14) humidifier; (15) effluent air (to fume hood).

sorbing and volatile but recalcitrant tracer and was shown neither to inhibit nor to enhance toluene biodegradation at the concentrations used (Allen 1991). Experiments were conducted in a stepwise fashion in each column by varying the source concentration or water flux (Table 1), or by adding mineral nutrients, and monitoring until a steady profile was achieved. A controlled airflow through the top of each column maintained near-atmospheric levels of oxygen and carbon dioxide at the soil surface. The mass of toluene removed by the airflow was subtracted from the total mass flux to obtain the mean applied concentration reported in Table 1. Soil gas and source and effluent water samples were monitored for toluene and chloroform at least once per day, oxygen and carbon dioxide and inorganic salts were monitored less frequently. Water content was monitored by time domain reflectometry (Topp et al. 1980). Soil cores were collected at the end of some of the experiments to determine the activity of toluene-degrading microorganisms.

Field Experiments

Field Site Description. The field experiments were conducted at the Canadian Forces Base (CFB) Borden in south-central Ontario, Canada, in a sand plain area. The soil profile was relatively undisturbed. Characteristics of the relatively poorly developed, medium-to-fine grained, sandy soil are listed in Table 2. The field column experiment site was covered with a mixture of birches, coniferous trees, ferns, and other mixed vegetation. The field plot experiment was conducted in a clearing adjacent to the forest. There was a slightly silty, fine sandy layer located between approximately 15 cm and 25 cm depth at the field plot experiment site only.

General. In the field experiments, as in the laboratory column experiments, water containing dissolved BTEX was applied continuously throughout each experiment. The soil remained unsaturated although water fluxes were relatively high. Plants were cleared to allow uniform distribution of water over the soil. Parameters monitored in the field were essentially the same as those monitored in the laboratory column experiments. Soil gas samples were collected through stainless steel soil gas probes screened at specific depths. The effluent was not collected in the field experiments. Groundwater samples, representative of the effluent from the system, were collected from multilevel piezometers constructed of 0.32-cm-OD Teflon™ tubing and screened near the water table depth. Water content was monitored using a neutron probe.

Unsaturated Columns. Field column experiments were conducted using a design similar to the laboratory column but at a larger scale, with BTEX concentration as the principal variable. In the field, a natural column of soil was isolated from the surrounding soil by pounding a 91-cm-OD steel pipe 230 cm into the ground. The pipe extended 30 cm above the ground surface. The water table varied seasonally between 180 and 230 cm below ground surface. Gasoline-contacted water was applied to the soil and total BTEX concentrations at the soil surface and water fluxes for the experiments are listed in Table 1. The concentration ratio of benzene : toluene : ethylbenzene : *p*- and *m*-xylene : *o*-xylene in the

TABLE 1. Source conditions for laboratory column and field experiments.

Experiment Type	No.	Soil Type	Source of Aromatics	Mean Applied BTEX Concentration [$\mu g L^{-1}$][(c)]	Mean Water Flux [$cm^3 cm^{-2} d^{-1}$]	Mean Total BTEX Flux [$cm^3 cm^{-2} d^{-1}$]	Nitrogen Addition	Water Source
Laboratory column experiments[(d)]	1	Layered-repacked	Toluene only	19,100	33.9	647.	None	Groundwater
	2A	Homogeneous-C horizon[(a)]	Toluene only	56,100	64.5	3,620.	None	Groundwater
	2B	Same as above	Same as above	5,810	20.2	117.	None	Groundwater
	2C	Same as above	Same as above	2,910	20.2	58.8	424 mg/L NH_4-N	Groundwater
	3	Homogeneous-mixed[(b)]	Toluene only	24,000	39.1	938.	None	Groundwater
Field column	A	Natural	Gasoline-contacted	7,000	32.2	225.	None	Groundwater
	B	Same as above	Same as above	19,000	32.2	612.	None	Groundwater
Prototype field	A	Natural	BTEX	9,700	58.	563.	None	Municipal supply
	B	Same as above	Same as above	9,600	58.	557.	800 g	Municipal supply

(a) C-horizon soil only; (b) mixture of A-, B-, and C-horizon soil from the field site, f_{oc}=0.32 to 0.39% (w/w); (c) concentration at the soil surface for the column experiments; (d) Each number represents a distinct column, letters designate sequential experiments with the same soil.

TABLE 2. Fraction organic carbon content (f_{oc}), nitrogen, and phosphorus in natural soil at/from the field site.

	Layered Laboratory Column				Field Column				Prototype-Scale Field Experiment				
Soil Horizon	Depth [cm]	f_{oc} [mg/mg]	NH_4-N [mg/kg]	NO_3-N [mg/kg]	Depth [cm]	f_{oc} [mg/mg]	NH_4-N [mg/kg]	NO_3-N [mg/kg]	Depth [cm]	f_{oc} [mg/mg]	NH_4-N [mg/kg]	NO_3-N [mg/kg]	P [mg/kg]
A	0-7	1.990	1.78	50.0	0-14	1.374	2.4 -3.9	16.7 -21.7	0-15	0.711	3.6	7.3	8
	7-13	1.374	3.94	16.7									
B	13-35	0.193	0.94	3.26	14-35	0.193	0.9	1.8 -3.3	15-25	1.065	2.4	6.9	11
C	35-165	0.021	0.71	0.	35-250	0.021	0-2.6	0	25-50	0.364	1.5	6.9	12
									50-75	0.189	1.9	6.8	7
									75-100	0.087	2.4	6.8	4
									100-150	0.068	3.2	6.9	3
									150-200	0.076	2.3	6.9	5
									200-250	0.059	1.7	7.0	5
									250-300	0.055	1.5	7.1	3
									300-350	0.088	2.1	7.5	4

applied water during field column experiment A was approximately 1.0 : 1.3 : 0.08 : 0.3 : 0.08, respectively. Water was applied through a drip applicator set atop the steel pipe, with a headspace between the soil and applicator. The headspace was not sealed in the field experiment. Additional water samples collected in vials at the soil surface inside the column were used to determine that about 40% of the dissolved BTEX in the source water was removed due to volatilization from the droplets in the headspace above the soil surface. As the removal above the soil was not the focus of this research, the values reported in Table 1 have been corrected for this removal.

Prototype Application. The prototype application was conducted on a 5 m × 5 m soil plot. A mixture of 1.0 : 0.58 : 0.075 : 0.044 : 0.19 : 0.12 benzene : toluene : ethylbenzene : *p*-xylene : *m*-xylene : *o*-xylene was added to the applied water. The water table was 300 to 375 cm below ground surface. Water was applied through drip irrigation tubing with drippers spaced on a 10 cm × 15 cm grid. Results from two experiments with source conditions listed in Table 1 are reported. In the second of the two tests, 3.2 mg•cm^{-1} (800 g total) N as ammonium nitrate fertilizer was applied to the test plot by spraying a solution over the site and adding dissolved fertilizer into the application water. Cores were collected prior to application of BTEX, and following the experiments, to determine the activity of toluene-specific degraders.

There was little opportunity for BTEX volatilization during drip application in these experiments because the drip emitters were in direct contact with the soil. However, diffusion of BTEX from below the soil surface to the atmosphere is one pathway by which BTEX could be removed from the soil. The diffusive flux from within the soil to the atmosphere was estimated by measurements and calculation of diffusive fluxes at the soil surface to account for less than 2 to 6% of the total BTEX applied (O'Leary 1991).

Analytical Methods

Gas-phase BTEX was analyzed immediately after collection by gas chromatograph (GC) with flame ionization detection (FID). Aqueous samples were preserved with sodium azide and analyzed for BTEX by microextraction with hexane or pentane (Patrick et al. 1985) followed by GC/FID analysis. Method detection limits for extracted samples were less than 1 µg/L for benzene and toluene, and 1 to 3 µg/L each for ethylbenzene and xylene isomers. Gas samples were analyzed for O_2 and CO_2 on a gas partitioner. Detailed analytical methods are given by Allen (1991) and O'Leary (1991).

The activity of specific toluene degraders was determined by measurement of $^{14}CO_2$ produced from [U-^{14}C]-toluene degradation in 20 mL respirometers. The heterotrophic activity method used by Jensen (1989) was modified to prevent volatile loss of toluene during incubation and to use a soil sample as the inoculum (Allen 1991). A nutrient-rich growth medium was chosen to supply nutrients in excess. The respirometers were incubated 6 to 9 hr.

RESULTS AND DISCUSSION

Laboratory Experiments

After several days of continuous application of water containing the compounds of interest, a steady profile of BTEX concentration with depth developed in the column experiments. Shown in Figure 2 are the approximately steady toluene, chloroform and oxygen profiles observed in laboratory column experiment 1. The results shown are the measured gas-phase concentrations with depth, except for the points at 0 and 165 cm. The point at 165 cm depth is the measured effluent water concentration, converted to the equivalent gas-phase concentration using the Henry's law constant. Regression coefficients to calculate Henry's law constants with temperature for BTEX were taken from Ashworth et al. (1988). In this paper, all concentration versus depth profiles were derived in this manner. In the laboratory column experiments, where the flow of gas through the headspace was controlled, the concentration of toluene at the soil surface (0 cm) was calculated using a simple mass balance approach. Decreased toluene concentrations with depth in the steady profiles indicate removal. Removal of toluene in the soil columns is attributable to one of three processes: sorption, transfer to the gas-phase followed by diffusion in the gas-phase to the atmosphere or headspace above the soil column, and/or transformation. The essentially uniform chloroform concentration with depth shows that volatilization and sorption are not major removal mechanisms at steady state and that assumption of solid/aqueous/gas-phase equilibrium is justified in this system. Volatile removal in the laboratory column system was shown to occur primarily from the droplets above the soil surface during application, with the diffusive flux from below the soil surface, calculated according to Fick's first law, contributing secondarily to removal (Allen 1991).

At steady state in laboratory experiment 1, toluene removal was rapid in the near-surface portion of the soil column, and the removal rate diminished with depth in the B horizon. The steady-state oxygen concentration profile was lower than in the pretoluene-addition background case, indicating additional utilization of oxygen. Additional CO_2 production with toluene addition and the observation of a lag phase prior to the development of the steady profile suggest a biologically mediated removal process for toluene. In other column experiments, the activity of toluene-degraders also was shown to increase with exposure to toluene from <0.03 to 0.1 $\mu g(g{\cdot}d)^{-1}$ (µg toluene per g dry soil per day) to 1 to 11 $\mu g(g{\cdot}d)^{-1}$, confirming a biological process.

The results of toluene additions to the microcosms are shown in Figure 3 as relative concentration over time, or measured gas-phase concentration normalized by the expected initial gas-phase concentration assuming instantaneous equilibrium between the aqueous and gas phases only. Transformation of toluene was rapid in the A microcosms, with complete removal of the initial high concentration occurring in less than 25 hr (Figure 3a). A corresponding decrease in O_2 concentration and increase in CO_2 concentration was observed in each of the microcosms. Observation of persistence of chloroform, added to the microcosms as a volatile, sorbing, recalcitrant tracer, in combination with the above observations of changes in O_2 and CO_2 concentrations, was used to confirm the biological degradation process.

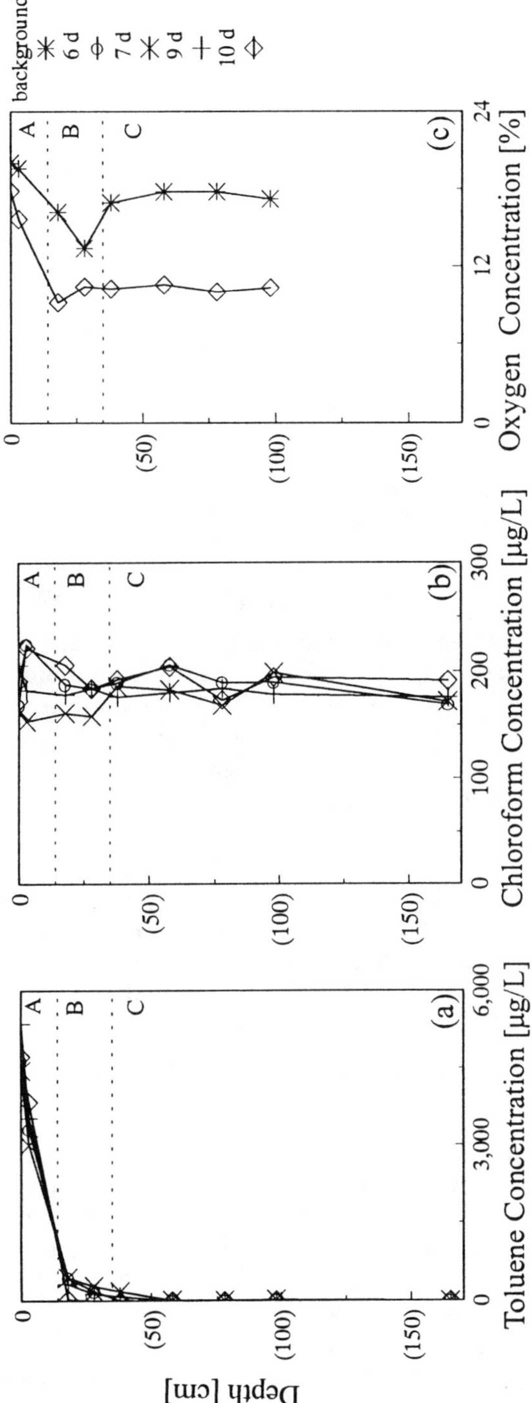

FIGURE 2. Steady gas-phase concentration versus depth profiles for layered laboratory column experiment on several days (d): (a) toluene; (b) chloroform; (c) oxygen.

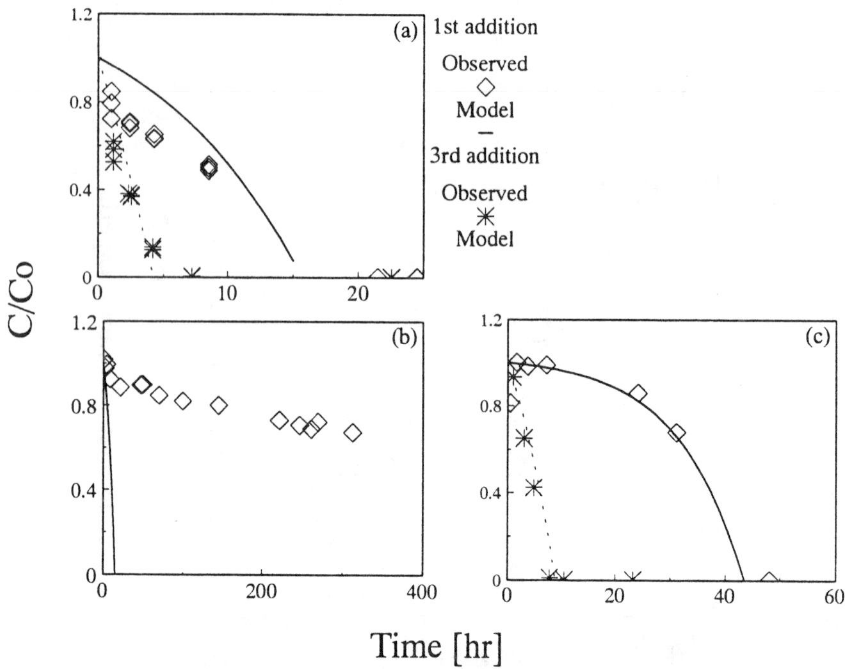

FIGURE 3. Gas-phase toluene concentrations in batch microcosms: (a) A microcosms, Co = 41,000 to 42,000 µg/L; (b) C microcosms without ammonium, Co = 27,000 µg/L; (c) C microcosms with ammonium, Co=24,000 to 27,000 µg/L. C/Co is the measured gas-phase toluene concentration normalized by the expected initial gas-phase concentration assuming equilibrium between the aqueous and gas phases.

In the A microcosms, the apparent rate of degradation increased with subsequent additions of toluene at the same concentration level. This type of behavior also was observed for benzene and *p*-xylene by English and Loehr (1990) and can be modeled by substrate-limited exponential-growth of the microorganisms carrying out the degradation reaction (Allen-King et al. 1993a). It was shown that the microcosm data from all three horizons could be adequately fit by this model using a maximum substrate utilization rate (μ_{max}) of 2.0 d^{-1} in circumstances where sufficient oxygen and nutrients were present.

In contrast to the rapid degradation observed in the A microcosms, biodegradation of toluene in the C microcosms was very slow (Figure 3b). Toluene persisted at about 80% of the initial concentration after more than 300 hr of incubation time when a relatively high concentration of toluene was added. These results could not be fit by the growth-limited model. It was hypothesized that a mineral nutrient was limiting the rate of degradation in the subsoil. Therefore, ammonium sulfate was added to the microcosms in the same concentration as used in the growth medium. Following nutrient addition, the degradation rate increased significantly such that all the toluene degraded in less than 50 hr

(Figure 3c). Subsequent additions of toluene were degraded more rapidly. With the addition of ammonium, the toluene removal could be fit by the substrate-limited growth model used for the A microcosms, suggesting that nitrogen was needed for biomass growth. The trends of accelerated degradation with ammonium addition when relatively high dissolved toluene concentrations were present and apparent substrate-limited degradation with sufficient oxygen and nutrients also were observed in the B microcosms (data not shown).

In microcosms prepared with uncontaminated C horizon soil from the field site and supplied with sufficient nitrogen, potassium, sulfate, and oxygen, the substrate-limited growth equation was found not to fit the results. Although the addition of inorganic nitrogen accelerated the biotransformation rate significantly, apparent zero-order behavior suggested that another nutrient also could be limiting biotransformation in this soil.

The apparent slower transformation rate in the deeper soil and the importance of inorganic nutrients on transformation rate were further investigated with a series of stepwise experiments conducted in a soil column into which homogenized soil from the C horizon only had been packed. Initially, a high flux of water and toluene were applied to the column (laboratory experiment 2A, Table 1). The profile at steady state was relatively uniform, and degradation was sufficiently slow that there was no significant toluene removal throughout the 167 cm depth of the soil column (data not shown).

In a later step (2B), the water flux and applied toluene concentration were reduced such that significant toluene removal was observed (Figure 4a). The rate of removal was much slower than the rate observed in the surface layers of the layered soil column. When nutrients of the growth medium were added to the

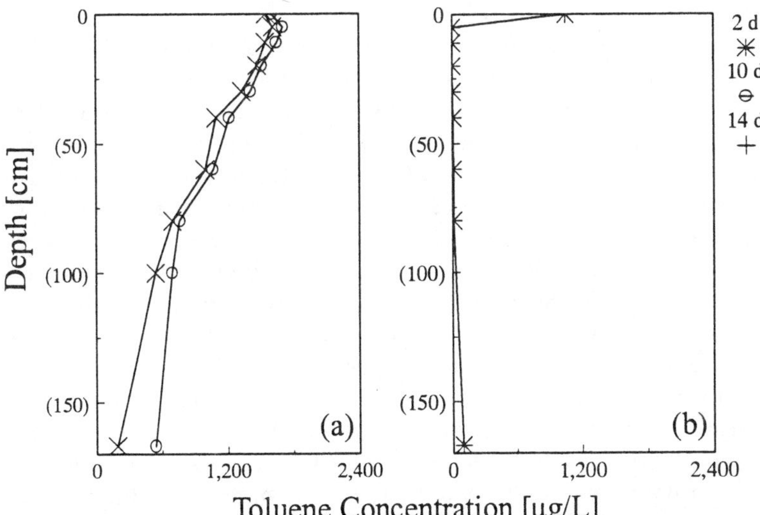

FIGURE 4. Almost steady gas-phase toluene concentration versus depth profiles for homogeneous C horizon soil column: (a) without nutrient addition; (b) with nutrient addition.

applied groundwater, the rate of degradation increased substantially, such that the toluene concentration was reduced to less than detectable below the surface of the column and above the water table after 2 days of nutrient addition (Figure 4b). The toluene concentration in the effluent water was reduced to nondetectable slightly later (less than 4 days), due to acclimation and the residence time of water in the column. Thus, nutrients enhanced the rate of transformation in the C horizon soil in both the dynamic column experiments and in the batch experiments.

The effect of limited oxygen supply on degradation in unsaturated soil was observed in laboratory column experiment 3, which was packed with a homogenized mixture of soil from all three soil horizons. In this column, as in the layered column, the background O_2 profile showed some consumption, suggesting active metabolism by aerobic microorganisms, before toluene was applied. After toluene was applied and a steady profile was achieved, oxygen was undetectable in the column below about the 40 cm depth (Figure 5b). Toluene removal occurred above the 40 cm depth only (Figure 5a), and there was apparently no degradation below 40 cm depth in this experiment. Limited dissolved oxygen availability has been shown to limit the apparent rate of dissolved BTEX degradation in the saturated zone (Barker et al. 1987, MacQuarrie et al. 1990). These results demonstrate that lack of oxygen availability can limit degradation of dissolved BTEX in unsaturated soil.

Toluene-degrader activity was determined from soil cores collected between 7 cm and 88 cm in the laboratory column experiment 3 described above. The activity measured (0.65 to 6.31 $\mu g(g{\cdot}d)^{-1}$) was significantly increased compared to samples from a background column that received the same water flux but no toluene (less than detectable to 0.11 $\mu g(g{\cdot}d)^{-1}$).

Field Experiments

Because the airflow above the field experiments was not controlled, as it was in the laboratory experiments, the component of BTEX removal attributable to volatilization cannot be determined directly from the field results. However, the diffusive flux from within the soil measured during one of the prototype field experiments was shown to account for less than about 2% of the total applied mass flux (O'Leary 1991). This is slightly lower than, but agrees well with, the maximum diffusive flux estimated using Fick's law for the same experiment. Based on these results, diffusion of BTEX from below the soil surface has been neglected as a significant removal mechanism in the following discussion.

The results of the two types of field experiments are shown as steady-state profiles in Figures 6 and 7. In both types of experiments, BTEX removal was rapid in the upper part of the soil profile, and in all cases, the concentration of BTEX groundwater was significantly decreased from the source concentration. This behavior also was demonstrated by the layered laboratory column experiment. The qualitative similarity of these results suggests that the same mechanisms are controlling BTEX or toluene removal in all the experiments. Toluene-degrader activity increased from 0.02 to 0.03 $\mu g(g{\cdot}d)^{-1}$ in background soil samples collected prior to initiation of the field plot experiment to 1.14 to 4.27 $\mu g(g{\cdot}d)^{-1}$ in samples from a depth of 20 cm collected after the experiments were completed.

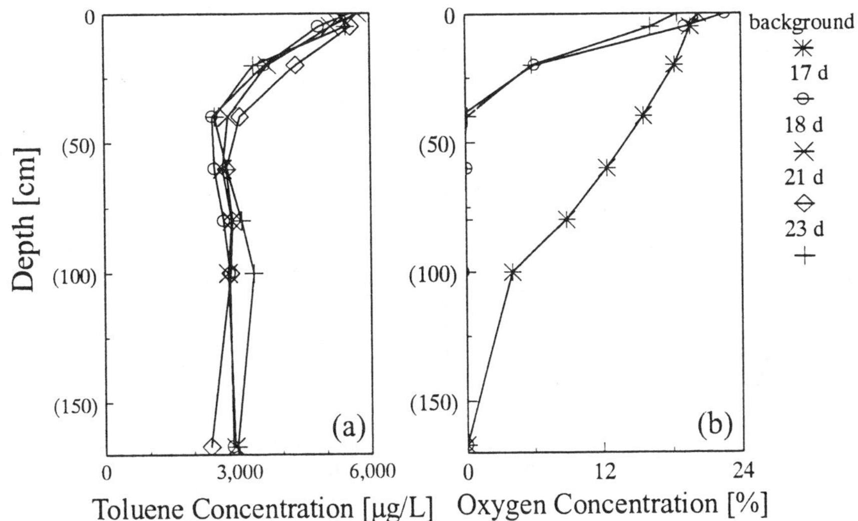

FIGURE 5. Steady gas-phase (a) toluene and (b) oxygen concentration versus depth profiles for homogeneous mixed soil column.

The observed increases are of the same order of magnitude as the increases observed in the laboratory column experiments.

In field column experiment A (lowest BTEX flux in the field experiments, Table 1), BTEX was completely removed above a depth of 35 cm after a steady profile had been achieved (Figure 6a). However, BTEX was persistent even after long exposure times when applied at a higher flux in the field column experiment (Figure 6b) or in the prototype experiment (Figure 7a). The variability shown between the profiles in these figures reflect the variation in the applied concentrations in these particular experiments. Although BTEX was reduced by 75 to 90% in these experiments, the concentrations present at the water table were significant with respect to drinking water standards, particularly for benzene. Oxygen was not limiting in any of the field experiments. Based on the rapid rate of biodegradation observed in the upper part of the column and complete removal of BTEX in field column experiment A, the behavior shown by the other two experiments might not have been expected. However, the laboratory experiments showed that nutrients, particularly nitrogen, can limit BTEX degradation in the subsoil at this site. Therefore, nitrogen was applied as ammonium nitrate fertilizer to the test plot before prototype experiment B was conducted. BTEX degradation was enhanced by the fertilizer application at this site such that essentially complete remediation of the applied water occurred in the soil profile (Figure 7b), thus demonstrating at the prototype scale that mineral nutrient availability can limit the rate of dissolved aromatic hydrocarbon degradation in soil and that nutrient addition can relieve the limitation and promote remediation. Nitrogen availability has been shown as a potentially limiting condition for aromatic hydrocarbon degradation in soil and groundwater at other sites (Aelion & Bradley 1991,

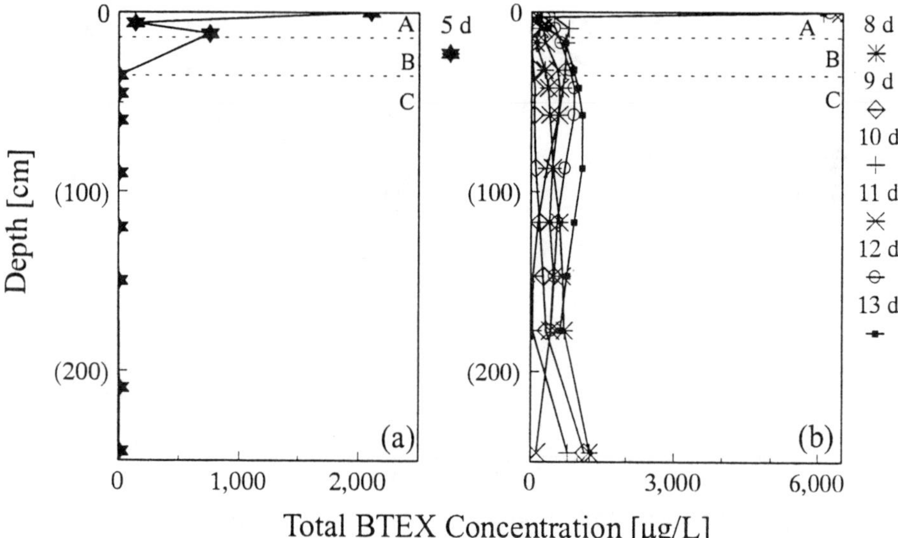

FIGURE 6. Steady gas-phase BTEX concentration versus depth profiles for field column experiment on several days (d) with water flux of 32.2 $cm^3cm^{-2}d^{-1}$: (a) 7,000 µg/L total BTEX applied to soil surface; (b) 19,000 µg/L total BTEX applied to the soil surface.

Armstrong et al. 1991). However, nitrogen addition alone may not enhance degradation at all hydrocarbon contaminated sites (Miller et al. 1990).

BTEX degradation rates estimated from a linear concentration gradient with depth in the upper A horizon in the layered laboratory column, in the upper 10 to 20 cm of field column experiments A and B, and in the upper 10 cm of prototype experiments A and B are 37.4, 9.9 to 20, 21 to 43, 43.4 to 55.5, and 33.0 to 34.1 $\mu g(g{\cdot}d)^{-1}$, respectively. The field column rates increased with applied mass flux, and thus agree qualitatively with the batch microcosm results. The degradation rates for the experiments that had similar BTEX mass fluxes (laboratory column, field column B, and the prototype experiments) are all within a narrow range. The agreement between the field and laboratory experiment results suggest that substrate-limited growth of microorganisms may have governed the fate of BTEX in the A horizon in the field experiments, as in the batch experiments.

SUMMARY AND CONCLUSIONS

Even without nutrient amendment in the poorly developed sandy soil at the CFB Borden, biodegradation by natural soil microorganisms resulted in removal of significant concentrations of BTEX from groundwater, at all scales of experimentation, demonstrating the potential for remediation by surface application to soil. The BTEX removal trends and biodegradation rates agree well for all laboratory and field systems, demonstrating that laboratory model systems can be created to be indicative of behavior in the field.

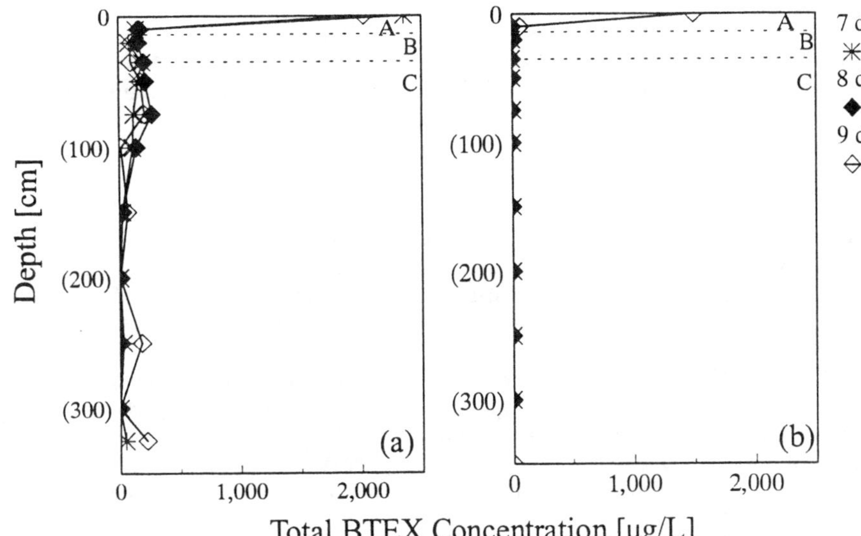

FIGURE 7. Steady gas-phase BTEX concentration versus depth profiles for prototype field experiment on several days (d) with average water flux of 58 $cm^3cm^{-2}d^{-1}$ and applied concentration of 9,600 to 9,700 µg/L total BTEX: (a) without nitrogen addition; (b) with nitrogen addition.

The results of batch experiments with A horizon soil can be modeled by substrate-limited growth kinetics when sufficient oxygen was present. However, biodegradation of dissolved BTEX was slow in subsoil when either nutrients or oxygen were in limiting supply. Under these circumstances, the growth-kinetic formulation was inadequate to describe degradation because it accounts only for the dependence on substrate and biomass and not on inorganic nutrient or electron acceptor availability. Nitrogen amendment significantly accelerated BTEX biodegradation in both laboratory- and field-scale experiments. Toluene removal in nitrogen amended B- and C-horizon microcosms fit the substrate-limited model.

Knowledge of the kinetic formulation for BTEX degradation allows predictive modeling of systems where natural or enhanced remediation is occurring. However, factors that can limit the degradation process, such as inorganic nutrient supply, often are not considered in kinetic studies. For the degradation of BTEX in soil, these factors can control the reaction rate, allowing rapid reaction rate when available, and essentially stopping or severely reducing the rate of the reaction when supply is limited. As a result, it is important to identify factors that may limit reaction rate as well as to determine reaction kinetics.

ACKNOWLEDGMENTS

The authors wish to thank the technical staff and students at the Waterloo Centre for Groundwater Research for their contributions to this research. The assistance of Tiffany Svenson, Marianne Vandergriendt, Shirley Chatten, Stephanie

O'Hannesin, Greg Friday, Ralph Dickhout, Paul Johnson, and Bob Ingleton is particularly appreciated. This research was financially supported by the American Petroleum Institute and the Ontario University Research Incentive Fund.

REFERENCES

Aelion, C. M. and P. M. Bradley. 1991. "Aerobic biodegradation potential of subsurface microorganisms from a jet fuel-contaminated aquifer." *Appl. Environ. Microbiol.* 57(1): 57-63.

Allen, R. M. 1991. "Fate and transport of dissolved monoaromatic hydrocarbons during steady infiltration through unsaturated soil." Ph.D. thesis, University of Waterloo, Waterloo, Ontario. 199 pp.

Allen, R. M., R. W. Gillham, and J. F. Barker. 1987. "Remediation of gasoline contaminated ground water by infiltration through soil." In *Proceedings NWWA Conference, FOCUS on Eastern Regional Ground Water Issues*, Burlington, VT, July 14-16.

Allen, R. M., R. W. Gillham, and J. F. Barker. 1989. *Soluble Petroleum Constituents: Rehabilitation of Groundwater Through Surface Application.* American Petroleum Institute Publication No. 4475, 90 pp.

Allen-King, R. M., J. F. Barker, R. W. Gillham, and B. K. Jensen. 1993a. "Substrate and nutrient-limited toluene biotransformation in sandy soil from three soil horizons." Submitted to *Environ. Toxicol. Chem.*

Allen-King, R. M., R. W. Gillham, J. F. Barker, and E. A. Sudicky. 1993b. *Rehabilitation of gasoline-contaminated groundwater through surface application: Part 1 - Laboratory Experiments.* American Petroleum Institute Publication, In press.

Alvarez, P. J. J., P. J. Anid, and T. M. Vogel. 1991. "Kinetics of aerobic biodegradation of benzene and toluene in sandy aquifer material." *Biodegradation* 2: 43-51.

Angley, J. T., M. L. Brusseau, W. L. Miller, and J. J. Delfino. 1992. "Nonequilibrium sorption and aerobic biodegradation of dissolved alkylbenzenes during transport in aquifer material: Column experiments and evaluation of a coupled-process model." *Environ. Sci. Technol.* 26(7): 1404-1410.

Armstrong, A. Q., R. E. Hodson, H. M. Hwang, and D. L. Lewis. 1991. "Environmental factors affecting toluene degradation in ground water at a hazardous waste site." *Environ. Toxicol. Chem.* 10: 147-158.

Ashworth, R. A., G. B. Howe, M. E. Mullins, and T. N. Rogers. 1988. "Air-water partitioning coefficients of organics in dilute aqueous solutions." *J. Hazard. Mat.* 18: 25-36.

Barker, J. F., G. C. Patrick, and D. Major. 1987. "Natural attenuation of aromatic hydrocarbons in a shallow sand aquifer." *Ground Water Monitoring Review* 7(1): 64-71.

Claus, D., and N. Walker. 1964. "The decomposition of toluene by soil bacteria." *J. Gen. Microbiol.* 36: 107-122.

English, C. W., and R. C. Loehr. 1990. "Removal of organic vapors in unsaturated soil. "*Proceedings of Petroleum Hydrocarbons and Organic Chemicals in Ground Water: Prevention, Detection and Restoration*, NWWA/API, Houston, TX, October 31-November 2.

Jensen, B. K. 1989. "ATP related specific heterotrophic activity in petroleum contaminated and uncontaminated groundwaters." *Can. J. Microbiol.* 35: 814-818.

MacQuarrie, K. T. B., E. A. Sudicky, and E. O. Frind. 1990. "Simulation of biodegradable organic contaminants in groundwater: 1. Numerical formulation in principal directions." *Water Resour. Res.* 26(2): 207-222.

McGill, W. B., M. J. Rowell, and D. W. S. Westlake. 1981. "Biochemistry, ecology and microbiology of petroleum components in soil." In E. A. Paul and J. N. Ladd (Eds.), *Soil Biochemistry*, pp. 229-297. Marcel Dekker, NY.

Miller, R. N., R. E. Hinchee, C. M. Vogel, R. R. Dupont, and D. C. Downey. 1990. "A field scale investigation of enhanced petroleum hydrocarbon biodegradation in the vadose-zone at Tyndall AFB, Florida." *Proceedings of Petroleum Hydrocarbons and Organic Chemicals in Ground Water: Prevention, Detection and Restoration*, NWWA/API, October 31 - November 2, Houston, TX.

O'Leary, K. E. 1991. "Remediation of dissolved BTEX through surface application: A prototype field investigation." M. Sc. Thesis, University of Waterloo, Waterloo, Ontario, Canada.

Patrick, G. C., C. J. Ptacek, R. W. Gillham, J. F. Barker, J. A. Cherry, D. Major, C. Il Mayfield, and R. D. Dickhout. 1985. *The Behavior of Soluble Petroleum Product Derived Hydrocarbons in Groundwater*. Petroleum Association for Conservation of the Canadian Environment, PACE 85-3, Phase 1, Ottawa, Ontario, Canada, 70pp.

Topp, G. C., J. L. Davis, and A. P. Annan. 1980. "Electromagnetic determination of soil water content: Measurements in coaxial transmission lines." *Water Resour. Res. 16*(3):574-582.

HETEROGENEITY IN CONTAMINANT CONCENTRATION AND MICROBIAL ACTIVITY IN SUBSURFACE SEDIMENTS

C. M. Aelion and S. C. Long

ABSTRACT

The association between sediment type and contaminant levels was investigated by collecting sediment at discrete depth intervals (15 to 20 cm) in the saturated zone from a jet fuel-contaminated aquifer near Charleston, South Carolina. Sediments from the site are unconsolidated, fine- to medium-grained sands, with interfingering lenses of clay, underlain by an impermeable clay confining layer. Gas chromatographic/mass spectrometric (GC-MS) analyses were carried out for JP-4 jet fuel and for benzene, toluene, ethylbenzene, and xylene (BTEX), and samples were analyzed for sediment grain size to determine percentages of sand versus clay and silt. Fine sands were predominant in the contaminated intervals; however, sediments containing 80% clay and silt were measured. Contamination was associated with sand and clay, but in two boreholes the sediments with the greatest concentrations of JP-4 jet fuel were 50% clay and silt. Direct counts of microorganisms were equal in both sediment types, but microbial activity, was observed only in sands. Heterogeneous sediments and large concentrations of contamination, in areas of low permeability and limited biological activity, should be addressed in field applications of in situ bioremediation.

INTRODUCTION

Petroleum hydrocarbon contamination is a significant problem. The U.S. Environmental Protection Agency (U.S. EPA) has estimated that oil spills alone contribute an estimated 10,000 to 15,000 newly contaminated sites each year in the United States (U.S. EPA 1990). In 1975, a leak of 300,000 L of JP-4 jet fuel from a fuel storage facility contaminated a shallow aquifer in Hanahan, South Carolina. This facility contains seven aboveground storage tanks that hold approximately 12.5 million L. The leak occurred from Tank 1, and contamination has migrated off site into the surrounding residential area. BTEX concentrations as high as 3 mg/L have been measured in groundwater both on and off site, and total petroleum hydrocarbons of 4,000 mg/kg dry sediment have been measured in sediments.

Subsurface sediment heterogeneity affects not only water and contaminant transport, but also contaminant horizontal and vertical distribution, and microbial activity. The microbial community at the Hanahan site has been shown to be active in highly contaminated and uncontaminated sediments (Aelion & Bradley 1991), but the influence of heterogeneity of sediment type on microbial activity has not previously been investigated. The purpose of this study was to examine the vertical association of contaminant concentration and microbial activity with sediment grain size. JP-4 and BTEX contamination was measured, and sediment wet sieving was carried out on aquifer sediments in 15- to 20-cm increments to assess heterogeneity in chemical contamination and sediment grain size with depth. In addition, microbial enumeration and activity were examined in predominantly sand and predominantly clay and silt fractions to assess the impact of sediment type on microbial numbers and amino acid respiration.

SITE DESCRIPTION AND CONTAMINANT CHARACTERISTICS

Previous site assessments determined the horizontal extent of contamination, estimated hydraulic conductivity, and examined site hydrogeology (McClelland Engineers, Inc. 1987; Residual Management Technologies, Inc. 1988). The upper aquifer sediment is characterized as an unconsolidated Pleistocene Formation consisting of fine sands with numerous clay and silt layers. The shallow aquifer is underlain by the confining layer of the Oligocene Cooper Formation, which consists of laterally alternating lenses of clay and interbedded sands, silts, and clays. Most of the monitoring wells previously installed at the site were screened over a 1.5- to 3-m interval, representing a significant portion of the saturated zone of the shallow aquifer. BTEX concentrations measured in these wells represented a vertically integrated value and did not reflect concentrations present in localized contaminated areas.

Petroleum hydrocarbons are less dense than water and tend to collect on the water table surface, potentially forming free product. However, in this area of South Carolina, heavy rain events are common and the shallow aquifer is recharged primarily from precipitation. Contamination can become smeared over larger vertical distances as the water table rises and falls seasonally. The water table elevation ranges from 3 to 4.5 m below land surface (bls) on site, and a decrease in topographic relief in the off-site residential area results in groundwater occurring at shallower depths, including at land surface, after severe rain events.

METHODS

In this study six boreholes were drilled using a Giddings Probe to assess the vertical extent of the contamination, and four were used to examine the heterogeneity of sediment grain size with depth. Using a sediment extruder, boreholes were continuously sampled every 15 to 20 cm in the saturated zone in the area of contamination near Tank 1. Subsamples were collected directly into VOA vials containing water and methylene chloride using a paste sampler. GC-MS analyses were carried out by the U.S. EPA laboratory in Ada, Oklahoma, for BTEX, JP-4, and fuel carbon concentrations (Vandegrift & Kampbell 1988). Sample depth ranged from 3 to 6 m below land surface (bls) and encompassed the vertical range of the majority of contamination. Samples from four of the boreholes were identified as E, F, H, and J. Samples E, H, and J were clustered within a 3-m radius, whereas sample F was approximately 21 m east of the other three samples (Figure 1).

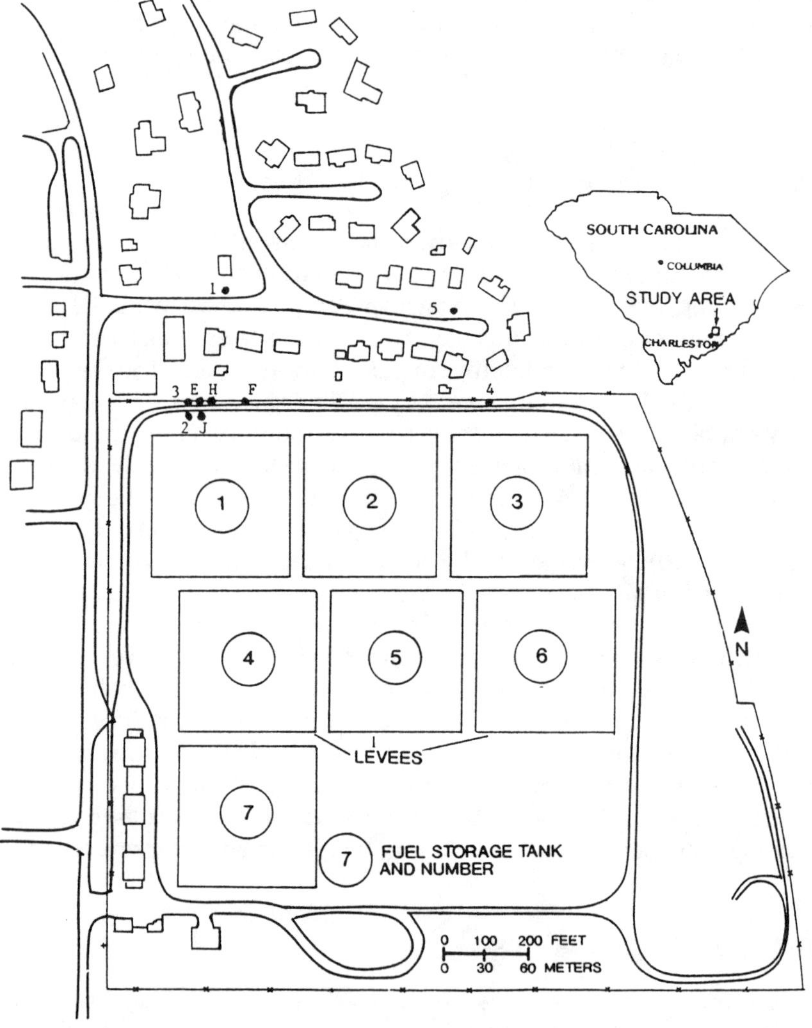

FIGURE 1. Site map showing location of aboveground fuel tanks and sediment samples.

A detailed examination was made of the distribution of the sand versus clay plus silt fraction in these four boreholes. The sediments were analyzed for grain size by wet sieving according to the standard procedure described by Lewis (1984). Sediments are classified according to grain size as follows: gravel (64.00 to 2.00 mm), sand (2.00 mm to 62.5 μ), silt (62.5 to 3.9 μ), and clay (3.9 to 0.25 μ) (Witkowski et al. 1987). Sediment grain sizes smaller than 62.5 μ were not further differentiated, and will be termed the clay fraction in this study. The sieve sizes and corresponding sediment grain diameters used in this study were –1.0 Φ (2.00 mm), –0.5 Φ (1.41 mm), 0.0 Φ (1.00 mm), 0.5 Φ (710 μ), 1.0 Φ (500 μ), 1.5 Φ (350 μ), 2.0 Φ (250 μ), 2.5 Φ (177 μ), 3.0 Φ (125 μ), 3.5 Φ (88 μ), and 4.0 Φ (62.5 μ). Negative Φ sizes correspond to larger sediment grain size.

Additional samples for microbial enumeration were collected on and off site in both uncontaminated and contaminated sediments. Microbial enumerations were carried out from direct microscopic counts using the acridine orange direct count (AODC) modification of the Ghiorse and Balkwill (1983) procedure (Swindoll et al. 1988). A serial dilution of sediment slurry was stained with 0.01% acridine orange, filtered through a 0.2 μ black counter-stained Nuclepore filter, and enumerated using a Leitz Ortholux epifluorescent microscope under 1250x oil immersion. Microbial plate counts were used to enumerate organisms capable of growth on standard high-nutrient and low-nutrient agars (Difco Laboratories, Detroit, Michigan) using the standard spread plate method (American Public Health Association 1989).

The most-probable-number (MPN) technique was used to measure the evolution of $^{14}CO_2$ from incubations with ^{14}C-radiolabeled mixed amino acids, a readily mineralized substance which should represent activity of the general microbial community (Aelion et al. 1987). Serially diluted samples are scored positively or negatively based on the magnitude of the $^{14}CO_2$ produced in the live samples compared to the dead control samples. Based on probability theory, an estimate of the MPN of degraders for the specific chemical used was calculated from the number of positive and negative vials at each dilution (Lehmicke et al. 1979).

RESULTS

Incremental additions in cumulative weight (expressed as a percent of the total sand fraction sample weight) for each sieve size (with increasing Φ size corresponding to decreasing sediment grain size) were plotted versus depth of the sample. These changes in sand fraction were calculated after wet sieving to remove the clay. The remainder of the sample, the difference between the final cumulative percent shown and 100% of the sample weight, is the clay fraction. The sediments at Hanahan showed distinct areas of sand and clay that changed sharply over small vertical distances in all cores. A representative sample is shown in Figure 2.

In general, large sediment grain sizes, larger than 1 Φ (500 μ) were not present in any of the sediments. Within the sand fraction, samples from different depths and different boreholes contained remarkably similar grain sizes, predominantly

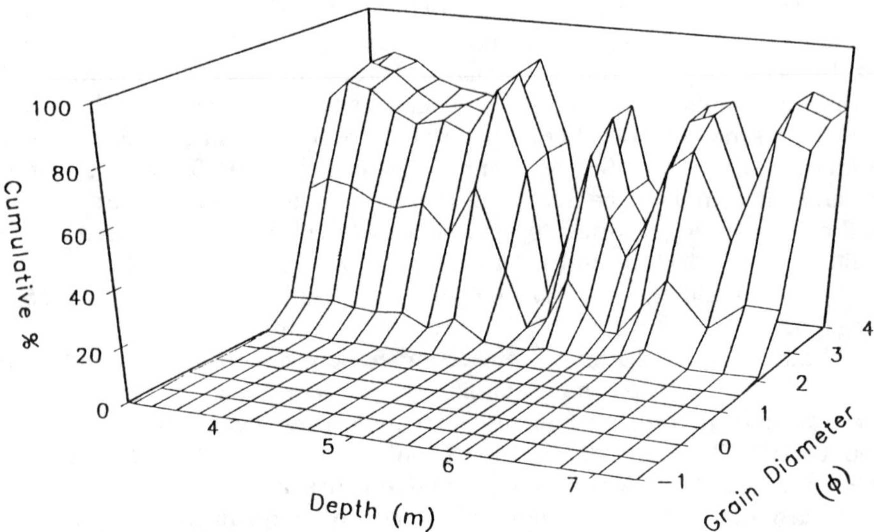

FIGURE 2. Distribution of sand fraction by sediment grain size in core H, expressed as the incremental addition in cumulative percent weight of sand fraction with depth for each sieve size used. Increasing Φ corresponds to decreasing grain size.

fine sands of grain sizes 2.5 to 2.0 Φ (177 to 250 μ) and 3.0 to 2.5 Φ (125 to 177 μ). Samples contained distinct sand and clay fractions, however, with no real patterns evident in their distribution. Samples with greater than 70% clay (grain sizes smaller than 4.0 Φ [63 μ]) were present in all four boreholes. Heterogeneity in sediment type was evident in all samples. Significant fluctuations in percent clay occurred over small vertical distances. For example, sample H (Figure 2) showed rapid changes in percent clay at approximately 5.2 and 5.8 m.

Figure 3 is representative of the relation between the amount of JP-4 contamination and the percent sand in each sample as a function of depth. Data are presented only for the area of contamination for the core. At the time of sample collection, the water table elevation was approximately 3.7 m bls, and no free product was found floating on the water table. JP-4 contamination was high in several of the sampled intervals. Sediments from E, H, and J contained approximately 1,000 mg/kg JP-4, an order of magnitude lower concentration than was present in core F (8,000 mg/kg dry weight sediment). In general, contamination was present over a vertical distance of 4 to 6.7 m, with the greatest concentrations found between 4.9 and 5.8 m (data not shown).

Sand and clay fractions within a core were highly contaminated with JP-4 jet fuel and BTEX. For example, the interval from 4.3 to 5.3 m bls of sediment F contained significant concentrations of JP-4 and BTEX (data not shown). Discrete samples from these depths contained both predominantly clay and predominantly sand fractions. The highest contamination in sample F contained 64% sand,

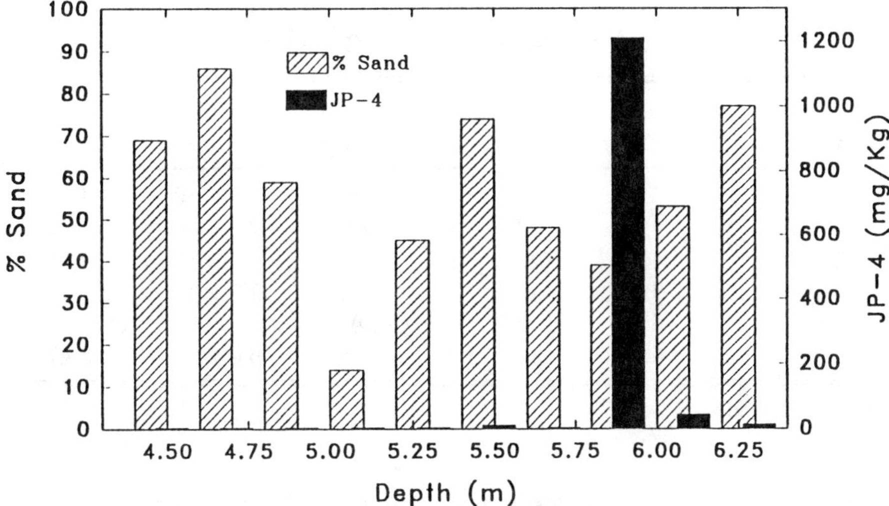

FIGURE 3. Percent sand of total sediment weight, and corresponding concentration of JP-4 (mg/kg dry sediment) as a function of depth (meters below land surface) in samples from core H.

whereas the second-most contaminated sediment contained 26% sand. A similar situation occurred for sediment sample E (data not shown). The majority of JP-4 and BTEX contamination was associated with sediments 4.4 to 5.8 m bls. Some of these sediments contained 80% sand, whereas others contained less than 50% sand.

Although contamination was associated with sediments containing significant amounts of sand and clay, microbial characteristics associated with the these fractions differed. Based on AODC results, significant numbers of organisms were present in all samples, although samples came from different depths and different levels of contamination. Five representative samples from both on and off site are shown in Table 1. Samples 1 and 3 were highly contaminated, but samples 2,

TABLE 1. Acridine orange direct counts of bacteria from aquifer sediments.

Sample ID	Depth (m)	Cells/g soil (replicates)		Contamination Range
1	6.1	2.4×10^8	(4.9)	mg/kg
2	9.2	1.0×10^8	(1.8)	BDL[(a)]
3	4.9	6.0×10^6	(13.7)	mg/kg
4A	5.2	1.1×10^7	(2.8)	μg/kg
4B	6.4	1.0×10^8	(1.0)	μg/kg
5	6.1	6.9×10^7	(3.6)	BDL

(a) Below detection limit.

TABLE 2. Plate counts on high- and low-nutrient agar of bacteria from aquifer sediments (colony-forming units/g soil dry weight).

Sample ID	Depth (m)	Cells/g soil High Nutrient (replicates)		Cells/g soil Low Nutrient (replicates)	
1	6.1	3.2×10^6	(2.5)	1.5×10^7	(1.2)
2	9.2	2.9×10^6	(2.8)	2.2×10^6	(2.7)
3	4.9	4.1×10^5	(4.7)	3.5×10^5	(4.0)
4A	5.2	1.7×10^5	(3.4)	2.2×10^5	(2.1)
4B	6.4	ND[(a)]	(ND)	ND	(ND)
5	6.1	1.3×10^5	(0.52)	5.0×10^6	(5.4)

(a) Not detected.

4A, and 4B were less contaminated on-site samples, and 5 was an uncontaminated off-site sample. Sample 4B was predominantly clay; the other samples were predominantly sandy material. Overall, AODCs of all samples contained between 10^7 to 10^8 cells/g soil dry wt. AODCs do not differentiate between live and dead cells, but simply give a count of the total microbial population (Atlas & Bartha 1993).

The results of the plate counts are presented in Table 2. In general, plate counts were 1 to 10% of the total cell counts. The contaminated sample 1 was the most active, whereas sample 4B that was predominantly clay showed no activity based on both high nutrient and low nutrient agar. Colony-forming units (CFUs) represent microorganisms that have the ability to grow on the agar provided, and do not preclude that growth could have occurred on an alternative medium. Therefore, although there were no platable organisms in sediment 4B, it is possible that cells could have grown on other types of growth medium.

Similarly, results of the MPNs of organisms based on incubation with a mixture of amino acids after 2 days of incubation show that approximately 1% of the total cell counts represented active amino acid-respiring bacteria. Overall, the number of amino acid respirers was smaller than the numbers derived from plate counts, suggesting that a portion of the platable bacteria were not actively respiring amino acids. Similar to the plate count results, the most contaminated sample 1 contained the greatest number of amino acid respirers, and no respiration was measured in the predominantly clay sample 4B (Table 3).

CONCLUSION

The coastal plain sediments in South Carolina are predominantly fine sand. Although fine sand represents the majority of the sand fraction in the sediments sampled, large fluctuations and variations in clay content occurred on a 15-cm vertical scale. Even in these sandy sediments, heterogeneities in subsurface sediments existed on both a small vertical and horizontal scale. In a bioremediation

TABLE 3. Most probable number of amino acid degraders from aquifer sediments (MPN/g soil dry weight).

Sample ID	Depth (m)	Cells/g soil (95% CI)[(a)]		Contamination Range
1	6.1	2.44×10^6	(0.8-7.4)	mg/kg
2	9.2	1.47×10^5	(0.5-4.8)	BDL[(d)]
3	4.9	1.89×10^4	(0.7-5.5)	mg/kg
4A	5.2	8.00×10^3	(2.9-21.8)	µg/kg
4B	6.4	ND[(b)]	(NA)[(c)]	µg/kg
5	6.1	5.17×10^3	(1.8-14.7)	BDL

(a) Confidence interval.
(b) Not detected.
(c) Not analyzed.
(d) Below detection limit.

system in which nutrients and water are added to an infiltration gallery, the sand fraction will receive preferential flow due to higher hydraulic conductivity, and thus be more readily remediated. Water and nutrient movement through the relatively impermeable clay fractions will be hindered. This heterogeneity must be incorporated into in situ bioremediation design.

The fate of petroleum hydrocarbons in aquifer systems is generalized as a floating free product containing the majority of contamination, a significant amount of contamination from the more readily water-soluble constituents dissolved in groundwater, and the volatile fraction in sediment pores of the vadose zone. In the area sampled for this study, free product was not observed during the time of sampling, and JP-4 jet fuel contamination was present over a significant vertical distance. This was due in part to seasonal variations in precipitation and a fluctuating water table. Although present in a significant portion of the saturated zone, JP-4 concentrations peaked strongly at discrete intervals.

In Hanahan sediments, both sand and clay fractions were contaminated, but clay sediments did not have significant microbial activity compared to sand sediments based on the two active enumeration methods used. The lack of growth on selected media used in this study does not indicate that potentially viable organisms are not present, because a response could be elicited on an alternative growth medium. Additional clay samples need to be examined to verify and determine the underlying reasons for limited microbial activity. Significant contamination and reduced microbial activity levels in these clay fractions must be addressed to promote the most efficient and effective use of in situ bioremediation.

ACKNOWLEDGMENTS

This research was supported by a grant from the Hazardous Waste Management Research Fund. The authors wish to thank Dr. Stephen Hutchins of the U.S. EPA, R. S. Kerr Environmental Research Laboratory in Ada, Oklahoma.

REFERENCES

Aelion, C. M., and P.M. Bradley. 1991. "Aerobic biodegradation potential of subsurface microorganisms from a jet fuel-contaminated aquifer." *Appl. Environ. Microbiol. 57*: 57-63.

Aelion, C. M., C. M. Swindoll, and F. K. Pfaender. 1987. Adaptation to and biodegradation of xenobiotic compounds by microbial communities from a pristine aquifer." *Appl. Environ. Microbiol. 53*: 2212-2217.

American Public Health Association. 1989. *Standard Methods for the Examination of Water and Wastewater*, 17th ed. Washington, DC.

Atlas, R. M., and R. Bartha. 1993. *Microbial Ecology*, 3rd ed. The Benjamin/Cummings Publishing Co., Inc., Redwood City, CA.

Ghiorse, W. C., and D. L. Balkwill. 1983. "Enumeration and morphological characterization of bacteria indigenous to subsurface environments." *Dev. Ind. Microbiol. 24*: 213-244.

Lehmicke, L. G., R. T. Williams, and R. L. Crawford. 1979. "^{14}C-most-probable-number method for enumeration of active heterotrophic microorganisms in natural waters." *Appl. Environ. Microbiol. 38*: 644-649.

Lewis, D. W. 1984. *Practical Sedimentology*. Hutchinson Ross Publishing Co., New York, NY.

McClelland Engineers, Inc. 1987. *Final Report, Confirmation Study, Characterization Step, Defense Fuel Supply Point, Charleston, South Carolina.*

Residual Management Technologies, Inc. 1988. *Final Report, Aquifer Evaluation, Defense Fuel Supply Point, Charleston, South Carolina.*

Swindoll, C. M., C. M. Aelion, and F. K. Pfaender. 1988. "Influence of inorganic and organic nutrients on aerobic biodegradation and on the adaptation response of subsurface microbial communities." *Appl. Environ. Microbiol. 54*: 212-217.

U.S. Environmental Protection Agency. 1990. *Bioremediation of Hazardous Wastes*. EPA/600/9-90/041. Office of Research and Development, Washington, DC.

Vandegrift, S. A., and D. H. Kampbell. 1988. "Determination of aviation gasoline and JP-4 jet fuel in subsurface core samples." *J. Chromatogr. Sci. 26*: 566-569.

Witkowski, P. J., J. A. Smith, T. V. Fusillo, and C. T. Chiou. 1987. *A Review of Surface-Water Sediment Fraction Interactions with Persistent Manmade Organic Compounds*. U.S. Geological Survey Circular 993.

NATURAL BIORECLAMATION OF ALKYLBENZENES (BTEX) FROM A GASOLINE SPILL IN METHANOGENIC GROUNDWATER

J. T. Wilson, D. H. Kampbell, and J. Armstrong

ABSTRACT

A spill of gasoline from underground storage tanks (USTS) at the Sleeping Bear Dunes National Lakeshore in Benzie County, Michigan, produced a plume of contamination that reached the banks of the Platte River. The plume was short (70) feet and it had a short residence time (5 to 53 weeks). The plume was in transmissive glacial sands and gravels. The groundwater is cold (10 to 11°C), hard (alkalinity 200 to 350 mg/L), and well buffered (pH 6.1 to 7.6). Ambient concentrations of oxygen, nitrate-N, and sulfate are 2.4, 15.3, and 20 mg/L, respectively. Along the most contaminated flow path, methanogenesis, nitrate reduction, sulfate reduction, iron reduction, and oxygen respiration accepted enough electrons to destroy 39, 14, 4.2, 1.1, and 0.8 mg/L of benzene, toluene, ethylbenzene, and xylenes (BTEX compounds) respectively. The actual concentration of BTEX compounds consumed was 42 mg/L. After correction for dilution, benzene was not bioattenuated. Bioattenuation of toluene ranged from 0.16 to 0.47 per week, ethylbenzene from 0.022 to 0.077 per week, *p*-xylene 0.017 to 0.067, *m*-xylene 0.026 to 0.10 per week, and *o*-xylene 0.028 to 0.11 per week.

PURPOSE AND STRATEGY OF THE STUDY

Fuel spills from USTs are a major source of groundwater contamination. Currently, more than 100,000 leaking tanks have been identified in the United States. Although the number of spills is large, the extent of contamination is less than would be expected.

Hadley and Armstrong (1991) compared the number of water supply wells in California that were contaminated with benzene, which is easily degraded in aerobic groundwater, to trichloroethylene or tetrachloroethylene, which are not easily degraded in aerobic groundwater. A great deal more benzene should enter the subsurface of California from gasoline spilled from leaking USTs, yet the solvents were encountered more frequently and at higher concentrations. Only nine wells were contaminated with benzene, compared to 188 with trichloroethylene and 199 with tetrachloroethylene. The median concentration of benzene in the contaminated wells was 0.0002 mg/L, compared to 0.0032 and 0.0019 mg/L for trichloroethylene and tetrachloroethylene. They suggest that natural biodegradation was removing the benzene from aerobic California groundwaters.

This study assesses the importance of natural bioattenuation of benzene, toluene, ethylbenzene, and the xylenes (BTEX) in groundwater at a typical UST spill. The following strategy was used to acquire information that would all a quantitative assessment at field scale:

1. Conduct a soil gas survey for hydrocarbon vapors to identify those areas with oily-phase hydrocarbons, which act as a source of groundwater contamination.
2. Acquire vertical cores in the area showing hydrocarbon vapors to define the vertical extent of gasoline contamination, and the cross section of the spill that was perfused by moving groundwater.
3. Install monitoring wells in the plume of contaminated groundwater. Well clusters were installed in the spill, at the point of discharge to surface water, and at an intermediate point.
4. Monitor concentrations of contaminants and potential electron acceptors over time.
5. Conduct a pumping test to measure hydraulic conductivity of the aquifer.
6. Monitor the hydraulic gradient over time to estimate the direction and velocity of groundwater flow.
7. Use information on groundwater flow to predict the average residence time of water sampled in the well clusters.
8. Use information on groundwater residence time and extent of attenuation to calculate bioattenuation rate constants.

The rate constants were calculated for portions of the groundwater plume that were not in contact with the oily-phase gasoline. Attenuation of a conservative constituent of the gasoline was used as a tracer to separate attenuation due to dilution from attenuation due to biotransformation.

HISTORY OF THE SITE

The study area is located in the Sleeping Bear Dunes National Lakeshore near Empire, Michigan. A gasoline service station had been operated on a corner lot where Michigan Highway M-22 crosses the Platte River. Figure 1 pictures the location of surface infrastructure, the three USTs, the highway, and the river. The National Park Service had acquired the land and removed the service station. On December 11, 1989, the three USTs were excavated and removed. Fill and excavated soil around the tanks smelled of gasoline. After the tanks were removed, the excavation was backfilled with the same soil that had been removed, to prevent injuries to visitors to the open pit.

The water table aquifer is in highly transmissive glacial outwash. In February 1989, monitoring wells were installed in the plume of contamination to monitor

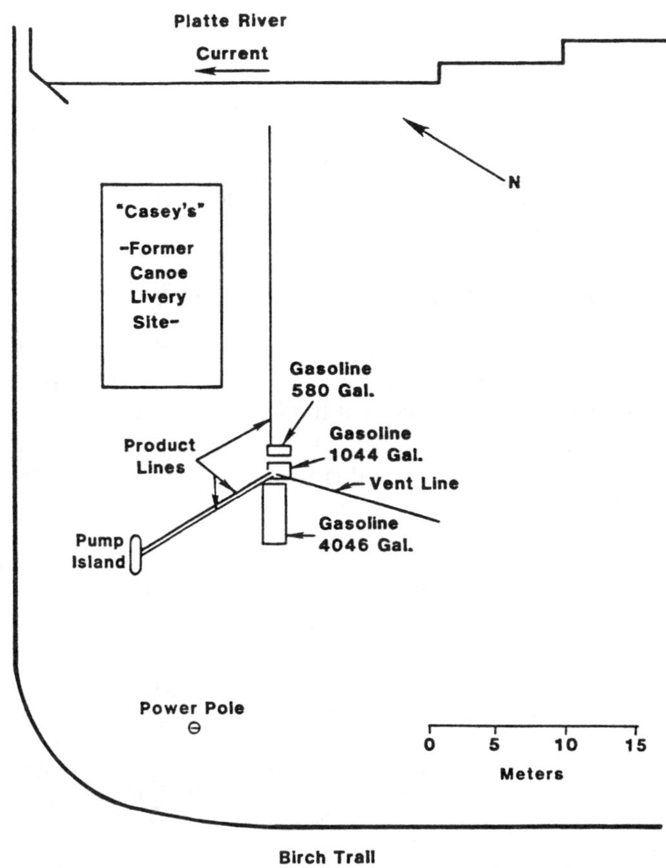

FIGURE 1. Location of underground storage tanks and associated infrastructure at the former "Casey's" Canoe Livery site prior to removal.

natural bioattenuation. The last samples were taken in November 1992. In December 1992, sheet piling was installed at the margin of the river, and the material contaminated with gasoline was excavated.

MATERIALS AND METHODS

Groundwater was sampled from 5-cm-I.D. wells and analyzed for volatile organic compounds by purge and trap using a flame ionization detector (FID). Core samples were extracted into methylene chloride, then analyzed by direct injection gas chromatography (GC) using an FID detector. Total petroleum hydrocarbons was estimated from the peak areas of the ten most abundant components of the spill. Methane was measured by headspace analysis using direct injection on a GC with an FID. Oxygen was measured with an electrode in the field. The wells contained a free water surface exposed to the atmosphere, and there was a possibility of oxygenation of the sample during purging of the wells. Measured oxygen concentrations should be considered an upper boundary on the concentration in the aquifers. Total organic carbon was measured by combustion to carbon dioxide, which was quantified by infrared absorbance. Hydraulic conductivity was measured from changes in head in passive monitoring wells during a 24-hour pumping test.

PROPERTIES OF THE GASOLINE SPILL

Figure 2 presents the soil gas survey used to locate material containing oily-phase contaminants. Figure 3 plots the results. There was an area with high concentrations of hydrocarbon vapors at the former location of the USTs. The area with oily-phase material extended from the boundary of the excavation toward the northeast, in the direction of the Platte River. The contaminated area was roughly

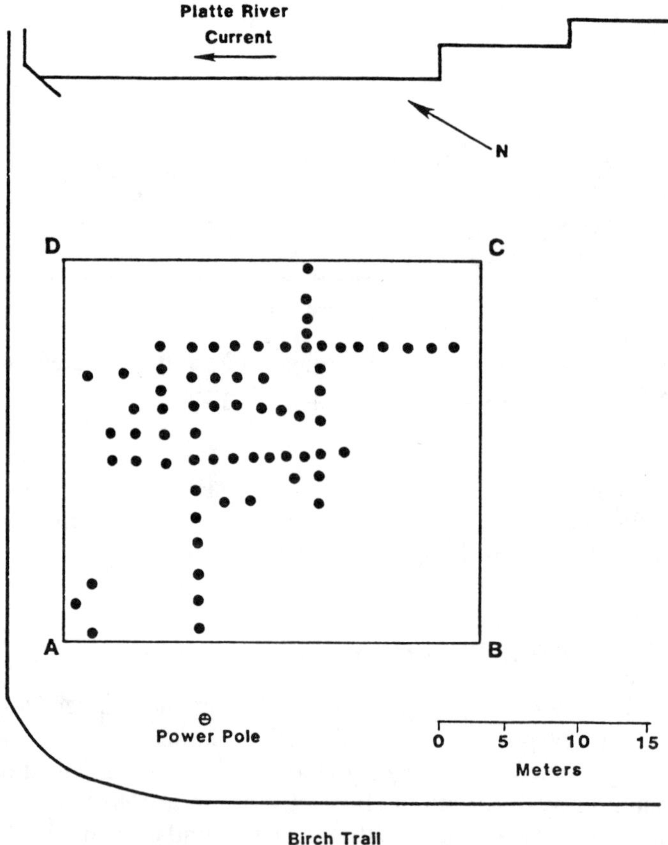

FIGURE 2. Soil gas survey used to identify regions containing oily-phase hydrocarbons after removal of the underground storage tanks.

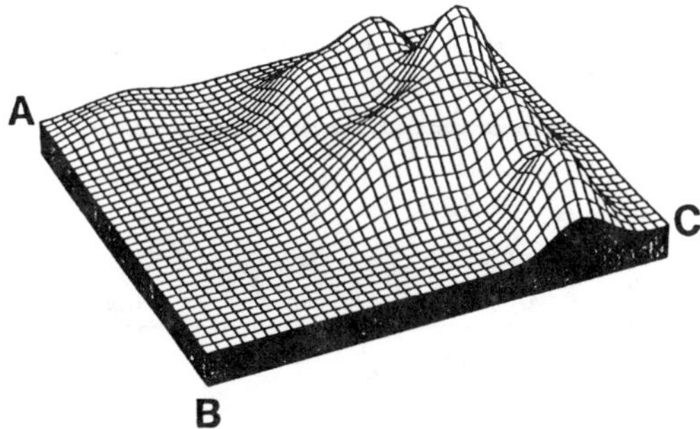

FIGURE 3. Relative concentrations of hydrocarbon vapors in the soil gas survey. See Figure 2 to identify the area contoured. The vertical axis is proportional to the logarithm of the vapor concentration.

circular, approximately 30 m in diameter; its margin was approximately 25 m from the river. Three series of cores were collected in the area contaminated with gasoline. All three series were in close agreement, the vertical extent of the oily-phase gasoline was only 0.64 m, most of it above the water table (see Table 1 for typical data). The gasoline in the contaminated interval averaged 9,200 mg/kg at borehole C and 10.2 mg/kg at borehole B. The concentration in the volume of soil corresponds to approximately 4,500 liters of gasoline left behind after the removal of the tanks.

The mean molecular weight of the gasoline was 128 daltons. It was 1.36 mole % benzene, 13.1 mole % toluene, 3.38 mole % ethylbenzene, 4.1 mole % *o*-xylene, and 12.6 mole % *m*+*p*-xylene.

GROUNDWATER FLOW

The gasoline spill served as the source area for a plume of groundwater contamination that extended toward the Platte River. The concentrations of organic contaminants in contact with oily phase materials should be controlled by partitioning between the gasoline and the water. As the plume moved away from the region with oily phase material, the concentration of organic contaminants should attenuate, due to dilution and biodegradation.

Water table elevations surveyed in July 1990, indicated that the gradient sloped 1.5 m/1,000 m to the northeast toward the Platte River. Monitoring wells with stainless steel screens were installed in the plume in clusters of seven. Each well was screened over a 0.91 m interval, and each cluster sampled groundwater from the water table to a depth of 6.4 m below the water table.

A monitoring well cluster was installed in the spill area between core boreholes 54C and 54B. A second well cluster was installed 21.3 m from the edge of the spill, adjacent to the river bank at the projected point of discharge. There was no evidence of oily phase material at this location. A third well cluster was installed at the midpoint between the first two clusters, 9.1 m from the edge of oily-phase spill.

TABLE 1. Vertical distribution of gasoline in the area of the spill after removal of the underground storage tank.

Core	Elevation (ft above sea level)[(a)]	Benzene (mg/L)	Toluene (mg/L)	Total BTEX (mg/L)	Total Petroleum Hydrocarbons (mg/L)
54C-4	587.5 - 587.2	<0.1	0.1	2.4	9.2
54C-3	587.2 - 586.9	0.1	0.6	6.1	5.0
54C-2	586.9 - 586.5	24	320	920	5,000
54C-1	586.5 - 586.2	93	930	2,400	19,000
54C-17	586.3 - 586.0	72	1,100	3,400	16,000
54C-16	586.0 - 585.8	48	650	2,000	9,900
54C-15	585.8 - 585.4	106	1,200	3,400	17,000
54C-14	585.4 - 585.1	3.8	35	130	730
Water table	585.3				
54C-13	585.1 - 584.8	7.2	93	108	1,400
54C-12	484.8 - 584.4	0.39	1.0	3.6	19

(a) 1.0000 ft equals 0.3048 m.

This arrangement is presented in plan view in Figure 4 and in cross section in Figure 5. There was good hydraulic connection between the wells within each cluster. At any sampling day, water table elevations in a cluster agreed within 0.01 m.

For purposes of the study, elapsed time was calculated from the date of UST removal. Figure 6 plots the hydraulic gradient against elapsed time, showing large variations in the gradient. In the fall and winter the gradient is toward the river. In late spring and early summer it reverses, and groundwater flows away from the river. The cyclical pattern is controlled by the balance between recharge to the aquifer and runoff to the river. In winter and early spring, precipitation in the area is held as snow and frozen soil water. At spring thaw the runoff increases, raising the stage of the river and reversing the hydraulic gradient.

The average hydraulic conductivity in the area containing the spill is 5×10^{-2} cm per second. A gradient of 3×10^{-3} corresponds to a plume velocity of 0.4 m per day. The wide variations in hydraulic gradient have a complex effect on the average residence time of the water sampled in the well clusters that are 9.1 and 21.3 m from the spill. To calculate average residence time in the period from week 50 to week 150, the instantaneous gradient was estimated by linear interpolation between the data points. At each sampling time, the gradient was integrated backward over time until the plume had traveled the distance from the source to the monitoring well cluster.

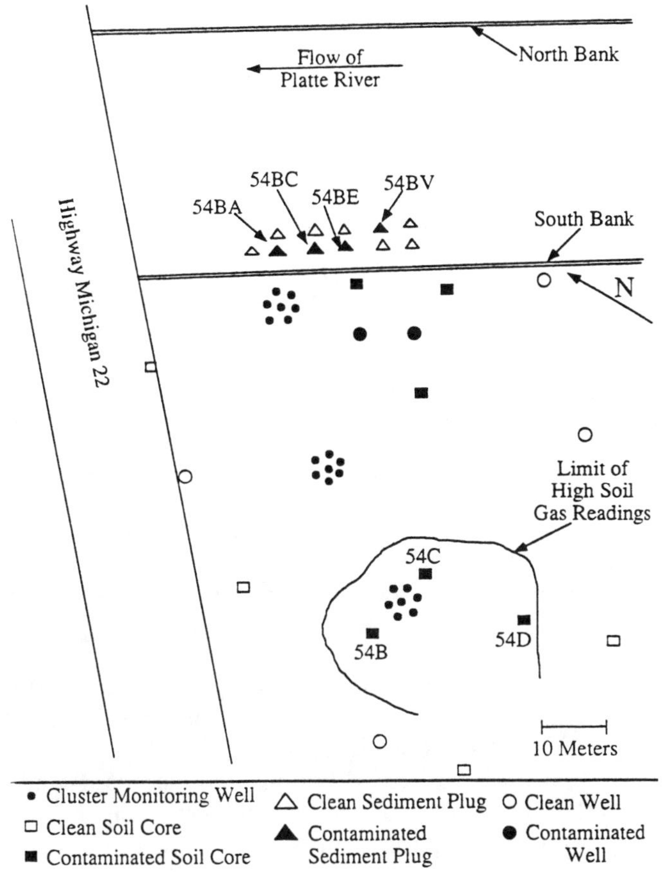

FIGURE 4. Location of monitoring wells with respect to the area containing oily-phase hydrocarbons and the Platte River.

The results of the calculations are presented in Figure 7. When the gradient is high and toward the river, the residence time of water is approximately 5 weeks in the 9.1 m well cluster, and 10 to 15 weeks in the 21.3 m cluster. As the gradient slowed and reversed in 1991, the plume swept past a well cluster, reversed and swept back across it, then reversed and swept past it a third time. The average residence time reached as high as 30 weeks before the plume was flushed out by the water that had not undergone the reversal. Obviously, such wide variations in residence time should have a profound effect on the extent of bioattenuation.

GEOCHEMISTRY OF BIOATTENUATION

A second profound influence on bioattenuation is the availability of electron acceptors for microbial metabolism. A third influence on apparent attenuation is a change in the geometry of the plume as it approaches the river. Flow lines tend

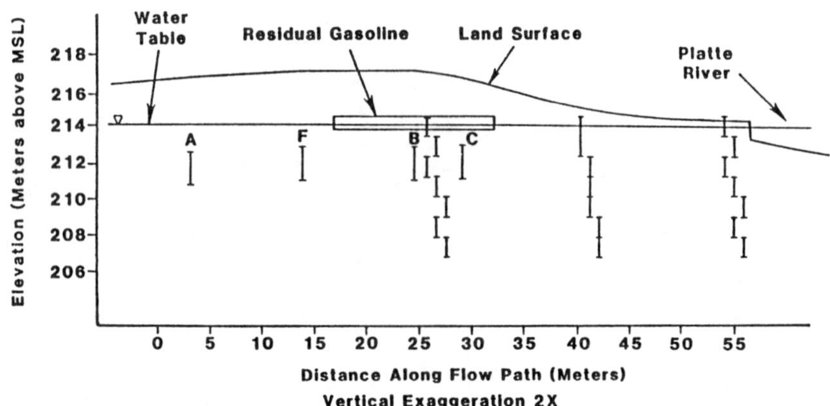

FIGURE 5. Vertical cross section showing the relation between the water table, the residual gasoline, and the screened interval of monitoring wells.

to converge as they approach a point of discharge. Both of these influences can be evaluated by examining the changes in the concentrations of contaminants and redox active species in the well clusters.

Table 2 presents the vertical distribution of contaminants and redox active materials under the spill area. Concentrations of benzene and total alkylbenzenes decreased with depth; the significant concentrations were confined to the interval

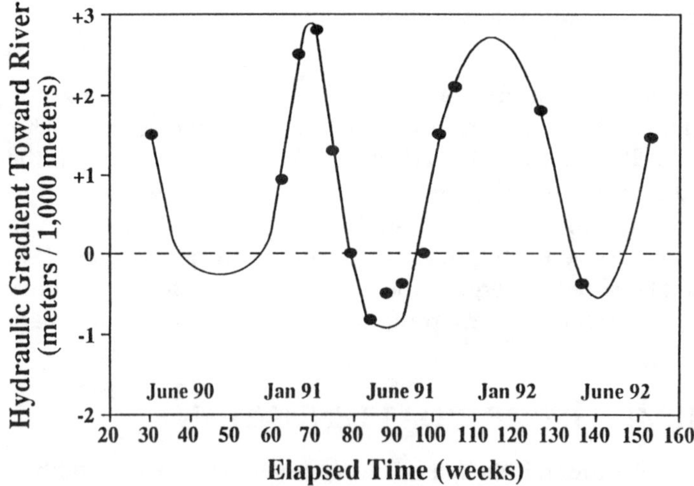

FIGURE 6. Changes in the hydraulic gradient at the site during the study.

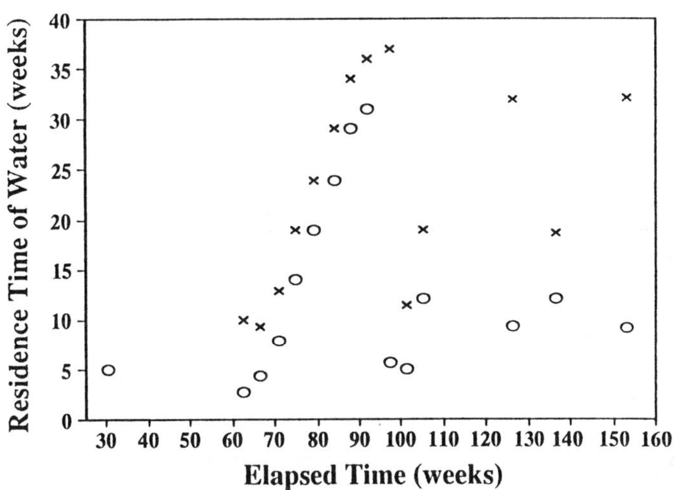

FIGURE 7. Changes in the residence time of water in wells from the cluster 9.1 m downgradient of the spill (O) and 21.3 m downgradient of the spill (X), as a result of changes in the hydraulic gradient.

from the water table to 3.7 m below the water table. The depth interval containing significant concentrations of BTEX compounds corresponds to an interval showing depletion of oxygen, sulfate, and nitrate, and accumulation of methane and iron II. The boundary between the contaminant plume and unimpacted groundwater underneath occurred at an elevation of 175.3 m.

The vertical distribution of material in the midpoint well cluster is presented in Table 3. The contaminants were confined largely to an interval between 0.91 m and 3.7 m below the water table. Oxygen was depleted and methane and iron II accumulated in this depth interval. Sulfate returned to ambient concentrations below the plume of contamination, but oxygen and nitrate were depleted through the entire depth interval sampled by the well cluster.

Table 4 presents the distribution of materials at the bank of the river. Only the interval from 0.91 to 1.8 m below the water table had significant concentrations of contaminants. As was the case with the other well clusters, methane and iron II accumulated, and oxygen, sulfate, and nitrate were depleted in the contaminated interval. As was the case with the midpoint well cluster, sulfate returned to ambient concentrations below the plume of contamination, but oxygen and nitrate did not.

To evaluate the relative importance of the various electron sinks, a balance on electrons transferred was attempted for a flow path that originated in uncontaminated water upgradient of the spill and passed through the most contaminated well in each cluster (Table 5). For purposes of calculation, BTEX compounds represented by the empirical formula CH were fermented to acetate and dihydrogen, sulfate was reduced to sulfide, and nitrate to dinitrogen. Methane expected from the acetate produced by fermentation of BTEX compounds was added to methane expected to accumulate from bicarbonate respiration of the excess

TABLE 2. Vertical distribution of materials in groundwater in the spill area. Data are an average of samples collected 136 and 153 weeks after removal of the tanks.

Elevation (ft above mean sea level)[a]	Benzene (mg/L)	Total BTEX (mg/L)	Methane (mg/L)	Nitrate (mg/L)	Sulfate (mg/L)	Oxygen (mg/L)[b]	Iron II (mg/L)
587 - 584	3.12	51.8	0.21	4.6	8.5	0.3	6.3
584 - 581	2.94	44.8	0.096	2.2	<0.05	0.1	7.7
581 - 578	0.83	14.9	3.5	0.50	0.12	0.7	6.5
578 - 575	0.37	4.5	3.0	0.40	<0.05	0.1	4.3
575 - 572	0.13	0.12	0.34	<0.05	20	1.1	0.3
572 - 569	0.0028	0.048	0.15	<0.05	20	1.0	0.05
569 - 566	0.00077	0.014	0.074	<0.05	10	0.4	0.05

(a) 1.0000 ft equals 0.3048 m.
(b) The groundwater samples may have be oxygenated during well purging. The measured concentrations should be considered upper boundaries on the actual concentration in the aquifer.

hydrogen to predict total methane production. Iron II accumulating in water was assumed to be from reduction of iron III, not mixed-valence iron minerals. As discussed in a later section, attenuation or accumulation was corrected for dilution by normalizing the concentrations to the concentration of 2,3-dimethylpentane, which was assumed to be conservative.

TABLE 3. Vertical distribution of materials in groundwater 9.1 m downgradient from the spill area. Data are an average of samples collected 136 and 153 weeks after removal of the tanks.

Elevation, ft above mean sea level[a]	Benzene (mg/L)	Total BTEX (mg/L)	Methane (mg/L)	Nitrate (mg/L)	Sulfate (mg/L)	Oxygen (mg/L)	Iron II (mg/L)
587 - 584	0.0064	0.18	0.75	1.8	9.5	0.1	4.2
584 - 581	0.33	3.3	1.71	0.3	6.0	<0.1	4.1
581 - 578	0.23	2.3	1.86	0.6	2.7	0.1	5.2
578 - 575	0.39	3.0	1.20	0.4	0.77	0.1	4.4
575 - 572	0.0019	0.0026	0.19	0.3	13	<0.1	1.0
572 - 569	0.00006	0.00015	0.048	0.3	19	0.3	0.5
569 - 566	0.00023	0.0021	0.048	0.3	13	2.0	0.3

(a) 1.0000 ft equals 0.3048 m.

TABLE 4. Vertical distribution of materials in groundwater 21.3 m downgradient of the spill area. Data are an average of samples collected 136 and 153 weeks after removal of the tanks.

Elevation, ft above mean sea level[a]	Benzene (mg/L)	Total BTEX (mg/L)	Methane (mg/L)	Nitrate (mg/L)	Sulfate (mg/L)	Oxygen (mg/L)	Iron II (mg/L)
587 - 584	0.103	0.17	1.55	<0.05	4.8	0.8	3.3
584 - 581	0.45	2.0	3.1	0.1	<0.05	0.4	5.2
581 - 578	0.0233	0.041	0.56	<0.5	8.9	0.7	5.1
578 - 575	0.049	0.086	0.47	0.2	18.4	0.7	3.0
575 - 572	0.019	0.037	0.087	0.2	16.2	0.5	0.17
572 - 569	0.00006	0.00006	0.035	0.4	12.9	0.7	0.05
569 - 566	0.00006	0.00006	0.0006	<0.05	6.0	1.3	0.05

(a) 1.0000 ft equals 0.3048 m.

Methanogenesis was the most important electron sink, followed by nitrate reduction, and sulfate reduction. Oxygen-based respiration and iron solubilization were not important. The actual electron acceptor demand was greater than the theoretical supply of electrons. Other compounds in the plume such as trimethylbenzenes and naphthalenes, also may have been biodegraded. The groundwater

TABLE 5. Balance of electrons transferred among redox active compounds in the most contaminated interval in the flow path from the spill to the Platte River. Data are an average of samples collected 136 and 153 weeks after removal of the tanks, corrected for dilution.

Compound	Upgradient	In Spill	9.1 m Downgradient	21.3 m Downgradient	BTEX Consumed
	-----(mg/L)------				
BTEX	<0.001	56.3	9.8	13.7	42.6
Acetate	<0.1	4.0	0.2	<0.1	1.7
Methane	0.08	0.09	24.1	29.8	39
Nitrate-N	15.3	0.25	<0.05	<0.05	14
Sulfate	20.0	8.5	6.0	<0.05	4.2
Iron II	3.5	6.3	17.5	27.8	1.1
Oxygen	2.4	<0.1	<0.1	<0.1	0.8

also contained large concentrations of nonvolatile total organic carbon, presumably of natural origin. Total organic carbon contents were reduced from 58 mg/L in the well cluster in the spill to 47 mg/L at the well cluster 9.1 m downgradient and 21 mg/L in the cluster 21.3 m downgradient.

There is little evidence that vertical dispersion was mixing electron acceptors into the contaminant plume. Concentration boundaries between the plume and unimpacted groundwater are sharp. Reduction of carbonate, nitrate, sulfate, oxygen, and iron III within the plume were adequate to explain the disappearance of hydrocarbons.

APPARENT ATTENUATION OF CONTAMINANTS

As discussed above, it is difficult to estimate bioattenuation at the study site if the only data available are concentrations of contaminants in monitoring wells. Data are presented to illustrate the variability in apparent attenuation through the annual cycle. To estimate attenuation of contaminants along the plume, the arithmetic average was taken of concentrations in the seven wells in each cluster. The same depth interval of aquifer is being compared, although the plume occupied different proportions of the interval at different clusters.

Concentrations of toluene in groundwater under the spill did not change detectably over the 100 weeks of the study (Figure 8), indicating no appreciable weathering of toluene from the oily phase hydrocarbons. Concentrations 9.1 m from

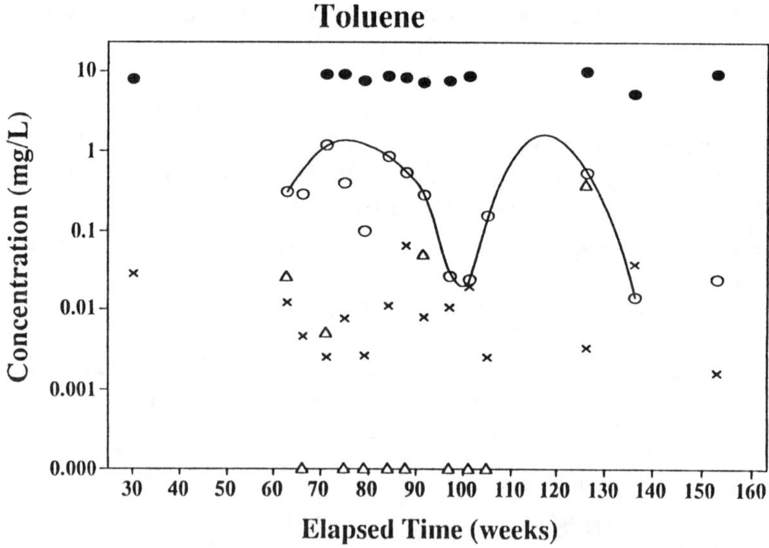

FIGURE 8. Apparent attenuation of toluene in monitoring wells 9.1 m downgradient (o) and 21.3 m downgradient (x), compared to monitoring wells in the spill area (•). A control well 6.1 m upgradient of the spill is indicated by (▲).

the spill were from 1 to 3 orders of magnitude lower; concentrations 21.3 m from the spill were from 2 to almost 4 orders of magnitude lower. Low concentrations in the midpoint cluster well corresponded to times when the residence time of the groundwater was great (compare 90 to 100 weeks in Figure 8 and Figure 7).

Concentrations of benzene in the groundwater under the spill declined almost an order of magnitude over the study period (Figure 9). This was probably due to simple leaching of benzene from the oily phase material in the spill area. Attenuation of benzene 9.1 and 21.3 m from the spill was less than the attenuation of toluene. Attenuations varied between 1 and 2 orders of magnitude. As was the case with toluene, the greatest attenuation occurred at times with the greatest residence time.

Accumulation of methane along the flow path did not follow the expected pattern (Figure 10). During fermentation of BTEX compounds to methane, roughly half the mass of contaminant removed by methanogenesis should ultimately be transformed to methane. Concentrations above 10 mg/L should have been produced for the destruction of the BTEX compounds (Table 5), much less than that was actually monitored. Further, concentrations of methane did not increase with distance along the flow path. Compared on a log scale, there was little fluctuation in methane concentrations over time compared to toluene and benzene.

Water from the well clusters was examined for volatile fatty acids up to C7. Only acetate accumulated to a significant extent (Table 5), and its concentrations were far below those that could have explained the limited accumulation of methane. The low concentration of methane suggested that there was significant attenuation due to dilution.

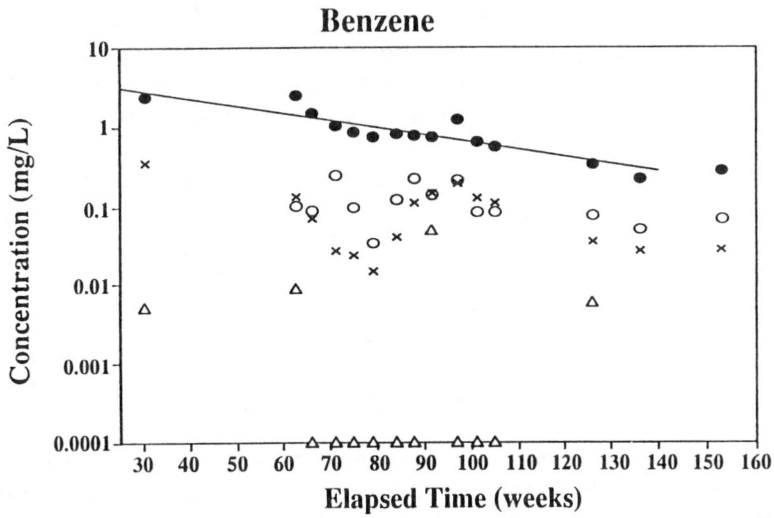

FIGURE 9. Apparent attenuation of benzene in monitoring wells 9.1 m downgradient (o) and 21.3 m downgradient (x), compared to monitoring wells in the spill area (•). A control well 6.1 m upgradient of the spill is indicated by (▵).

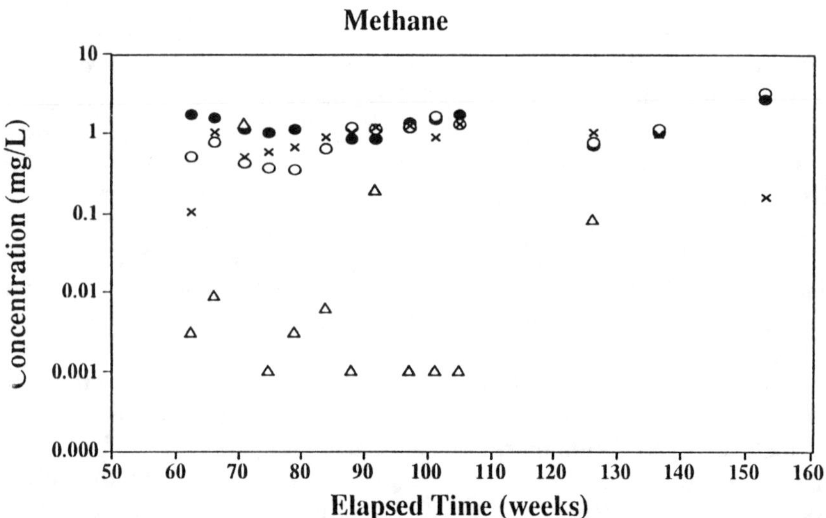

FIGURE 10. Lack of attenuation of methane in monitoring wells 9.1 m downgradient (o) and 21.3 m downgradient (x), compared to monitoring wells in the spill area (•). A control well 6.1 m upgradient of the spill is indicated by (△).

CORRECTION FOR DILUTION

Analysis of the preliminary data from the study indicated that it would be difficult to offer a rigorous interpretation of bioattenuation based on the data being collected. The best estimate of attenuation due to dilution would be reduction in concentration of a hydrocarbon in the plume that should not degrade under reducing conditions. Concentrations of 2,3-dimethylpentane were great enough to be measured accurately, and mass spectral analysis of extracts indicates that there was little interference from co-eluting peaks. Table 6 presents the distribution of 2,3-dimethylpentane in the well clusters during the last two sampling periods of the study. The distribution of 2,3-dimethylpentane corresponds to the distribution of total petroleum hydrocarbons (compare Table 6 with Tables 2, 3, and 4) and is reasonably consistent between the last two sampling periods.

BIOATTENUATION OF CONTAMINANTS

In the last two sampling times, attenuation or production of organic compounds was corrected for dilution by normalizing to the concentration of 2,3-dimethylpentane. Any attenuation or production remaining was presumed to be the result of biological processes (Table 7). Methane production was consistent with concentrations that would be expected from the disappearance of BTEX compounds. Concentrations of benzene were not significantly altered, whereas

TABLE 6. Vertical distribution of 2,3-dimethylpentane in groundwater in the spill area, and 9.1 m and 21.3 m downgradient of the spill.

Elevation (ft above MSL)[a]	In the spill area (mg/L)		9.1 m downgradient (mg/L)		21.3 m downgradient (mg/L)	
	After 136 weeks	After 156 weeks	After 136 weeks	After 153 weeks	After 136 weeks	After 153 weeks
587 - 584	0.082	0.070	<0.001	<0.001	<0.001	0.0021
584 - 581	0.079	0.069	0.0173	0.038	0.0118	0.0075
581 - 578	0.036	0.0178	0.0108	0.0074	0.0015	0.0019
578 - 575	0.0114	0.0073	0.0081	0.0057	<0.001	<0.001
575 - 572	<0.001	<0.001	<0.001	<0.001	<0.001	<0.001
572 - 569	<0.001	<0.001	<0.001	<0.001	<0.001	<0.001
569 - 566	<0.001	<0.001	<0.001	<0.001	<0.001	<0.001

(a) 1.0000 ft equals 0.3048 m.

concentrations of toluene were reduced 1 to 2 orders of magnitude. Attenuation of the other BTEX compounds was slightly less than 1 order of magnitude.

Bioattenuation corrected for dilution was converted to first-order rate constants by taking the natural logarithm, then dividing by the average residence time (Figure 7). Results of the calculations are presented in Table 8. The rates vary about threefold between cluster wells on the same sampling date, and between dates. The rate of bioattenuation contributed considerably less to variation in contaminant concentration than effects of groundwater flow.

DISCUSSION

Wilson et al. (1990) described the methanogenic bioattenuation of benzene, toluene, and xylenes (BTX) in groundwater contaminated by a spill of aviation gasoline. Attenuation of total BTX was measured quarterly over a 4-year period in wells along a flow path running through the centerline of the plume. Removal followed first-order kinetics, varying from 0.10 to 0.34 per week. When a purge well field went on line, one of the monitoring wells was isolated from the source area. Alkylbenzene concentrations dropped rapidly in the isolated well. Interestingly, toluene disappeared more than twice as rapidly as benzene. Shortly after the BTEX compounds disappeared, core material was acquired for a laboratory microcosm study. The kinetics of bioattenuation were very similar during flow along a flow path, in stagnant water near the isolated monitoring well, and in the microcosm study. The rate of toluene degradation was 0.3 to 1.3 per week, the xylenes degraded at 0.03 to 0.1 per week, and in the gasoline plume, benzene degraded at 0.05 to 0.17 per week.

TABLE 7. Bioattenuation or production of organic compounds in the plume of contaminated groundwater, corrected for attenuation due to dilution.

Elapsed time (weeks)	Position	Methane (mg/L)	Benzene (mg/L)	Toluene (mg/L)	Ethylbenzene (mg/L)	*p*-Xylene (mg/L)	*m*-Xylene (mg/L)	*o*-Xylene (mg/L)
136	Spill	0.98	0.230	5.3	1.7	2.1	2.5	2.3
	9.1 m	1.07	0.220	0.063	1.3	1.35	1.7	1.1
	21.3 m	14.7	0.403	0.281	0.410	0.60	0.85	0.55
153	Spill	1.74	0.253	8.2	1.5	1.4	3.4	2.4
	9.1 m	7.7	0.287	0.109	0.79	0.76	1.3	0.87
	21.3 m	13.0	0.410	0.025	0.62	0.79	1.5	0.97

TABLE 8. Apparent first-order rates of anaerobic transformation of monoaromatic compounds. The rates were calculated from the residence time of the water at the time of sampling and the attenuation in concentration after correction for dilution.

Elapsed time (weeks)	Position	Benzene (per week)	Toluene (per week)	Ethylbenzene (per week)	*p*-Xylene (per week)	*m*-Xylene (per week)	*o*-Xylene (per week)
136	9.1 m	+0.003	0.37	0.022	0.036	0.032	0.060
	21.3 m	-0.003	0.16	0.077	0.067	0.058	0.0078
153	9.1 m	-0.014	0.47	0.069	0.066	0.10	0.11
	21.3 m	-0.015	0.18	0.017	0.017	0.026	0.028

The independent field-scale estimates of bioattenuation agreed with each other within a factor of 3, and with the microcosm within a factor of 10. The groundwater flow velocity was known within a factor of 2.

Cozzarelli et al. (1990) reported the anaerobic bioattenuation of alkylbenzenes in a plume of contamination from a spill of crude oil in Minnesota. The groundwater was actively methanogenic and accumulated long-chain volatile fatty acids. They report bioattenuation in monitoring wells in a transect along the centerline of the plume. Toluene and *o*-xylene disappeared without lag. Both were depleted within 12 m of the source area. Ethylbenzene degradation began after the toluene and *o*-xylene had disappeared. Benzene depleted without a lag, but when toluene and *o*-xylene disappeared, the rate of benzene depletion slowed greatly. The average seepage velocity in the plume is 0.1 m per day (personal communication, Philip Bennett, University of Texas, Austin, Texas). This value can be used to express the attenuation between wells as a first-order rate constant. Depletion of benzene, toluene, *o*-xylene, and ethylbenzene would be 0.12, 0.50, 0.40 and 0.19 per week, respectively.

Bioattenuation of toluene, ethylbenzene, and the xylenes at the study site was consistent with that seen in other methanogenic aquifers contaminated with petroleum hydrocarbons. The rate constants across all three sites do not vary more than 1 order of magnitude. The agreement is remarkable, considering uncertainty introduced into these field-scale estimates from variation in groundwater flow and changes in plume geometry.

The failure of benzene to degrade at the Sleeping Bear site is inconsistent with the other two field studies, but is very consistent with a number of laboratory microcosms studies. The plume at Sleeping Bear was short (less than 30 m and the residence time was short (5 to 35 weeks). There may not have been adequate opportunity for anaerobic degradation of benzene.

RISK ASSESSMENT

The source area was 30 m wide. The plume will be considered to be 30 m wide. It was contained with in vertical interval of 6.4 m. The highest gradient toward the river was 3.0 m/1,000 m, resulting in a Darcy flow of 0.13 m per day. Multiplying depth by width, then by flow, the plume would contribute at a maximum 25 m^3 of water to the river each day. The discharge of the Platte when the stage is low is approximately 300,000 m^3 a day (personal communication, R. K. Popp, measurements taken by the Michigan Department of Natural Resources in 1990, 1991, and 1992). Under the worst conditions, the plume would be diluted at least 12,000-fold in the river.

Benzene is the compound of regulatory concern in the plume. When first monitored 30 weeks after removal of the tanks, the average concentration in the spill area was 2.4 mg/L. The average concentration of benzene in the spill area after 153 weeks was 0.25 mg/L.

After 153 weeks, assuming no bioattenuation of benzene as the plume moves from the spill area to the river, the expected concentration of benzene in the river due to the contribution of the plume would be 0.00002 mg/L. The drinking water standard for benzene in Michigan is 0.0006 mg/L. The analytical detection limit is near 0.0001 mg/L.

In addition to simple dilution, there is a possibility for aerobic biodegradation of benzene as the plume enters the aerobic benthic layer of the river. A plastic syringe with the end cut away was used to take plug samples of the river bottom (see Figure 4 for locations). A brown oxidized layer 1- to 4-cm thick was easily distinguished from anaerobic sediments deeper in the bed of the river. The samples were analyzed for BTEX compounds by purge and trap. Although benzene was detected in the anaerobic samples, no samples of the oxidized shallow sediment contained benzene above the detection limit of 0.010 mg/kg, corresponding to a concentration in the pore water of no more than 0.06 mg/L. There was at least a 4-fold reduction in benzene concentrations before the groundwater moved from the sediment to the water column.

REFERENCES

Cozzarelli, I. M., R. P. Eganhouse, and M. J. Baedecker. 1990. "Transformation of Monoaromatic Hydrocarbons to Organic Acids in Anoxic Groundwater Environment." *Environmental Geology and Water Science* 16(2):135-141.

Hadley, P. W., and R. Armstrong. 1991. "Where's the Benzene? Examining California Ground-Water Quality Surveys." *Ground Water* 29(1):35-40.

Wilson, B. H., J. T. Wilson, D. H. Kampbell, B. E. Bledsoe, and J. M. Armstrong. 1990. "Biotransformation of Monoaromatic and Chlorinated Hydrocarbons at an Aviation Gasoline Spill Site." *Geomicrobiology Journal* 8:225-240.

THE USE OF INTERNAL CHEMICAL INDICATORS IN PETROLEUM AND REFINED PRODUCTS TO EVALUATE THE EXTENT OF BIODEGRADATION

G. S. Douglas, R. C. Prince, E. L. Butler, and W. G. Steinhauer

ABSTRACT

An analytical approach based on internal petroleum biomarkers is used to quantitatively determine the extent of oil weathering from biological or physical processes. The triterpane C_{30} 17α(H),21β(H)-hopane is used as an internal biomarker to monitor the degradation of both specific petroleum compounds and total oil in crude oil, whereas for refined petroleum products that do not contain the hopane, C_4-phenanthrenes/anthracenes may be substituted. Provided all concentration data are presented on an oil-weight basis, the increase in the internal biomarker concentration relative to the concentration in the source oil is proportional to the amount of oil lost due to weathering and biodegradation. Field data indicate the use of an internal chemical indicator reduces spatial variability of oil data when compared to other mass balance approaches, allowing degradation to be monitored effectively and reducing the number of samples required to monitor the remediation effectiveness. Field data also demonstrate the utility and limitations of internal chemical indicators for fingerprinting fresh and weathered oils.

INTRODUCTION

The use of remediation techniques to treat petroleum-contaminated soils has increased steadily in the past decade (U.S. Congress 1991). Many new chemical, biological, and physical approaches have been developed to enhance hydrocarbon weathering rates in these soils. The effectiveness of remediation agents in soils may be dependent on a variety of environmental factors, including oxygen concentration, populations of indigenous bacteria, water content, temperature, nutrient concentration, and type of contamination (U.S. Congress 1991). Studies monitoring degradation often focus on the total oil remaining in soil determined by standard analytical methods and often are confounded by inhomogeneity of the contaminants in the affected soil. Studies relying on standard methods provide little data on the degradation of individual classes of compounds, data that may be crucial to characterize true effectiveness of the biodegradation processes occurring in the soil (Douglas et al. 1992).

A rigorous analytical method based on internal chemical markers is used to evaluate the relative efficiency of bioremediation agents in degraded field samples. The method relies on gravimetric methods, gas chromatography using flame ionization detection (GC/FID), and gas chromatography using mass spectrometry (GC/MS) to determine total oil or specific analyte concentrations. The internal chemical marker C_{30} 17α(H),21β(H)-hopane in crude petroleum is identified and quantified by GC/MS. The concentration of this chemical marker in degraded samples relative to their concentrations in original products is used to estimate analyte and total oil depletion in the weathered soils.

ANALYTICAL METHODS

Sediment Sample Extraction

A 30:50 g sediment/soil sample is mixed with a sufficient excess of sodium sulfate to dry the soil. The mixture is spiked with the appropriate surrogate compounds (Table 1) and serially extracted three times with 60 mL of methylene chloride using orbital shaker and sonication techniques [EPA Method 3550 (EPA 1986)]. The extracts are combined, dried with sodium sulfate, filtered through a 293-mm Gelman Type A/E glass-fiber filter, and concentrated to 1 mL using Kuderna-Danish and nitrogen solvent evaporation techniques. A small aliquot of the extract is dried and weighed to determine total methylene chloride extractables [analogous to an Oil and Grease measurement, EPA Method 413.1 (EPA 1983)]. The remaining extract, containing up to 300 mg of extractables, is passed through a 30-mm × 100-mm column containing 10 g of neutral alumina (activity grade I) following modified EPA Method 3611 (EPA 1986). This procedure removes polar compounds that interfere with the chromatographic analyses. The saturated (F_1) and unsaturated/aromatic (F_2) hydrocarbon fractions are eluted with 100 mL of methylene chloride. The combined F_1 and F_2 fraction is then concentrated using Kuderna-Danish and nitrogen solvent evaporation techniques to an appropriate pre-injection volume (PIV). An aliquot of the final extract is dried and weighed to determine total saturated and unsaturated/aromatic hydrocarbons. The remaining extract is spiked with appropriate quantitation internal standards (Table 1) and analyzed by GC/FID and GC/MS.

Water Sample Extraction

A 1-L water sample is spiked with appropriate surrogate compounds (Table 1) and serially extracted with methylene chloride [EPA Method 3510 (EPA 1986)]. The combined extracts are dried with sodium sulfate and concentrated as above. The final extract is spiked with the appropriate quantitation internal standards (Table 1), and analyzed by GC/FID and GC/MS. The alumina column cleanup and gravimetric procedure is generally not required for most water samples.

TABLE 1. Target analyte list includes: PAH and PHC target analytes, PAH and PHC surrogate and quantitation internal standards, primary and secondary ions used for GC/MS-SIM analysis, and method reporting limits.

Compound	Abbreviation	#Rings	Primary Ion (M/z)	Secondary Ion (M/z)
Polycyclic Aromatic Hydrocarbons (PAH)				
naphthalene[a]	N	2	128	127
C_1-naphthalenes	N1	2	142	141
C_2-naphthalenes	N2	2	156	141
C_3-naphthalenes	N3	2	170	155
C_4-naphthalenes	N4	2	184	169
biphenyl	BI	2	154	152
acenaphthylene[a]	AE	3	152	153
dibenzofuran	DI	3	168	169
acenaphthene[a]	AC	3	154	153
fluorene[a]	F	3	166	165
C_1-fluorenes	F1	3	180	165
C_2-fluorenes	F2	3	194	179
C_3-fluorenes	F3	3	208	193
anthracene[a]	A	3	178	176
phenanthrene[a]	P	3	178	176
C_1-phenanthrenes/anthracenes	P1	3	192	191
C_2-phenanthrenes/anthracenes	P2	3	206	191
C_3-phenanthrenes/anthracenes	P3	3	220	205
C_4-phenanthrenes/anthracenes	P4	3	234	219
dibenzothiophene	D	3	184	152
C_1-dibenzothiophenes	D1	3	198	184
C_2-dibenzothiophenes	D2	3	212	197
C_3-dibenzothiophenes	D3	3	226	211
fluoranthene[a]	FL	4	202	101
pyrene[a]	PY	4	202	101
C_1-fluoranthenes/pyrenes	FP1	4	216	215
benz[*a*]anthracene[a]	B	4	228	226
chrysene[a]	C	4	228	226
C_1-chrysenes	C1	4	242	241
C_2-chrysenes	C2	4	256	241
C_3-chrysenes	C3	4	270	256
C_4-chrysenes	C4	4	284	269
benzo[*b*]fluoranthene[a]	BB	5	252	253
benzo[*k*]fluoranthene[a]	BK	5	252	253
benzo[*e*]pyrene	BE	5	252	253
benzo[*a*]pyrene[a]	BA	5	252	253
perylene	PER	5	252	253
indeno[*1,2,3-c,d*]pyrene[a]	IP	6	276	277
dibenz[*a,h*]anthracene[a]	DA	5	278	279
benzo[*g,h,i*]perylene[a]	BP	6	276	277
Biomarker Compounds				
diterpanes	DI	NA[b]	191	NA
hopanes (i.e. C_{30} 17α(H)21β(H)-hopane)	TRI ($\alpha\beta$)	NA	191	NA
steranes	STER	NA	217	NA

TABLE 1. (continued)

Compound	Abbreviation	#Rings	Primary Ion (M/z)	Secondary Ion (M/z)
PAH Surrogate and Quantitation Internal Standards				
naphthalene-d_8 (SIS)	Nd_8	2	136	134
fluorene-d_{10} (SIS)	Fd_{10}	3	176	174
chrysene-d_{12} (SIS)	Cd_{12}	4	240	236
acenaphthene-d_{10} (QIS)	Ad_{10}	3	164	162
phenanthrene-d_{10} (QIS)	Pd_{10}	3	188	184
benzo[*a*]pyrene-d_{12} (QIS)	$BAPd_{12}$	5	264	260
C_{30} 17β(H)21β(H)-hopane (QIS)	$\beta\beta$	NA	191	NA
Normal (linear) Alkanes				
n-C_7 - n-C_{36}[c]	NA	0	57	NA
Isoprenoid Hydrocarbons				
pristane[c]	NA	0	55	NA
phytane[c]	NA	0	55	NA
1380[c]	NA	0	55	NA
1470 (farnesane)[c]	NA	0	55	NA
1650[c]	NA	0	55	NA
PHC Surrogate and Quantitation Internal Standards				
ortho-terphenyl[c] (SIS)	NA	0	230	NA
5α-androstane[c] (QIS)	NA	0	245	NA
QA/QC Standards				
crude oil control samples				
Reporting Limits (waters)				
alkanes, 0.2 μg/L				
PHC, 50 μg/L				
PAH, 10 ng/L				
Reporting Limits (sediments)				
alkanes, 0.1 mg/kg				
PHC, 10 mg/kg				
PAH, 0.001 mg/kg				
Reporting Limits (oils)				
alkanes, 20 mg/kg				
PHC, 80,000 mg/kg				
PAH, 5 mg/kg				

[a] Target analyte is a priority pollutant PAH.
[b] Not applicable
[c] Target analyte concentration determined by GC/FID analysis (GC/MS M/z is generic for saturated hydrocarbon type).

Oil Sample Processing

Oil samples are diluted to 5 mg/mL in methylene chloride and spiked with appropriate surrogate compounds (Table 1). The diluted oil is then processed through an alumina column cleanup procedure (described above) to remove polar compounds. The combined F_1 and F_2 fraction is concentrated using Kuderna-Danish and nitrogen solvent evaporation techniques to the appropriate pre-injection volume. An aliquot of the combined hydrocarbon fraction (F_1 + F_2) is dried and weighed to determine total gravimetric petroleum hydrocarbons. The remaining extract is spiked with appropriate quantitation internal standard compounds (Table 1) and analyzed by GC/FID and GC/MS.

Individual and Total Petroleum Hydrocarbons by GC/FID – Modified EPA Method 8100

Analysis of the oil, sediment, and water samples for total petroleum hydrocarbons (PHC), individual C_7 to C_{36} *n*-alkanes, and isoprenoid hydrocarbons (Table 1), is performed by GC/FID. A 2-µL aliquot of the sample extract is injected into a gas chromatograph equipped with a split/splitless injection port operated in the splitless mode. A high-resolution capillary column (J&W fused-silica DB-5 column, 30 meters, 0.32-mm internal diameter, and 0.25-µm film thickness) and optimal temperature program is employed to achieve near-baseline separation of all of the saturated hydrocarbons listed in Table 1.

The GC/FID conditions are:

Initial column temperature–35°C
Initial hold time–10 minutes
Oven temperature program rate–3°C/minute
Final oven temperature–320°C
Final hold time–10 minutes
Injection port temperature–275°C
Detector temperature–325°C
Column flowrate (hydrogen)–1 mL/minute.

Prior to sample analysis, a five-point calibration is established to demonstrate the linear range of the analysis and to determine the relative response factors (RRFs) for individual compounds [EPA Method 8100 (EPA 1986)]. The calibration solution contains C_7 through C_{36} *n*-alkanes as well as the isoprenoid hydrocarbons pristane and phytane. A midlevel calibration verification sample is analyzed every 10 samples and must meet the calibration criteria. Check standards of suspected source oil or reference oil are analyzed with every analytical batch (approximately 15 samples) to provide an additional level of quality assurance (Douglas et al. 1992). Methylene chloride blanks are analyzed every 10 samples to evaluate baseline drift. The RRF of *n*-C_{36} and *n*-C_{21} are monitored to minimize mass discrimination in the GC/FID analysis. In order to proceed with the sample analysis, the ratio of *n*-C_{36} RRF to *n*-C_{21} RRF in the calibration verification sample analysis must be greater than 0.80.

Quantitation of the individual hydrocarbons in samples is performed by the method of internal standards using the RRF for the individual hydrocarbons relative to the quantitation internal standard, 5α-androstane. Total PHC (defined operationally as the resolved plus unresolved hydrocarbons eluting between the C_7 and C_{36} *n*-alkanes) is also quantified by the method of internal standards. The mean RRF of all hydrocarbon compounds in the calibration standard serves as the RRF for PHC. The total baseline-corrected area of resolved plus unresolved compounds determined over the entire analytical range corrected for baseline drift and internal standard area response is used for the PHC response (Figure 1a). Reporting limits for this method are listed in Table 1. All alkane and PHC data are corrected for the recovery of ortho-terphenyl.

Polycyclic Aromatic Hydrocarbons and C_{30} 17α(H),21β(H)-hopane Analysis by GC/MS-SIM — Modified EPA Method 8270

Analysis of target PAH compounds, including the EPA priority pollutant PAHs, petroleum-specific alkylated PAHs, selected heterocyclic PAHs, and polycyclic aliphatic hydrocarbons (Table 1) is performed by GC/MS (Hewlett Packard HP5970B GC/MS). The mass spectrometer is operated in the selected ion monitoring mode (SIM) to increase sensitivity and selectivity relative to full-scan operation. A 2-µL aliquot of the sample extract is injected into a GC equipped with a split/splitless injection port operated in the splitless mode. A high-resolution capillary column (J&W fused-silica DB-5MS column, 30 m, with a 0.25-mm internal diameter and 0.25-µm film thickness) and optimal temperature program is used to achieve near-baseline separation of all the PAHs listed in Table 1.

The GC conditions are:

Initial column temperature – 40°C
Initial hold time – 1 minute
Oven temperature program rate – 6°C/minute
Final oven temperature – 290°C
Final hold time – 20 minutes
Injection port temperature – 300°C
Detector temperature – 280°C
Column flowrate (helium)–1 mL/minute.

The mass spectrometer conditions are:

Ionization mode – positive ion, 70 eV
Source pressure – 1 to 5 × 10^{-5} Torr
Interface temperature – 280°C
Interface – capillary direct

Prior to sample analysis, a five-point calibration is established to demonstrate the linear range of the analysis and to determine the mean RRF for individual

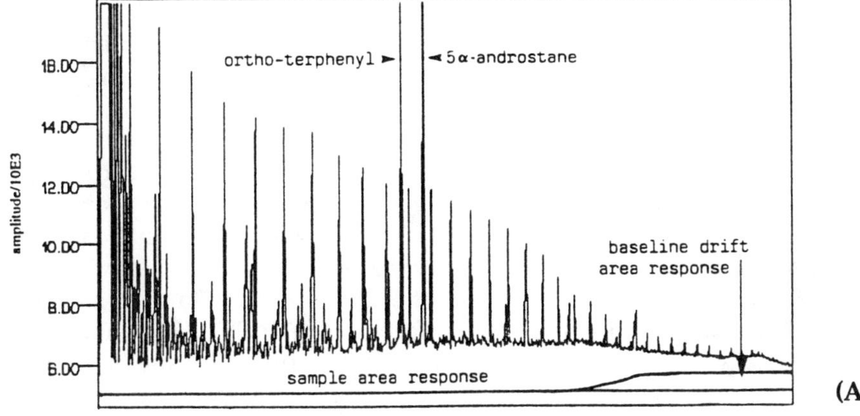

(A)

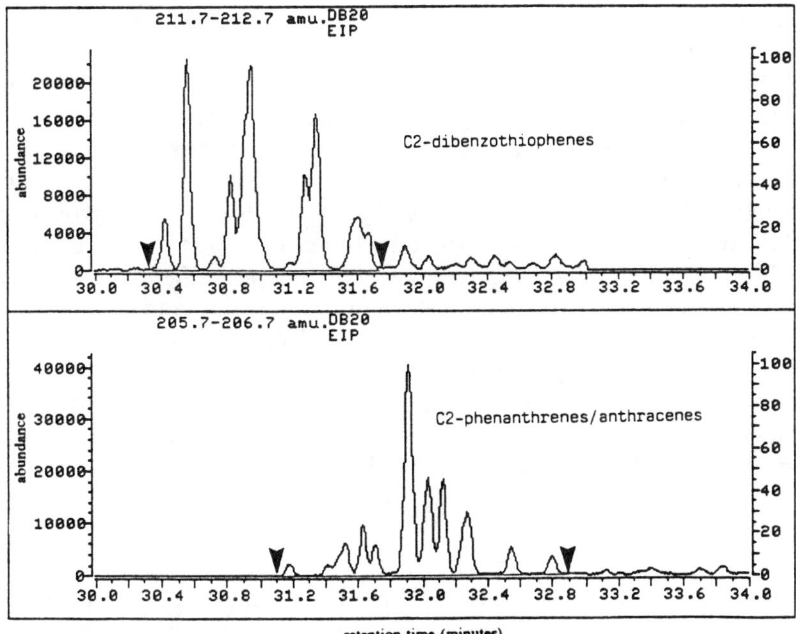

(B)

FIGURE 1. A. GC/FID chromatogram of the source crude oil identifying the baseline integration method. The baseline drift area was determined from a methylene chloride blank that was analyzed every 10 samples. B. GC/MS-extracted ion profiles of C_2-dibenzothiophenes and C_2-phenanthrenes/anthracenes demonstrating the method of straight baseline integration used to quantify the alkylated homologues. The baseline was drawn between the two arrows. Compounds outside the arrows were not selected because the appropriate confirmation ions were not present in the initial full-scan GC/MS analysis of the standard oil.

compounds [EPA Method 8270 (EPA 1986)]. The calibration solution contains the 16 priority pollutant PAHs, dibenzothiophene, biphenyl, dibenzofuran, perylene, benzo[*e*]pyrene, C_{30} 17β(H),21α(H)-hopane, and the appropriate internal standards listed in Table 1. A midlevel calibration verification sample is analyzed every 10 samples and must meet the required calibration criteria. Check standards of suspected source oil or reference oil are analyzed with each analytical batch (approximately 15 field samples and 5 quality control samples) to provide an additional level of quality assurance (Douglas et al. 1992) and to provide batch-specific analyte data.

All target analytes are quantified by the method of internal standards using the RRF for the individual unsubstituted PAHs relative to four quantitation internal standards (Table 1). Due to the fact that C_{30} 17α(H),21β(H)-hopane is not commercially available, the RRF for C_{30} 17β(H),21α(H)-hopane is determined relative to the internal standard C_{30} 17β(H),21β(H)-hopane (a triterpane not commonly detected in petroleum or its refined products) and used to quantify C_{30} 17α(H),21β(H)-hopane in the sample. The retention time of C_{30} 17α(H),21β(H)-hopane is easily determined by pattern recognition of the triterpane pattern in the crude oil control sample (Butler et al. 1991, Peters & Moldowan 1993). In addition to monitoring the molecular ion for C_{30} 17α(H),21β(H)-hopane (m/*z* 191), the principal biodegradation products (25-nor-hopane m/*z* 171, Peters & Moldowan 1991) are monitored to detect hopane degradation.

Each multicomponent alkylated PAH homologous series is defined by the total appropriate m/*z* response within a defined retention time window, and the characteristic pattern for each series of isomers. Each isomer grouping is defined by analyzing a crude oil sample by full-scan GC/MS prior to GC/MS SIM analysis. Once the homologous group has been identified by the analyst, the area response is determined by a straight baseline integration (Figure 1b). The RRF of the respective unsubstituted parent PAH compound serves as the RRF for all alkylated homologues in a given series (i.e., the RRF for naphthalene is used to quantify all of the C_1-, C_2-, C_3-, and C_4-naphthalenes). Reporting limits for this method are listed in Table 1. All PAH and biomarker data are corrected for the recovery of the surrogate fluorene-d_{10}.

Analytical Quality Control

The reliability of the above analytical methods is dependent on the quality control procedures followed. The relative standard deviation of the RRFs for the analytes in the initial five-point calibration should not exceed 25%, and the calibration verification standard percent difference from the initial five-point calibration must not exceed 20%. A control chart of the standard oil should be prepared and monitored. Variations of analytes in the control chart should be no more than 25% from historical averages. RRF stability for the PHC analysis is a key factor in maintaining the quality of the analysis. RRFs for *n*-C_7 through *n*-C_{36}, pristane, and phytane should be 0.90 ±0.1 relative to the internal standard 5α-androstane. Mass discrimination (i.e., the loss of high-molecular-weight compounds) must be carefully monitored; the ratio of RRF *n*-C_{36}/RRF *n*-C_{21} alkanes should not be allowed to fall below 80%.

With each analytical batch (approximately 15 samples), one procedural blank, two matrix spike samples, one duplicate, and one standard oil are analyzed. Procedural blanks should contain no analytes at greater than three times the target compound method detection limit, surrogate and matrix spike recoveries for the PAH and PHC analyses should be within 50 to 120%, and duplicate relative percent difference values should be ±25%. The relative percent difference for the matrix spike and matrix spike duplicate samples should be less than 25%. When the data fall within the above limits, then the surrogate-corrected results provide reliable estimates of analyte concentration (e.g., ± 5% precision for C_{30} 17α(H),21β(H)-hopane).

DISCUSSION

The detailed chemical evaluation of oil degradation in environmental samples requires that a conservative (e.g., compound that is not degraded in the oil), source-specific reference compound be identified. If the selected compound is the most or nearly the most conservative chemical species in the oil, then the increase in indicator concentration with time relative to the concentration in the initial source oil is a measure of the amount of oil degraded. This type of data analysis requires that the analyte data be reported on an oil weight basis (Butler et al. 1991, Douglas et al. 1992, Prince et al. 1994) so that samples with different oil loadings may be compared.

The application of an internal chemical indicator to monitor oil weathering or degradation is dependent on the following assumptions: (1) the source of the oil contamination has been identified and is primarily a single source, (2) the chemical indicator is not formed during weathering or biodegradation (Prince et al. 1993), (3) the indicator is not degraded during the weathering process (Prince et al. 1993), and (4) the extraction efficiency of the chemical marker is the same as the rest of the oil. If we assume some indicator degradation with time, then the amount of oil degraded estimated by this method represents a minimum value (i.e., more degradation may be occurring). The percent depletion of the oil can be estimated by the following equation (Douglas et al. 1992):

$$\text{percent Total Oil Depletion} = (1 - H_0/H_1) \times 100$$

where H_1= conservative analyte concentration in the degraded oil
H_0= conservative analyte concentration in the source oil

Similarly, individual analyte depletion is estimated by the following equation:

$$\text{percent Analyte Depletion} = [1 - (C_1/C_0) \times (H_0/H_1)] \times 100$$

where C_1= target analyte concentration in the degraded oil
C_0= target analyte concentration in the source oil

For crude oil, Prince et al. (1990), Butler et al. (1991), Prince et al. (1994), and Bragg et al. (1993) found that the triterpane, C_{30} 17α(H),21β(H)-hopane is an ideal reference compound for biodegradation monitoring. Triterpanes, in general, are resistant to biological and chemical degradation (Kennicutt 1988; Peters & Moldowan 1991, 1993) on time scales associated with recent oil spills (Prince et al. 1993). In addition, triterpane distributions in crude oil are characteristic of the source formation and thus provide an excellent fingerprint of the oil (Peters & Moldowan 1993). The triterpane C_{30} 17α(H),21β(H)-hopane is relatively abundant and easily identified in most crude oils, and using the GC/MS-SIM analysis there are few, if any, analytical interferences with this compound (Butler et al. 1991).

Triterpanes are not found in measurable concentrations in mid-range refined petroleum products, such as diesel fuel and fuel oil #2, because they are lost during the distillation of the product. Based on compounds measured in these products, Douglas et al. (1992) found that C_4-phenanthrenes/anthracenes were the most biodegradation-resistant class of compounds measured and can be substituted for C_{30} 17α(H),21β(H)-hopane in the above equation. Data from biodegradation studies of crude oil demonstrated that C_4-phenanthrenes/anthracenes were degraded slowly, but at a greater rate than C_{30} 17α(H),21β(H)-hopane. Because we can verify that these compounds are being degraded, percent oil loss and percent analyte depletion calculated from C_4-phenanthrenes/anthracenes represent conservative estimates of actual oil depletion. For historical studies of crude oil degradation where hopane was not measured and alkylated PAHs were, C_3-chrysenes will provide percent depletion results that are similar to those derived from hopane-based values.

Case Study

As an example of the utility of the proposed approach for evaluating oil and compound-specific degradation, we present the analysis of a sediment sample from a major oil spill that was collected from an impacted beach more than 1 year after the spill. The full data set is described elsewhere in detail (Bragg et al. 1993, Prince et al. 1994, Prince et al. 1990). In this paper we present the aliphatic and PAH data and calculated percent depletion for each target compound (Tables 2 and 3). Figure 2a suggests that the percent depletion of individual alkanes decreases with increasing chain length (e.g., n-C_{19} versus n-C_{34}) and increasing degree of branching (e.g., phytane versus n-C_{18}). The calculation of analyte depletion is based on the individual analyte/hopane ratios within the source and degraded oil in the sediment, and is therefore independent of the units. However, the calculation of percent depletion of total oil is dependent on the concentration of the marker in the residual oil, and, therefore, the total extractable oil in the sample must be carefully measured.

Three separate determinations of extractable oil were made in this study:

1. Total methylene chloride extractables (PREOIL on Figure 2a) corresponding to a modified EPA 413.1 oil and grease gravimetric measurement.

TABLE 2. Target aliphatic hydrocarbons: Concentrations and percent depletion of alkanes in the degraded crude oil and source crude oil. Concentration units are reported as milligrams of alkane per kilogram (mg/kg) of PREOIL.

Aliphatic Hydrocarbons	Units:	Degraded Crude[a] Oil mg/kg Oil	Source Crude Oil mg/kg Oil	Percent Depleted
n-C_7		NM[b]	NM	NA[c]
n-C_8		NM	NM	NA
n-C_9		NM	NM	NA
n-C_{10}		ND	3680	100
n-C_{11}		ND	3760	100
n-C_{12}		ND	3950	100
n-C_{13}		ND	3300	100
isoprenoid 1380		NM	NM	NA
n-C_{14}		130	3870	98
isoprenoid 1470 (farnesane)		NM	NM	NA
n-C_{15}		209	3540	97
n-C_{16}		186	2810	97
isoprenoid 1650		NM	NM	NA
n-C_{17}		370	3650	95
pristane		1030	1750	72
n-C_{18}		230	2520	96
phytane		828	1110	64
n-C_{19}		177	2290	96
n-C_{20}		166	2210	96
n-C_{21}		230	2100	95
n-C_{22}		303	2030	93
n-C_{23}		284	1910	93
n-C_{24}		306	1800	92
n-C_{25}		219	1560	93
n-C_{26}		202	1350	93
n-C_{27}		198	988	90
n-C_{28}		97	702	93
n-C_{29}		92	590	93
n-C_{30}		188	546	84
n-C_{31}		320	527	71
n-C_{32}		180	273	69
n-C_{33}		202	240	60
n-C_{34}		290	323	57
n-C_{35}		NM	NM	NA
n-C_{36}		NM	NM	NA
PHC[d]		519000	606000	59
PREOIL[e]				52
POSTOIL[f]				60

[a]Concentration based on pre-alumina column oil weight (PREOIL).
[b]Not measured
[c]Not detected
[d]GC/FID detectable hydrocarbons.
[e]Total methylene chloride extractables - gravimetric analysis.
[f]Post alumina column oil weight - gravimetric analysis.

TABLE 3. Target polycyclic aromatic hydrocarbons: Reports concentrations and percent depletion of PAHs, dibenzothiophenes, and C_{30} 17α(H),21β(H)-hopane in the degraded crude oil and the source crude oil. Concentrations are reported as milligrams of PAH per kilogram of PREOIL. The concentration of 17α(H),21β(H)-hopane is reported based on PREOIL, POSTOIL, and PHC oil weights to calculate total oil depletion using the three different values.

Polycyclic Aromatic Hydrocarbons	Units:	Degraded[a] Crude Oil mg/kg Oil	Source[a] Crude Oil mg/kg Oil	Percent Depleted
naphthalene		ND[b]	631	100
C_1-naphthalenes		ND	1450	100
C_2-naphthalenes		13.1	1930	100
C_3-naphthalenes		52.3	1370	98
C_4-naphthalenes		385	876	79
biphenyl		NM[c]	NM	NA[d]
acenaphthylene		ND	ND	NA
dibenzofuran		NM	NM	NA
acenaphthene		ND	ND	NA
fluorene		ND	94.9	100
C_1-fluorenes		ND	249	100
C_2-fluorenes		193	430	79
C_3-fluorenes		372	442	60
phenanthrene		ND	255	100
anthracene		ND	ND	NA
C_1-phenanthrenes/anthracenes		25.8	625	98
C_2-phenanthrenes/anthracenes		377	799	78
C_3-phenanthrenes/anthracenes		637	659	54
C_4-phenanthrenes/anthracenes		402	359	47
dibenzothiophene		ND	217	100
C_1-dibenzothiophenes		27.9	388	97
C_2-dibenzothiophenes		452	736	71
C_3-dibenzothiophenes		740	710	50
fluoranthene		ND	3.22	100
pyrene		10.9	8.07	36
C_1-fluoranthenes/pyrenes		99.0	88.0	46
benz[*a*]anthracene		6.76	ND	NA
chrysene		66.3	39.5	20
C_1-chrysenes		134	81.8	22
C_2-chrysenes		222	123	14
C_3-chrysenes		169	102	21
C_4-chrysenes		ND	ND	NA
benzo[*b*]fluoranthene		8.70	4.48	8
benzo[*k*]fluoranthene		ND	ND	NA
benzo[*e*]pyrene		NM	NM	NA
benzo[*a*]pyrene		ND	ND	NA
perylene		NM	NM	NA
indeno[1,2,3-*cd*]pyrene		ND	ND	NA
dibenzo[*ah*]anthracene		ND	ND	NA
benzo[*g,h,i*]perylene		6.79	3.33	3
Total PAH		4400	12700	83
C_{30} 17α(H) 21β(H) - hopane - PREOIL		453	216	
C_{30} 17α(H) 21β(H) - hopane - POSTOIL		640	256	
C_{30} 17α(H) 21β(H) - hopane - PHC		869	364	

[a]Based on PREOIL weight
[b]Not detectable
[c]Not measured
[d]Not applicable

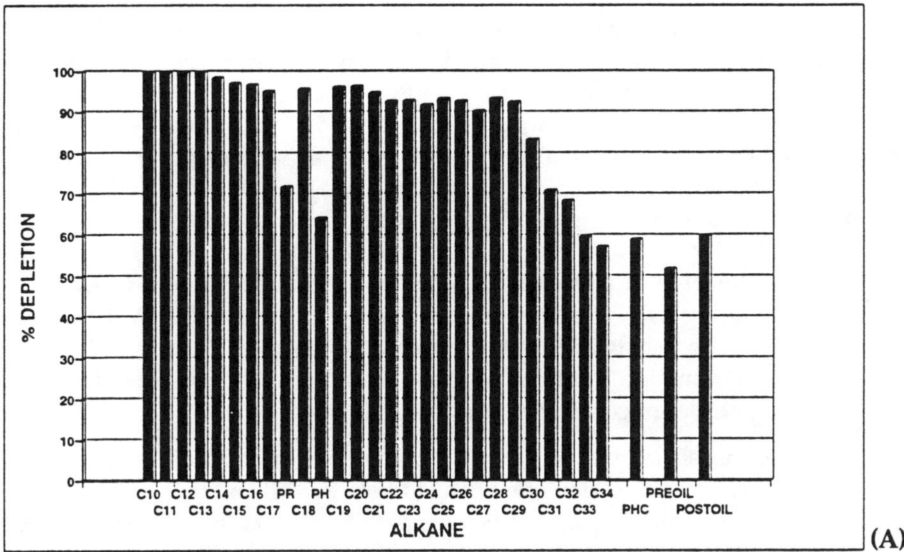

(A)

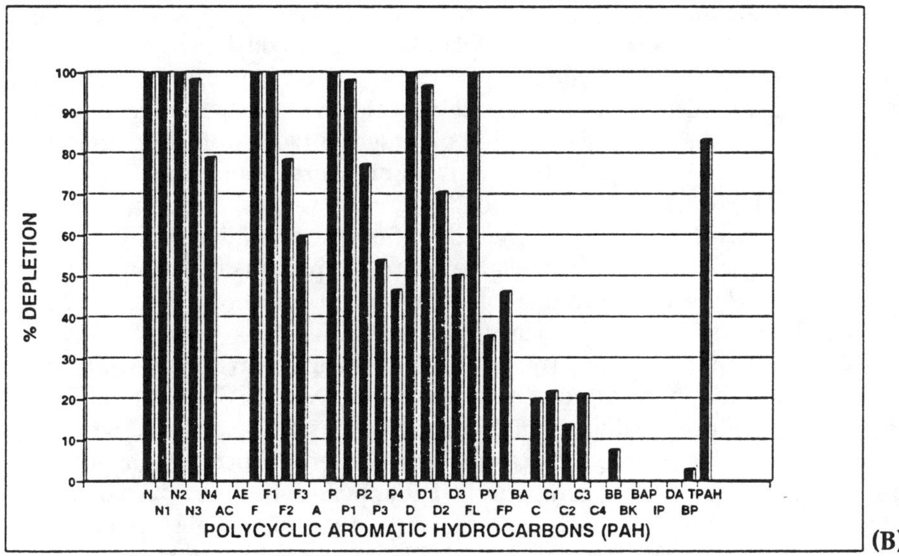

(B)

FIGURE 2. A. Alkane and total oil percent depletion plot for the degraded crude oil. Total oil depletion was estimated three ways using the PREOIL, POSTOIL, and PHC C_{30} 17α(H),21β(H)-hopane data. B. PAH and total PAH percent depletion plot for the degraded crude oil.

This extract contains saturated and aromatic hydrocarbons, asphaltenes, and polar compounds. It is likely that this fraction contains all of the oil in the sample but also likely includes methylene chloride-extracted biomass and other biogenic compounds.

2. Post-alumina column oil weight (POSTOIL on Figure 2a) corresponding to a modified EPA 418.2 total petroleum hydrocarbon gravimetric measurement. This extract contains saturated and aromatic hydrocarbons, perhaps some asphaltenes, and probably relatively little extracted biomass.
3. GC/FID-detectable hydrocarbons in the post-alumina column fraction (PHC in Figure 2a). This measurement also represents saturated and aromatic hydrocarbons and asphaltenes eluting from n-C_7 through n-C_{36}. Gross contamination from any biomass would probably be detectable by a nonpetroleum chromatographic signature.

All three measurements of percent oil depletion produced similar results for this sample; percent oil depletion ranged between 50 and 60%. The percent oil depletion based on PHC and POSTOIL data (based on the concentration of C_{30} 17α(H),21β(H)-hopane in oil) were almost identical, whereas the PREOIL data resulted in slightly lower percent oil depletion. This presumably results from an increase in relative amount of the polar compounds in the degraded oil compared to the source oil. In general, the use of total methylene chloride extractable weight data (PREOIL) for the calculation of percent oil depletion will underestimate the percent oil depletion and the greater the amount of polar compound material (e.g., biogenic compounds), the greater will be the underestimation. We have found that heavily degraded oils such as those from biodegradation flask studies (Elmendorf et al. 1994) or from biodegradation agent field testing (Prince et al. 1990, Prince et al. 1994) tend to have more polar compound material than the source oil prior to biodegradation. Some remediation agents also may contain methylene chloride-extractable polar components (e.g., Inipol) which would also lead to an underestimation of percent oil depletion if the calculations are based on total extractable measurements.

We have found that percent oil depletion calculated from gravimetric PREOIL measurements may also be underestimated (compared to POSTOIL or PHC basis) for high asphaltic or weathered oils. This may occur because as oil degradation proceeds, the relative amount of asphaltenes generally increases (asphaltenes are high-molecular-weight compounds that appear to be resistant to biodegradation) compared to the source material. For low asphaltic oils such as diesel and light crude oils, the chromatographic PHC data appear to be more accurate than POSTOIL data for estimating product degradation because losses of the light-end hydrocarbons during the POSTOIL gravimetric analysis bias these data. For highly asphaltic and degraded (e.g., loss of volatile compounds) crude oils, the POSTOIL gravimetric data may provide a more accurate measure of total oil than the chromatographic PHC data because the asphaltenes may not be measured by GC/FID. The agreement in percent depletion for the POSTOIL and PHC measurements suggests that the crude oil used in this study has a relatively low asphaltene content and either one could be used for this oil. We have found that it is advantageous to collect total oil data from all three methods and calculate the percent oil depletion from each. In this manner, additional information that may be valuable for interpretation (e.g., increase in percent of polar compounds versus percent depletion, or substantial influence from biomass) may be gained at little additional cost.

Prince et al. (1993) and Elmendorf et al. (1994) performed a series of laboratory studies to evaluate crude oil biodegradation in a closed system. The use of a closed system prevents compound losses due to differential compound solubilities. They concluded that the approximate preference for microbial degradation was *n*-alkanes and naphthalene > pristane > phytane > fluorene > dibenzothiophene > phenanthrene > chrysene > hopane. Prince et al. (1993) also observed that alkylated substituted forms were more resistant to biodegradation than the unsubstituted parent compounds. Figure 2b contains the calculated percent depletion of individual and alkylated PAH for the petroleum-contaminated sediment sample mentioned above. Due to the fact that many of the unsubstituted compounds are 100% depleted in these samples, the relative rates of depletion for naphthalene, fluorene, phenanthrene, and dibenzothiophene cannot be directly determined. However, because the parent/alkylated PAH percent depletion relationship is clearly demonstrated by the phenanthrene data (percent depletion of parent > P1 > P2 > P3 > P4), and the relationship of ring size to percent depletion is also demonstrated (percent depletion of N3 > F3 > P3 = D3 > C3), then one can infer from the data that the percent depletion for the parent compounds follows the same pattern as the alkylated PAH compounds with the same number of aromatic rings (N > F > P = D > C > hopane). This relationship has been confirmed in samples from the same oil spill that were not as degraded and had measurable concentrations of parent PAH compounds (N, F, P, D, C).

An understanding of the biodegradation characteristics of hydrocarbons in the environment will assist in the selection of source-specific diagnostic compounds that can be used to fingerprint the source and to evaluate the fate and transport of the spilled oil in the environment. An acceptable oil source indicator must (1) provide source-specific chemical information, (2) be resistant to the effects of weathering and biodegradation, (3) remain associated with the bulk oil, and (4) be extracted from the weathered and aged soils/sediments in a similar fashion as the other associated compounds (including total oil). The dibenzothiophene and alkylated dibenzothiophene distributions provide source-specific chemical information in oil (Overton et al. 1981, Page et al. 1993, Bence & Burns 1993). Figure 2a clearly demonstrates that these individual compounds are degraded in the environment and therefore their effectiveness as unique source indicators diminishes with time. However, by normalizing the concentration of these diagnostic compounds with other compounds in the oil that have similar physical/chemical/biological weathering properties, the effects of oil degradation on the utility of the source indicator are minimized. For example, the similarity in the depletion rates for the C_2- and C_3-phenanthrenes/anthracenes (P_2 and P_3) and C_2- and C_3-dibenzothiophenes (D_2 and D_3) can be generalized. This is an important observation because the D_2:P_2 and D_3:P_3, important source oil ratios

(Overton et al, 1981, Page et al. 1993, Bence & Burns et al. 1993), may also be used to characterize weathered oils.

Relative differences in target compound degradation rates can be used to construct a series of oil weathering ratios of varying sensitivity. For example, the ratio of C_3-naphthalenes/C_3-chrysenes is useful to monitor light to moderate crude oil degradation, whereas the ratio of C_3-dibenzothiophenes/C_3-chrysenes has more utility for moderate to extensive oil degradation. When a weathering ratio (e.g., C_3-dibenzothiophenes/C_3-chrysenes) is plotted versus a source oil indicator (C_3-dibenzothiophenes/C_3-phenanthrenes) the stability of the source oil indicator over the oil degradation range can be evaluated. Figure 3 is a plot of C_3-dibenzothiophenes/C_3-chrysenes versus C_3-dibenzothiophenes/C_3-phenanthrenes for fresh and degraded crude oil samples collected from three separate oil spills. This plot demonstrates the stability of the oil source ratios over a wide range of weathering, biodegradation and oil types making these ratios critical analytical tools for hydrocarbon fingerprinting studies.

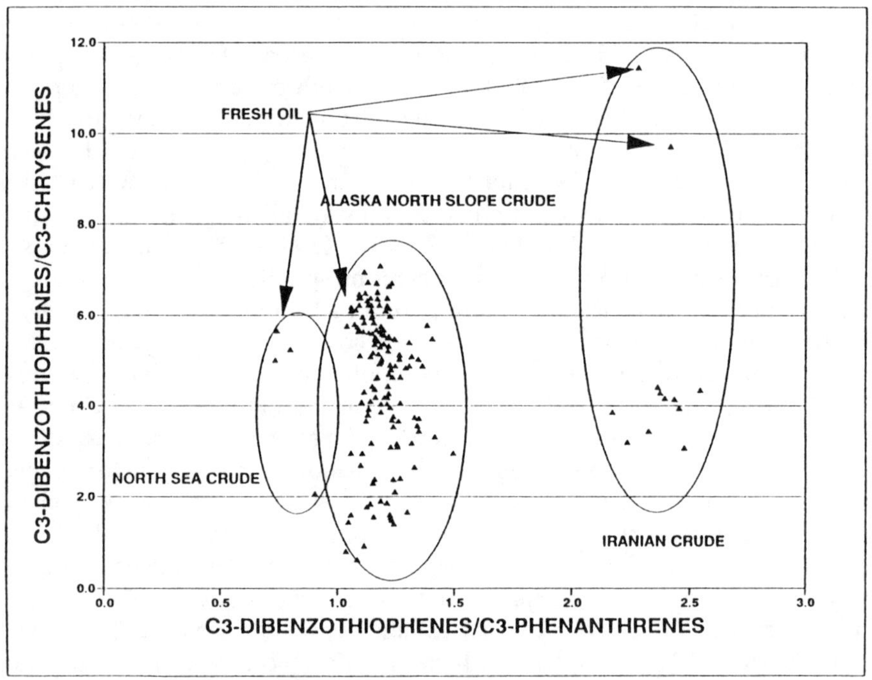

FIGURE 3. A plot of C_3-dibenzothiophenes/C_3-chrysenes (weathering ratio) versus C_3-dibenzothiophenes/C_3-phenanthrenes (oil source ratio) for fresh and degraded crude oil samples collected from three separate oil spills. This plot demonstrates the stability of the oil source ratio over a wide range of weathering and biodegradation.

CONCLUSIONS

An analytical method has been presented to isolate and quantify total petroleum hydrocarbons, alkanes, PAHs, and the triterpane C_{30} 17α(H),21β(H)-hopane in soil, water, and oil samples. An interpretive approach has been described to evaluate oil and analyte depletion based on the relative increase on an oil weight basis of the refractory compound C_{30} 17α(H),21β(H)-hopane in the sample. The importance of the method used to quantify the total oil concentration in the sample has been reviewed. This approach has been applied to both field (Bragg et al. 1993, Prince et al. 1990, Prince 1993, 1994) and laboratory biodegradation studies (Elmendorf et al. 1994) and has been demonstrated to reduce the inherent variability associated with traditional quantitative approaches that are often used to evaluate oiled environmental samples. The use of this analytical method can provide valuable quantitative information that can be used to monitor the degradation of petroleum in highly variable environments and enhance our understanding of the rates and mechanisms responsible for oil biodegradation. Finally, an improved understanding of the degradation properties of potential source indicator compounds in petroleum and its refined products will improve our ability to develop new hydrocarbon fingerprinting tools that will provide unique source information even after the sample has been extensively degraded.

REFERENCES

Bence, A. E., and Burns, W. A. 1993. "Fingerprinting Hydrocarbons in the Biological Resources of the *Exxon Valdez* Spill Area." Presented at the Third Symposium on Environmental Toxicology and Risk Assessment, Atlanta, GA, April 26-29, 1993.

Bragg, J. R., R. C. Prince, E. J. Harner, and R. M. Atlas. 1993. "Bioremediation Effectiveness Following the *Exxon Valdez* Spill." 1993 International Oil Spill Conference, Tampa, FL, March 29 - April 1, 1993.

Butler, E. L., G. S. Douglas, W. G. Steinhauer, R. C. Prince, T. Aczel, C. S. Hsu, M. T. Bronson, J. R. Clark, and J. E. Lindstrom. 1991. "Hopane, a New Chemical Tool for Measuring Oil Biodegradation." In R. E. Hinchee and R. F. Olfenbuttel (Eds.), *On-Site Bioreclamation*, pp. 515-521. Butterworth-Heinemann, Stoneham, MA.

Douglas, G. S., K. J. McCarthy, D. T. Dahlen, J. A. Seavey, W. G. Steinhauer, R. C. Prince, and D. L. Elmendorf. 1992. "The Use of Hydrocarbon Analysis for Environmental Assessment and Remediation." *Journal of Soil Contamination* 1(3): 197-216.

Elmendorf, D. L., C. E. Haith, G. S. Douglas, and R. C. Prince. 1994. "Relative Rates of Biodegradation of Substituted Polyaromatic Hydrocarbons." In R. E. Hinchee, A. Leeson, L. Semprini, and S. K. Ong (Eds.), *Bioremediation of Chlorinated and Polycyclic Aromatic Hydrocarbon Compounds*. Lewis Publishers, Ann Arbor, MI.

Kennicutt, M. C. 1988. "The Effect of Biodegradation on Crude Oil Bulk and Molecular Composition." *Oil Chem. Pollut.*, pp. 4, 89-112.

Overton, E. B., J. A. McFall, S. W. Mascarella, C. F. Steele, S. A. Antoine, I. R. Politzer, and J. L. Laseter. 1981. "Identification of Petroleum Residue Sources after a Fire and Oil Spill." *1981 International Oil Spill Conference*, pp. 541-546.

Page, D. S., P. D. Boehm, G. S. Douglas, and A. E. Bence. 1993. "Identification of Hydrocarbon Sources in the Benthic Sediments of Prince William Sound and the Gulf of Alaska Following the *Exxon Valdez* Oil Spill." Presented at the Third Symposium on Environmental Toxicology and Risk Assessment, Atlanta, GA, April 26-29, 1993.

Peters, K. E., and J. M. Moldowan. 1991. "Effects of Source, Thermal Maturity, and Biodegradation on the Distribution and Isomerization of Homohopanes in Petroleum." *Org. Geochem.* 17 and 47-61.

Peters, K. E., and J. M. Moldowan. 1993. *The Biomarker Guide.* Prentice-Hall, Inc., Englewood Cliffs, NJ.

Prince, R. C., J. R. Clark, and J. E. Lindstrom. 1990. *Bioremedation Monitoring Program.* Submitted to U.S. Coast Guard Federal On-Scene Coordinator, Anchorage, AK, December 1990.

Prince, R. C., S. M. Hinton, J. R. Bragg, D. L. Elmendorf, J. R. Lute, M. J. Grossman, W. K. Robbins, C. S. Hsu, G. S. Douglas, R. E. Bare, C. E. Haith, J. D. Senius, V. Minak-Bernero, S. J. McMillen, J. C. Roffall, and R. R. Chianelli. 1993. "Laboratory Studies of Oil Spill Bioremediation; Toward Understanding Field Behavior." Paper presented at Proceedings of the ACS Meeting, Denver, CO, April 1993.

Prince, R. C., J. R. Clark, J. E. Lindstrom, E. L. Butler, E. J. Brown, G. Winter, M. J. Grossman, P. R. Parrish, R. E. Bare, J. F. Braddock, W. G. Steinhauer, G. S. Douglas, J. M. Kennedy, P. J. Barter, J. R. Bragg, E. J. Harner, and R. M. Atlas. 1994. "Bioremediation of the *Exxon Valdez* Oil Spill; Monitoring Safety and Efficacy." In R. E. Hinchee, B. C. Alleman, R. E. Hoeppel, and R. N. Miller (Eds.), *Hydrocarbon Bioremediation.* Lewis Publishers, Ann Arbor, MI.

U.S. Congress, Office of Technology Assessment. 1991. *Bioremediation for Marine Oil Spills.* Government Printing Office, Washington, DC, OTA-BP-0-70, U.S. May.

U.S. Environmental Protection Agency. 1983. *Methods for Chemical Analysis of Water and Wastes.* U.S. Environmental Protection Agency, Washington, DC.

U.S. Environmental Protection Agency. 1986. "Organic Analytes." Chapter 4 in *Test Methods for Evaluating Solid Waste.* 3rd ed. Document SW-846, U.S. Environmental Protection Agency, Office of Solid Waste and Emergency Response, Washington, DC.

THE ROLE OF SURFACTANTS IN ENHANCED IN SITU BIOREMEDIATION

J. Ducreux, D. Ballerini, and C. Bocard

INTRODUCTION

Although bioremediation techniques have proved to be effective for restoring oil-contaminated soils and aquifers, some limitations still exist, especially related to the duration of field operations and to the level of residual contaminant concentrations that can be achieved in low-permeability soils. The use of surfactants, either synthetic or biogenic, has been considered as a way of improving the bioremedial efficiency by enhancing the accessibility of contaminants to microorganisms, nutrients, and possibly oxygen. However, surfactant use is still controversial. Surfactant injection must be carefully controlled to be ecologically acceptable and to offer more economical and physical advantages than disadvantages.

Contradictory results on the activation of in situ biodegradation of PAH in soil-water laboratory systems have been published (Aronstein et al. 1991, Laha & Luthy 1991). In a study on soil columns, a selected surfactant failed to enhance jet fuel biodegradation under simulated soil-venting conditions, compared to venting alone (Marsh et al. 1992).

The development of field application of surfactant-aided bioremediation needs further laboratory studies in order to characterize and quantify the effects of surfactants, especially by distinguishing the mobilizing effect (aided extraction) and the bioavailability-enhancing effect which operate simultaneously to an extent depending on the concentration of the surfactants. It should be noted that, specially in clay soils, the surfactant concentration in water may be lowered by adsorption and, in the case of anionic compounds, by precipitation.

Based on preliminary results (Ducreux et al. 1990), the present study was carried out in the case of diesel fuel, which is not extractable by air, to make a complete mass balance and to quantify the bioactivating effect of a selected nonionic surfactant.

METHOD

The biodegradation experiments of a diesel fuel (GOR 86) in aerobic conditions were carried out in glass columns, 4.5 cm i.d. × 17.0 cm long, packed with fine or coarse silica sands, with permeabilities respectively of 6×10^{-5} and 3×10^{-3} m/s. After creating an unsaturated zone, a mixture of hydrocarbonoclast bacterial strains, isolated from media contaminated with different types of hydrocarbons, and grown on liquid Trypticase soja medium, during 24 hours at 30°C, was injected into the column from both the top and the bottom. The sands were contaminated by oil at residual saturation. Either an aqueous solution containing mineral nutrients (nitrates and phosphates) as main components, or the same solution added with a nonionic surfactant (NIS) at a concentration of 0.05 wt % was passed through the porous medium during 60 days in four soaking/drainage cycles per day with oxygen supplied during the drainage phase. The aqueous effluents were collected and analyzed as cumulated fractions at 8, 22, 36, and 60 days. At the end of the tests, the residual oil in sand was extracted simultaneously by *n*-hexane and dichloromethane to control the efficiency of *n*-hexane extraction. The extracts were analyzed, for both concentration and composition, by gas chromatography on nonpolar capillary column. Hydrocarbons of *n*-hexane extracts were separated into aliphatics and aromatics, and total metabolites were determined in the polar compounds fraction after separation of the added surfactant. These three fractions were separated by using low-pressure liquid chromatography on silica gel.

RESULTS

Tables 1, 2, and 3 show the results obtained for the fine sand. The global effect of the surfactant is characterized in Table 1 by the increase of the percentage of hydrocarbons or oil (hydrocarbons + polar metabolites) eliminated from the

TABLE 1. Quantitative evolution of diesel oil in biodegradation experiments.

	Time (d)	Without surfactant	0.05 % NIS solution
Residual oil in sand (g/kg)	0	58.6	43.9
	60	44.6	9.4
Oil[(a)] removed from the sand (%)	60	23.9	78.6
% Displaced oil	0-8	0.17	29.9
	8-22	0.07	36.7
	22-36	0.18	5.5
	36-60	0.58	3.8
	0-60	1.00	71.9

(a) Oil : hydrocarbons (aliphatics + aromatics) + polar compounds.

TABLE 2. Chemical composition of the effluents.

Compounds		Original diesel oil	Without NIS				0.05 % NIS			
	Time (d)	0	0-8	8-22	22-36	36-60	0-8	8-22	22-36	36-60
Aliphatics (%)	**Total**	65.3	69.4	30.8	43.4	53.8	64.4	52.5	15.2	16.5
	n-alkanes	17.0	3.14	2.73	1.69	0.96	13.3	3.4	0.99	0.43
Aromatics (%)	**Total**	33.6	22.6	34.5	36.5	40.4	32.8	35.8	14.8	14.2
	Toluene	0.05	-	0.05	-	-	-	-	-	-
	Ethyl benzene	0.05	-	0.01	< 0.01	< 0.01	< 0.01	< 0.01	-	-
	Xylenes	0.22	< 0.01	0.07	0.01	< 0.01	0.03	< 0.01	-	-
	Naphthalene	0.16	0.06	2.27	1.10	0.26	0.09	0.10	0.09	0.09
	C1 naphthalenes	0.70	0.32	0.51	0.22	0.11	0.66	0.36	0.09	0.06
	C2 naphthalenes	1.28	0.16	1.06	0.69	0.61	1.27	1.40	0.33	0.26
Polar (%)	Total	1.1	8.0	34.7	20.1	5.8	2.8	11.7	70.0	69.4

sand, from approximately 24 to 80%. When the surfactant was applied, 71.9% of the initial oil, i.e., 91.5% of the removed oil, was found in the effluents. As shown by the oil composition in the effluents (Table 2), a significant amount of polar compounds, resulting presumably from biooxidation of both aliphatics and aromatics, was found from day 8 to day 22; the concentration of polar compounds was much higher from day 22 to day 36, roughly corresponding to the same weight amount as during the previous period. Taking into account the polar compound content in the accumulated effluents from day 8 to day 60 and in the residual oil in the sand at the end of the test (Table 3), a 22.7% rate of hydrocarbon biodegradation is calculated based on the initial oil in the sand. This rate corresponds to a 32.5% based on hydrocarbons remaining in the sand after 8 days when 30% of oil has been displaced before biodegradation takes place, compared to 23.7% in the test without surfactant.

Furthermore, in the test with surfactant, the residual oil in sand (9.4 g/kg) was more chemically modified than in the test without surfactant (44.6 g/kg), as shown in Table 3, for both on aliphatics and aromatics but mainly with regard to the concentration of water-soluble aromatics (up to C2 naphthalenes). The use of surfactant is thus beneficial both for lowering the residual oil concentration and making it less contaminating.

The oil composition of the effluent corresponding to the last period in the surfactant test (Table 2) is very different from the composition of the residual oil in sand (Table 3), mainly with regard to the polar compounds. That suggests that the trapped oil is made progressively more bioavailable, the surfactant probably both enhances the interfacial exchanges and continuously displaces the mobilized contaminant fraction.

TABLE 3. Chemical composition of the residual diesel oil entrapped in sand after 60-day biodegradation.

Compounds		Original diesel oil	Without NIS	0.05 % NIS
Aliphatics (%)	Total	65.3	60.50	40.20
	n-alkanes	17	6.41	0.69
Aromatics (%)	Total	33.6	37.30	30.70
	Toluene	0.053	-	-
	Ethyl benzene	0.054	-	-
	Xylenes	0.217	< 0.01	-
	Naphthalene	0.159	0.08	0.04
	C1 naphthalenes	0.696	0.25	0.14
	C2 naphthalenes	1.281	0.92	0.61
Polar (%)	TOTAL	1.1	2.20	29.10

TABLE 4. Hydrocarbons biodegradation rates within 60 days.

		% Aliphatics			
		Total	n-alkanes	% Aromatics	% Total
Fine sand	Without NIS	28.7	71.3	14.4	23.7
	NIS 0.05 %	40.2	89.5	17.0	32.5
Coarse sand	Without NIS	44.8	95.5	32.8	40.8
	NIS 0.05 %	86.1	97.4	76.1	83.6

Hydrocarbons Biodegradation Rate (%) = (HCo-HCm) - (HCw+HCr) x100 / HCo-HCm

HCo: Initial hydrocarbons entrapped in sand at to
HCm: Non-degraded quickly mobilized oil (< 8days)
HCw: Hydrocarbons in aqueous effluents
HCr: Residual hydrocarbons entrapped in sand at t60 (end of experiment)

Without surfactant, the oil displaced in the effluents at very low concentrations is less degraded, and the polar compound content of the effluents steadily decreases after a maximum value corresponding to the 8- to 22-day period when the most accessible oil is degraded (Table 2). On the contrary, the polar compound content of the effluents remains high and constant during the last two periods in the surfactant test. Thus the biodegradation rate would be significantly increased by carrying on the test. This conclusion is supported by the mass balance of accumulated polar compounds in the effluents which shows that, for the same amount of degraded hydrocarbons, more polar compounds are produced and displaced when using the surfactant. These results are due to the biological attack of the more recalcitrant hydrocarbons, both aliphatic and aromatic.

The data in Table 4 show that a more permeable porous medium enhances the oil biodegradation by increasing the oxygen exchange, the bioavailability of

TABLE 5. Residual oil concentration in sand in sterile and biodegradation tests.

	Residual oil in sand (g/kg)			
Time (days)	Sterile tests		Biodegradation tests	
	Without NIS	0.05 % NIS	Without NIS	0.05 % NIS
0	38.8	40.6	58.6	43.9
60	30.2	1.4	44.6	9.4

the entrapped oil, and interfacial exchanges when surfactant was used. Thus, by using the NIS solution, the total hydrocarbon biodegradation rate is much higher, i.e. 83.6% compared to 32.5% in the fine sand experiment. This difference corresponds to the a more pronounced degradation of the aromatic fraction, producing more polar compounds.

Without surfactant biodegradation also is enhanced in the coarse sand (40.8%), but remains much lower than when using the NIS solution.

Two sterile control tests were carried out on fine sand to estimate the role of surfactant on oil mobilization. The data in Table 5 show that residual oil concentrations in sand were always lower in the case of control tests: 30.2 and 1.42 g/kg compared to 44.6 and 9.4 g/kg in biodegradation experiments. It appeared that the colonization of porous media by microorganisms seems to have increased the oil retention on sand in both cases.

CONCLUSION

In conclusion, laboratory results show that surfactants can enhance the bioavailability of low-accessibility residual hydrocarbons when added to circulating water. However when optimizing their surfactant use, its mobilizing effect must be considered by making a mass balance on the water effluents, the composition of which indicates the rate of the different processes. To use surfactant-aided biodegradation to its greatest advantage, the production of biomass, especially in low-permeability soils, must be controlled.

ACKNOWLEDGMENT

This work has been funded in part by Agence de l'Eau Seine-Normandie.

REFERENCES

Aronstein, B. N., Y. M. Calrillo, and M. Alexander. 1991. "Effect of Surfactants at Low Concentrations on the Desorption and Biodegradation of Sorbed Aromatic Compounds in Soil." *Environ. Sci. Technol. 25*: 1728-1731.

Ducreux, J., C. Bocard, C. Gatellier, and L. Minssieux. 1992. "Contamination of Aquifers by Hydrocarbons and Restoration with Surfactants." In A. A. Balkema Publisher, K. U. Weyer (Eds.), *Subsurface Contamination by Immiscible Fluids*, pp. 337-350. Rotterdam, The Netherlands.

Laha, S., and R. G. Luthy. 1991. "Inhibition of Phenanthrene Mineralization by Nonionic Surfactants in Soil-Water Systems." *Environ. Sci. Technol. 25*: 1920-1930.

Marsh, S. S., T. C. Zwick, M. F. Arthur, and G. K. O'Brien. 1992. "Evaluation of Innovative Approaches to Stimulate Degradation of Jet Fuels in Subsoils and Groundwater." National Technical Information Service, Springfield, VA, AD-A252 359/5/xAD.

VENTING AND BIOVENTING FOR THE IN SITU REMOVAL OF PETROLEUM FROM SOIL

J. van Eyk

INTRODUCTION

The development and application of the technique of venting and bioventing for in situ removal of petroleum from soil have been reported before (van Eyk & Vreeken 1991, van Eyk 1991). This technical note reports the results of an ongoing demonstration project at a retail gasoline station (van Eyk & Vreeken 1991), which is expected to be finalized in the fall of 1993.

RESULTS

Soil Venting. A plan view of the retail gasoline station outlining the cleanup system is shown in Figure 1. The removal of gaseous hydrocarbons (HC) from soil by venting is shown in Table 1 and Figure 2. The results show that the removal of HC by venting is essentially completed after 1 year. For this reason and because

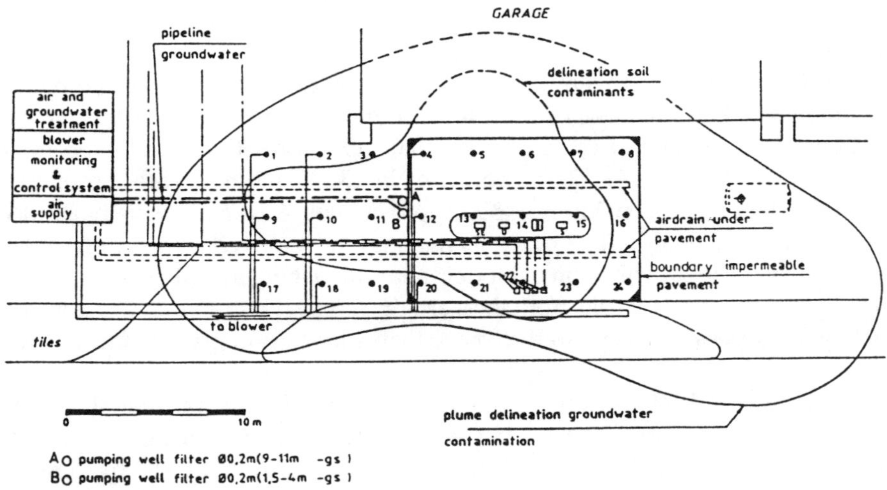

FIGURE 1. Plan view of retail gasoline station outlining the installation of the venting system, pumping wells, cleanup, and monitoring systems.

TABLE 1. Concentrations of hydrocarbons removed by venting in mg/L air extracted.

Week no.	Flux m^3/h	Benzene	Toluene	Xylenes	Total HC	Total HC removed in kg
0	19.2	2.37	4.71	1.84	33.28	117
1	24.8	0.44	1.64	1.23	9.32	190
2	32.0	0.16	1.00	0.87	5.11	227
3	35.0	0.12	0.64	0.50	3.88	253
7	29.8	0.09	0.37	0.52	3.00	331
14	33.0	0.10	0.41	0.45	2.90	436
30	30.2	0.034	0.21	0.41	1.60	573
38	47.7	0.043	0.169	0.296	1.69	677
43	39.0	0.016	0.103	0.22	1.14	718
48	41.5	0.06	0.054	0.196	0.987	751
56	39.9	0.007	0.008	0.075	0.354	771
63	37.8	0.0003	0.006	0.049	0.247	782
67	36.6	0.00018	0.003	0.024	0.121	785
76	9.7	0.0001	0.0014	0.018	0.085	786
80	8.5	0.00005	0.0011	0.020	0.100	787
105	6.9	0.00006	0.00017	0.0052	0.033	788

the emissions for benzene, toluene, and xylene (BTX) (0.18 mg/m^3, 3 mg/m^3, and 24 mg/m^3, Table 1) are far below the maximum allowable discharge values for BTX, the airflux was reduced by 75% on December 9, 1991 (after 67 weeks). As Table 1 shows, the total amount of volatile HC removed from soil by venting after 67 weeks is 785 kg.

Soil Bioventing. The removal of HC from soil as a result of biodegradation was monitored on the basis of carbon dioxide (CO_2) production. Oxygen (O_2) consumption was measured to confirm that CO_2 production was linked to O_2 consumption (see Table 4). The carbon dioxide (CO_2) data shown are cumulative. Assuming that 14 g HC yields 44 g CO_2, approximately 430 kg HC were biograded after 67 weeks, about half as much as the amount removed by venting in that same period. The contribution of biodegradation to the removal of HC from soil initially is nil. After 14 weeks from the start, the removal rate amounts to approximately 330 g per day, compared to more than 2 kg carbon per day for removal by venting.

Approximately 40 weeks after the start, both removal rates are equal (around 1.3 kg carbon per day), and after 48 weeks the removal rate for biodegradation (2 kg carbon per day) is twice as high as the removal rate for HC by venting. As the concentration in the soil continues to decline (see below), the removal rate for HC as a result of biodegradation also starts to decline after approximately 60 weeks (Figure 2). In the previous section we mentioned the reduction of the

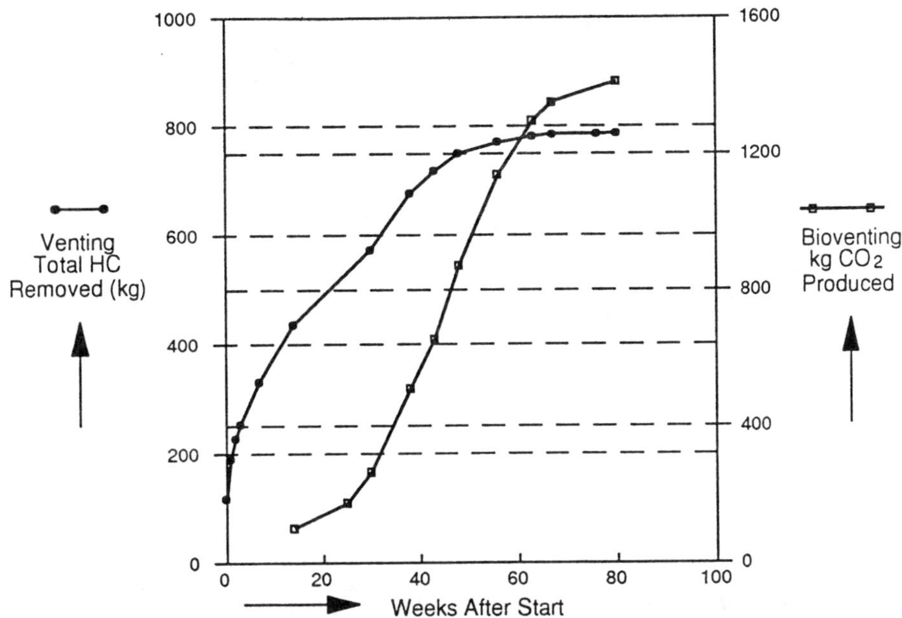

FIGURE 2. Graphic representation of (bio)venting results (cumulative).

airflux after week 67. Calculations show that this reduced flow was still more than adequate to sustain biodegradation.

Assuming a maximum CO_2 production of 3,000 ppm, then based on an average airflux of 34 m^3/h and equating 1 ppm to 1.964 mg CO_2 per m^3, it can be shown that the amount of HC oxidized equals:

$$3{,}000 * 1.964 \text{ mg/m}^3 * 34 \text{ m}^3\text{/h} * 24 \text{ h/d} = 4.8 \text{ kg d}^{-1} \quad (1)$$

Based on $[CH_2] + 3O \rightarrow CO_2 + H_2O$, 4.8 kg per day equals 4,800/14 moles per day, which requires 3/2 * 4,800/14 moles * 0.0224 m^3/mole = 11.52 m^3 O_2 per day or 58 m^3 air per day. The actual reduced airflux amounted to approximately 200 m^3 per day.

Annual Bacterial Carbon Dioxide Production. The data in Figure 3 show the production of CO_2 in ppm. From week 25 (February 13, 1991) to week 56, the CO_2 production increased 8-fold. Although the soil temperature was not monitored, the effect shown is most likely the result of both bacterial growth and temperature. The observation that the CO_2 produced after 1 year (11/28/91, week 66) is two times higher than observed in week 14 (11/28/90) may be caused in part by the presence of high concentrations of BTX components at the start of the remediation. Low-molecular-weight HCs generally are considered to be toxic to microorganisms because of their interactions with lipoidal membranes. Benzene in particular is known to be cytotoxic to many bacteria.

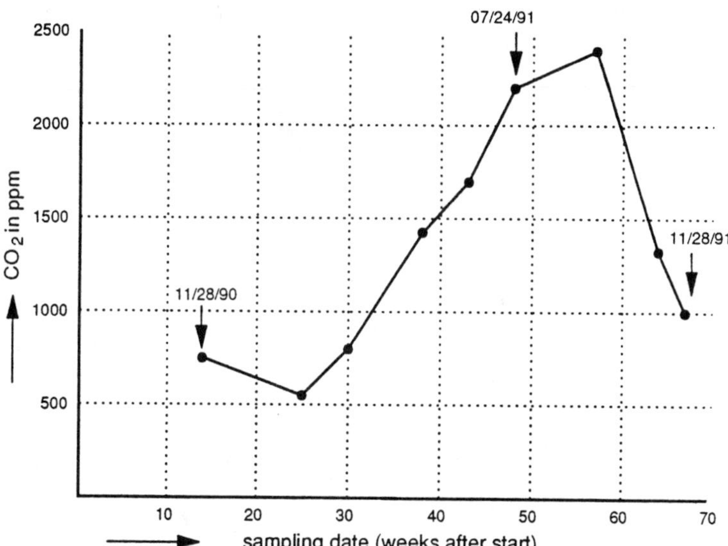

FIGURE 3. Graphic representation of CO_2 production in ppm.

Enumeration of Bacteria. In addition to monitoring the CO_2 production (and O_2 consumption, see below), an epifluorescence counting was carried out on soil samples obtained from two different soil horizons: one from the groundwater table, which was still polluted, and one from a soil horizon, which had been cleansed of HC. These measurements were carried out on soil smears, in which bacteria were made fluorescent with a protein dye. The enumeration results are based on 30 different microscopic views, carried out in triplicate for both samples with the help of confocal laser scan microscopy, equipped with automatic image processing. Tables 2 and 3 summarize the results. The number of the bacteria and the biovolume in μm^3 per gram soil in sample 3029 are, respectively, 3.2 and 3.9 times higher than in sample 3027. The average volume per cell is 20% higher. The differences are still more evident (data not shown) if the mean values are used rather than the average numbers. The mean values are less influenced by the effects of small number of cells with extreme values. Based on the values of the mean, the biovolume is 5 times higher, whereas cell volume, length, and width are, respectively, 55%, 27%, and 9% higher. These findings—higher numbers, higher biovolume, and larger cells—indicate higher microbial activity in sample 3029. This supports the view that the observed CO_2 production is coupled to bacterial growth on HC.

Monitoring of Pumping Wells. For insight into how closely CO_2 production is coupled to O_2 consumption, both parameters were determined for each individual airwell (Table 4). Although, apart from the stack gas the stoichiometry is not perfect (due in part to the inaccuracy of the O_2 monitor), the results clearly show that CO_2 production is accompanied by O_2 consumption. These results support the view that CO_2 production is linked to HC oxidation. Assuming that the airflux for

TABLE 2. Enumeration and morphological characterization of subsurface bacteria in a hydrocarbon-free zone: Sample 3027 (0.8 to 1.1 m BGS).

Sample number	Bacteria/ gram soil	Bio- [a] volume ($\mu m^3/g$)	Cell length (μm)	Cell width (μm)	Length/ width
3027-1	1.8 E+08	2.8 E+07	0.80	0.56	1.52
3027-2	1.9 E+08	4.5 E+07	1.10	0.57	1.98
3027-3	1.9 E+08	3.8 E+07	0.98	0.60	1.62
Average	1.9 E+08	3.7 E+07	0.96	0.58	1.71
SD [b]	0.01 E+08	0.8 E+07	0.15	0.02	0.24
CV [c]	4	23	16	3	14

(a) The biovolume is the number of bacteria per gram soil × the volume per cell.
(b) Standard deviation.
(c) Coefficient of variance in %.

individual airwells does not vary significantly, the CO_2 production rates also should give an indication of the degree of pollution within the radius of influence of each individual airwell. This assumption is not too unreasonable, in view of the homogeneity of the soil (van Eyk & Vreeken 1991) and the balanced position of each well in the grid.

This view is supported by the observation that the highest CO_2 values measured correspond to the areas where initially the highest concentrations of HCs were detected. Indeed, these values are found in wells 4 and 5 (Figure 1), which are installed in the area where floating HCs were detected in 1989.

TABLE 3. Enumeration and morphological characterization of subsurface bacteria in a hydrocarbon-contaminated zone: Sample 3029 (1.4 to 1.7 m BGS).

Sample Number	Bacteria/ gram soil	Bio- [a] volume ($\mu m^3/g$)	Cell length (μm)	Cell width (μm)	Length/ width
3029-1	5.1 E+08	12.7 E+07	1.03	0.61	1.77
3029-2	5.6 E+08	12.4 E+07	0.94	0.64	1.59
3029-3	7.3 E+08	18.5 E+07	1.03	0.64	1.73
Average	6.0 E+08	14.5 E+07	1.00	0.63	1.70
SD [b]	0.11 E+08	3.4 E+07	0.05	0.02	0.10
CV [c]	19	24	5	3	6

(a) The biovolume is the number of bacteria per gram soil × the volume per cell.
(b) Standard deviation.
(c) Coefficient of variance in %.

TABLE 4. CO_2 production and oxygen consumption for each individual pumping airwell (Figure 1).

Airwell number	CO_2 in %[(a)]	O_2 in %[(b)]	Airwell number	CO_2 in %[(a)]	O_2 in %[(b)]
1	1.50	2.6	13	0.34	1.0
2	1.76	3.0	14	0.16	0.6
3	0.95	1.9	15	0.28	0.7
4	3.10	4.6	16	0.20	0.4
5	2.40	3.6	17	0.86	1.6
6	0.13	0.6	18	0.91	1.9
7	0.31	0.8	19	1.01	2.2
8	0.43	0.8	20	0.34	0.9
9	0.53	1.2	21	0.10	0.6
10	0.36	1.0	22	0.68	1.1
11	0.20	0.8	23	0.42	1.0
12	0.56	1.4	24	0.22	0.7

(a) Background zeroed.
(b) Background 21%.
Off-gas from gas stack CO_2 : 0.675%
O_2 : 1.0%

Soil Sampling. In addition to monitoring the vented air samples for HCs (Table 1) and CO_2 (Figures 2 and 3), soil samples were taken annually to confirm the progress of remediation. To minimize the number of holes that had to be drilled through the 2-inch-thick (5-cm-thick) slabs of watertight pavement, the number of hand borings was restricted. The borings (numbers 22 and 25, Figures 4 and 5) were taken, however, from the center of the (no longer existing) floating oil layer (approximately the location of airwell no. 4). The results of these borings for BTEX and mineral oil, respectively, are presented in Figures 4 and 5, together with analysis results for BTEX and mineral oil for Begemann boring B3, taken in 1989 before the start of the in situ remediation. (The Begemann boring is a continuous sampling device that, in contrast to hand borings, can be applied below the groundwater table.) When the results of these borings are compared, it appears that the soil has been cleansed from the ground surface to the groundwater table downwards. This is not unexpected because, as the moisture content toward the capillary fringe increases, and thus air permeability decreases, both venting and biodegradation become less efficient. In view of the somewhat disappointing results for boring number 25, the sample between 1.4 and 1.7 below ground surface (BGS) was quantitatively analyzed by gas chromatography (GC). The results (data not shown), show that the GC scan contains, in addition to diesel oil, approximately 50% HC in the boiling range beyond that of diesel oil (860 to 990°C), which most likely represents lube oil.

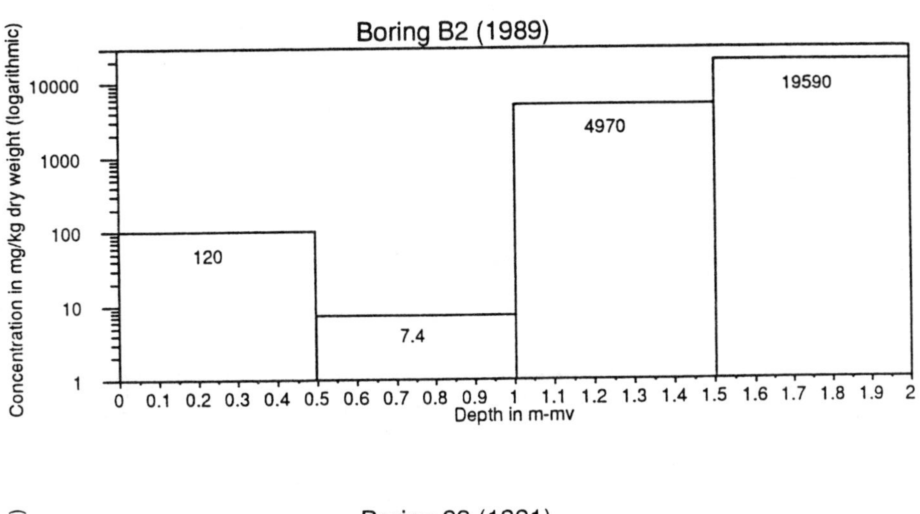

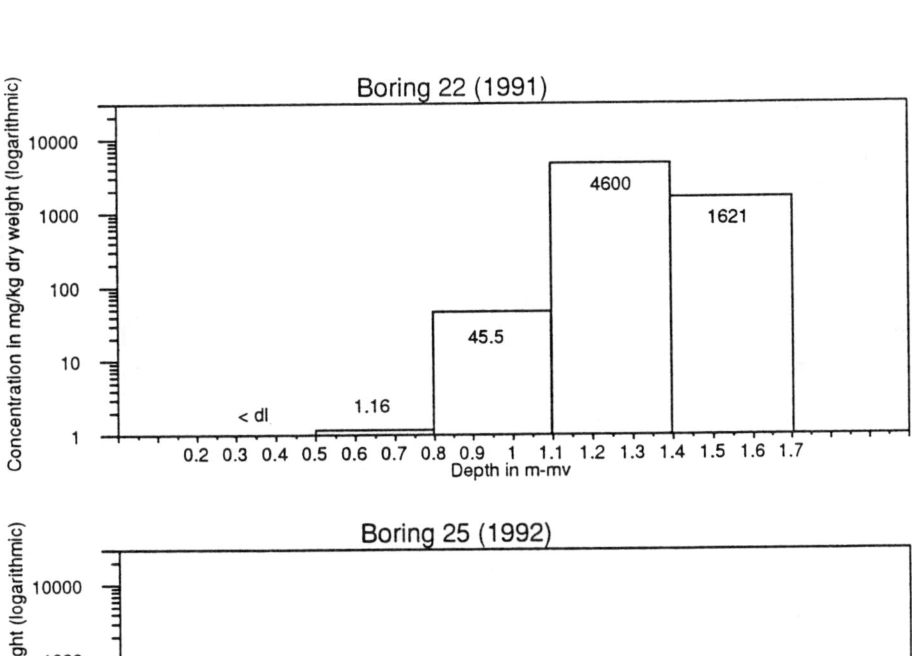

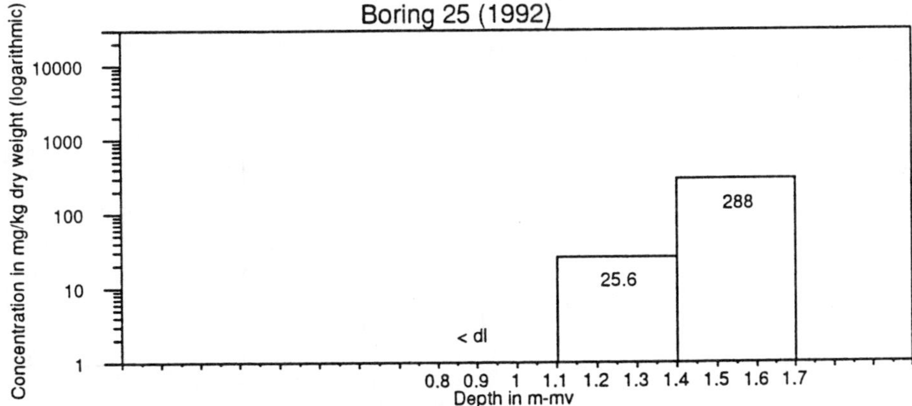

FIGURE 4. BTEX levels in borings B2, 22, and 25 after 0, 1, and 2 years of remediation by applying (bio)venting.

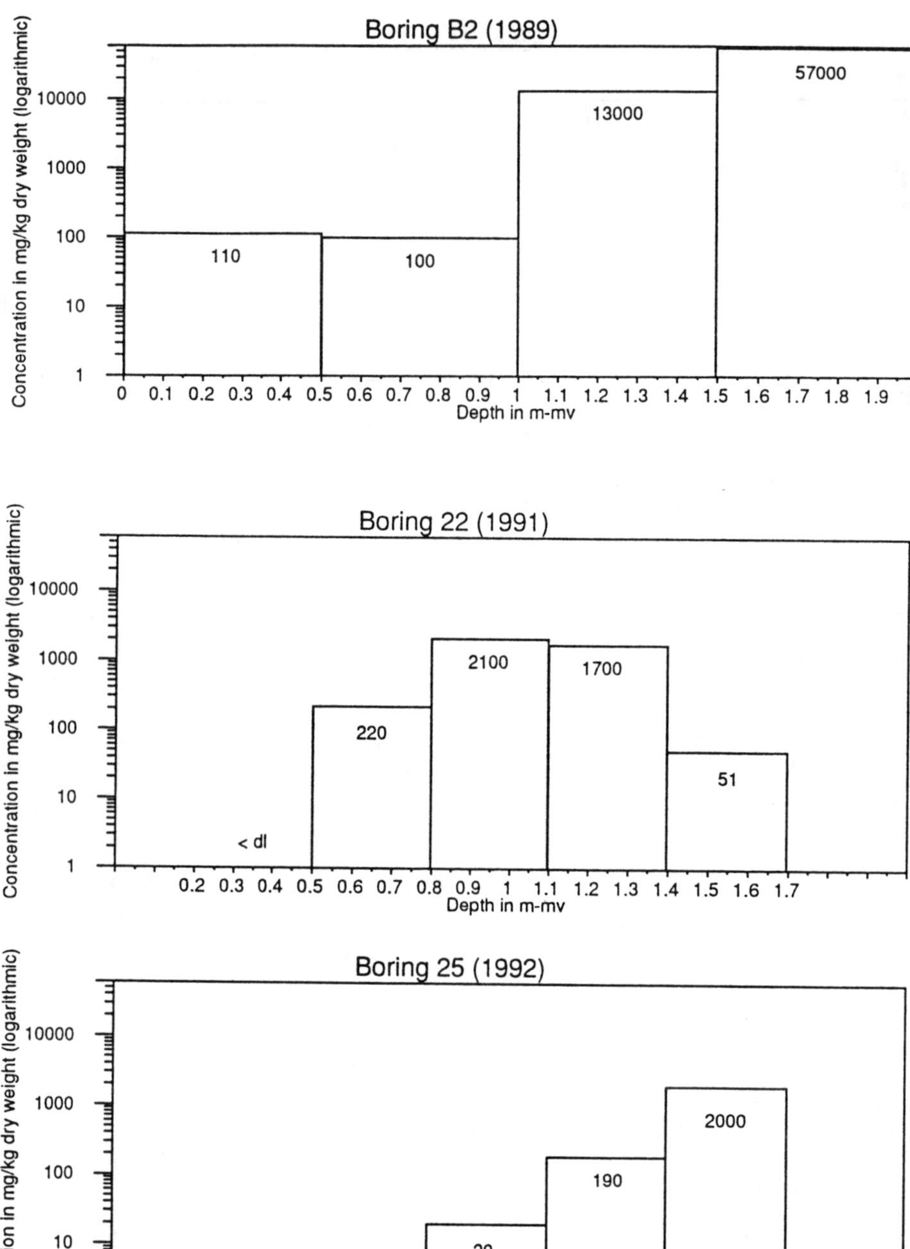

FIGURE 5. Mineral oil levels in borings B2, 22, and 25 after 0, 1, and 2 years of remediation by applying (bio)venting.

CONCLUSIONS AND OUTLOOK

After 2 years of in situ remediation, approximately 800 kg of HCs were removed from soil by venting, and approximately 572 kg by biodegradation. Soil samples show that soil cleansing progresses from the ground surface downwards. As the concentration of HCs in the higher soil horizons reaches background values, it is not unreasonable to assume that similar values will eventually be obtained near the capillary fringe; achievement of these values will just take longer. The project is expected to be successfully completed in fall 1993.

REFERENCES

van Eyk, J., and C. Vreeken. 1991. "In Situ and On-Site Subsoil and Aquifer Restoration at a Retail Gasoline Station." In R. E. Hinchee and R. F. Olfenbuttel (Eds.), *In Situ Bioreclamation: Applications and Investigations for Hydrocarbon and Contaminated Site Remediaton*, pp. 303-320. Butterworth-Heinemann, Boston, MA.

van Eyk, J. 1991. "Bioventing. An In Situ Remedial Technology: Scope and Limitations." Paper presented at the Symposium on Soil Venting. April 29-May 3, 1991, Houston, TX.

OPTIMIZING IN SITU BIOREMEDIATION AT A PILOT SITE

H. Bonin, M. Guillerme, P. Lecomte, and M. Manfredi

INTRODUCTION

In 1990, BRGM and ESYS, in the framework of a French "Research in Industrial Partnership" project and with the collaboration of Elf Aquitaine Production (SNEA(P)), developed a process for the decontamination of aquifers polluted by hydrocarbons. The principle is that of in situ bioremediation by bacteria that are naturally present in the soil. Such bacteria are activated by the nutrients nitrogen and phosphorus, as well as by oxygen in the form of hydrogen peroxide. The SNEA(P) laboratories contributed their competence in the field of bioremediation, ESYS was responsible for the engineering aspects, and BRGM intervened in its capacity of specialist in the Earth Sciences and in particular hydrogeochemistry and hydrodynamics. The use of hydrogen peroxide and its monitoring in the environment were the responsibility of Atochem. As a result of the first phase of the project, which demonstrated the feasibility of in situ bioremediation, a complementary program was proposed. The aims of this Phase 2 program were threefold: to improve our knowledge of the mechanisms involved, to optimize the development of bacteria, and to test variants of the process developed during Phase 1. This paper summarizes the main results obtained on the test site during Phase 2 of the work.

THE PILOT SITE AND PHASE 1 OF THE WORK

Phase 1 effectively took place from 15 July to 15 November 1990. The selected test site lies in the Petroleum Harbor of Strasbourg, where pollution was caused by a spill of 65 m^3 of heating oil. The layer that was decontaminated corresponds to the upper part of the sand and gravel aquifer, whose groundwater level lies at a depth of about 3 m. Locally, the entire aquifer can be 90 m thick.

The industrial pilot installation consisted of a treatment unit at the surface, and eight injection wells around a central pumping well. The surface area affected by this radial flow pattern was about 500 m^2. The three objectives of this installation were to ensure a good hydraulic circulation of the treatment fluid in the zone at 3 to 3.5 m depth that is impregnated with hydrocarbons, to recycle most of the water charged with additives, and to skim off the floating hydrocarbons (Figure 1).

The pumping well was designed for the installation of two pumps, which enabled the differentiation of the pumped water. A shallow pump recovered the

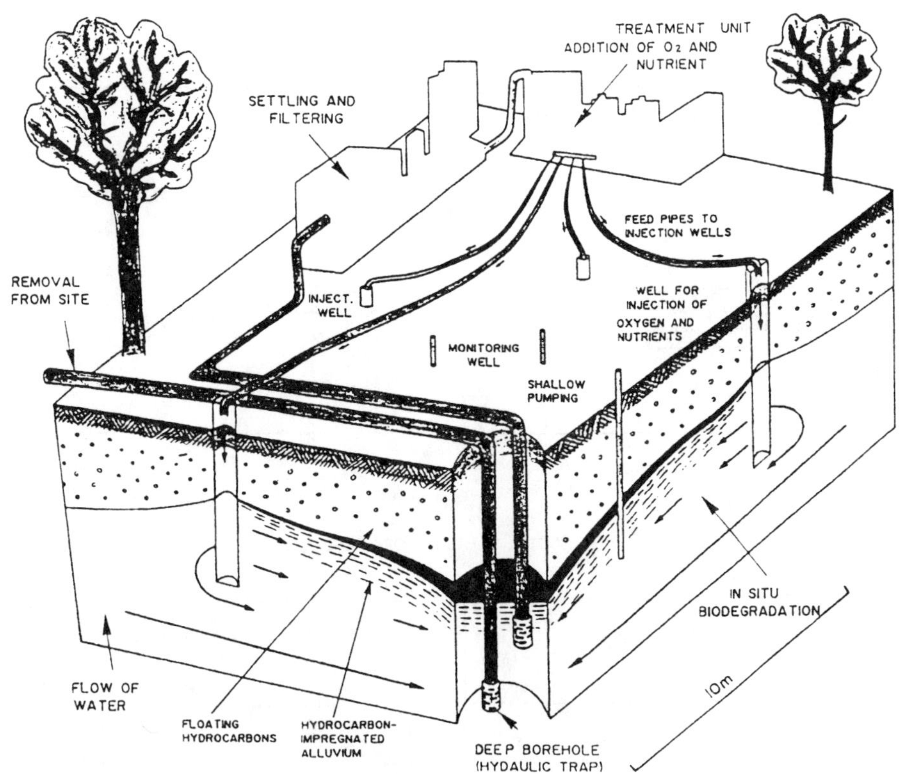

FIGURE 1. Block diagram of pilot site.

most heavily charged water and sent it to the treatment unit. The additional pump, installed at the bottom of the well, was needed to maintain an efficient hydraulic trap as well as complete recycling. The injection of nutrients and hydrogen peroxide, as well as the recycling of pumped water, required a minimum of surface installations that were mounted on two separate skids.

As the soil initially contained only a few microorganisms, the subsurface was seeded with autochthonous cultured bacteria. The overall efficiency of the treatment was evaluated by analyzing soil samples (cores) taken before and after treatment. This showed that the hydrocarbon concentrations in the soil generally were reduced by about 50% (Figure 2). Analysis of chromatograms showed the systematic disappearance of linear fractions, i.e., paraffins up to C_{22}, from the heating oil.

PHASE 2 – OPTIMIZATION OF THE PROCESS (1991)

Objectives. After Phase 1, which demonstrated the feasibility of the process, a Phase 2 program included the use of surfactants for homogenizing the distribution of bacteria in the subsurface, or for checking their movements, and optimization of the oxygenation from hydrogen peroxide addition.

AREA 1 - SOIL SAMPLES S25 AND C25

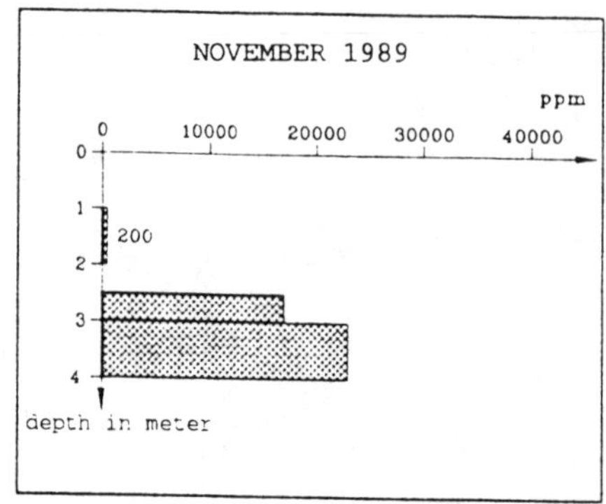

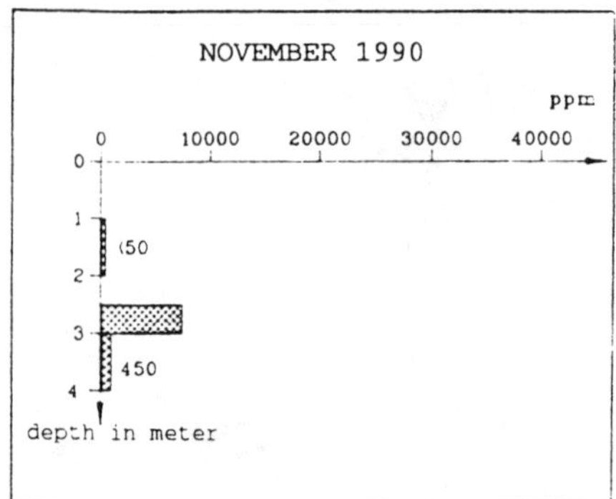

AREA 2 - SOILS SAMPLES S24 AND C24

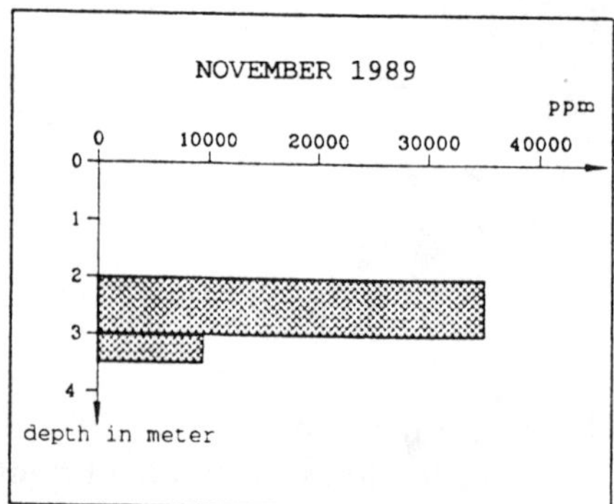

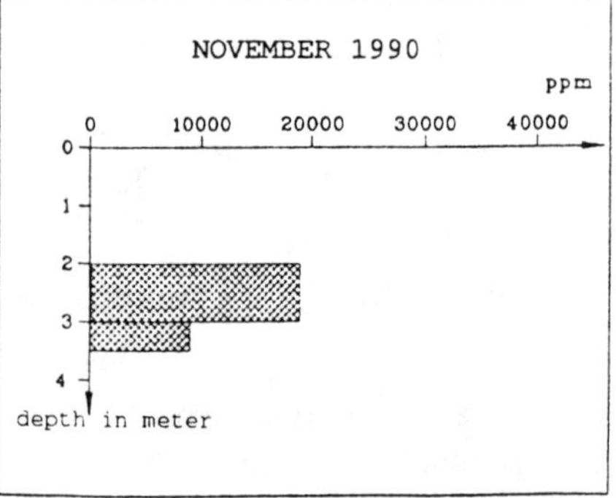

FIGURE 2. Bioremediation effectiveness evaluated by soil sampling.

Surfactant Tests. Preliminary tests were carried out to select an efficient, biodegradable, anionic surfactant that would be minimally toxic for the environment. The tests were made on soil columns of varying grain size, and clearly showed the efficacy of some surfactants in improving bacteria transfer after optimization of the injected concentrations.

The flooding of the porous substratum with anionic surfactants in the concentrations used, i.e. 20 to 100 ppm, did not lead to a remobilization of the hydrocarbons impregnated in the sand. In addition, the curves obtained with a biophotomer showed that the selected organic compounds were not toxic. No bacteriostatic of bactericidal effects was observed.

After these preliminary tests, the test work on the pilot site included: (1) conditioning of the bacteria selected during Phase 1, with pre-cultures in a wet medium and incubation; (2) injection of bacteria into one of the eight injection wells (number 1 on Figure 1); (3) monitoring of bacteria migration through the phreatic groundwater and in the injection wells, the pumping well (number 3, Figure 1), and intermediate observation wells; (4) two tests with different concentrations (50 to 150 mg/L) of the surfactants selected during the laboratory tests. The results (Figure 3) indicate that the quantities of bacteria in the groundwater significantly increased after injection of the surfactant. This variation, which is caused by the elution of bacteria fixed on soil particles, shows that the tested product is efficient. Monitoring of the behavior of the surfactant in the natural environment showed that this organic compound is subject to strong absorption to the aquifer sediments.

Oxygenation Tests. The objective was to inject a minimum amount of hydrogen peroxide in the ground, while maintaining a dissolved-oxygen content of more than 1 mg/L that guarantees aerobic conditions. Both dissolved oxygen and hydrogen peroxide were regularly measured while entering into and exiting from the system, as well as within two flow axes below the pilot site.

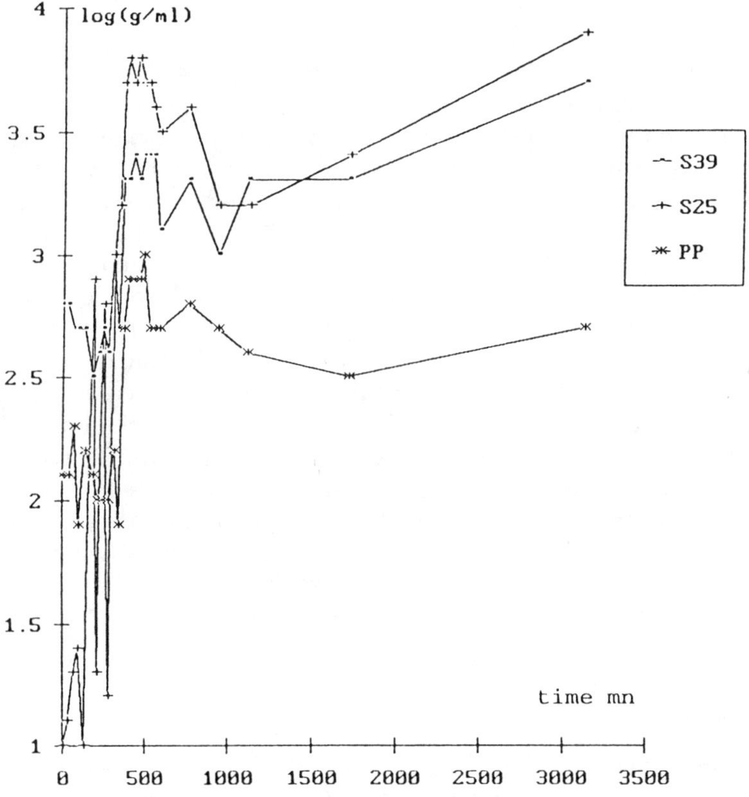

FIGURE 3. Evolution of organisms during the first test (50 mg/L).

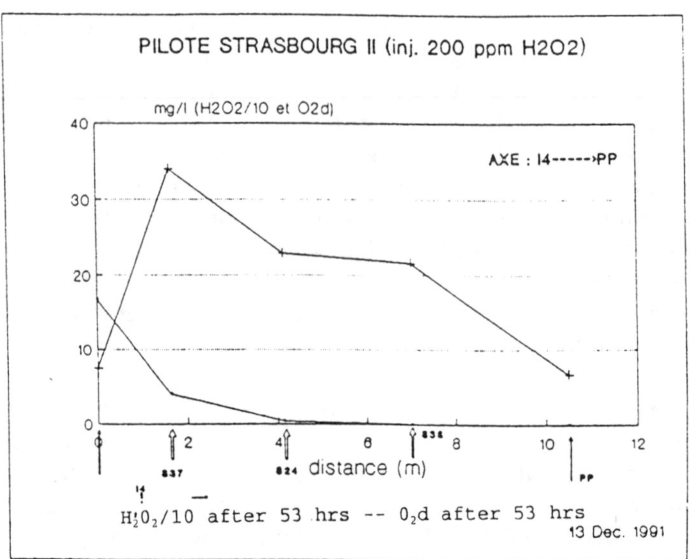

FIGURE 4. H_2O_2 and O_2 evolution.

During the first period, from 17 September to 26 November 1991, an average injection value of 55 mg/L H_2O_2 was maintained, which caused a strong increase of between 2 and 10 mg/L in the dissolved O_2 content. The background values of dissolved oxygen, measured upstream from the site during the same time, varied between 0.1 and 2 mg/L. H_2O_2 rapidly disappears after its injection; on average, only 1 mg/L remains at a distance of 3 m from the injection point (Figure 4).

During the second step, various injection tests of hydrogen peroxide were programmed for checking the behavior of this product. The first of these aimed at correlating the dissolved H_2O_2 and O_2 contents in the groundwater upon injection. Doubling the peroxide content from 115 to 230 ppm did not result in a doubling of the dissolved oxygen content, but in an increase of only 1.75. However, the interest of this operation lies in the fact that the impact of this oxygenation can be felt over a greater distance downstream. The second test consisted in the injection of pure H_2O_2, with a stabilizer based on phosphoric acid. The aim of this test was to limit the spontaneous breakdown of peroxide in the first few meters after having been injected in the aquifer. The concentration of this injection was set at 200 ppm, but the results (Figure 4) did not show any difference from the behavior of unstabilized peroxide.

CONCLUSION

This program has combined competence in biological treatment with hydrogeochemical/hydrodynamic and peroxide behavior know-how to study the implementation of in situ bioremediation of hydrocarbons. Independently of the success of the operation, which was indicated by an overall reduction of the pollutants by 50% after 4 months of treatment, the greatly improved understanding of operating mechanisms and a much better technical know-how constitute an appreciable body of experience. In this respect, we can mention (1) definition of a protocol for adapting bacteria that "specialize" in degradation of the pollutant, (2) understanding of the relationships between bacteria present in water and on sediment surfaces, (3) optimization of the process by adding surfactants, and (4) optimization of the addition of H_2O_2 for oxygenation of the subsurface.

The success of the pilot operation has led to the adoption of this process on an industrial scale: it will be used to decontaminate a former refinery site of 20 hectares in Eastern France.

DISSOLUTION KINETICS OF TOLUENE POOLS IN SATURATED POROUS MEDIA

M.-F. Yeh and E. A. Voudrias

INTRODUCTION

Nonaqueous-phase liquids (NAPLs) are immiscible with water and may consist of single components, such as toluene, or mixtures of a large number of components, such as gasoline and aviation fuel. In recent years, groundwater contamination by NAPLs has reached significant proportions. This contamination has occurred due to leaking underground storage tanks, ruptured pipelines, surface spills, hazardous landfills, and disposal sites. Although NAPL solubilities are generally low (a few hundred to a few thousand mg/L), they are frequently several orders of magnitude higher than the maximum contaminant level for drinking water. Therefore, a small amount of NAPL may contaminate a large volume of groundwater (Anderson et al. 1992, Mackay & Cherry 1989).

The rate of biodegradation of NAPLs in groundwater depends on oxygen and nutrient concentrations and the availability of dissolved NAPL components. The latter may be limited by the slow dissolution kinetics of the NAPL. The objective of this work was to study the dissolution kinetics of toluene pools floating at the capillary fringe above the water table in a two-dimensional experimental aquifer.

MATERIAL AND METHODS

Toluene pools, approximately 60 cm long, 20 cm wide, 3 cm thick, were formed by applying 2 L of toluene containing 1 g/L of Oil-Red EGN dye above the water table in a two-dimensional aquifer, 100 cm long by 50 cm high by 20 cm wide (Figure 1). Medium sand, with particle size between 0.8 mm and 0.5 mm, was used as the porous medium in all experiments. Tap water was used in all experiments. All experiments were conducted in a 20°C constant temperature room.

Pore water velocities between 10 cm/day and 100 cm/day were used. Sampling ports were located in one of the side glass walls of the experimental aquifer. Groundwater samples were taken using 10.2 cm long, 18-gauge stainless-steel needles, which were inserted into the ports and pushed into the porous medium. Interstitial water samples were withdrawn from selected sampling ports of the aquifer with a 1-mL gastight syringe (model #1005, Hamilton Co., Reno,

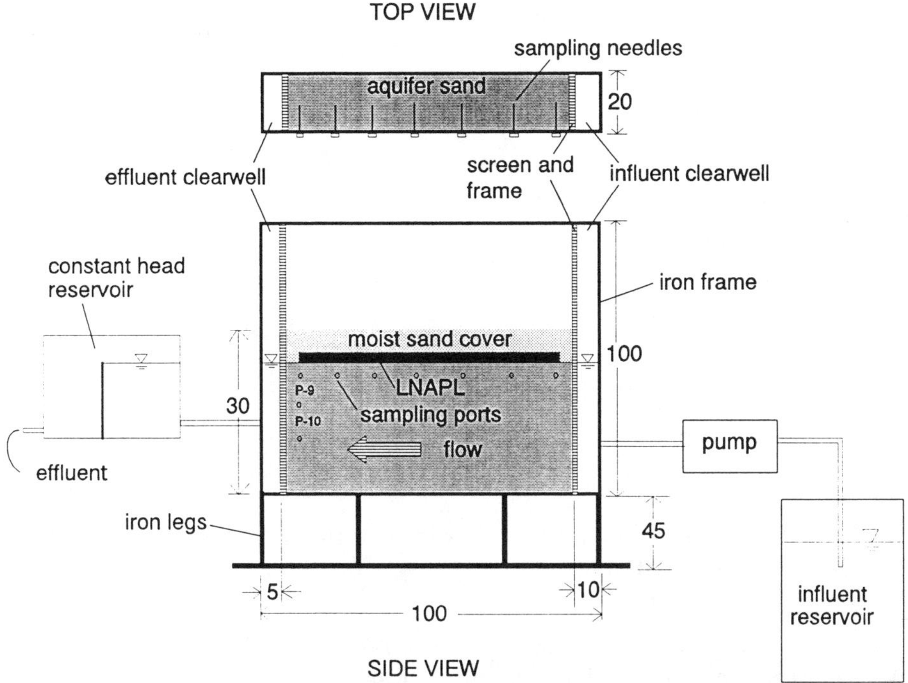

FIGURE 1. Schematic diagram of experimental aquifer and tank assembly with influent and effluent appurtenances (dimensions in cm).

Nevada). The samples were analyzed using a purge-and-trap procedure with a 4460A Sample Concentrator (O.I. Corporation) and a Hewlett Packard Gas Chromatograph (Model 5710A) equipped with a flame ionization detector.

RESULTS AND DISCUSSION

The evolution of toluene concentration with time, at sampling port P-9, which was located at the downstream end of the pool 4 cm below the pool surface, is shown in Figure 2. The results indicate that the dissolution of the toluene pool required approximately 100 hours to reach steady state at 57.2 cm/d pore velocity. The highest toluene concentration detected at sampling port P-9 was 24 mg/L, which is approximately 5% of the toluene solubility of 515 mg/L at 20°C (Verschueren 1983). Dissolved toluene concentrations increased with increasing distance along the pool. Figure 3 presents the steady-state concentration versus distance profile for the sampling ports located 4 cm below the pool surface.

As the vertical distance from the pool increased, concentration levels decreased sharply, resulting in a relatively thin plume of dissolved toluene. For example, the concentration in port P-10 (Figure 1) detected under steady state was 0.03 mg/L.

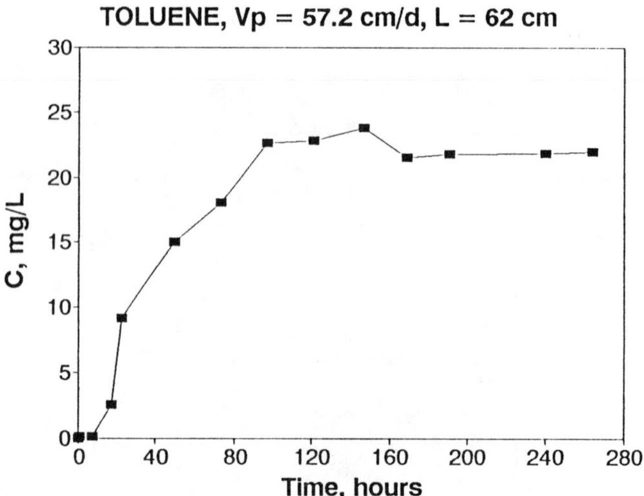

FIGURE 2. Concentration (C) of dissolved toluene vs. experimental time (Vp is pore velocity).

This is 800 times lower than that detected at P-9 (24 mg/L), which was located only 4 cm above P-10. Within 8 cm of vertical distance from the pool surface, the dissolved toluene concentration decreased from solubility (515 mg/L) to 0.03 mg/L at P-10, i.e., a decrease by 99.994%. Figure 3 illustrates the effect of vertical distance from the pool on dissolved toluene concentrations under steady-state conditions.

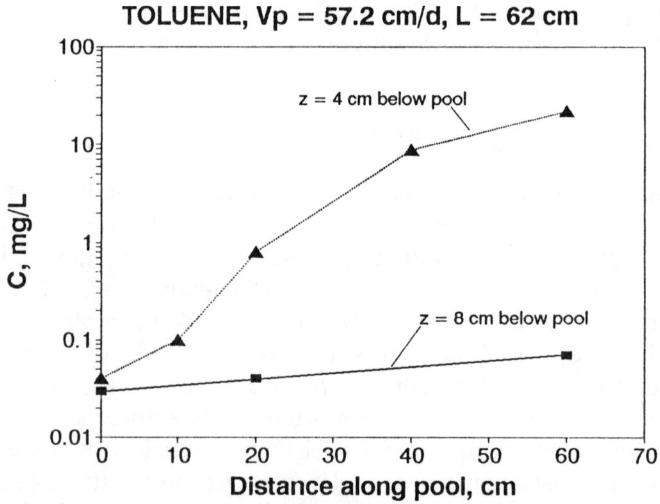

FIGURE 3. Steady state concentration profile vs. distance along the pool for sampling ports located at distances 4 and 8 cm below the pool.

The above data suggest that cleanup of NAPL-contaminated groundwater by the pump-and-treat procedure may be an ineffective and extremely long approach, producing massive volumes of dilute wastestreams. This occurs because NAPL dissolution kinetics result in low dissolved-NAPL concentrations on the absolute scale, although these concentrations may be high relative to drinking water standards.

One of the factors affecting in situ biological treatment of NAPL-contaminated groundwater is the rate at which NAPL components are dissolved in groundwater. However, the rate of biodegradation may be limited by low dissolved-NAPL concentrations, even when oxygen and nutrients are abundant. Development of a reliable approach for in situ biorestoration will require an accurate description of the NAPL dissolution process.

ACKNOWLEDGMENT

This research was supported by the National Science Foundation, under Grant No. BCS-9022205.

REFERENCES

Anderson, M. R., R. L. Johnson, and J. F. Pankow. 1992. "Dissolution of dense chlorinated solvents into groundwater: 1. Dissolution from a well-defined residual source." *Ground Water 30*: 250-256.

Mackay, D. M., and J. A. Cherry. 1989. "Groundwater contamination: Pump-and-treat remediation." *Environmental Science and Technology* 23: 630-636.

Verschueren, K. 1983. *Handbook of Environmental Data of Organic Chemicals,* 2nd ed. Van Nostrand Reinhold Company, New York, NY.

ENHANCED AEROBIC BIOREMEDIATION OF A GASOLINE-CONTAMINATED AQUIFER BY OXYGEN-RELEASING BARRIERS

C.-M. Kao and R. C. Borden

Contamination of groundwater by gasoline and petroleum-derived hydrocarbons has been recognized as a serious and widespread environmental problem. BTEX (benzene, toluene, ethylbenzene, *m+p*-xylenes, and *o*-xylene) are major components of gasoline and of great concern because they are both toxic and relatively soluble in water. All of these BTEX compounds are readily biodegradable under aerobic conditions and have been successfully remediated using enhanced in situ aerobic remediation (Wilson et al. 1986). The aim of this study is to design and field-test a barrier system as an alternative method for treating contaminated groundwater. This system is expected to be less expensive to construct and simpler to maintain than the pump and treat or other biological techniques currently in use. The permeable barrier system includes a line of fully screened polyvinyl chloride (PVC) wells installed downgradient of the hydrocarbon spill and perpendicular to the groundwater flow direction. Each well contains a column of oxygen-releasing chemical (ORC) concrete that acts as a diffusion source for the oxygen. Oxygen is released from the ORC columns at a controlled rate, enhancing the biodegradation of dissolved hydrocarbons by native microorganisms within the barriers and the downgradient aquifer.

MATERIALS AND METHODS

Laboratory studies have been conducted to identify an appropriate ORC mixture and determine the oxygen release rate from ORC concrete. Three different oxygen-releasing compounds (calcium peroxide, urea peroxide, and magnesium peroxide) were investigated in batch manometer experiments (Page et al. 1982). From our laboratory data, magnesium peroxide was selected for field testing because it is stable and maintains a uniform oxygen-releasing rate. Oxygen is given off when MgO_2 reacts with water in the following reaction:

$$2\ MgO_2 + 2\ H_2O \rightarrow 2\ Mg(OH)_2 + O_2$$

Laboratory results indicate that a 37% by weight mixture of commercially available MgO_2 is suitable. Low-strength concrete was prepared by blending cement, sand, water, and commercial MgO_2 together at the ratio of 1.79:1:1.66:2.58 by weight. In the laboratory test, the release rate from a 4 × 4-in. (10 × 10-cm) cylinder of this mixture was approximately 0.17 mg O_2 per day per gram of ORC-concrete and was almost constant over a 2-month monitoring period. During this period, approximately 37% of the oxygen-releasing potential was liberated.

SITE DESCRIPTION

Soil and groundwater contamination due to release of gasoline from an underground storage tank was initially identified and investigated at the Jennifer Mobile Home Park by the North Carolina Division of Environmental Management. Additional site characterization has been completed by North Carolina State University to further define the plume and aid in designing the permeable barrier system. The total soluble BTEX concentration is approximately 8 mg/L in the middle of the plume (150 ft [46 m] downgradient of the source). At this location, the maximum width of the plume is 120 ft (37 m) and the maximum depth is approximately 10 ft (3 m) from the water table. Within the plume, the oxygen concentration is zero and the Eh is below −50 mV. The background wells (F-1, MW-1, ORCMW-2, and ORCMW-6) give positive oxygen concentrations (from 2 to 4.5 mg/L) and Eh values (larger than 50 mV). Figure 1 shows the locations of the monitoring wells. The groundwater flow direction is shown on Figure 1 and the conductivity of this aquifer is around 76 ft/d (23 m/d).

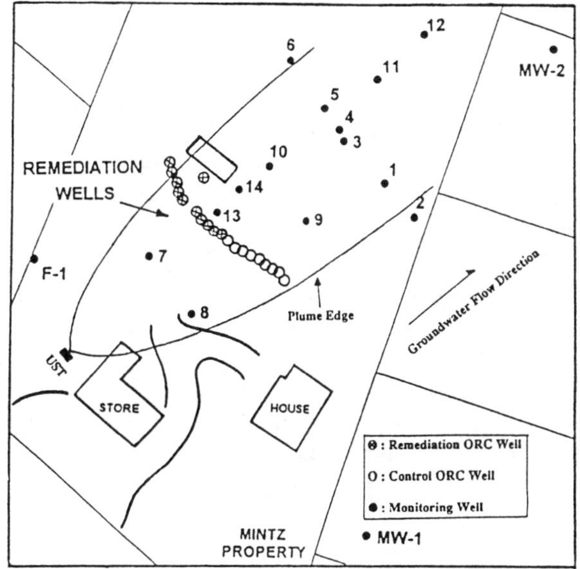

FIGURE 1. Site map and ORC barrier system.

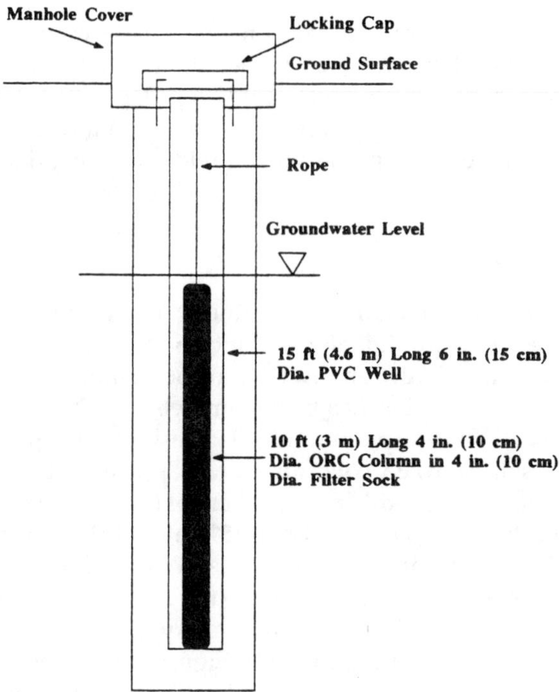

FIGURE 2. Schematic of ORC well.

ORC BARRIER SYSTEM

The full-scale ORC barriers were constructed intersecting the plume and 150 ft (46 m) downgradient from the spill source. The permeable barrier system consists of 22 fully screened PVC wells in a line perpendicular to the plume located 5 ft (1.5 m) on center. These wells are 6 in. (15 cm) diameter and 15 ft (4.6 m) deep. The ORC concrete was installed as 4-in.-diameter (10-cm-diameter) by 10-ft-long (3-m-long) columns in filter fabric socks that were hung inside the 6-in.-diameter (15-cm-diameter) PVC wells (Figure 2). Each ORC concrete column weighs approximately 80 lb (36 kg), and includes 35 lb (16 kg) of ORC. Half of the PVC wells are used as control wells without ORC columns inside. A schematic diagram of the ORC barrier system is presented in Figure 1. Monitoring wells were installed upgradient and downgradient of the barriers on both the ORC and control sections. Groundwater samples are monitored for dissolved oxygen, individual BTEX concentrations, pH, redox potential, and other relevant parameters. The BTEX concentrations are determined by purge and trap gas chromatography (Kao 1993).

MONITORING RESULTS

Dissolved oxygen concentrations and total BTEX from monitoring wells ORCMW-7, 13, 14, 10, and F-1 are plotted versus time (Figures 3 and 4). The

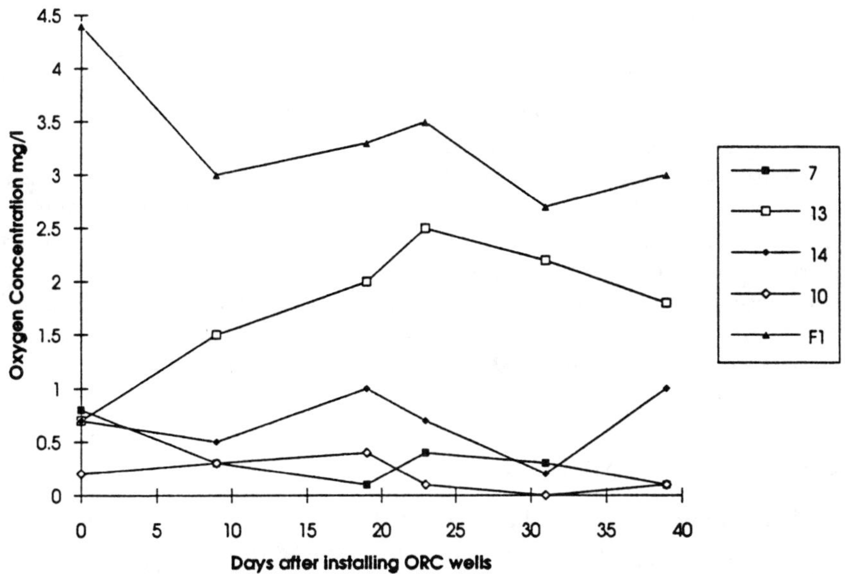

FIGURE 3. Oxygen concentration versus time in ORCMW-7, 13, 14, 10, and F-1.

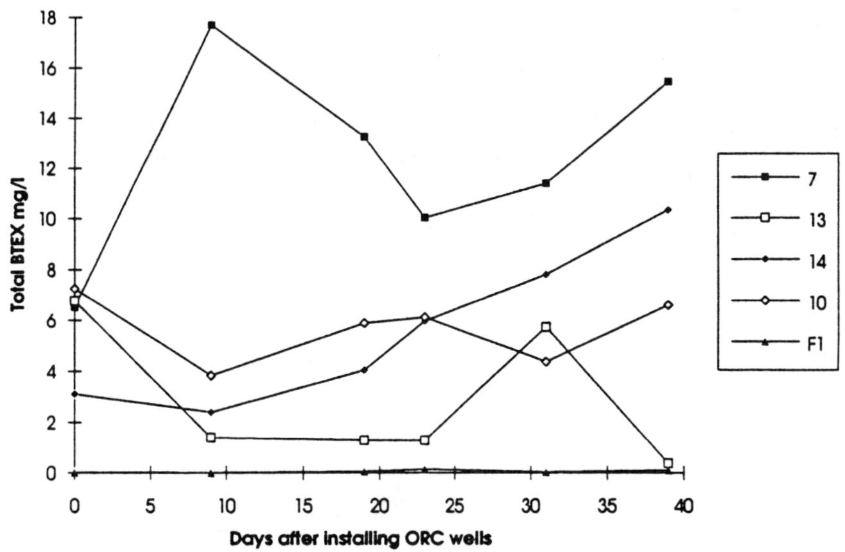

FIGURE 4. Total BTEX versus time in ORCMW-7, 13, 14, 10, and F-1.

oxygen concentration in ORCMW 13 began to gradually increase 9 days of the installation of ORC wells. Approximately 80% of the total BTEX is removed based on the difference in BTEX concentration between monitoring wells immediately upgradient and downgradient of the barrier. At wells further downgradient, the removal efficiency is much more variable. We believe this variation is due to changes in groundwater flow direction. If there are minor changes in direction, a well may move into an out of the treated zone of groundwater downgradient from an ORC well. We expect these treated zones will eventually merge and the entire plume will be controlled.

Construction cost for the system of 22 ORC barrier wells was $15,000. The ORC concrete will be replaced every 5 to 6 months to obtain an efficient oxygen-releasing rate. This is estimated to cost $3,000 per year for the ORC.

ACKNOWLEDGMENTS

This work was supported in part by the U.S. Environmental Protection Agency (EPA) through cooperative agreement No. CR820468-01-0. It has not been subject to EPA review and therefore does not necessarily reflect the views of the EPA and no official endorsement should be inferred. We would also like to thank Plant Research Laboratories, Inc., for providing the ORC used in these studies.

REFERENCES

Kao, C.-M. 1993. Ph.D. Dissertation, Department of Civil Engineering, North Carolina State University, Raleigh, NC.

Page, A. L., R. H. Miller, and D. R. Keevey (Eds.). 1982. *Methods of Soil Analysis*. Part 2, 2nd ed., Chap. 41, pp. 831-872. American Society of Agronomy, Inc., Madison, WI.

Wilson, J. T., L. E. Leach, M. Henson, and J. N. Jones. 1986. "In Situ Biorestoration as a Groundwater Remediation Technique." *Ground Water Monitoring Review 7(4)*: 56-64.

MAXIMIZATION OF *METHYLOSINUS TRICHOSPORIUM* OB3b SOLUBLE METHANE MONOOXYGENASE PRODUCTION IN BATCH CULTURE

J. P. Bowman and G. S. Sayler

INTRODUCTION

The type II methanotroph, *Methylosinus trichosporium* OB3b, oxidizes methane to methanol either with particulate methane monooxygenase (pMMO) or with soluble methane monooxygenase (sMMO). Biochemically, these enzymes differ significantly (Dalton 1992, Nguyen et al. 1992); however both require nicotinamide adenine dinucleotide (NADH). In the last 5 years sMMO has become regarded as a potentially highly useful enzyme for bioremediation applications. The enzyme has an exceptionally wide substrate range. For instance, sMMO can epoxidate trichloroethylene (TCE) resulting in a cascade of reactions that essentially leads to complete mineralization of this pollutant. This process is orders of magnitude more efficient than other mono- or dioxygenase systems. Many other chlorinated aliphatic compounds also can be degraded by sMMO (Ensley 1991). By comparison, pMMO has a much narrower substrate range and can oxidize TCE only very slowly (DiSpirito et al. 1992). Consequently, conditions favoring sMMO synthesis are critical if sMMO systems are to be applied to bioremediation.

The objective of this study was to define experimental conditions under which the rate of sMMO synthesis in the methanotroph *Methylosinus trichosporium* OB3b was maximized. This study is part of a larger effort involving the development of an innovative bioreactor design for the biotreatment of TCE and other halogenated aliphatics in groundwaters using a methanotrophic sMMO system.

METHODOLOGY

Methylosinus trichosporium OB3b was grown in a copper-free nitrate mineral salts (NMS) medium (Cornish et al. 1984) at 25°C. Unless otherwise specified, OB3b was cultivated under a 1:4 methane:air atmosphere in vials or flasks. The sMMO-specific enzyme activity was quantified by a modification of a naphthalene oxidation assay (Brusseau et al. 1990). The sMMO activity was expressed as nmoles of naphthol formed per h per mg of protein. The sMMO naphthalene oxidation rate directly correlates to TCE degradation rate in OB3b and in other methanotrophs (unpublished data). TCE degradation and methane levels were quantified using gas chromatography (Brusseau et al. 1990).

An initial series of experiments, determined the effects of methane availability on sMMO activity. Subsequent experiments were directed towards media optimization and the observation of the effects of supplementary carbon compounds and a variety of growth factors on sMMO activity. Using the data obtained in these experiments, OB3b batch cultures were set up to and incubated for several weeks in order to observe sMMO stability and to determine what was necessary to maintain sMMO activities at a stable level.

RESULTS AND DISCUSSION

When methane headspace concentrations (5 to 50% v/v) were increased, a corresponding increase in sMMO activity was detected (Figure 1). Upon the onset of methane limitation, enzyme activity rapidly declined. The sMMO activity was

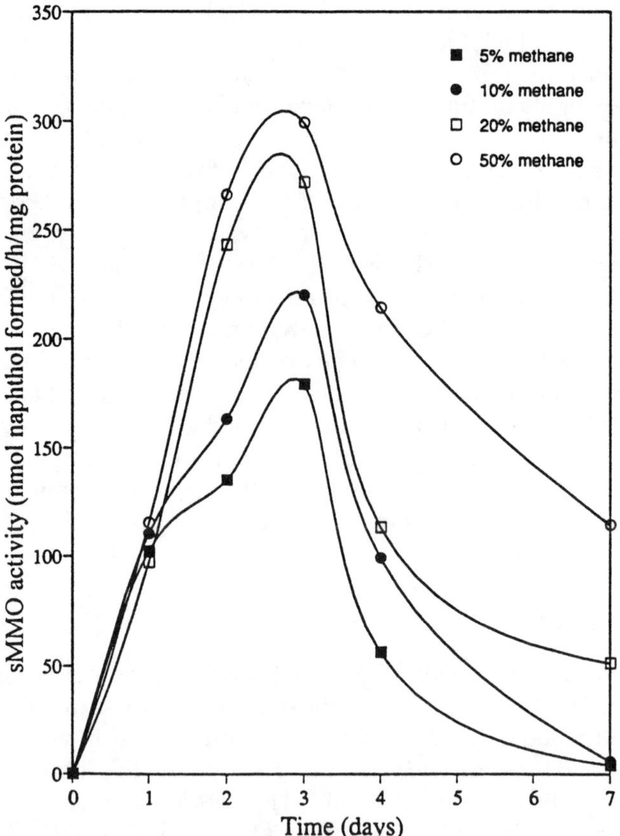

FIGURE 1. The effect of methane headspace levels on sMMO-specific activity in *Methylosinus trichosporium* OB3b.

retained to some extent in cultures cultivated with 20% and 50% methane headspace levels. Significant levels of residual methane was detected in these cultures, whereas no methane could be detected in cultures grown with 5% and 10% methane (Figure 1). It has been shown that cell-free sMMO extracts switch from an oxygenase (for methane oxidation) to a hydrolase when a suitable oxidizable substrate is not present (Dalton 1992). It is thought that this switch is employed in methanotrophs as an NADH conservation mechanism. Thus, without a suitable substrate, the enzyme shuts down but at least in the short term is not degraded. Another possible reason for the decline in sMMO activity could be exhaustion of the cellular NADH pools. Enzyme NADH requirements and NADH pools are maintained primarily by the oxidation of methane to CO_2 by a dissimilatory pathway. With the continual supply of methane, sMMO activity is maintained.

A series of experiments were performed to assess the importance of various constituents of the NMS medium. Physicochemical parameters affecting sMMO activity including temperature, pH, and NaCl concentration have been reported previously (Brusseau et al. 1990). Nitrate and phosphate, and to a lesser extent iron and magnesium, were found to be the most critical factors for both biomass development and sMMO expression. Nitrate and phosphate limitation led to a distinct sMMO synthesis decline (Figure 2). Presumably sMMO turnover is

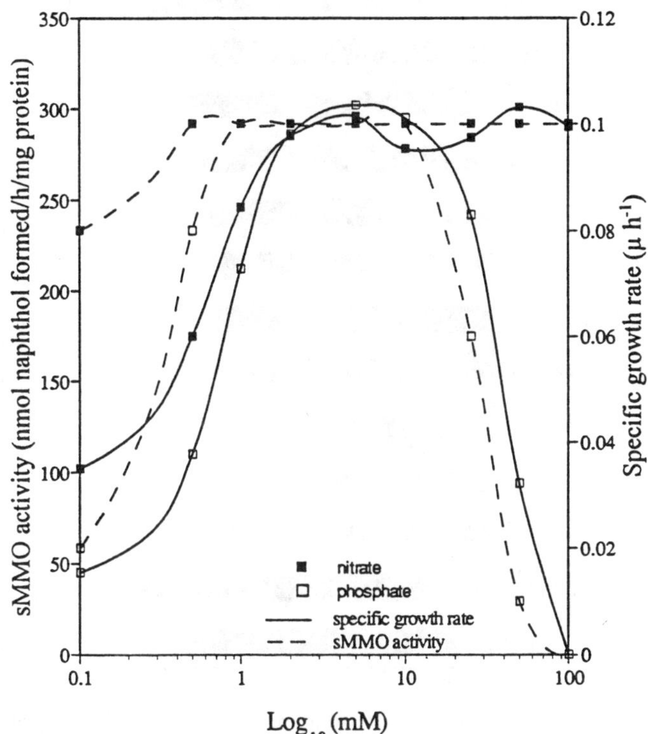

FIGURE 2. The effect of nitrate and phosphate concentration on specific growth levels and sMMO-specific activity in *Methylosinus trichosporium* OB3b.

affected directly by the lack of nitrate and phosphate. In nitrogen-free NMS medium, sMMO levels were much lower than compared to NMS prepared with nitrate as the combined nitrogen source (data not shown). This reduction was perhaps due to channeling of NADH from the methane oxidation dissimilatory pathway to nitrogen fixation. Other constituents in the NMS medium (including Ca(II), Co(II), Mn(II), Zn(II), and Mo(VI)) when absent seemed to have no effect on sMMO synthesis or specific growth rates.

The addition of supplementary carbon sources (including amino acids, organic acids, alcohols, and carbohydrates) to the NMS medium generally had little effect OB3b sMMO synthesis and specific growth rates. A significant increase in sMMO synthesis and TCE degradation was seen when a vitamin solution (diluted 1:2000) was added to the NMS medium; the vitamins mostly responsible for the increase were d-biotin, vitamin B_{12}, and pyridoxine (Figure 3). These growth factors did not appear to affect sMMO synthesis or growth as no change in protein levels

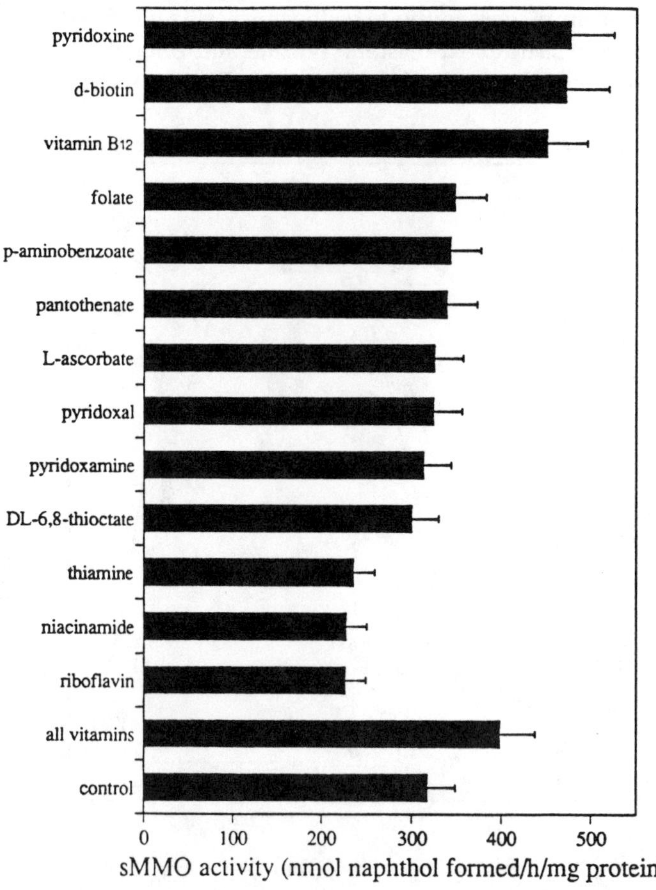

FIGURE 3. The effect of various vitamins on sMMO-specific activity in *Methylosinus trichosporium* OB3b.

or specific growth rate was detected. Nor did they increase activity by acting as artificial electron donors as is the case for formate which stimulates sMMO by indirectly providing NADH via formate dehydrogenase (Brusseau et al. 1990). The increase in sMMO activity occurred only if OB3b was cultivated in the presence of these vitamins. The addition of the vitamins to resting cell suspensions did not stimulate sMMO activity.

In a final series of experiments OB3b was cultivated for several weeks in a simplified NMS medium which consisted of 2 mM $NaNO_3$, 2 mM phosphate buffer (pH 6.8), 50 M $FeCl_3.6H_2O$, and 50 M $MgSO_4.7H_2O$. In addition, separate flasks were supplemented with d-biotin, pyridoxine, and vitamin B_{12}. Specific growth rates were only slightly less than what was usually obtained with the complete NMS medium. When methane and the nutrients were maintained in flask cultures (approximating the initial levels) the sMMO levels, after peaking in the early stationary growth phase, declined over time at an increasingly slower rate until a plateau level was obtained (Figure 4). When starved of methane, a much more rapid decline in activity was observed. Within 9 to 10 d no residual methane was detectable in these cultures. After 20 d of incubation, the addition of methane recommenced for cultures deprived of methane. A recovery of sMMO activity ensued (Figure 4), indicating cellular NADH pools can be renewed if given time.

CONCLUSION

In order to obtain and then maintain a high level of sMMO activity in bioreactor system cultures, methane is required continually. To avoid reduction of TCE degradation rates due to competition of the sMMO active site with methane, a dual-stage bioreactor system such as that modeled by Alvarez-Cohen and McCarty (1991) would be ideal. Cells are cultivated in a continually—stirred through tank reactor (CSTR) and are supplied with sufficient methane to stably maintain cellular NADH pools. Cells are then pumped to a plug flow reactor in which TCE degradation takes place. It is possible to make TCE degradation economical in "pump and treat" systems by simplifying the medium. Ostensibly, groundwater supplemented with nitrate, phosphate, and iron could sufficiently maintain methanotroph growth in the CSTR. In addition higher sMMO levels could be attained if biomass was initially cultivated in the presence of d-biotin, pyridoxine, and vitamin B_{12}. As sMMO levels can be maintained at relatively high levels in the long term, it could be possible to recycle cells back into the CSTR after TCE exposure in the plug flow reactor. This assumes groundwater TCE is completely degraded and TCE toxicity is not a significant factor. The results of the batch culture work presented here will be adapted to a bench-scale bioreactor designed for methanotrophic biotreatment of TCE.

ACKNOWLEDGMENTS

The work presented here and in other ongoing studies represent a demonstration by the Air Force Civil Engineering Support Agency (AFCESA), Oak Ridge

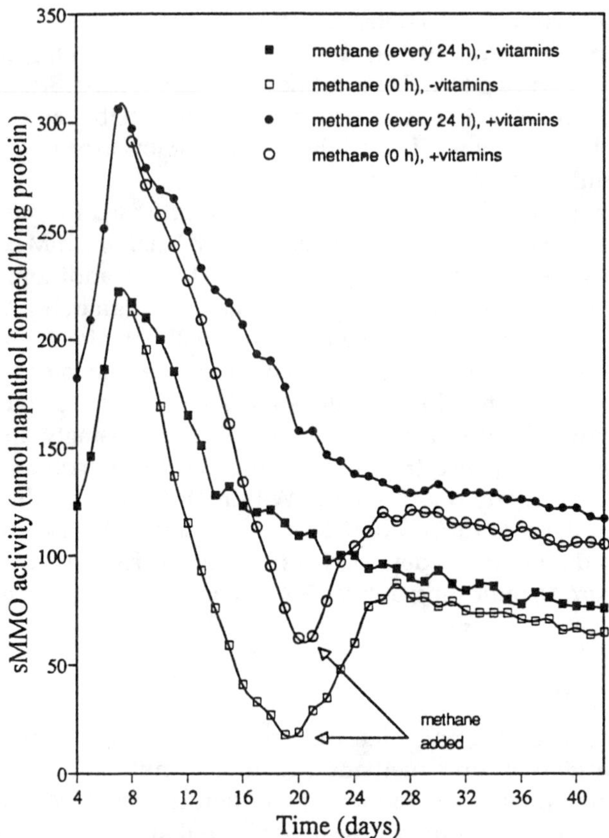

FIGURE 4. Maintenance of sMMO-specific activity in *Methylosinus trichosporium* OB3b in simplified NMS medium (see text) and in vitamin-supplemented (0.5 g/L vitamin B_{12}, 10 g/L d-biotin, and 10 g/L pyridoxine), simplified NMS medium. In one set of flasks, methane was supplied every 24 h, whereas in a second set methane was only added initially at 0 h. After 20 d incubation, addition of methane to the methane-starved cultures recommenced.

National Laboratory, and The University of Tennessee for the U.S. Department of Energy (Contract No. DE-AC05-84OR21400).

REFERENCES

Alvarez-Cohen, L., and P. L. McCarty. 1991. "Two-Stage Dispersed Growth Treatment of Halogenated Aliphatic Compounds by Cometabolism." *Environ. Sci. Technol. 25*:1380-1386.

Brusseau, G. A., H. C. Tsien, R. S. Hanson, and L. P. Wackett. 1990. "Optimization of Trichloroethylene Oxidation by Methanotrophs and the Use of a Colorimetric Assay to Detect Soluble Methane Monooxygenase Activity." *Biodegradation* 1:19-29.

Cornish, A., K. M. Nicholls, D. Scott, B. K. Hunter, W. J. Aston, I. J. Higgins, and J. K. M. Sanders. 1984. "*In-Vitro* ^{13}C-NMR Investigations of Methanol Oxidation by the Obligate Methanotroph *Methylosinus trichosporium* OB3b." *J. Gen. Microbiol. 130*: 2565-2575.

Dalton, H. 1992. "Methane Oxidation by Methanotrophs: Physiological and Mechanistic Implications." In J. C. Murrell and H. Dalton (Eds.), *Methane and Methanol Utilizers*, pp. 85-114. Plenum Press, New York, NY.

DiSpirito, A. A., J. Gulledge, J. C. Murrell, A. K. Shiemke, M. E. Lidstrom, and C. L. Krema. 1992. "Trichloroethylene Oxidation by the Membrane Associated Methane Monooxygenase in Type I, Type II and Type X Methanotrophs." *Biodegradation* 2: 151-164.

Ensley, B. D. 1991. "Biochemical Diversity of Trichloroethylene Metabolism." *Ann. Rev. Microbiol. 45*: 283-299.

BIODEGRADATION OF NAPHTHENIC ACIDS AND THE REDUCTION OF ACUTE TOXICITY OF OIL SANDS TAILINGS

D. C. Herman, P. M. Fedorak, M. D. MacKinnon, and J. W. Costerton

INTRODUCTION

Since 1978, Syncrude Canada Ltd. has extracted bitumen from the Athabasca oil sands deposit located in northeastern Alberta, Canada, using a caustic hot water flotation method. The bitumen is then upgraded to a synthetic crude oil. The separation of bitumen from the oil-bearing sands requires large volumes of water and results in the production of fluid tailings composed of water, solids (sand and clays), and nonextracted bitumen. The oil sands industry follows a "zero discharge" policy; all tailings are contained within a tailings pond and must eventually be reclaimed by either wet or dry landscape options.

At present, the tailings pond covers more than 12 km^2, is approximately 45 m deep, and contains 250 to 300 × 10^6 m^3 of tailings. As the tailings slurries enter the pond, they stratify into a "free-water" zone of low-suspended-solids content (less than 0.1%) and a fine tails zone of higher solids content (10 to 60%) which is slowly densifying. The fine tails solids are mainly clays (mostly kaolinite and illite) mixed with unrecovered bitumen (1 to 2%) and a slightly saline water (MacKinnon 1989).

Water within the tailings pond has been shown to be acutely toxic, with a 96-hour rainbow trout LC_{50} of < 10% (v/v) and Microtox EC_{50} of < 30% (v/v) (MacKinnon & Boerger 1986). The major source of acute toxicity has been identified as a complex group of polar organic acids, believed to be naphthenic acids, that are leached into the process water during bitumen extraction and, at the pH of the tailings (8 to 8.5), will be present as naphthenates.

RECLAMATION OF OIL SANDS TAILINGS

The ability of natural processes to detoxify tailings has been investigated by transferring water from the surface zone of the tailings pond into a shallow pit (3 m at deepest point) where the water was isolated from the constant input of fresh tailings (Boerger & Aleksiuk 1987, MacKinnon & Boerger 1986). Monitoring of water quality (Figure 1) revealed that acute toxicity of the water was reduced

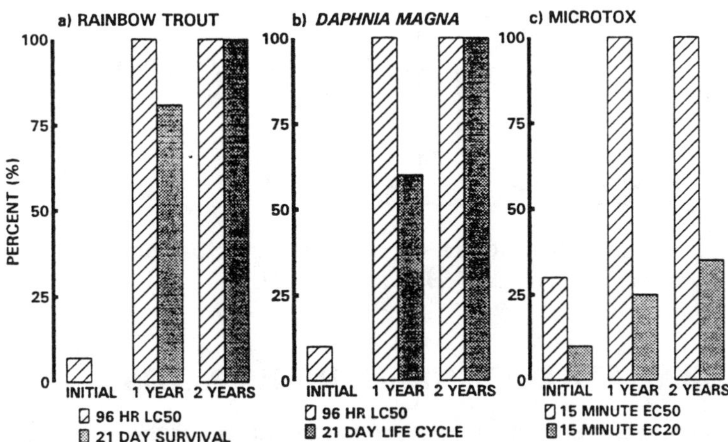

FIGURE 1. Toxicity testing of tailings pond surface water after 0, 1, and 2 years storage in shallow pits. Toxicity testing included static 96-hour rainbow trout and *Daphnia* LC_{50} tests, longer term tests (21-day) in which percent survival is indicated, and Microtox analysis.

within 1 year, as indicated by an increase in Microtox EC_{50} values from an initial level of 35% (v/v) to 100% (v/v) and by rainbow trout and *Daphnia* tests that revealed no mortality during 96-hour exposure periods. After 2 years of containment in pits, 100% survival in longer-term 21-day *Daphnia* life-cycle tests and 21-day rainbow trout survival tests corroborated the loss of acute toxicity of the tailings water.

The strategy for reclamation of tailings fines involves covering or "capping" fine tails with a layer of natural surface water of sufficient depth to isolate and contain the fine tails in the bottom layer and allow the development of an aquatic ecosystem in the capping layer (Boerger et al. 1992, MacKinnon & Boerger 1991). A series of experimental field pits (15 m × 50 m × 7 m) were constructed to test the feasibility of the capping pond approach. During the first 3 years of their development, no resuspension of fine tailings in the capping layer was observed, although changes in water chemistry indicated a slow release of pore water into the capping water. Boerger et al. (1992) and MacKinnon and Boerger (1991) reported that both Microtox and rainbow trout toxicity bioassays revealed no acute toxicity associated with the capping water, even though water from the fine tails layer remained acutely toxic.

An investigation of microbial activity within the capping ponds determined that both total cell number (determined using viable plate counts) and heterotrophic activity (determined from the percent mineralization of carbon-14 labeled glutamic acid) was greatest at the capping water/fine tails interface (data not shown). This layer of biological activity may provide a barrier to the movement of organics, such as the naphthenates, released from tailings fines by the diffusion of interstitial waters during tailings densification.

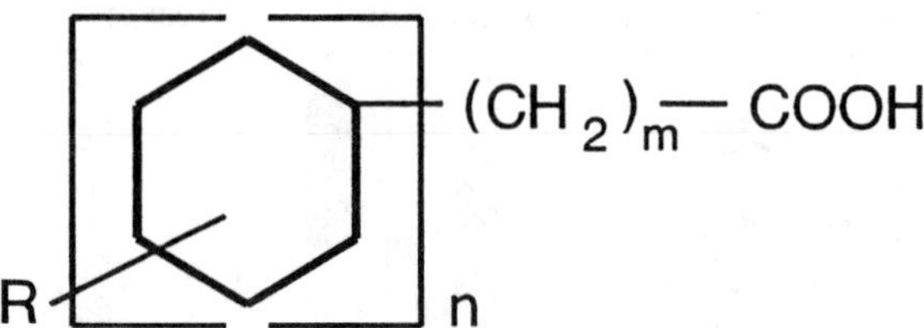

FIGURE 2. Proposed structure of naphthenic acids, modified from the *Encyclopedia of Chemical Technology* (1981).

BIODEGRADATION OF NAPHTHENIC ACIDS

Naphthenic acids are a mixture of predominantly mono- and polycycloalkane (cyclohexane and cyclopentane) carboxylic acids with aliphatic side chains of various lengths. A structural formula (Figure 2) shows the carboxylic acid group at the end of an aliphatic side chain where "m" is greater than one. "R" is a small aliphatic group, such as methyl, and "n" equals one or more.

The biodegradation pathway of naphthenic acids is expected to begin with the β-oxidation of the aliphatic side chain leading to the oxidation of the cycloalkane ring. Therefore, the potential for naphthenic acids biodegradation was investigated by determining the degradation of hexadecane and carboxylated cycloalkanes within oil sands tailings (Herman et al. 1993). Cyclohexane carboxylic acid, cyclopentane carboxylic acid, and a methyl-substituted cyclohexane carboxylic acid were used to represent the carboxylated cycloalkane backbone of naphthenic acids, while hexadecane was used to reveal the presence of n-alkane-degrading bacteria. The combined results indicated that naphthenic acids can be biodegraded within oil sands tailings.

Current studies examining the biodegradation of naphthenic acids use a commercially available naphthenic acid sodium salts (NAS) mixture (Kodak Chemicals, Rochester, New York). A naphthenic acids-degrading mixed bacterial culture was enriched from oil sands tailings using 100 mg/L of NAS dissolved in a minimal salts medium. Microbial activity reduced the toxicity of 100 mg/L of NAS from an initial Microtox EC_{50} level of 30% (v/v) to an EC_{50} of greater than 100% after 9 days of incubation.

The naphthenic acids-degrading enrichment culture was also examined for its capacity to degrade and detoxify organic acids extracted directly from oil sands tailings. The tailings extract contained lower molecular weight organic acids associated with acute toxicity (290 to 600 molecular weight range), plus higher molecular weight acids that showed little toxic character (MacKinnon & Boerger 1986). Incubation of the tailings extract with the naphthenic acids-degrading enrichment culture resulted in a reduction in acute toxicity from an initial Microtox EC_{50} of 40% (v/v) to an EC_{50} between 84 and 95% (v/v) after 20 days of incubation.

These results indicated that bacteria indigenous to oil sands tailings degraded naphthenic acids and reduced acute toxicity of organic acids extracted from tailings. The residual toxicity of the tailings extract may be due to organic acids that are highly recalcitrant to microbial degradation. Further studies will focus on the biodegradation of these compounds and on the biodegradation of naphthenic acids within the oil sands tailings pond under environmental conditions.

REFERENCES

Boerger, H., and M. Aleksiuk. 1987. "Natural detoxification and colonization of oil sand tailings water in experimental pits." In J. H. Vandermeulen and S. E. Hrudey (Eds.), *Oil in Freshwater: Chemistry, Biology, Countermeasure Technology*, pp. 379-387. Pergamon Press, New York, NY.

Boerger, H., M. MacKinnon, T. Van Meer, and A. Verbeek. 1992. "Wet landscape option for reclamation of fine tails." In R. K. Singhal (Ed.), *Environmental Issues and Waste Management in Energy and Minerals Production*, Second International Conference on Environmental Issues and Management of Waste in Energy and Minerals Production, Calgary, Alberta, Canada, September 1992, pp. 1249-1261. Bakkema, Rotterdam.

Encyclopedia of Chemical Technology, 3rd. ed. 1981. pp. 749-753. John Wiley and Sons, New York, NY.

Herman, D. C., P. M. Fedorak, and J. W. Costerton. 1993. "Biodegradation of cycloalkane carboxylic acids in oil sand tailings." *Can. J. Microbiol.*, in press.

MacKinnon, M. D. 1989. "Development of the tailings pond at Syncrude's oil sands plant: 1978 - 1987." *AOSTRA J. Res.* 5: 109-512.

MacKinnon, M., and H. Boerger. 1986. "Description of two treatments for detoxifying oil sands tailings pond water." *Water Poll. Res. J. Canada* 21: 496-512.

MacKinnon, M., and H. Boerger. 1991. "Assessment of a wet landscape option for disposal of fine tails sludge from oil sands processing." *Proc. Pet. Soc. of CIM and AOSTRA Tech. Conf.*, Paper 91-124. Banff, Alberta. April, 1991.

OXIDATION CAPACITY OF AQUIFER SEDIMENT

G. Heron, J. C. Tjell, and T. H. Christensen

INTRODUCTION

The leaching of organic matter and pollutants into shallow, aerobic aquifers constitutes a major groundwater pollution risk. Leachates rich in organic matter typically lead to the development of highly reduced redox environments as oxidized aquifer species are reduced during microbial oxidation of the organics. The dominating aquifer redox reactions have been reviewed by Baedecker & Back (1979), McFarland & Sims (1991) and Stumm & Morgan (1981). The redox half-reactions listed in Table 1 will be key electron-accepting reactions as electrons are donated by the reduced organic matter leaching into the groundwater.

The total oxidation capacity (OXC) of an aquifer volume and subsequently the aquifer sediment can be written as (modified after Barcelona & Holm 1991):

$$OXC = 4[O_2] + 5[NO_3^-] + [Fe(III)_{available}] + 2[Mn(IV)_{available}] + 8[SO_4^{2-}] + x[TOC] \quad (1)$$

$$OXC_{sediment} = [Fe(III)_{available}] + 2[Mn(IV)_{available}] + x[TOC] \quad (2)$$

Where OXC = oxidation capacity
Fe(III) = iron at oxidation state +3 at the sediment
Mn(IV) = manganese at oxidation state +4 at the sediment
TOC = total organic carbon

All concentrations are converted to mol per cubic meter of the aquifer. The availability index is added because not all the oxidized iron and manganese will be available for redox processes. The unknown factor x for the organic matter is the fraction; of organic matter multiplied by the oxidation state change during the reduction of this fraction; x should be small because aquifer organic matter is relatively refractory.

The sediment content of oxidized iron oxides and hydroxides should be very important in controlling the development of aquifer redox zones, as it constitutes the major redox pool to meet the reducing power of the pollution plume. In support of iron reduction as an important redox-buffering mechanism in aquifers, large iron-reducing zones were identified in polluted aquifers by Lyngkilde & Christensen (1992b) and Nicholson et al. (1983).

TABLE 1. Governing electron-accepting reactions in aquifers.

Reaction	Phase
$O_2 + 4H^+ + 4e^- \rightarrow 2H_2O$	(aq)
$NO_3^- + 6H^+ + 5e^- \rightarrow ½N_2 + 3H_2O$	(aq)
$Fe(OH)_3 + 3H^+ + e^- \rightarrow Fe^{2+} + 3H_2O$	(solid)
$FeOOH + 3H^+ + e^- \rightarrow Fe^{2+} + 2H_2O$	(solid)
$Fe_2O_3 + 6H^+ + 2e^- \rightarrow 2Fe^{2+} + 3H_2O$	(solid)
$MnO_2 + 4H^+ + 2e^- \rightarrow Mn^{2+} + 2H_2O$	(solid)
$SO_4^{2-} + 9H^+ + 8e^- \rightarrow HS^- + 4H_2O$	(aq)
$CH_2O + 4e^- + 4H^+ \rightarrow CH_4 + H_2O$	(aq + solid)

The oxidation capacity (OXC) represents the pool of oxidized species governing the redox zone formation. An OXC of 0 will allow for large zones of methanogenesis and sulfate reduction in areas of oxygen and nitrate depletion. Dilution and diffusion of oxygen and nitrate will then determine the boundaries of the reduced zones. High levels of OXC related to iron species as expected in reddish, iron-rich, oxic aquifers may diminish the reduced zones and allow for larger zones of iron and manganese reduction, as the thermodynamically less favorable redox reactions are suppressed. Therefore, the OXC will be an important geochemical parameter in controlling both plume formation and, indirectly, the fate of organic contaminants in polluted aquifers.

This paper presents the anaerobic titanous EDTA extraction technique designed to quantify the oxidation capacity related to oxidized iron and manganese species of aquifer sediment.

METHODOLOGY

Fresh sediments were collected from the contaminant plume downgradient of the Vejen landfill in Denmark previously described by Lyngkilde & Christensen (1992a, 1992b). A series of well-described redox zones was identified in the aquifer. For this study, sediments were collected from the highly reduced methanogenic zone (M samples), the active iron-reducing zone (F samples), and the weakly polluted aerobic parts of the sandy aquifer (A samples).

Sediments were sampled anaerobically as intact cores with the preservation of the pore water and kept anaerobically at 10°C for a maximum of 10 days prior to transfer to an anaerobic glovebox.

For the oxidation capacity determinations, the anaerobic titanous EDTA extraction was developed. This technique is a refinement of the TiE method by Ryan & Gschwend (1991). Briefly, a 8×10^{-3} M solution of Ti^{3+} in 5×10^{-2} M ethylenediaminetetraacetic acid (EDTA) was prepared anaerobically from $TiCl_3$ (15% W/v, low in iron) dissolved in deionized water. The pH was adjusted to 6.0 with NaOH and remained between pH 6 and 5.6 during sediment extraction. The extractant was stored under oxygen-free headspace inside an anaerobic chamber to prevent oxidation of the Ti^{3+} by atmospheric oxygen. Anaerobically, 10 mL of extractant was transferred to the extraction vessel containing 0.6 to 1.0 g of wet sample. The vessels were rotated (2 rpm) at 20°C in the dark for 24 hours.

During the extraction, aquifer oxidized species were reduced in combination with this half-reaction:

$$Ti^{3+}\text{-EDTA} \rightarrow Ti^{4+}\text{-EDTA} + e^- \tag{3}$$

Next, the vessels were centrifuged to remove particles greater than 0.25 µm. The remaining nonreacted Ti^{3+} in the extract was quantified by immediate anaerobic titration with a solution of 10^{-3} M $K_2Cr_2O_7$ in 10% H_2SO_4 using the redox indicator Neutral Red (E_H(pH 7)= −0.29V). During titration, Ti^{3+} was oxidized via the reaction

$$6Ti^{3+} + Cr_2O_7^{2-} + 14H^+ \rightarrow 6Ti^{4+} + 2Cr^{3+} + 7H_2O \tag{4}$$

The maximum OXC detected (75% reductant consumption) corresponded to 110 µeq/g. The interference of Fe^{2+} with the titration was found to be negligible by additions of $FeCl_2$ to the extractant (data not shown). The extracted Fe and Mn were determined from split samples of the extracts by atomic absorption spectrometry. A detection limit of 4 µeq/g was achieved for reduced samples by increasing the sample size to 1 g.

The amount of 0.50 M HCl-extractable Fe(II) was determined using a slight modification of the method described by Lovley & Phillips (1986b). Briefly, 0.60 g of fresh sediment was extracted anaerobically for 24 hours by 0.50 M HCl. After centrifugation, the amount of Fe^{2+} in the extract was determined by the ferrozine method.

As an estimate of the total non structural iron and manganese content, sediments were extracted with 5 M HCl at 90°C for 8 hours. The extracted amounts of iron and manganese was determined by atomic absorption spectrometry.

RESULTS AND DISCUSSION

OXC of Standard Oxidized Aquifer Species. To test the OXC procedure, standard amounts of oxidized species were extracted either as single samples or as additions to aquifer sediments. The Ti^{3+}-EDTA instantly reduced oxygen. Figure 1 lists the results for synthetic aquifer minerals. The OXC recovery was determined as the measured OXC divided with the added amounts of oxidation equivalents. The electron transfer quantification was in close agreement with the amount of reduced species found in solution. In conclusion, the Ti^{3+}-EDTA solution extracted the minerals ferrihydrite (100%), akageneite (100%), goethite (93%), hematite (93%), magnetite (8 to 10%), and pyrolusite (99%) when these were added as powder. The low degree of reduction of magnetite was in context with related studies on anaerobic sediments showing magnetite resistance to

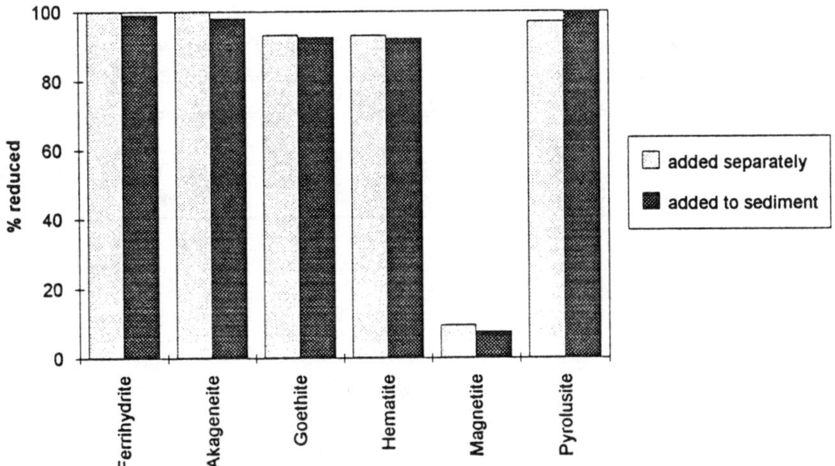

FIGURE 1. Recovery of oxidation equivalents from synthetic minerals following 24-hr anaerobic extraction of minerals added as single samples and as additions to fresh oxic sediments. Values are mean of 5 replicates.

microbial reduction and magnetite formation by iron-reducing bacteria (Lovley & Phillips 1986a; Lovley et al. 1987).

Organic matter, represented by glucose, hydroquinone, and humic acid, was insignificantly reduced by the anaerobic titanous EDTA extraction technique (data not shown).

Oxidation Capacities of Aquifer Sediments. A well-defined fraction of the oxidized aquifer species was reduced in single 24-hr extractions. Larger extraction periods or multiple extractions only slightly increased the measured oxidation capacity (data not shown). The oxidation capacities of two strongly reduced (methanogenic), two iron-reducing, and two aerobic sediments were determined (Table 2 and Figure 2). The oxidation capacity were significantly higher for the oxic sediments than for the reduced sediments. This pattern was supported by the analysis of more than 40 other sediments from the Vejen landfill site (data not shown), yielding high oxidation capacities (20 to 110 μeq/g) for oxidized, reddish sediments and low oxidation capacities (0 to 10 μeq/g) for the reduced sediments.

For the aerobic sediments, the major contributors to the oxidation capacity were oxidized iron species. This balance is based on the assumption that only minor amounts of reduced iron and manganese species were extracted by the solution as confirmed by control experiments (data not shown). The iron extracted by the Ti^{3+}-EDTA solution typically made up for 90 to 100% of the transferred electron equivalents, whereas the manganese species typically contributed 2 to 5%. In these sediments, the organic matter contributed insignificantly to the measured

TABLE 2. Oxidation capacity and extracted amounts of iron and manganese for reduced and oxic sediments from the aquifer contaminated by the Vejen landfill, Denmark (mean of 5 replicates). Units are μeq/g of dry sediment.

Redox zone		OXC	st. dev.	Fe in TiE extract	Mn in TiE extract	Total Fe in 5M HCl extr.	Total Mn in 5M HCl extr.
Oxic	A1	31	4.2	32	0.7	56	1.0
Oxic	A2	23	2.5	22	1.2	56	1.6
Iron reducing	F1	5	2.5	2.4	0.0	36	0.8
Iron reducing	F2	10	1.9	8.1	0.1	107	1.4
Methanogenic	M1	4	4	1.6	0.0	29	0.6
Methanogenic	M2	4	4	2.1	0.0	36	0.5

oxidation capacity of the sediment. About 40 to 70% of the total acid-extractable iron contributed to the oxidation capacity of aerobic sediments.

For the reduced sediments, oxidation capacities were significantly lower and approached the detection limit of the method. The low amounts of iron and manganese reduced during the extraction further indicated that these redox pools were depleted in the reduced sediments. Due to the very low levels, the contributors

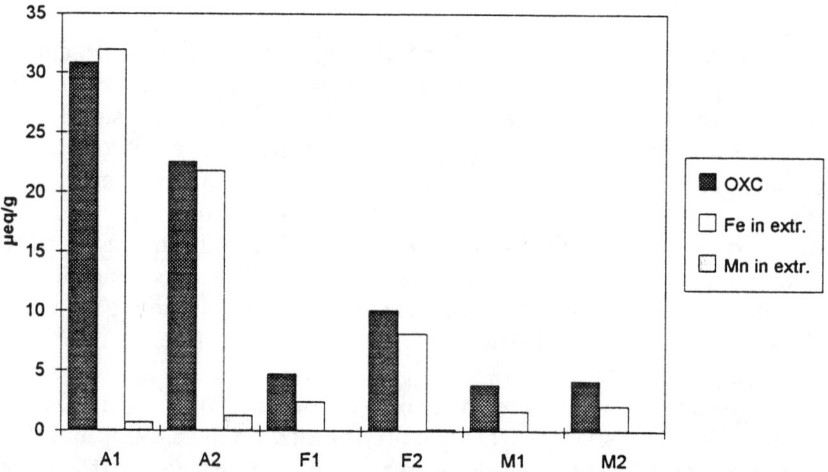

FIGURE 2. Oxidation capacity and extracted amounts of iron and manganese equivalents for 6 selected sediments from the redox zones at the Vejen landfill site. A1, A2: aerobic; F1, F2; iron reducing; M1, M2: methanogenic. Values (μeq/g) are mean of 5 replicates. Standard deviations are listed in Table 2.

TABLE 3. Oxidation capacity (OXC) and 0.50 M HCl-extractable Fe(II) determined for fresh, freeze-dried, and anaerobically stored sediments. OXC units are μeq/g, Fe(II) units are μmol/g.

	OXC			Fe(II)		
Sediment	Fresh	Freeze dried	Stored anaerobically	Fresh	Freeze dried	Stored anaerobically
F1	5	8	5	6	3	6
F2	10	24	9	25	14	26
M1	4	4	4	7	3	7
M2	4	5	4	7	4	8

to the measured oxidation capacities cannot be separated. However, it was indicated that iron compounds made up for at least half of the oxidation capacity in these sediments. The majority of the iron content of the reduced sediments consisted of either Fe(II) compounds as pyrite or very stable crystalline Fe(III) and mixed Fe(III)-Fe(II) oxides.

TOC concentrations of typically 0.2 mg/g were recorded for the sediments, which leads to the conclusion that organic matter was reduced in a very minor degree. In neither the aerobic nor the reduced sediments could significant oxidation capacity be attributed to the reduction of organic matter.

Supplementary tests revealed the need for redox-intact sediment sampling and handling in an oxygen-free atmosphere, as Fe(II) was oxidized to Fe(III) during freeze-drying leading to an overestimation of the oxidation capacities (Table 3). A 30-day holding period inside an anaerobic chamber was accepted, as oxidation capacities and the content of Fe(II) remained constant.

CONCLUSIONS

An anaerobic titanous EDTA extraction technique to determine the oxidation capacity of aquifer sediment was developed. The selected extraction procedure proved reliable for quantifying electron transfer to oxidized aquifer species via the oxidation of Ti^{3+}-EDTA to Ti^{4+}-EDTA. The extractions yielded a well-defined fraction of the oxidized sediment species. For aerobic sediments, the oxidation capacity (on the order of 20 to 110 μeq/g) was related to a well-defined fraction of the Fe(III) species. Significantly lower oxidation capacities (0 to 10 μeq/g) close to the detection limit were determined for reduced sediments.

REFERENCES

Baedecker, M. J., and W. Back. 1979. "Modern Marine Sediments as a Natural Analog to the Chemically Stressed Environment of a Landfill." *J. of Hydrol.* 43: 393.

Barcelona, M. J., and T. R. Holm. 1991. "Oxidation-Reduction Capacities of Aquifer Solids." *Environ. Sci. Technol. 25*: 1565.

Lovley, D. R., and E. J. P. Phillips. 1986a. "Availability of Ferric Iron for Microbial Reduction in Bottom Sediments of the Freshwater Tidal Potomac River." *Appl. & Environ. Microbiol.* 52: 751.

Lovley, D. R., and E. J. P. Phillips. 1986b. "Organic Matter Mineralization with Reduction of Ferric Iron in Anaerobic Sediments." *Appl. & Environ. Microbiol. 51*: 683.

Lovley, D. R., J. F. Stolz, G. L. Nord, Jr., and E. J. P. Phillips. 1987. "Anaerobic Production of Magnetite by a Dissimilatory Iron-Reducing Microorganism." *Nature 330*: 252.

Lyngkilde, J., and T. H. Christensen. 1992a. "Fate of Organic Contaminants in the Redox Zones of a Landfill Leachate Pollution Plume (Vejen, Denmark)." *J. Contam. Hydrol. 10*: 291.

Lyngkilde, J., and T. H. Christensen. 1992b. "Redox Zones of a Landfill Leachate Pollution Plume (Vejen, Denmark)." *J. Contam. Hydrol. 10*: 273.

McFarland, M. J., and R. C. Sims. 1991. "Thermodynamic Framework for Evaluating PAH Degradation in the Subsurface." *Ground Water 29*: 885.

Nicholson, R. V., J. A. Cherry, and E. J. Reardon. 1983. "Migration of Contaminants in Groundwater at a Landfill: A Case Study." *J. of Hydrol. 63*: 131.

Ryan, J. N., and P. M. Gschwend. 1991. "Extraction of Iron Oxides from Sediments Using Reductive Dissolution by Titanium(III)." *Clays and Clay Minerals 39*: 509.

Stumm, W., and J. J. Morgan. 1981. *Aquatic Chemistry*. John Wiley & Sons, New York, NY.

BIOREMEDIATION: A SOUTH AFRICAN EXPERIENCE

Z. M. Lees and E. Senior

INTRODUCTION

As in the rest of the world, treatment and disposal of organic wastes in South Africa are being addressed. The rising costs of existing options, as well as the severe lack, or inadequacy, of facilities, planning and legislation, have led to chronic contamination of our soils and groundwater by indiscriminate dumping and by accumulation of wastes. Unfortunately, due to the absence of legislation, there has been a lack of systematic monitoring. As a consequence, the extent of soil and groundwater pollution is not known.

There is an urgent need to develop cost-effective technologies to treat contamination in South Africa. These technologies must be both simple and flexible, facilitating treatment in situ. Bioremediation, as an emerging technology that can compete effectively in the waste management arena, holds tremendous potential for contaminated site remediation.

In March 1991, a study was initiated by Shell S. A. (Pty) Ltd to evaluate in situ bioremediation of an oil-contaminated site. The central aim of this study was to make a comprehensive laboratory evaluation of the efficacy of indigenous microbial species to decontaminate the soil. The preliminary results are reported here.

FIELD INVESTIGATION

The site (2 hectares) comprises an oil recycling plant that ceased operation in 1990, after operating for more than 2 decades. Test pits (37, each <1.2 m) were excavated across the site at locations considered representative of the different areas of contamination. The pits were profiled in accordance with standard practice, and the soil horizons were sampled and analyzed for oil and heavy metals (As, Zn, Pb, Cd, Ni, and Cr).

Six 30-m-deep, and one 70-m-deep monitoring boreholes were installed to investigate subsurface contamination and geology. Core samples for oil analysis were taken at regular intervals and stored at 4°C. Due to the highly permeable nature of the colluvium overlying the residual clay, the perched water table in this area appeared to be highly polluted. Shallow wells (2 to 5 m) were installed adjacent to each deep monitoring well to test the quality of the perched water. Water pressure tests were carried out at different depths, and at three pressure increments, to determine rock mass permeability. Water table levels were also measured.

LABORATORY INVESTIGATIONS

Enrichment of Hydrocarbon-Oxidizing Associations. Liquid enrichment cultures were prepared in Erlenmeyer flasks, each containing 100-mL mineral salts medium, by inoculating with 2.5 g of lightly contaminated soil. Individual cultures were then overlaid with 0.05%, 0.10%, 0.50%, or 1.0% (v/v) of pristane and *n*-hexadecane. The flasks were closed with cotton wool and incubated at 29-30°C on a rotary shaker (150 rpm), in the dark, for approximately 21 days. Then 10-mL aliquots were withdrawn and subcultured. This procedure was repeated after 21 days. The pH was regularly monitored during the first 24 hours of growth, using a calibrated ion analyzer (Crison Micro pH 2000).

Isolation. After serial dilution (10^{-1} to 10^{-8}), 0.1-mL samples of the enrichment cultures were inoculated onto soil extract plates overlaid with the respective alkane and incubated at 29 to 30°C for 21 to 28 days. Potential isolates were subcultured onto nutrient agar overlaid with hydrocarbon.

Scanning Electron Microscopy (SEM). Cellular aggregates were removed, aseptically, from the enrichment cultures of both pristane and *n*-hexadecane, for SEM examination. After conventional alcohol dehydration, the granules were subjected to critical point drying, and then viewed under the SEM (Hitachi 570-S).

Surfactant Screening. Mineral salts medium (100-mL) was dispensed into each Erlenmeyer flask, containing 10 g of highly contaminated soil (900 to 1000 mg oil kg^{-1} soil). Three nonionic chemical dispersants (Arkopal N-050, Arkopal N-060 and Emulsifier 2491) (Hoechst S.A.(Pty) Ltd.) were added individually to give concentrations of 0.1% to 1.0% (v/v). The flasks were incubated in a shaking water bath (150 rpm) at 30°C for 3 days. The emulsifying capacity of the dispersants was estimated by measuring the turbidity (545 nm) of the supernatant after centrifugation (Mulkins-Phillips & Stewart 1974). To examine the toxicity of the dispersants to the microorganisms, solutions were withdrawn aseptically from each flask, filtered through 0.2-µm filters, and subjected to critical point drying prior to SEM examination. Agar plates, overlaid with 50 µL dispersant, also were inoculated to determine possible carbon source utilization.

Soil Microcosms. Next, 16 glass columns (300 mm × 43 mm inner diameter) were constructed with sampling ports at 50-mm intervals and a Teflon™ tap at the base. Temperature regulation (30°C) is via coiled PVC tubing in conjunction with a circulating water bath (Lauda RMT 6). The soil moisture content, porosity, bulk density, field capacity, total pore space, nutrient status, exchange capacity and pH were determined. Soil samples for abiotic controls were γ-irradiated (25 Mrad). Liquid samples are withdrawn into syringes via miniature tensiometers inserted at the last sampling port.

Chemical Analyses. The oil concentrations of samples are determined using a classical 6- to 8-cycle Soxhlet extraction procedure, followed by evaporation of the CS_2 solvent and analysis by gas chromatography: SPI: 80°C, 100°C min^{-1} to 320°C, hold; Column: 46 m x 0.2 μm OV 73 (Chromtech), 80°C, 20°C min^{-1} to 320°C, hold; Detector: FID at 320°C, range 10^{-12}, attenuation 256; carrier gas nitrogen at 41 cm sec^{-1}. Before injection (0.5 μL), samples are filtered and centrifuged.

Treatments. Four combinations of N, P, and K are used to treat both soil microcosms and field plots. These treatments are based on an analysis of the nutrients and organic carbon present in the contaminated soil. Treatment combinations of C:N (made up with a stock solution of NH_4NO_3, NaH_2PO_4, and KCl, respectively) are (1) 20:1, (2) 15:1, (3) 10:1, (4) 5:1, (5) 5:1 + Emulsifier 2491 (1.0% v/v), and (6) 5:1 + 40 mgL^{-1} H_2O_2. Abiotic microcosm controls receive identical treatments, and a single biological control receives only water. Treatments are administered as evenly as possible over the field plots and to the soil columns, at 2-week intervals for a period of up to 6 months. Prior to each treatment, samples are taken for analysis.

FIELD PLOTS

Field sites consist of 8 vegetation-free plots (7 experimental, 1 control), each 1 m × 1 m, with a border of 1 to 1.5 m. Three different subplots (10 × 10 cm) are sampled regularly (random number table) to monitor changes in oil concentrations. No two subplots are sampled twice during the study. The sampling grid has an elevated plank platform so that the plots are not disturbed during sampling. A small hand-operated auger is used to sample the plots (depth 15 to 20 cm). After the samples are composited and the temperature is monitored, residual soil is returned to the holes, prior to the next treatment application.

RESULTS AND DISCUSSION

Field Investigations: Extent of Contamination. The geotechnical investigation coupled with visual inspection clearly identified the major zones of contamination. The GC traces of samples confirmed that the soil was contaminated (60 to 15,800 mg oil kg^{-1} soil) by oils of the paraffinic and naphthenic types. These concentrations exceed the South African limit for of 50 mgL^{-1} for mineral oil in water. Concentrations of heavy metals in all samples were very low (<0.0001 mgL^{-1}).

The borehole cores revealed that the groundwater may be seriously contaminated as a result of the perched water table, the permeability of the sandy soil, and the degree of shattering in the highly weathered sandstone, and the geology of the area suggested that there may be a pollution plume emanating from the site in a northerly direction. The regional water table, however, appears to be uncontaminated.

Laboratory Investigations: Hydrocarbon Degradation. Depending on concentration, oil may be a nutrient or a potential toxicant (Pfaender & Buckley 1984). Nevertheless, the aerobic enrichment of indigenous *n*-hexadecane- and pristane-catabolizing microorganisms was successful.

SEM examination of the microbial aggregates revealed that pristane and *n*-hexadecane were metabolized by remarkably different species of microorganisms. The *n*-hexadecane batch cultures revealed a dominance of filamentous forms, including at least three predominant species of fungi and their spores. In contrast, the pristane aggregates were dominated by bacterial species. Both pristane- and *n*-hexadecane-degrading populations existed together in a honeycomb-like matrix, resembling layers of glycocalyx, and cells were frequently shown to be held closely together by an intricate network of extracellular fibers. This polysaccharide-fatty acid polymer in alkane transport is thought to increase cellular hydrophobicity, which allows passive binding (partitioning) of alkane to the cell surfaces (Käppeli & Fiechter 1980).

It was apparent in both culture types that the bacteria tend to respond more rapidly to the alkane, while the colonization by fungi is somewhat delayed (Pinholt et al. 1979). Conversely, the activity of fungi tended to persist long after bacterial activity diminished. The changes in pH during growth concurred with this: The pH initially was >7 (bacterial dominance), becoming <7, possibly as a result of hydrocarbon emulsification or the accumulation of metabolic products (organic acids), thereby facilitating fungal dominance.

This study confirms the existence of an active microbial association capable of catabolizing two representative alkanes.

Surfactant Screening. Only one of the three dispersants (Emulsifier 2491) supported growth. Filters visualized under the SEM revealed only clay particles and cellular debris for the Arkopal N-050 and N-060 types. This was also observed when plates were overlaid with dispersant as the sole carbon source. Although Emulsifier 2491 did not have the best emulsifying properties of the three dispersants (Arkopal N-060 displaced the most oil from the soil, forming a stable emulsion), it did show potential and was, therefore, chosen for further study.

In summary, although this study is still in its early stages, the results have been encouraging. Viable hydrocarbon-degrading microorganisms are present in the contaminated soil, and are stimulated by the addition of nutrients, demonstrated by a rapid change in pH. The effects of the treatments on oil degradation rates in the soil microcosms and the field plots will facilitate the development of an in situ bioremediation program.

ACKNOWLEDGMENTS

The financial assistance of Shell S.A. (Pty) Ltd., the Foundation for Research Development, and the University of Natal is gratefully acknowledged.

REFERENCES

Käppeli, O., and A. Fiechter. 1980. "Partition of hexadecane to the cell surface of *Candida tropicalis*: Mechanism for the transport of water-insoluble substrates." *Biotechnol. Bioeng.* 22:1829-1841.

Mulkins-Phillips, G. J., and J. E. Stewart. 1974. "Effect of 4 dispersants on biodegradation and growth of bacteria on crude oil." *Appl. Microbiol. 28*:547-552.

Pfaender, F. K., and E. N. Buckley, III. 1984. "Effects of petroleum on microbial communities." In R. M. Atlas (Ed.), *Petroleum Microbiology*, pp. 507-537. Macmillan Publishing Company, NY.

Pinholt, Y., S. Struwe, and A. Kjoller. 1979. "Microbial changes during oil decomposition in soil." *Holarctic Ecol.* 2:195-200.

NATURAL BIOREMEDIATION OF A GASOLINE SPILL

R. C. Borden, C. A. Gomez, and M. T. Becker

Groundwater contamination was discovered at the Arvida Research Site in 1987 when a property owner downgradient of the site complained of objectionable tastes and odors in their water. A preliminary investigation indicated that an underground storage tank (UST) present at the site had leaked gasoline and contaminated the shallow aquifer. In 1990, North Carolina State University with support from the American Petroleum Institute began an extensive field study to aid in understanding the physical, chemical, and biological processes controlling the rate and extent of natural bioremediation of the gasoline plume at this site. In this work, we define "natural bioremediation" as using the capacity of naturally occurring microorganisms to degrade contaminants that have been released into the subsurface while minimizing risks to public health and the environment. This approach requires an assessment of those factors that influence the biodegradation capacity of an aquifer and the potential human and environmental risks.

HYDROGEOLOGY AND CONTAMINANT SOURCE

The primary aquifer at the site consists of dark gray and green micaceous, glauconitic fine sand that is overlain by a confining layer of 1.5 to 4.5 m of silts, clays, and clayey sands. The water table has a relatively gentle, constant slope of 0.006 in an almost due south direction. The hydraulic conductivity of the aquifer varies from 0.004 to 0.03 cm s^{-1} with a mean value of 0.006 cm s^{-1}. Using the average water table gradient, and a porosity of 0.25, the average groundwater velocity is estimated to be 46 m yr^{-1}.

The dissolved BTEX plume results from the steady release of dissolved BTEX from the residual nonaqueous-phase liquids (NAPLs) present in the aquifer beneath the former UST. Detailed soil coring (data not shown) in the unsaturated and saturated zones has demonstrate that while some weathering of the NAPL has occurred, large amounts of residual BTEX are still present in the source area.

DISSOLVED BTEX DISTRIBUTION

Dissolved benzene, toluene, ethylbenzene, and xylene isomers (BTEX) and related parameters were monitored seven times from July 1991 to December 1992. Monitoring well locations are shown in Figure 1. Average concentrations for

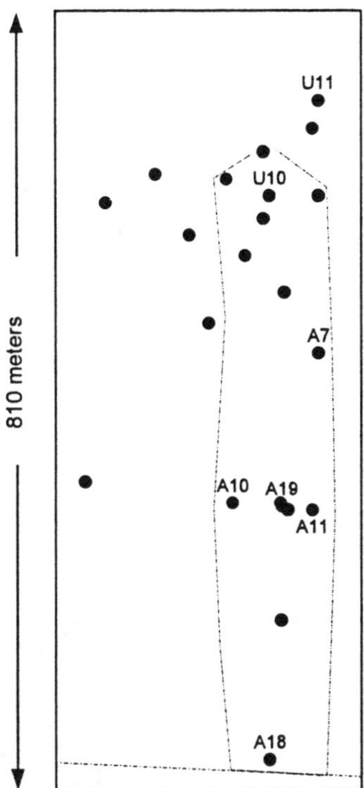

FIGURE 1. Monitoring well location map.

BTEX components and indicator parameters are shown in Table 1 for representative wells. Analytical and sampling procedures are described by Gomez (1993). A preliminary statistical analysis indicated that the dissolved BTEX plume is stable with no significant increase or decrease in concentration over the monitoring period. Based on these results, we have chosen to use average concentrations in all further analyses.

The BTEX plume has a long, narrow shape and appears to become narrower as it migrates downgradient. Although the concentrations in individual wells varied significantly from one sampling event to the next, the plume shape and distribution of contaminants was very consistent. The concentrations of toluene and *o*-xylene decline most rapidly in a downgradient direction, followed by benzene and then *m+p*-xylene and ethylbenzene. This pattern is somewhat surprising, because *o*-xylene and *m+p*-xylene have very similar physical and chemical properties. The only apparent explanation for this pattern is preferential biodegradation of the *o*-xylene isomer by subsurface microorganisms. The observed loss of BTEX is believed to be primarily a result of biodegradation. Dispersion cannot explain the observed decline in BTEX since the plume actually becomes narrower with travel distance. While volatilization could result in some loss of BTEX, this mechanism would not result in the preferential removal of toluene and *o*-xylene.

TABLE 1. Average parameter concentrations in representative monitoring wells.

Parameters/MW	U11	U10	A7	A10	A19	A11	A18
Benzene (mg/L)	BQL[a]	1,326	316	42	628	88	615
Toluene (mg/L)	BQL	10,436	117	8	80	9	26
Ethylbenzene (mg/L)	BQL	1,814	1,824	96	1,925	158	42
m+p-Xylene (mg/L)	BQL	8,179	4,501	199	2,752	152	335
o-Xylene (mg/L)	BQL	3,873	111	81	38	38	13
Total BTEX (mg/L)	BQL	25,629	6,869	425	5,423	444	1,030
Oxygen (mg/L)	3.1	0.3	0.3	0.4	0.2	0.4	0.7
Nitrate-N (mg/L)	1.4	0.1	0.6	1.1	0.1	1.2	0.1
Dissolved Iron (mg/L)	0.2	29.3	65.5	84	52.4	86	1.8
Sulfate (mg/L)	19.3	34.2	3.3	5	4	3	12.8
Methane (mg/L)	BQL	BQL	BQL	BQL	BQL	BQL	0.6
Carbon Dioxide (mg/L)	60	228	259	267	273	270	93
Redox Potential (mV)	196	–132	–181	–166	–118	–161	–187
pH	4.6	5.8	6.1	6.4	6.1	6.4	7.1
Alkalinity (mg/L)	6	54	97	67	122	72	148
Calcium (mg/L)	5	7	13	6	7	6	75

(a) BQL is below quantitation limits.

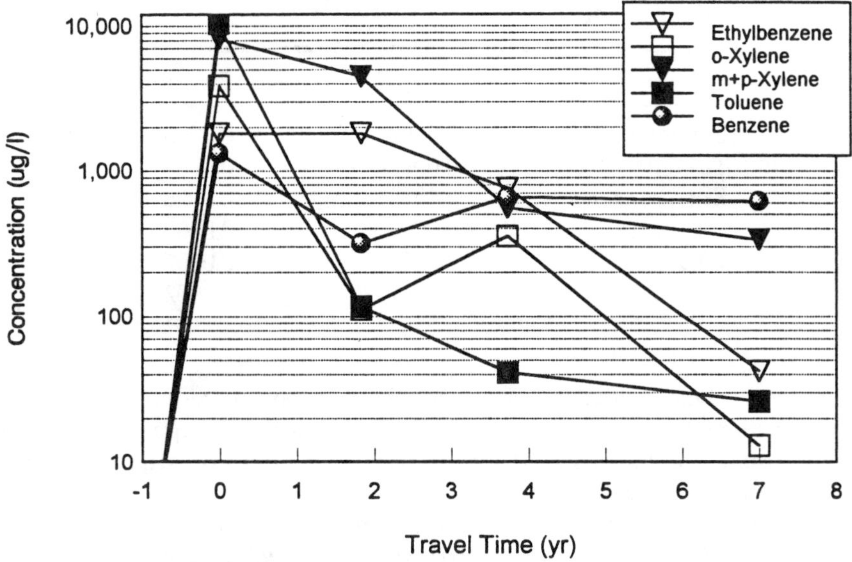

FIGURE 2. Variation in BTEX components with distance.

Figure 2 shows the observed variation in the average concentrations of BTEX components in monitoring wells along the plume centerline. Toluene declines most rapidly, followed by *o*-xylene, benzene, and then ethylbenzene and *m+p*-xylene. The total BTEX curve appears to follow a roughly exponential decay, suggesting that the change in total BTEX concentration may be approximated by the equation $C = Co \exp(-Kt)$, where K is the apparent first-order decay rate (d^{-1}), t is travel time from the source, and Co is the concentration at the source area. K was estimated by plotting concentration versus time on a logarithmic scale and then determining the best fit line by linear regression. Travel time was estimated as the distance from the source divided by the average groundwater velocity. Estimated first-order decay rates are shown in Table 2 along with the standard error of the decay rate. The apparent degradation rates at the Arvida site are significantly lower than have been observed at other sites (Kemblowski et al. 1987). The cause of this difference is not yet clear. Several factors could result in a lower biodegradation rate at this site, including (1) the overlying clay confining layer, which would reduce oxygen exchange with the unsaturated zone; and (2) a low background pH, which could inhibit methanogenic biodegradation of some BTEX components.

GEOCHEMICAL INDICATORS

Average concentrations of electron acceptors and indicator parameters are shown in Table 1. Dissolved oxygen and redox potential are both reduced in the plume due to biodegradation of BTEX. The background dissolved oxygen concentration varies from 2 to 3 mg/L, while in the center of the plume dissolved oxygen

TABLE 2. Apparent first-order decay rates for BTEX.

Parameter	K (day^{-1})	Std. Error of K
Benzene	0.0002	0.0003
Toluene	0.0021	0.0009
Ethylbenzene	0.0015	0.0006
m+*p*-Xylene	0.0013	0.0002
o-Xylene	0.0021	0.0007
Total BTEX	0.0011	0.0002

is below the field detection limit. Dissolved carbon dioxide is highest in the downgradient portion of the plume, indicating that significant mineralization of BTEX components has occurred. An electron balance indicates that the dominant electron acceptors are iron and sulfate followed by nitrate and dissolved oxygen. Methanogenesis does not appear to be an important process at the site, as only a small amount of methane is produced at the most downgradient monitoring well. Immediately upgradient of this well, the aquifer geochemistry changes (carbonate appears in significant amounts) as evidenced by the sudden rise in calcium in the most downgradient well. The increase in carbonate and accompanying rise in pH appears to result in conditions more favorable for methane production.

SUMMARY

Field monitoring indicates that a gasoline plume in the shallow aquifer at the Arvida Research Site is undergoing natural bioremediation. The dissolved BTEX plume does not spread out and concentrations of all compounds are decreasing in a downgradient direction. The apparent degradation rates for BTEX components are lower than at other sites, presumably due to the low pH in the aquifer and a restriction on oxygen transport through the overlying clay confining layer.

ACKNOWLEDGMENTS

The research described in this article was supported in part by the American Petroleum Institute under Grant No. GW-25A-0400-38. We would also like to thank the North Carolina Division of Environmental Management for their assistance in obtaining access to the site and in monitoring well installation.

REFERENCES

Gomez, C. 1993. "Characterization of a Dissolved Hydrocarbon Plume." Master of Science Thesis, Department of Civil Engineering, North Carolina State University, Raleigh, NC.

Kemblowski, M. W., Salanitro, J. P., Deeley, G. M., and Stanley, C. C. 1987. "Fate and Transport of Residual Hydrocarbon in Groundwater — A Case Study." *Proceedings, Petroleum Hydrocarbons and Organic Chemical in Ground Water — Prevention, Detection and Restoration*, pp. 207-231. Nat. Water Well Assoc., Houston, TX.

AGGREGATION OF OIL- AND BRINE-CONTAMINATED SOIL TO ENHANCE BIOREMEDIATION

D. H. McNabb, R. L. Johnson, and I. Guo

INTRODUCTION

Solid-phase bioremediation of contaminated soil and soil-like wastes requires that the material be permeable to water and air in order to leach soluble salts, supply water and nutrients to microorganisms, and exchange soil gases with the atmosphere. Tillage is often used to achieve the necessary porosity but is effective only to a depth of about 15 cm and must be repeated frequently. Tillage of wet, plastic soil can destroy soil structure and cause the soil to form into large clods. Clods with a diameter of 1 to 2 cm have a lower rate of bioremediation than do aggregates with a diameter of 1 to 2 mm (Mott et al. 1990).

Small, stable aggregates of contaminated soil are likely to remain permeable to water and air during handling, water infiltration, and compression over time, as well as providing a near-optimum environment for bioremediation. Our objective was to evaluate various amendments for aggregating an oil- and brine-contaminated soil in the laboratory. The evaluation consisted of measuring the aggregate size distribution, the stability of aggregates, and the compressibility and water retention of the contaminated soil that had been aggregated with the preferred amendment.

MATERIALS AND METHODS

The soil was a Black Chernozem that had been contaminated by crude oil and brine from a break in a pipeline transporting unprocessed oil. The soil, located in central Alberta, was contaminated to a depth of about 20 cm. The texture of the soil was a silty clay loam (39% sand, 27% silt, and 34% clay), and the organic matter content was 4.1%. The contaminated soil had an electrical conductivity of 26.8 mS/cm and an oil content of 6.7% (toluene extraction of oil and grease). The water content of the contaminated soil was about 35%; this caused the soil to form into clods between 4 and 10 cm in diameter when tilled. Contaminated soil was severely water repellent, requiring a 9.2 N solution of ethanol to penetrate air-dry soil (Yeung 1990). A bulk sample of noncontaminated soil from the spill site was also collected for tests of aggregate stability and soil consolidation.

Several amendments were considered for aggregating contaminated soil, including straw, peat moss, fly ash, hydrated lime, gypsum, and starch. The straw (wheat straw) and the peat moss (a commercial sphagnum moss) were air-dried and ground to pass a 3-mm sieve. Fly ash was from a local coal-fired electrical generating plant. Starch was a precooked, corn starch. The amendments were added as a percentage of the wet weight of contaminated soil. Amendments were mixed with the soil using a small commercial mixer with an attached paddle; the soil and amendments were mixed for 10 min at low speed. The aggregate size distribution was determined by sieving the amended soil immediately following mixing into 6 size fractions and reporting each fraction on a wet weight basis. The stability of fresh aggregates was measured in triplicate by wet sieving the 1.4- to 2.36-mm-diameter aggregates in a standard aggregate stability test (Kemper & Rosenau 1986).

The optimal aggregate size distribution was assumed to be that with the greatest mass of aggregates between 0.85 to 4.76 mm in diameter. Smaller aggregates were expected to have a lower stability and a reduced air and water permeability; larger aggregates were assumed to have a less favorable soil environment for biodegradation (Mott et al. 1990).

The compressibility of oil- and brine-contaminated soil that had been amended and aggregated, and stored at 2°C for five months, was tested in a one-dimensional consolidation test in the laboratory (Lambe 1951). The compressibility of non-contaminated, air-dried soil that had been ground and sieved to less than 2 mm in diameter, and oil- and brine-contaminated that had been dried to a water content of approximately 25% and then crushed and sieved to less than 4.76 mm in diameter was also measured. Three samples of each soil was gently packed into plastic rings with a diameter of 74 mm and a height of 35 mm by tapping the ring on a flat surface. The ring of soil was then soaked in water for 24 h. Similar samples were prepared and compressed to one of three normal stresses, 0.88, 2.84, and 9.71 kPa, to determine soil water retention at water potentials of −10, −30, and −100 kPa (Klute 1986).

RESULTS AND DISCUSSION

Mixing contaminated soil without amendments produced a range of aggregate sizes that depended on the soil water content at the time of mixing (Table 1). The stability of contaminated aggregates without amendments was less than the stability of aggregates of noncontaminated soil (Figure 1). The percentage of aggregates less than 1.4 mm in diameter increased markedly at water contents less than 25%. Contaminated soil also became water repellent at a water content less than 25%. The difficulty in controlling the rate of drying of contaminated soil, the release of fines during tillage of dry soil, and the difficulty of rewetting water repellent soil led a need to evaluate amendments for aggregating contaminated soil when wet.

Except for hydrated lime, all amendments produced relatively large aggregates (Table 1). Hydrated lime produced the widest range of aggregate sizes, but increasing the percentage of lime increased the fraction of aggregates less than 0.85 mm in diameter.

TABLE 1. Aggregate size distribution of oil- and brine-contaminated soil following mixing with and without amendments.

Amendment	Water Content[a]	Aggregate Size Distribution, mm <0.85	0.85-1.4	1.4-2.36	2.36-4.76	4.76-9.5	>9.5
		-------------		%		-------------	
None	2.2	45.0	6.6	9.6	31.4	7.4	0
None	22.2	--- 34.3 ---		--- 43.5 ---		--- 22.2 ---	
None	34.0	--- 0 ---		--- 5.8 ---		--- 94.2 ---	
Peat Moss-3%	31.5	0	2.3	7.2	26.1	44.6	19.8
Straw-3%	28.0	0	4.8	11.0	62.7	21.5	0
Starch-5%	27.8	0	0.7	2.4	17.1	52.8	27.0
Fly ash-10%	28.0	0	1.3	3.6	16.1	44.6	34.5
Gypsum-5%	27.8	0	4.8	11.3	31.7	41.1	11.1
Lime-5%	29.8	10.8	43.1	18.3	20.2	6.8	0.8
Lime-2%	32.0	0.7	8.7	26.6	45.6	16.4	2.0
Lime-1%	31.0	0	3.1	10.5	31.5	39.7	15.2
Lime-1% + Straw-1%	32.5	0	2.3	10.3	47.0	29.0	11.4
Lime-1% + Starch-1%	31.0	9.1	28.7	20.6	28.2	13.0	0.5
Lime-1% + Straw-1% + Starch-1%	30.5	5.2	22.6	31.7	35.2	5.3	0

(a) Water content was determined by oven drying soil at 105°C for 24 h.

All inorganic amendments increased aggregate stability while the organic amendments decreased aggregate stability (Figure 1). Peat moss decreased aggregate stability the most. From these tests, we concluded that organic amendments should be used only sparingly if required to improve the aggregate size distribution.

The initial testing of amendments identified hydrated lime, straw, and starch as the most promising amendments for creating an optimum aggregate size distribution. We then determined the best combination and amount of amendments needed to create an optimal size distribution. This combination of amendments was 1% each of hydrated lime, straw, and starch (Table 1 and Figure 1). The straw was mixed into the soil before the other two materials. These amendments are assumed to cause several complementary changes in the soil that would enhance aggregation. We observed that the straw was well distributed throughout the wet clods of contaminated soil. The value of the straw in aggregation is probably the formation of fracture planes, or planes of weakness, along which the clods broke when the lime and starch were incorporated. The hydrated lime increased soil pH which may have increased the solubility of silica and the formation of silicate bonds (Papadakis et al. 1992). The hydrated lime also absorbed water which made the wet soil less plastic and more susceptible to fracturing during mixing. The starch probably formed bonds between soil particles (Tisdall & Oades

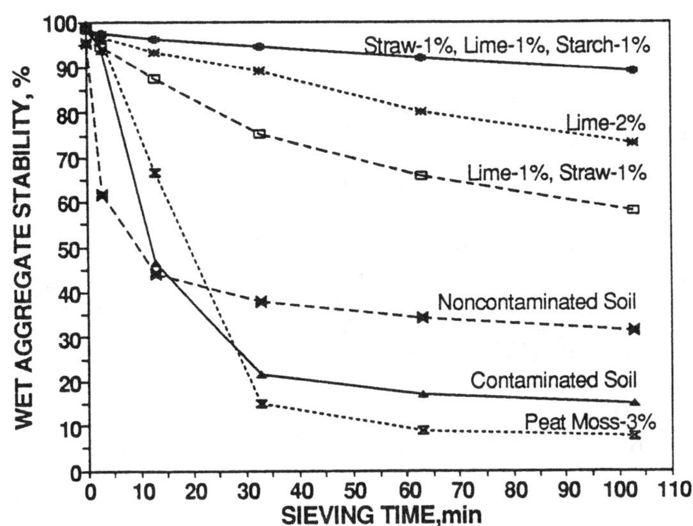

FIGURE 1. Wet aggregate stability of noncontaminated soil, and an oil- and brine-contaminated soil before and after amendments were added.

1982), and did reduce the amount of lime needed to achieve an equivalent aggregate size distribution.

The amended soil had a lower bulk density and a lower compression index than either the noncontaminated or the unamended, oil- and brine-contaminated soil (Figure 2). The low bulk density of the amended soil reflected the large increase in the macropores created by the amendment and aggregation process. The compression index (that portion of the compression curve where the relationship between bulk density and the logarithm of the normal stress is linear) of amended and aggregated soil was only 0.217 (Figure 2). In contrast, the noncontaminated soil and unamended, contaminated soil both had a compression index of 0.32. The compression index of the aggregated soil was less than half that predicted for agricultural soils with a similar clay content (Larson et al. 1980).

The consolidation of the soil was measured when the soil was most compressible because the tests were conducted on saturated soil that drained during the test. This test criterion simulate the conditions that will likely occur in the bioreactor. The effective stresses that control the compression of the soil placed in the bioreactor will be similar to that of a saturated soil whenever the soil is leached or irrigated to maintain a high soil water content. The effective stresses will be similar to those in saturated soil because the high porosity of the amended soil reduces the effect that the matric potential of partly saturated soil has on effective stress (Bishop & Blight 1963).

Gravimetric water content decreased as soil water potential decreased but the water content was unaffected when bulk density increased from 0.86 to 1.06 Mg/m^3 (Table 2). However, volumetric water content increased as a result of soil compression because of the increase in bulk density. As a result, soil

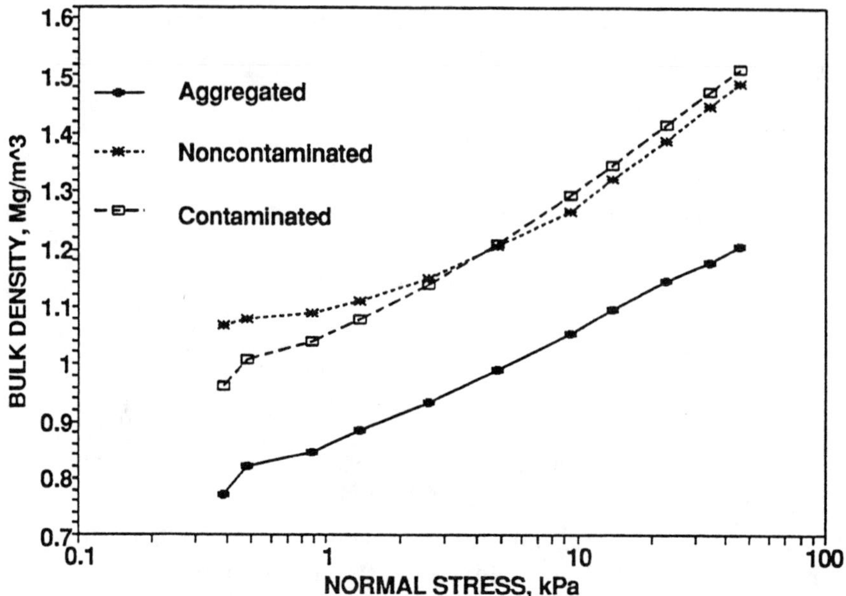

FIGURE 2. Consolidation of noncontaminated soil, oil- and brine-contaminated soil, and contaminated soil amended and aggregated with hydrated lime, straw, and starch, at 1% each.

compression had a much greater effect on air-filled porosity than was apparent from the measurement of gravimetric water content.

The data on compressibility and air-filled porosity are important for determining the maximum depth of soil placement in a solid-phase bioreactor. The data in Figure 2 and Table 2 can be used to estimate the bulk density, volumetric water content, and air-filled porosity of any depth of soil that is placed in a bioreactor. Air-filled porosity will be lowest at the bottom of the soil layer where the compressive stress produced by the weight of the wet soil will be highest. For example, if the normal stress at the bottom of a layer of soil was 9.7 kPa, the maximum bulk density would be 1.11 Mg/m^3, including an estimate for secondary consolidation of 0.05 Mg/m^3 (Table 2). This soil would have a wet weight of 1.43 Mg/m^3. Therefore, a 69-cm-thick layer of aggregated soil could be placed in the bioreactor before a stress of 9.7 kPa is reached. The air-filled porosity at the bottom of this soil layer would be 0.23 kg/kg at a water potential of −10 kPa. The actual stress will be less because soil near the surface will be less dense and retain less water. However, additional compression may occur because of a deterioration of the aggregates as the amendments decompose, or wetting and drying cycles cause volumetric changes in the aggregates.

The minimum air-filled porosity required to maintain an aerobic environment depends on the depth of soil and rate of biological activity. As the air-filled porosity decreases, proportionally more of the pore space is in discontinuous

TABLE 2. Bulk density, water content at three water potentials, and water and air-filled porosity at −10 kPa of aggregated, contaminated soil at three normal stresses.

Normal Compressive Stress	Bulk Density	Volume Soil Solids[a]	Soil Water Content −10 kPa	Soil Water Content −30 kPa	Soil Water Content −100 kPa	Volume of Soil Water at −10 kPa	Air-filled Porosity at −10 kPa
kPa	Mg/m^3				%		
0.88	0.86	36.4	29.6	28.9	24.7	23.2	40.4
2.84	0.94	39.6	29.2	26.8	24.0	28.9	31.5
9.71	1.06	44.4	29.4	26.5	23.4	32.6	23.0

(a) Assumes a particle density of 2.50 Mg/m^3, and a 0.05-Mg/m^3 increase in bulk density over time as a result of secondary consolidation.

pores that do not contribute to soil aeration (Dexter 1988). As a result, the diffusion of gases in soil approaches zero as an air-filled porosity of 10% (Xu et al. 1992). Furthermore, the 0.23-kg/kg air-filled porosity for this soil was calculated at a water potential of −10 kPa. While the soil layer near the surface will drain to this water potential in a few days, deep soil layers will take a much longer period to drain to this water potential without the benefit of vegetation to transpire water from deeper in the soil. Therefore, air-filled porosity in deep soil layers is expected to be much less for several days after irrigation.

CONCLUSIONS

A combination of amendments used to aggregate an oil- and brine-contaminated soil increased aggregate stability, decreased bulk density, and reduced the compressibility of the soil relative to the noncontaminated and unamended contaminated soil. The compressibility and water retention characteristics of aggregated soil are two parameters needed to determine the maximum depth that material can be placed in a solid-phase bioreactor; additional data on water retention and gas diffusion are needed to complete the evaluation.

ACKNOWLEDGMENTS

Funding for this research was provided by Canadian Petroleum Association, Gulf Canada Resources, Environment Canada, Development and Demonstration of Soil Remediation Technology Program (DESRT), and Alberta Environment Research Trust.

REFERENCES

Bishop, A. W., and G. E. Blight. 1963. "Some Aspects of Effective Stress in Saturated and Partly Saturated Soils." *Geotechnique 13*:177-197.

Dexter, A. R. 1988. "Strength of Soil Aggregates and of Aggregate Beds." *Catena Supplement 11*: 35-52.

Kemper, W. D., and R. C. Rosenau. 1986. "Aggregate Stability and Size Distribution." In A. Klute (Ed.), *Methods of Soil Analysis, Part 1 – Physical and Mineralogical Methods*, 2nd ed., pp. 425-442. Amer. Soc. Agronomy, Madison, WI.

Klute, A. 1986. "Water Retention: Laboratory Methods." In A. Klute (Ed.), *Methods of Soil Analysis, Part 1 - Physical and Mineralogical Methods*, 2nd ed., pp. 635-662. Amer. Soc. Agronomy, Madison, WI.

Lambe, T. L. 1951. *Soil Testing For Engineers*. John Wiley & Sons, New York, NY.

Larson, W. E., S. C. Gupta, and R. A. Useche. 1980. "Compression of Agricultural Soils from Eight Soil Orders." *Soil Sci. Soc. Amer. J. 44*: 450-457.

Mott, S. C., P. H. Groenevelt, and R. P. Voroney. 1990. "Biodegradation of a Gas Oil Applied to Aggregates of Different Sizes." *J. Environ. Qual. 19*: 257-260.

Papadakis, V. G., M. N. Fardis, and C. G. Vayenas. 1992. "Hydration and Carbonation of Pozzolanic Cements." *ACI Materials J. 89*(2): 119-130.

Tisdall, J. M., and J. M. Oades. 1982. "Organic Matter and Water-Stable Aggregates in Soils." *J. Soil Sci. 33*: 141-163.

Yeung, P. 1990. "A Method for Measuring Water-Repellent Soil." In *Proc. 27th Ann. Alberta Soil Sci. Workshop*, Feb. 21-22, 1990, pp. 59-64. Edmonton, Alberta, Canada.

Xu, X., J. L. Nieber, and S. C. Gupta. 1992. "Compaction Effect on the Gas Diffusion Coefficient in Soils." *Soil Sci. Soc. Amer. J. 56*: 1743-1750.

EVALUATION OF LIPOSOME-ENCAPSULATED CASEIN AS A NUTRIENT SOURCE FOR OIL SPILL BIOREMEDIATION

R. T. Herrington, G. D. Sayles, C. E. Furlong, R. J. Richter, and A. D. Venosa

INTRODUCTION

Approximately 1.5 million tons per year of crude and refined oils are introduced into the marine environment through transportation-related releases (Farrington 1985). For releases catastrophic enough to warrant remedial action bioremediation is often considered as a cleanup strategy because it has the potential for reducing the acute toxicity of oil in a more cost-effective and benign manner compared to other physical and chemical treatment technologies.

Marine oil spill biodegradation can be difficult because the oceans typically have low concentrations of nitrogen and phosphorus. Researchers have attempted to amend nutritive and bacterial deficiencies by adding water-soluble and oleophilic fertilizers to oil spills (Atlas & Bartha 1973, Goldstein et al. 1985, Horowitz & Atlas 1977). Water-soluble amendments were easily washed away by surface agitation and mixing, whereas oleophilic nutrients, such as paraffin-supported urea, exhibited greater ability to stimulate the biodegradation of floating oil. However, the use of paraffin-supported urea and other hydrocarbon-based nutrients can exacerbate an oil spill disaster by increasing the quantity of hydrocarbons added to the environment.

An alternative delivery system for applying nutrients without additional petroleum hydrocarbons uses liposomes. Liposomes (Figure 1) are comprised of amphipathic phospholipids aligned to form bilayered structures. This bilayer orients the hydrophobic ends of the lipids toward each other, resulting in an organic structure that can encapsulate water-soluble nutrients (Lasic 1992). Liposomes exist as unilamellar or multilamellar vesicles and range in size from 0.1 to 1.5 µm. Multilamellar vesicles have been shown to be effective dispersants of oil and may serve an additional purpose as nutrient-delivery vehicles (Gatt et al. 1990).

We are evaluating liposome-encapsulated casein (milk protein) as a nutrient delivery system to enhance oil spill bioremediation. Encapsulation of simple nutrient salts (e.g., nitrates) is impractical because of rapid leakage rates from the multilamellar vesicles (Furlong et al. 1993). Larger molecules such as casein (molecular weight of ~50,000) leak at very low rates because the vesicles are subject to lower osmotic pressures. Further, casein is an inexpensive source of essential

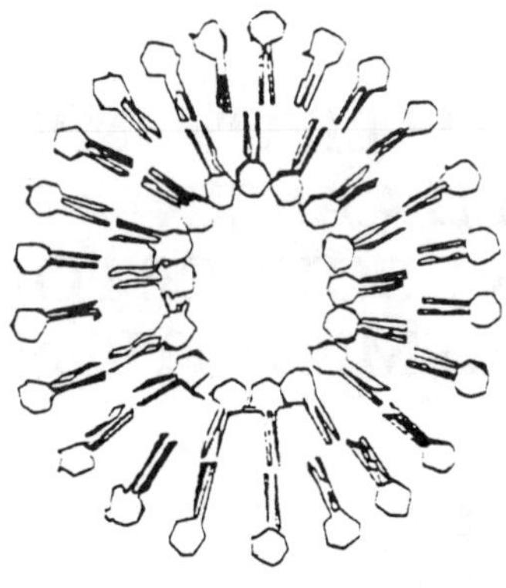

FIGURE 1. Drawing of a unilamellar phospholipid vesicle (liposome). Aqueous solutions can be encapsulated in the center of the liposome.

nitrogen for microorganisms. Casein is 14% N by weight, and would cost much less in bulk than the ~ $2/lb available from research biochemical suppliers.

The objective of this study is to show that encapsulated casein as a nitrogen source can promote crude oil biodegradation. Preliminary experiments (Sayles et al. 1993) have shown that an aqueous solution of (unencapsulated) casein can provide the necessary nitrogen to support oil biodegradation. Future studies will demonstrate the effectiveness of encapsulated casein to promote biodegradation of an oil spill in an open-sea simulation.

MATERIALS AND METHODS

Seawater. Seawater was prepared using an artificial mix at 3.5% total saline (Macleod & Onofrey 1956). This mix contained all essential trace nutrients.

Liposomes. Liposomes were prepared (Furlong et al. 1993) using l-α-phosphatidylcholine from soybean extract (Sigma Chemical Co., catalog #P5638, lot #129f0618) by homogenization at 1,100 rpm in 0.01 M phosphate buffer with either 180.5 mg/L KNO_3 or 178.1 mg/L casein (Sigma Chemical, catalog #C0376, lot #20h0263) followed by a 24-hour hydration period. The liposomes were separated from structural debris by brief centrifugation and from unencapsulated casein by passage through a Sepharose CL-4B column coupled to an ultraviolet monitor.

Oxygen Uptake Measurement. The oil biodegradation studies were performed in 500-mL respirometer vessels connected to an N-Con Comput-Ox WB512 Respirometer. The respirometer allowed for the automated measurement of oxygen uptake in each flask while maintaining a constant headspace oxygen concentration and constant temperature.

Carbon Dioxide Evolution Measurement. The CO_2 evolved in the flask due to biodegradation was scrubbed in the headspace by an internally mounted KOH trap. The amount of CO_2 produced during a time interval was quantified by measuring the drop in KOH pH over that time interval and calculating CO_2 concentrations with an in-house computer algorithm.

Total Heterotrophic Plate Counts. Total heterotrophic populations were measured by diluting samples from the controls with 2% sterile saline water and transferring to sterile plates containing marine bactoagar (Difco 2216). The plates were incubated for 6 days at room temperature, then the colonies were counted.

Total Petroleum Hydrocarbon (TPH) Measurement. TPHs were collected by a hexane extraction procedure adapted from U.S. Environmental Protection Agency (EPA) Method 8270. A total trace analysis of the aliphatics and aromatics was performed on a Hewlett Packard GC5890/FID with a dB-5 capillary column (0.25 mm ID). The injector temperature was 290°C, the detector temperature was 320°C, the initial oven temperature was 55°C, and the oven was ramped at 5°C/min to 290°C and then 3°C/min to 320°C.

EXPERIMENTAL DESIGN

The ability of liposomes to retain encapsulated casein in an uninoculated system was determined spectrophotometrically by the release of a specially marked casein. Casein was covalently tagged with 0.3% dabsyl-chloride (Sigma Chemical Co.) a colored, chromophoric label that absorbs at 445 nM. Marked, liposome-encapsulated casein was then added to three 250-mL Erlenmeyer flasks with 100 mL of sterile, artificial seawater in each. The flasks were then treated with mineral oil, Arabian light crude oil, or no oil. Aliquots of each treatment were extracted at intervals and the liposomes in each extract were purified on a sepharose 4B-CL column. The purified liposomes were burst with a 1% solution of Triton-X 100 detergent, then measured for casein content by absorbance at 445 nM.

The effectiveness of liposome-encapsulated casein to biodegrade oil was measured by comparing O_2 uptake, CO_2 evolution, TPH loss, and biomass increase in three separate treatments. Nitrogen source was the primary variable between the three treatments and was added in two predominant forms: as casein or as potassium nitrate (KNO_3). KNO_3 is commonly used as a nitrogen source in biodegradation experiments; thus, the KNO_3 treatment was selected to be used as a comparative control. Casein was added at an equivalent nitrogen concentration to KNO_3 and was either encapsulated in liposomes ("encapsulated casein") or added concomitantly with liposomes that carried no nutrients ("nonencapsulated casein"). Comparison of encapsulated and nonencapsulated controls would show differences, if any, in biodegradation rates as a result of a slow release of casein from the liposome. Note that in the casein treatments phosphatidylcholine (PC) in the form of liposomes was a secondary source of nitrogen (~1% nitrogen by weight). The fractional contribution of casein, PC, and KNO_3 to the total nitrogen in each treatment is tabulated in Table 1.

Phosphorus requirements were satisfied by the phosphate buffer (0.01 M KH_2PO_4) in the simulated seawater. The initial pH was adjusted to 7.5.

Arabian light crude oil was loaded at 500 mg/L in each flask containing an artificial seawater mix. The theoretical nitrogen demand of the oil is 25 mg/L and all treatments provided this concentration of nitrogen as KNO_3 or as casein. All controls were inoculated with 6×10^6 bacteria/mL from mixed cultures derived from Galveston Bay, Texas, and acclimated to Arabian light crude oil. A summary of the important contents of each treatment is shown in Table 1.

RESULTS

Figure 2 illustrates that liposomes can encapsulate casein for substantial periods of time. All three treatments (2 with oil, one without oil) retained at least 80% of their original casein cargo after 300 hours of gentile mixing, whereas, the no-oil treatment had no measurable casein loss after 400 hours. These results suggested that the liposomes were osmotically stable and could be used for casein delivery in the oil biodegradation studies.

Figure 3 demonstrates that encapsulated casein can provide the necessary nitrogen to promote oil biodegradation at a significant rate. The encapsulated casein treatments biodegraded 78% of the measurable TPH after 16 days and averaged 9.0% TPH/day initial rate during the first 5 days. The nonencapsulated casein treatment exhibited a similar initial rate of 8.1% TPH/day and biodegraded 83% of the oil after 16 days. Lack of significant differences in biodegradation

TABLE 1. Outline of experiment showing specific treatments for each of the three controls.

	Encapsulated Casein	Nonencapsulated Casein	KNO_3
Oil (mg/L)	500	500	500
Inoculum (cell/mL)	4.7×10^6	4.7×10^6	4.7×10^6
Casein (mg/L)	178	178	0
PC (mg/L)	6,800	6,800	0
KNO_3 (mg/L)	0	0	180
%N as Casein-N	64%	64%	0
%N as PC-N	36%	36%	0
%N as KNO_3-N	0	0	100

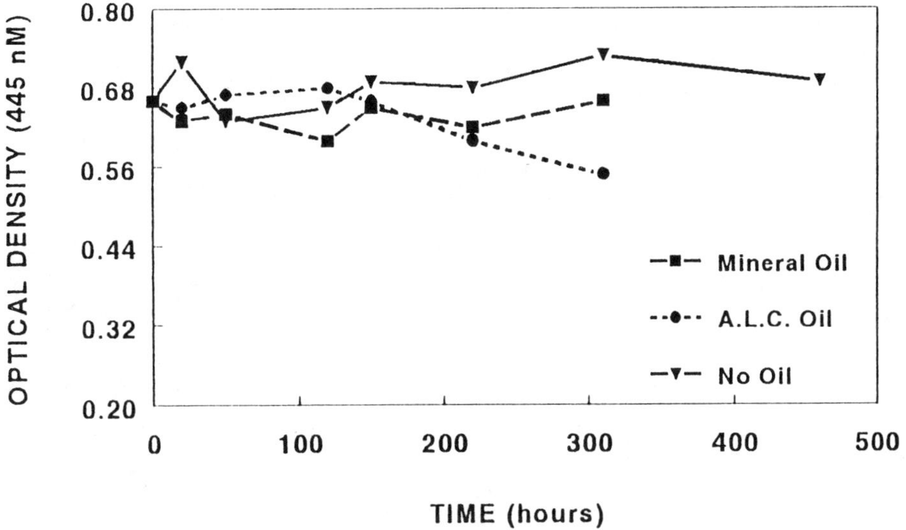

FIGURE 2. Retention of encapsulated casein in liposomes. Two oil types used to monitor leakage in presence of a hydrocarbon substrate.

rates of the encapsulated and nonencapsulated casein treatments suggests that the supply of nitrogen to the microorganisms was not limited by a slow release of casein-N from the liposome vesicles.

As expected, KNO_3 also stimulated biodegradation of oil very well (Figure 3). In the first 100 hrs, the rate of oil biodegradation in all three treatments was roughly the same. After 100 hours, however, the rate of degradation in the KNO_3 treatment increased to 23% TPH/day while the rate in the casein treatment slowed. The increase in the rate in the KNO_3 control is likely a result of the accumulation of additional cells with time (Figure 4).

The differences in the rates of oil biodegradation between the casein and KNO_3 treatments can be explained by comparing the efficiency of oil biodegradation with respect to oxygen consumption. Figure 5 displays oil degradation versus oxygen consumption for the three treatments. The overall efficiency of oxygen use for oil degradation in the KNO_3 treatment was about 7 times that in the casein treatments (~ 0.2 and 0.03% TPH/mg O_2/L for the KNO_3 and casein treatments, respectively). Presumably, this large difference existed because the casein treatments used casein, phospholipid, and oil as competing biodegradable carbon sources, whereas the KNO_3 treatment used oil as a sole carbon source. Also, bacteria must expend more energy to utilize the NH_3 nitrogen source from the casein than the KNO_3 nitrogen source of the mineral salt possibly slowing the mineralization of oil in the casein treatments.

The presence of easily degradable carbon sources in the casein treatments, although decelerating the rate of use of oil relative to the rate in the KNO_3 treatment, promoted a higher rate and extent of biomass production. Figure 4 depicts

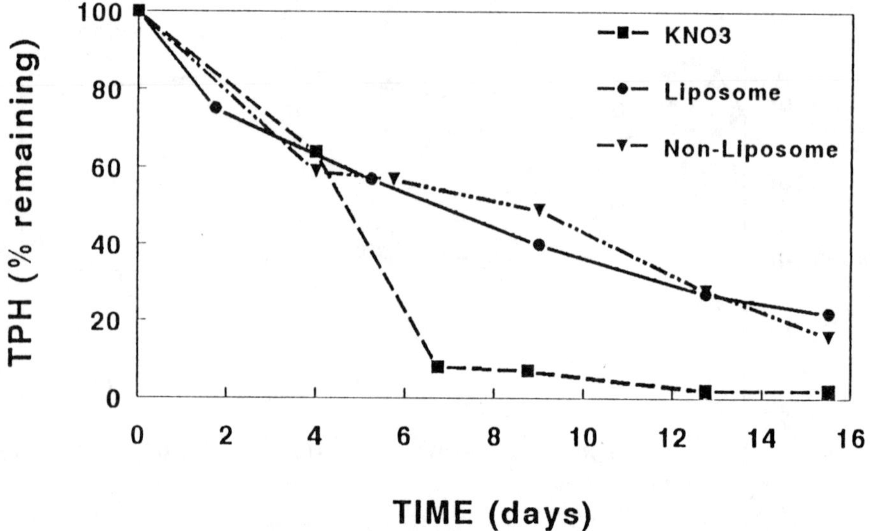

FIGURE 3. Percent TPH loss (hexane-soluble components) as a function of time for all three controls.

heterotrophic plate counts of microorganisms as a function of time for all three treatments. Total cell production is about 10 times higher in the casein treatments than in the KNO_3 treatment. In addition, cell growth in the casein treatments did not display the initial lag seen in the KNO_3 treatment.

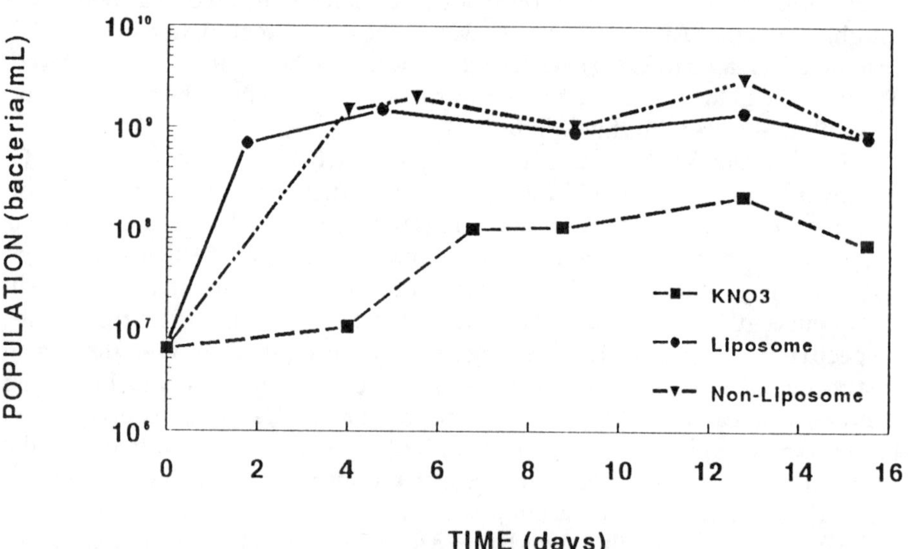

FIGURE 4. Bacterial populations (total heterotrophic plate counts) as a function of time for all three controls.

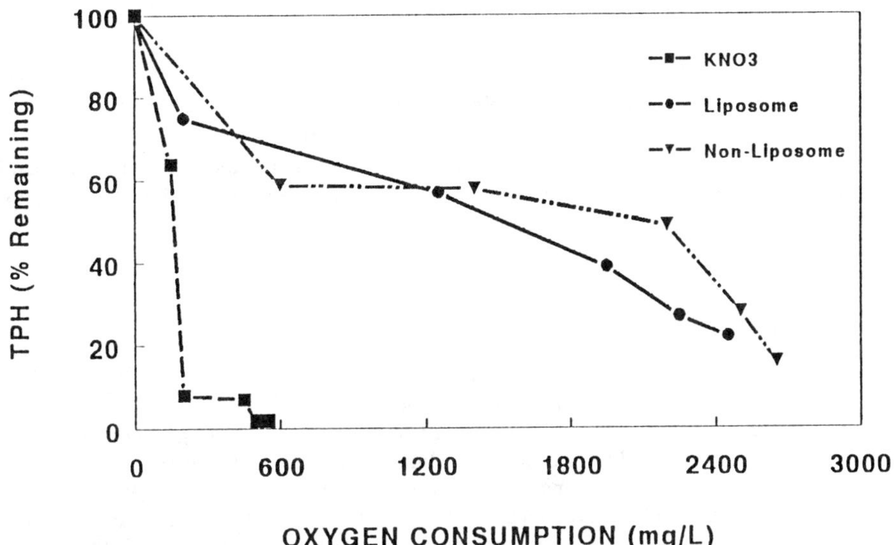

FIGURE 5. Percent TPH loss (hexane-soluble components) as a function of oxygen consumption for all three controls.

CONCLUSIONS

Relative to aqueous solutions of mineral nutrients, liposome-encapsulated casein has the potential to deliver essential nitrogen and phosphorus to an oil spill and continue to supply nutrients in the neighborhood of the spill over an extended period of time. A simple diffusion calculation shows that water-soluble nutrients would be in contact with the oil for no more than several hours.

This study has demonstrated several important characteristics that a nutrient delivery system must possess. The results demonstrated that the liposome-encapsulated casein successfully encapsulated and provided the essential nitrogen needed to stimulate biodegradation of crude oil in a closed vessel. It was also shown that liposome-encapsulated casein was a supplemental and easily biodegradable carbon source. The liposome-encapsulated casein carbon sources allowed for cell growth with no lag and high final biomass levels.

Based on these results, application of liposome-encapsulated casein on an oil spill would provide rapid initial cell growth with simultaneous oil degradation that would continue until the nitrogen, in the form of casein and phospholipid, or the oil, was depleted. At this time, cell numbers would be high, so, if necessary, an additional application of this or a related product would stimulate further oil degradation at a higher initial rate that during the first application. Demonstrating the applicability of the liposome-encapsulated casein in the continuous-flow environment of the open sea is the subject of our next studies.

DISCLAIMER

Although the research described in this article has been funded in part by the U.S. Environmental Protection Agency under assistance agreement CR-818852 to the University of Washington and CR-816700 to the University of Cincinnati, it has not been subjected to the EPA's peer and administrative review and therefore may not necessarily reflect the views of the EPA, and no official endorsements should be inferred.

REFERENCES

Atlas, R. M., and R. Bartha. 1973. "Simulated Biodegradation of Oil Slicks Using Oleophilic Fertilizers." *Environmental Science & Technology 7*: 528-541.

Farrington, J. W. 1985. "Oil Pollution: A Decade of Research and Monitoring." *Oceanus 28*: 2-12.

Furlong, C.R., R.J. Richter, G.D. Sayles, and R.T. Herrington. 1993. "A Comparison of the Stability of Liposomes Encapsulating Casein and Mineral Nutrients." In preparation.

Gatt, S., H. Bercovier, and Y. Barenholz. 1990. "Use of Liposomes for Combating Oil Spills and Their Potential Application to Bioreclamation." In R. E. Hinchee and R. F. Olfenbuttel (Eds.), *In Situ Bioreclamation: Applications and Investigations for Hydrocarbons and Contaminated Site Remediation*, pp. 293-312. Butterworth-Heinemann, Boston, MA.

Goldstein, R. M., L. M. Malloy, and M. Alexander. 1985. "Reasons for Possible Failure of Inoculation to Enhance Biodegradation." *Applied & Environmental Microbiology 50*: 977-983.

Horowitz, A., and R. Atlas. 1977. "Continuous Flow Through System as a Model for Oil Biodegradation in the Arctic Ocean." *Applied Environmental Biology 33*: 647-653.

Lasic, D. 1992. "Liposomes." *American Scientist 80*: 20-31.

Macleod, R. A., and E. Onofrey. 1956. "Nutrition and Metabolism of Marine Bacteria. II. Observation on the Relation of Seawater to the Growth of Marine Bacteria." *Journal of Bacteriology 72*: 661-667.

Sayles, G. D., R. T. Herrington, C. E. Furlong, and R. J. Richter. 1993. "Casein Provides Nitrogen Needed for Oil Biodegradation." In preparation.

BIOREMEDIATION AND PHYSICAL REMOVAL OF OIL ON SHORE

P. Sveum and C. Bech

INTRODUCTION

Although several experiments have been conducted on biological treatment methods of oil-polluted marine habitats, this topic had been given limited practical attention until the grounding of the *Exxon Valdez* in Alaska in March 1989. The main reason for the lack of interest is that the removal that could be attributed to biological processes will be minor when the total oil spill mass balance is considered. Following contamination of a shoreline, the spilt oil will be influenced by both biotic and abiotic factors. Biological activity can mineralize oil into CO_2, or increase mobility of the oil due to the production of biosurfactants by microbes. Abiotic factors include all types of physical transport processes (advection, diffusion, evaporation, etc.), as well as photochemical degradation. The rates of both the biotic and the abiotic processes depend on the distribution of the oil within the contaminated area. These processes interact and can result in an increased or decreased rate for any of the processes. When bioremediation of an oil-polluted shoreline site is considered an option, it is important to keep in mind all of the processes that are likely to alter oil concentration on the habitat.

The results presented were obtained from field and outdoor laboratory experiments done on Spitsbergen in 1987 and 1988 that were designed to evaluate how the rate of biodegradation is affected by the rate of physical self-cleaning. The results are discussed from the point of view of the role of physical self-cleaning with regard to strategies for bioremediation based on a conceptual model for beach self-cleaning presented by Sveum and Bech (1990).

MATERIALS AND METHODS

The field experiment consisted of eight plots, each with an area of 1 m × 2 m, located on a gravel beach close to Sveagruva on Spitsbergen. The plots consisted of homogeneous beach material, with similar exposure to waves. Oil (Statfjord crude oil) was added at the first low tide, and fertilizer (Inipol EAP22) was added at the second low tide. The amount of oil and the treatment used in each plot are presented in Table 1.

In addition to the on-site biodegradation experiment, outdoor laboratory experiments were performed in a series of five enclosures containing gravel material (Table 2). The enclosures had no wave exposure, and thus no or very

TABLE 1. Experimental setup for the on-site gravel beach experiment.

Plot #	Crude oil added (L/m^2)	Inipol EAP22 added (L/m^2)
1	10	0
2	8	0
3	6	0
4	4	0
5	10	1
6	8	0.8
7	6	0.6
8	4	0.4

little physical removal of oil took place. The experiments lasted 94 days. The tide was varied, and seawater was circulated but was not exchanged. The dimension of each sediment enclosure was 120 × 60 × 30 cm (l × w × d), and the volume of the reservoir was 200 L. The upper 5 cm of each enclosure and shoreline test plot was sampled at five randomly selected locations, where 5-cm-long core samples (ID = 3 cm) were collected, and the upper 1 cm of each was discarded. In the in situ experiment, each plot was divided into 10-cm squares. Samples were taken after 36 hours, 10 days, 30 days, 60 days, and 90 days. In the enclosure experiment, a grid system with 5-cm squares was used. In this experiment, sampling was done at the start of the experiment and after 24, 57, and 94 days. The samples from each plot or enclosure were thoroughly mixed in a Waring blender.

Oil was extracted from the mixed samples using n-hexane and measured gravimetrically. Gas chromatographic (GC) analysis of extracted hydrocarbons was done on an HP5730A GC, applying a temperature program from 100°C (3 min) to 300°C at 4°C/min. The GC was filled with a 15-m DB fused silica column. Helium was used as carrier gas, and 2-µL samples were injected.

TABLE 2. Experimental design of the outdoor laboratory enclosure experiment. Initial oil concentration is given as the amount of oil added.

Encl. #	Initial oil conc. (mg/kg)	Inipol added (L/m^2)
1	42,000	1
2	36,200	0.8
3	23,500	0.6
4	12,400	0.4
5	6,400	0.2

RESULTS AND DISCUSSION

The nC_{17}/pristane and nC_{18}/phytane ratios are conventionally used as indices of biodegradation (e.g., Atlas 1991), based on the assumption that pristane and phytane are more or less recalcitrant to biodegradation. The decrease in these ratios does not necessarlily reflect biodegradation of oil in a linear manner, it is considered to be a valid indication of oil biodegradation. The nC_{17}/pristane ratios and the nC_{18}/phytane ratios agree well ($r^2 = 0.929$, $P > F = 0.0001$, df = 19). Thus, only pristane is used to evaluate biodegradation in this presentation.

The decrease in the nC_{17}/pristane ratio for the outdoor laboratory enclosures, plotted against time, is given in Figure 1. The ratio decreases both with decreasing initial oil concentration and with time. It should be noted that the temperature and nutrient conditions in these experiments did not simulate natural arctic shoreline conditions. The temperature varied between 5 and 11°C, and added nutrient could not escape. In the field experiment, the shoreline sediment temperature varied between 2.5 and 5°C.

In all shoreline plots, the initial rate of self-cleaning was high (Figure 2), and after 25 to 30 days the total oil concentration in all shoreline plots was about 10,000 mg/kg or lower. In Plot 8, which had the lowest initial loading, the oil concentration was notably lower than in the other plots.

The biodegradation of oil in the shoreline plots is expressed as the decrease in nC_{17}/pristane ratio (Figure 3). In Plots 1 to 4 (i.e., the plots not treated with

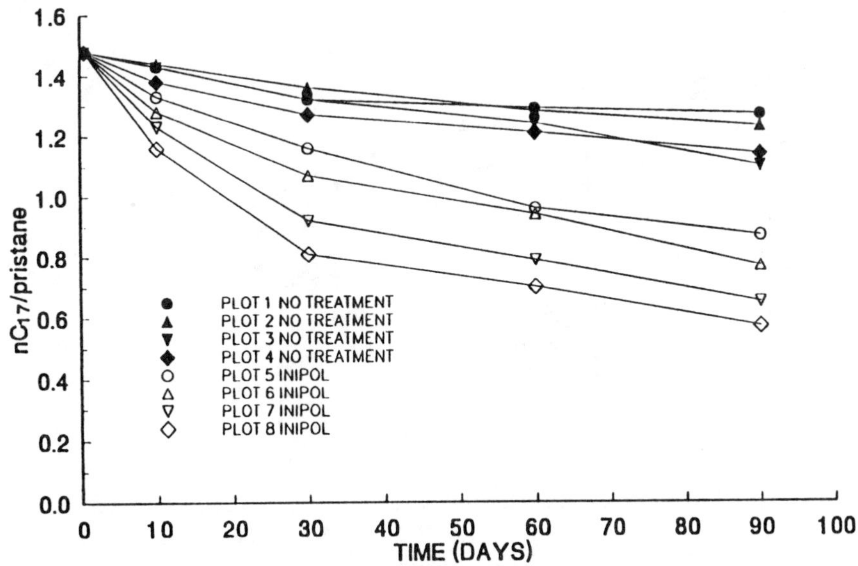

FIGURE 1. Biodegradation of Statfjord crude oil in the outdoor laboratory experiments given as decrease in the nC_{17}/pristane ratio, plotted against time and initial concentration of hydrocarbons in the sediment.

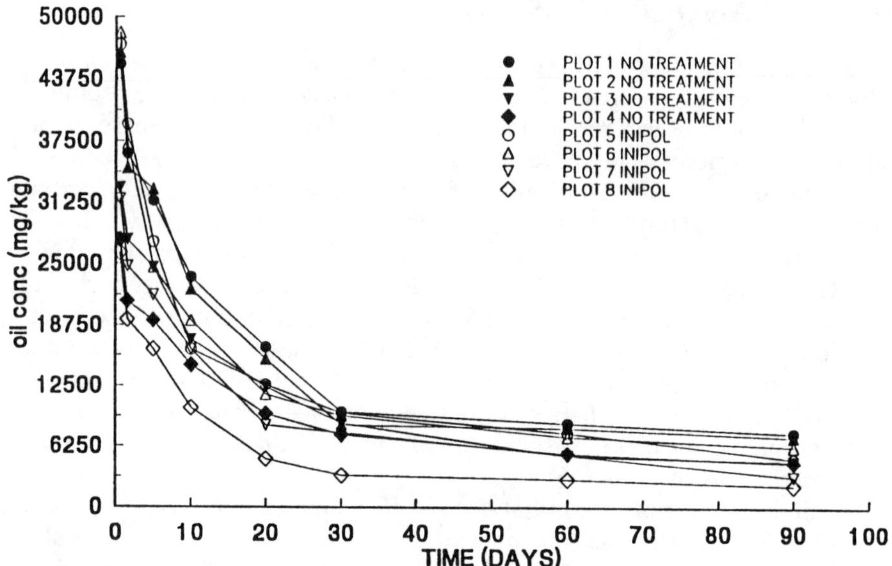

FIGURE 2. Removal of Statfjord crude oil in the on-site shoreline experiments given as decrease in the nC_{17}/pristane ratio, plotted against time and initial concentration of hydrocarbons in the sediment.

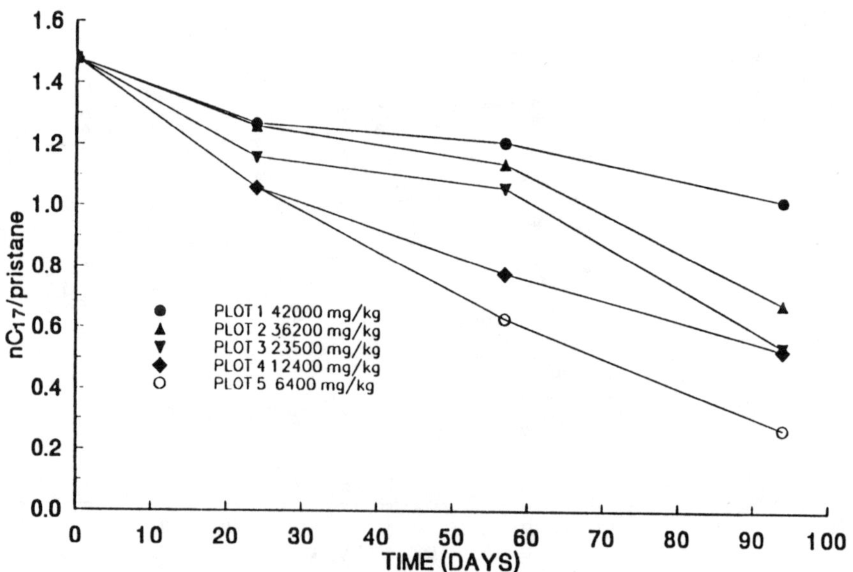

FIGURE 3. Biodegradation of Statfjord crude oil in the on-site shoreline experiments given as decrease in the nC_{17}/pristane ratio, plotted against time and initial concentration of hydrocarbons in the sediment.

fertilizer) the biodegradation increased with decreasing oil concentration (see Figure 1). The same tendency was found in those treated with Inipol. Plot 4, which had the lowest oil concentration of the nontreated plots, had a lower nC_{17}/pristane ratio than Plot 5 which had the highest oil concentration of the plots that received fertilizer treatment. After 90 days the nC_{17}/pristane ratio in the fertilizer-treated plots varied between 0.9 and 0.45, with the lowest ratios found in the plots with the lowest initial oil concentrations.

The results from both these experiments indicate that biodegradation of crude oil is directly related to the hydrocarbon concentration in the sediment. The present observation is in accordance with that of Atlas (1981) and Fusey & Oudot (1984). During the first 25 to 30 days following contamination physical removal predominates. However, the difference in removal rate between Plots 4 and 8 indicates that the fertilizer treatment increased the physical removal rate for this particular combination of experimental factors.

Biological degradation is one of several processes contributing to the self-cleaning of oil-polluted shorelines. The different components of self-cleaning are influenced by the distribution and interaction of the oil with the sediment at the time of contamination. The availability of oil to biological and nonbiological processes influences the rate of each individual process. The distribution of oil in a shoreline sediment indirectly affects the biological processes, due to its influence on important processes such as water and gas transport.

When a shoreline is inundated by an oil with a given viscosity O_v, the horizontal distribution of the oil, F_h, after contamination will depend on the exposure E, tidal range T, oil weathering O_f, and shoreline topography S_t at the time of inundation:

$$F_h = f(E, T, O_f, S_t) \qquad (1)$$

The vertical distribution of oil, F_v, will depend on the amount of oil O_m (i.e., the oil loading), the oil viscosity O_v, the internal properties of the shoreline sediment S_p, the organic content of the sediment S_o, and the solubility of oil in water O_s:

$$F_v = f(O_m, O_v, S_p, S_o, O_s) \qquad (2)$$

The sum of the horizontal and vertical contamination comprises the total contamination, F, of the shoreline:

$$F = F_h + F_v = F_h + F_o + Fs \qquad (3)$$

The vertical partition will change in time into two components:

$$F_v = F_o + F_s \qquad (4)$$

F_o is the vertical distribution of oil that is not and will never be bound to the sediment. This component can be a major part of F_v, depending, in particular, on initial oil loading:

$$F_o = f(O_m, O_v, S_p, O_s) \tag{5}$$

F_s on the other hand represents the oil saturation fraction (i.e., amount of oil that the sediment will hold after initial redistribution of the oil) of the shoreline sediment and is related to oil sorption:

$$F_s = f(O_m, O_v, S_o) \tag{6}$$

As demonstrated by the decrease in oil concentration in the present on-site shoreline experiment (Figure 2), a major fraction of the oil initially present in the sediment was easily removed after a few tidal cycles. The rate of decrease in oil concentration increased with initial oil loading. Thus, an increase in the component O_m does not increase the relative extent of bound oil, F_s.

Differences in mechanisms whereby the oil will be removed are the basis for choosing to differentiate between the different distribution components. The expression $F_h + F_o$ is described by a reciprocal exponential function (Seip et al. 1985), if the individual components are related to self-cleaning potential. This fraction of the contaminating oil is available for immediate distribution. In the experimental data (Figure 3), $F_h + F_o$ is described by the decrease in oil concentration between day 0 and day 25, which can be fitted to a reciprocal exponential function.

F_h respresents the horizontal contamination of the oil on the shoreline surface, i.e., the oil that does not penetrate the sediment. On solid rock, where penetration is not possible, all of the oil will be represented by this factor. On shorelines with a low organic content, or with oils of limited stickiness, the immediate contamination will be determined by F_o and F_h:

$$F \approx F_h + F_o \tag{7}$$

In cases when $F_h + F_o >> F_s$, self-cleaning is mainly a redistribution process. If the oil-contaminated shoreline is considered an isolated unit, a decrease in the concentration of oil in the sediment can be observed; however, an increase in concentration will occur in nearby sites. Redistribution of oil, which is part of F_h or F_o, can be prevented only by manual or mechanical means. If bioremediation is selected for cleanup, it will not reduce redistribution of the oil to any great extent during the removal of the F_h or F_o components. This should be considered when bioremediation is evaluated as a possible cleanup strategy.

Biodegradation is slow for a highly contaminated shoreline, when compared to the rates of other naturally occurring processes. As these processes cannot be controlled, they will have to be considered as an important factor within any bioremediation operation on shorelines. It can be expected that physical removal will be the most important factor in oil removal during the initial phases of an oil spill on a shoreline. Biodegradation also will take place during this phase; however, if the shoreline is heavily contaminated, the conditions for biodegradation will be suboptimal, due to problems with both oxygen availability and nutrient distribution. On shorelines, the oil concentration will reach steady state when the easily available oil has been removed by physical action. The concentration level at steady state will depend on the interaction between the spilled oil and the sediment, and on the remaining oil sorbed by the sediment. This is the oil that should be targeted for bioremediation.

REFERENCES

Atlas, R. M. 1981. "Microbial Degradation of Petroleum Hydrocarbons: An Environmental Perspective." *Microbial Rev. 45*: 180-209.

Atlas, R. M. 1991. "Bioremediation of Fossil Fuel Contaminated Soils." In R. E. Hinchee and R. F. Olfenbuttel (Eds.), *In Situ Bioreclamation: Applications and Investigations for Hydrocarbon and Contaminated Site Remediation*, pp. 14-32. Butterworth-Heinemann, Stoneham, MA.

Fusey, P., and J. Oudot. 1984. "Relative Influence of Physical Removal and Biodegradation in the Depuration of Petroleum-Contaminated Seashore Sediments." *Marine Pollution Bulletin 15*: 136-141.

Seip, K. L., K. A. Brekke, K. Kveseth, and H. Ibrekk. 1987. "Models for Calculating Oil Spill Damages to Shores." *Oil and Chemical Pollution 3*: 69-81.

Sveum, P., and C. Bech. 1990. "Oil Spill on Arctic Shoreline Sediments." *Esarc Report 20*(1). STF21 F90060, SINTEF Applied Chemistry, Trondheim, Norway.

THE ROLE OF BIOSURFACTANTS IN BIOTIC DEGRADATION OF HYDROPHOBIC ORGANIC COMPOUNDS

W. P. Hunt, K. G. Robinson, and M. M. Ghosh

INTRODUCTION

Hydrophobic organic compounds (HOCs) tend to partition onto soil. Consequently, their biodegradation may be severely limited. One group of hydrophobic compounds, polycyclic aromatic hydrocarbons (PAHs), are of particular concern. Widely distributed in the subsurface, many of these compounds are toxic and potentially carcinogenic. To foster in situ biodegradation, environmentally safe methods must be found to desorb PAHs from contaminated soils. Synthetic surfactants have been shown to dramatically increase the aqueous solubility of HOCs (Kile & Chiou 1989) and to be effective in extracting HOCs from contaminated soils (McDermott et al. 1989). However, the high concentrations of these chemicals required to extract HOCs may inhibit biodegradation (Laha & Luthy 1991). Furthermore, synthetic surfactants may adversely affect the permeability of the microbial cell membrane, thus reducing or eliminating the biodegradative potential of indigenous microorganisms.

Biosurfactants are synthesized and excreted into the environment by microorganisms while metabolizing water-insoluble substrates and in response to environmental stresses (Ramana & Karanth 1989). Soil bacteria such as *Pseudomonas, Rhodococcus,* and *Arthrobacter* have been shown to produce biosurfactants (Itoh & Suzuki 1972). Recent studies have demonstrated the effectiveness of biosurfactants produced by *Pseudomonas aeruginosa* in removing sparingly soluble HOCs from gravel (Harvey et al. 1990) and in enhancing the biomineralization rate of a sparingly soluble HOC (Zhang & Miller 1992).

The goal of the present research is to determine the effectiveness of biosurfactants in promoting bioavailability of HOCs in soil-water systems. In this paper, preliminary results outlining the purification and use of a biosurfactant produced by *Pseudomonas aeruginosa* are discussed. The increase in apparent water solubility of a PAH, phenanthrene, in the presence of biosurfactant and the effect of such increase on biomineralization also are reported.

MATERIALS AND METHODS

Pseudomonas aeruginosa PRP652 in 1-L cultures consisting of 30 g glycerol and mineral salts (Ramana & Karanth 1989) was used to produce the biosurfactant.

Surfactant was harvested after 7 to 9 days of growth at 31°C by centrifugation (10,000 rpm for 20 minutes), acidification of the supernatant (pH 2.5), and extraction of biosurfactant from it using ether. The extract was dried under vacuum, and the solids were redissolved in chloroform before chromatographic purification.

To evaluate the enhancement of the aqueous solubility of phenanthrene by the biosurfactant, 0.175 mg of phenanthrene (dissolved in acetone) was first added to triplicate 10-mL screw-cap test tubes. The acetone was then volatilized, leaving phenanthrene crystals, and 2 mL of sterile solution containing various amounts of purified biosurfactant were added to resolubilize the phenanthrene. Capped tubes were continuously mixed for 5 days (160 rpm) at ambient temperature (23°C). Preliminary results indicated that equilibrium was reached in about 2 days (data not shown). Samples were then centrifuged at 500 × g for 45 minutes to separate the nonsolubilized phenanthrene crystals. Aliquots (100 µL) of the aqueous phase were analyzed for phenanthrene by reverse-phase high-performance liquid chromatography (HPLC). Some samples were filtered through preconditioned glass fiber filters (0.7 µm). These results agree with those obtained by centrifugation to within a one standard deviation.

To measure biomineralization, aqueous suspensions of phenanthrene and biosurfactant were spiked with ^{14}C-phenanthrene (550,000 counts/min) and mixed on a shaker platform for 1 day. Duplicate vials containing phenanthrene were inoculated with washed cells from a log-phase culture of *Pseudomonas fluorescens* DFC50 grown on naphthalene as the sole carbon source. Mineralization was evaluated by measuring the production of $^{14}CO_2$ over time.

RESULTS AND DISCUSSION

The surface tension of the supernatant in a culture of *Pseudomonas aeruginosa* decreased from 66 dyne/cm to a constant minimum value of 31.8 dyne/cm during the first 2 days of growth. Reduction of surface tension to a constant value indicates biosynthesis of extracellular surfactant(s) at a concentration exceeding the critical micelle concentration (CMC). To maximize the amount of biosurfactant produced, the cultures were harvested at 5 to 7 days after reaching the stationary phase.

Pseudomonas aeruginosa synthesizes and excretes primarily two different surface active glycolipids and a yellow pigment under the growth conditions described. Silicic acid chromatography, employing chloroform and progressively higher ratios of acetone to chloroform, was used to separate these species. Most of the biosurfactant mass of interest (estimated from the intensity of anthrone color developed) was eluted in the acetone fraction. Thin-layer chromatographic analysis of this fraction showed the presence of only one anthrone-positive (glycosylate) species, in agreement with similar results reported in the literature (Hisatsuka et al. 1971). Reverse-phase HPLC of this fraction with absorption maxima at 254 nm further confirmed the presence of a single species.

A white, solid material was obtained upon vacuum evaporation of the acetone extract containing the biosurfactant. It was soluble in aqueous solutions buffered at pH greater than 6.5 but insoluble in acidic solutions (pH <6.5). The surface tension of serial dilutions of a 0.01% (w/v) solution of this biosurfactant in 0.1 M $NaHCO_3$ was determined; the surface tension at CMC (20 mg/L) was 35.1 dyne/cm. The purified biosurfactant is a glycosylated, amphipathic compound, produced and excreted by *Pseudomonas aeruginosa*. Furthermore, it is apparently anionic because it is insoluble in solutions below pH 6.5 and in the presence of divalent cations. Based on the results obtained during the production, purification, and characterization of this compound it is tentatively identified as Rhamnolipid R1 (Figure 1).

Biosurfactant R1, at concentrations of 0, 450, 900, 1,750, and 3,500 mg/L in 0.03 M phosphate buffer (pH 7.2), was used to increase the aqueous solubility of phenanthrene (Figure 2). The apparent solubility increased from 1.2 mg/L (no biosurfactant added) to 34.4 mg/L (3,500 mg/L biosurfactant). Maximum solubility was increased by approximately a factor of 28 using pure R1 biosurfactant at a dosage of 175 × CMC or 5.45×10^{-3} M assuming a molecular weight of 649 (Hisatsuka et al. 1971). The apparent phenanthrene solubility in this solution is 1.98×10^{-4} M resulting in a molar ratio of biosurfactant to phenanthrene of approximately 28.

Mineralization of phenanthrene in the presence of biosurfactant was evaluated by measuring the production of $^{14}CO_2$ over time after the addition of *Pseudomonas fluorescens* DFC50. This organism does not significantly reduce the surface tension of the test solution under these conditions. Vials were incubated at 31°C while shaking (160 rpm), samples were purged with air, and the evolved $^{14}CO_2$ was trapped in 0.5 N NaOH. The results indicate that mineralization was enhanced in samples containing biosurfactant (Figure 3). The enhancement of mineralization increased with incubation time such that after 42 hours an average increase of 75% was noted in samples containing the biosurfactant relative to those having none. The effect of surfactant concentration on the degree of mineralization is not amenable to meaningful interpretation; further work is warranted.

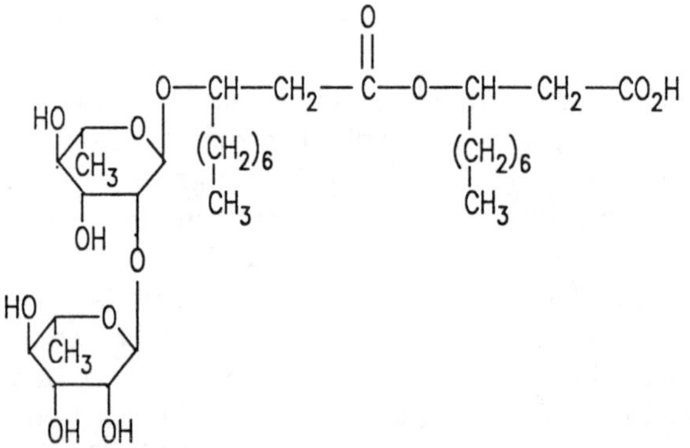

FIGURE 1. Molecular structure of the biosurfactant Rhamnolipid R1 produced by *Pseudomonas aeruginosa* (adapted from Hisatsuka et al. 1971).

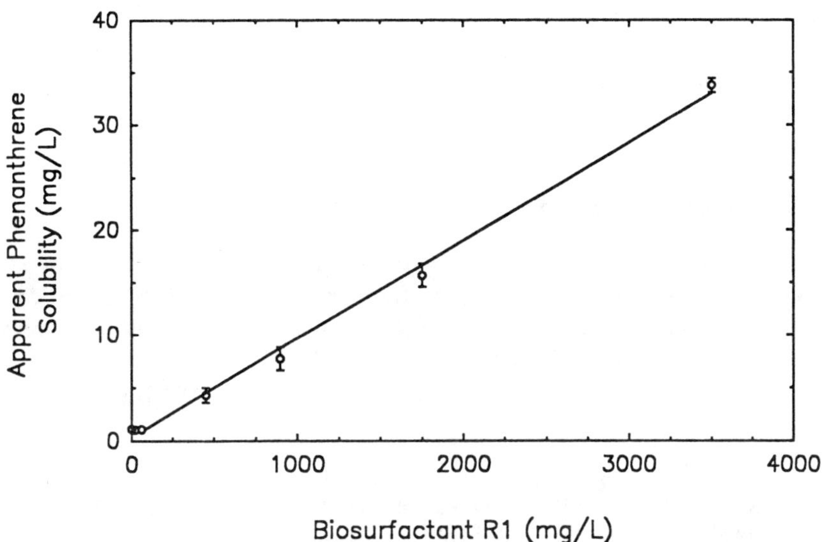

FIGURE 2. Effect of biosurfactant R1 concentration on the apparent aqueous solubility of phenanthrene.

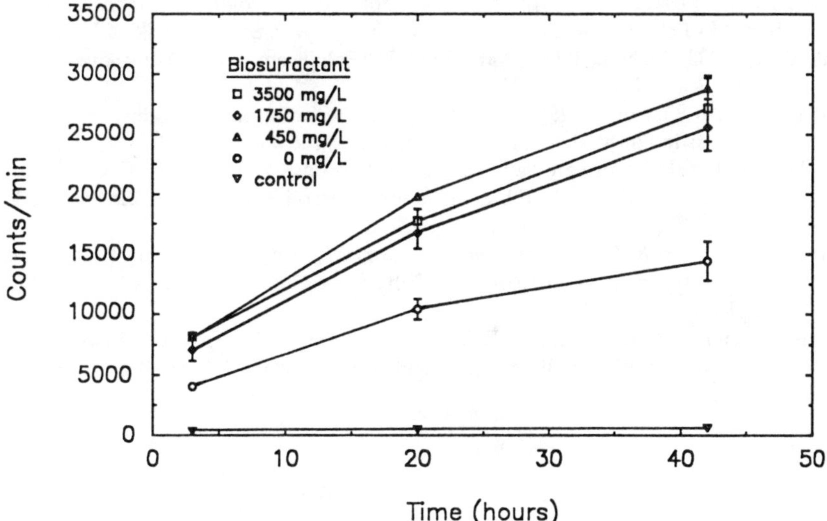

FIGURE 3. Effect of biosurfactant R1 on biomineralization of phenanthrene by *Ps. fluorescens* DFC50.

CONCLUSIONS

Biosurfactant Rhamnolipid R1 can be successfully extracted from viable cultures of *Pseudomonas aeruginosa*, a ubiquitous soil microorganism. Purified

Rhamnolipid R1 can effectively lower the surface tension of aqueous solutions to a constant minimum of 35.1 dyne/cm at a concentration of ≈ 20 mg/L. The compound is able to enhance the aqueous solubility and overall mineralization of phenanthrene.

ACKNOWLEDGMENTS

This project was funded by the Center for Environmental Biotechnology (CEB), University of Tennessee, Knoxville, Tennessee, and the U.S. Environmental Protection Agency, Grant No. R819168010. Bacterial strains used in this study were environmental isolates provided by the CEB.

REFERENCES

Harvey, S., I. Elashvili, J. J. Valdes, D. Kamely, and A. M. Chakrabarty. 1990. "Enhanced Removal of *Exxon Valdez* Spilled Oil from Alaskan Gravel by a Microbial Surfactant." *Bio/Technol. 8*: 228-230.

Hisatsuka, K., T. Nakahara, N. Sano, and K. Yamada. 1971. "Formation of Rhamnolipid by *Pseudomonas aeruginosa* and its Function in Hydrocarbon Fermentation." *Agr. Biol. Chem. 35*:686-692.

Itoh, S., and T. Suzuki. 1972. "Effect of Rhamnolipids on Growth of *Pseudomonas aeruginosa* Mutant Deficient in n-Paraffin-Utilizing Ability." *Agr. Biol. Chem. 36*:2233-2235.

Kile, D. E., and G. T. Chiou. 1989. "Water Solubility Enhancement of DDT by Some Surfactants." *Environ. Sci. Technol. 23*:832-838.

Laha, S., and R. G. Luthy. 1991. "Inhibition of Phenanthrene Mineralization by Nonionic Surfactants in Soil-Water Systems." *Environ. Sci. Technol. 25*:1920-1929.

McDermott, J. B., R. Unterman, M. J. Brennan, R. E. Brooks, D. P. Mobley, C. C. Schwartz, and D. K. Dietrich. 1989. "Two Strategies for PCB Soil Remediation: Biodegradation and Surfactant Extraction." *Environ. Prog. 8*:46-51.

Ramana, K. V., and N. G. Karanth. 1989. "Factors Affecting Biosurfactant Production Using *Pseudomonas aeruginosa* CFTR-6 under Submerged Conditions." *J. Chem. Tech. Biotechnol. 45*: 249-257.

Zhang, Y., and R. M. Miller. 1992. "Enhanced Octadecane Dispersion and Biodegradation by a *Pseudomonas* Rhamnolipid Surfactant (Biosurfactant)." *Appl. Environ. Microbiol. 58*:3276-3282.

IN SITU BIOREMEDIATION IN ARCTIC CANADA

J. G. Carss, J. G. Agar, and G. E. Surbey

BACKGROUND

Transport Canada (TC) operates a weather and radio station at the community of Reliance, Northwest Territories, Canada. In July 1989 a petroleum product sheen was observed on the surface of adjacent Great Slave Lake and TC was ordered to investigate and mitigate the impact of the contamination on the fish-bearing waters of the lake.

SITE INVESTIGATION

A site investigation was carried out in August 1989. Vapor concentrations were measured in approximately 150 probe holes drilled to a depth of 600 mm below grade and test holes were excavated at seven locations. Standpipe monitoring piezometers were installed in the test holes.

The stratigraphy observed at the locations of the test holes was comprised generally of angular, platy gravel underlain by fractured, indurated Cretaceous shale bedrock. The shale bedding planes dipped into the slope at approximately 15°. Visible ground ice, indicating the presence of permafrost, was observed in the shale.

Samples of groundwater and lake water were obtained from two of the piezometers and from Great Slave Lake for laboratory analyses of benzene, toluene, ethylbenzene and xylenes (BTEX) and total petroleum hydrocarbons (TPH) concentrations. The results of the hydrochemical analyses indicated a TPH concentration of 47 mg/L in the groundwater adjacent to the shoreline. Phase-separated liquid petroleum products were not detected in the piezometers. However, droplets of liquid hydrocarbons were observed to rise from the lake bottom and disperse on the lake surface during calm conditions.

HYDROCARBON CONTAINMENT AND REMEDIATION

A hydrocarbon containment system consisting of a sump and a 175-m-long recovery trench located along the shoreline of Great Slave Lake, a coalescing separator, a treatment tank, and a 100-m-long reinjection gallery was constructed in the summer of 1990. Contaminated groundwater is purged from the sump, pumped uphill to the coalescing separator and the treatment tank, enriched with nutrients and reinjected to the subsurface through the reinjection gallery. The treated and enriched water then cycles through the fractured bedrock back to the recovery trench. The components of the system are shown in profile on Figure 1.

The containment system was operated without the addition of nutrients from September 1990 to December 1990. Laboratory biotreatability testing conducted during the same period suggested that populations of indigenous hydrocarbon-degrading microorganisms could reach 10^3 CFU/mL to 10^4 CFU/mL. In the summer of 1991 approximately 20,000 L of 50% hydrogen peroxide and 22 tonnes of dry nutrients including 30% nitrogen (NH_4NO_3), 5% phosphorus (H_2PO_4), 3% potassium (KCl), and 1.5% other micronutrients were transported to the site. The oxygen objective was a concentration of at least 10 mg/L in the recovery system. A maximum concentration of 10 mg/L of nitrate at the recovery well was established, corresponding to the drinking water criterion. A field-monitoring program was implemented to gauge the quantities of nutrients added to the water.

In June 1991 the containment system was reactivated and was operating effectively, recovering groundwater at a rate of approximately 450 L/min. In August 1991 the addition of nutrients was initiated. By mid-September 1991, it was discovered that the maximum pumping rate had decreased to 120 L/min. The decrease in the pumping performance subsequent to the initiation of nutrient addition suggested that microbial growth was impeding the flow of groundwater to the recovery well. To remove the microbial growth, the recovery well was treated with successive additions of hydrogen peroxide and backflushed with lake water. Within 2 days the maximum pumping rate had increased to 410 L/min.

The operation of the recovery well pump was maintained by adding hydrogen peroxide to the recovery well and occasionally backflushing the well until the winter shutdown in December 1991. In September 1992 the system was reactivated and remained operational until November 1992. The total volume of groundwater pumped from August 1991 to December 1992 was 19 million L.

The temperatures of the recovered groundwater are summarized on Figure 2. In September 1991 the groundwater temperature reached 8.3°C but decreased to approximately 0.2°C in December 1991. In 1992 the temperature of the groundwater decreased from 2.0°C to 0.3°C during the same interval.

RESULTS

The concentrations of dissolved oxygen measured in the recovery system throughout the 1991 and 1992 seasons are presented on Figure 3. The data indicate that the concentrations of dissolved oxygen in the groundwater generally exceeded the target concentration of 10 mg/L.

In 1991 the initial addition of nutrients produced groundwater nitrate concentrations that exceeded the objective. The volume of nutrients added to the system was decreased in successive increments and nitrate concentrations of less than 10 mg/L were generally obtained during the final 10 weeks of system operation.

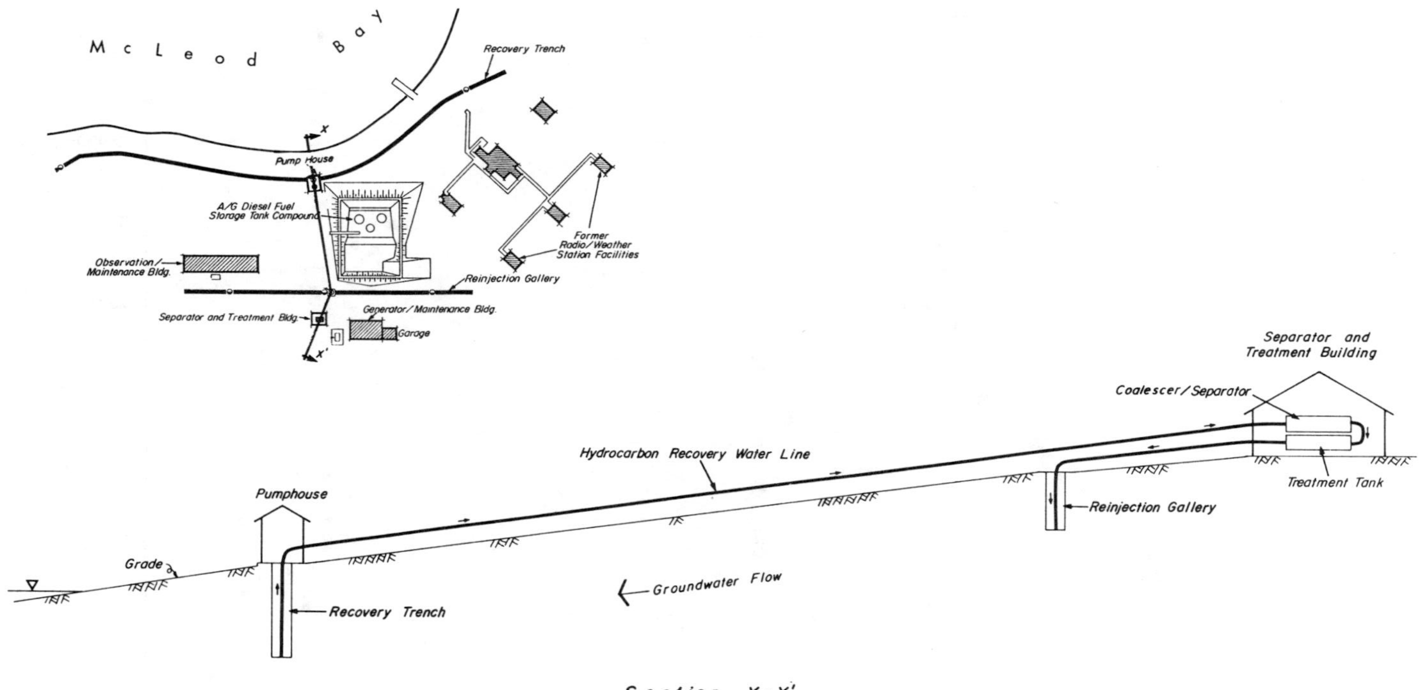

FIGURE 1. Hydrocarbon containment system cross-section.

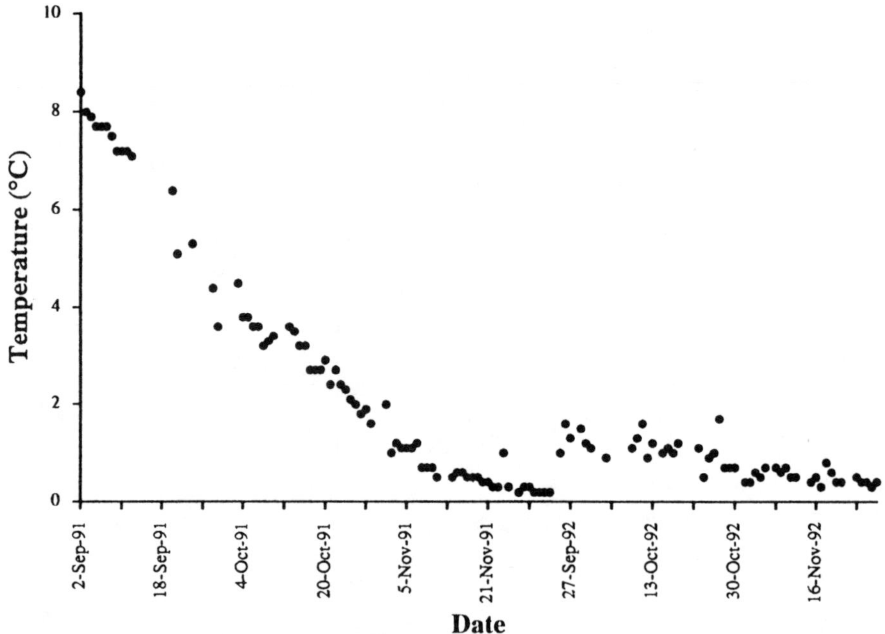

FIGURE 2. Groundwater temperatures.

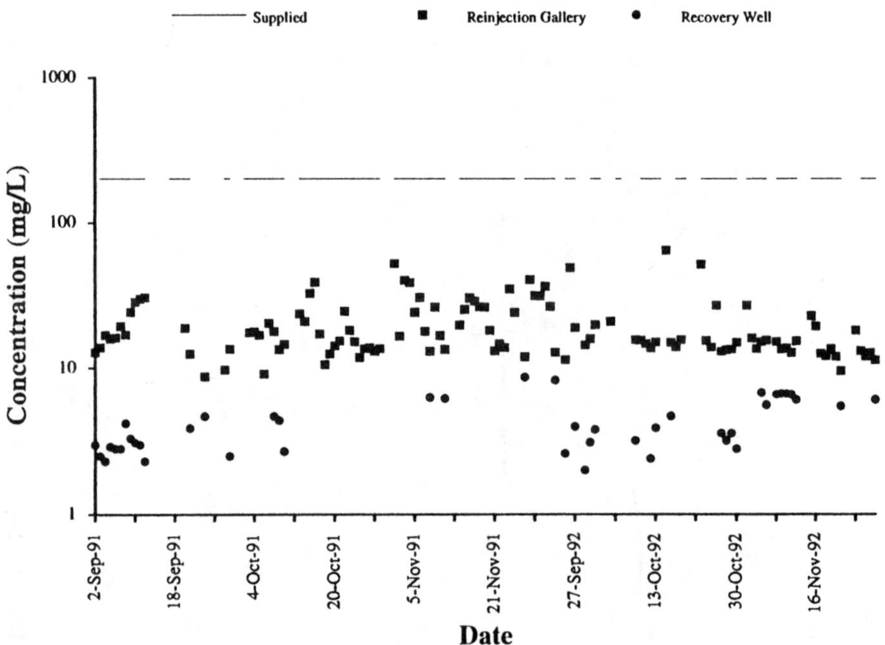

FIGURE 3. Dissolved oxygen concentrations.

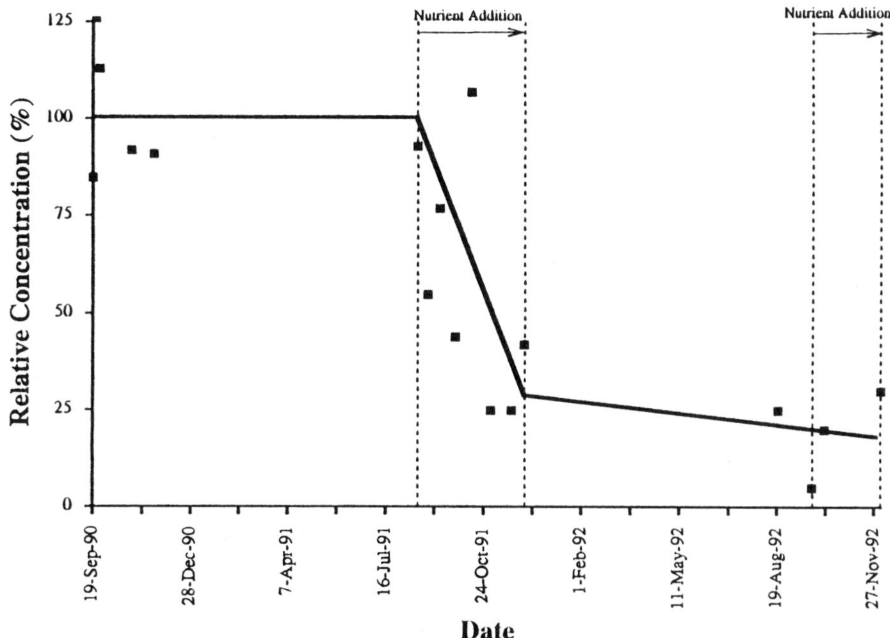

FIGURE 4. Total petroleum hydrocarbon concentrations.

In 1992 the nitrate concentrations were less than 10 mg/L until an unscheduled addition of nutrients resulted in elevated nitrate concentrations which persisted for the remainder of the season.

The results of the laboratory analyses of TPH in the water samples collected during the 1990, 1991 and 1992 operational seasons are summarized on Figure 4. The data indicate a decrease in the concentration of TPH in samples of the recovered water. In 1990 the concentration of TPH in the pumped groundwater reached 4.6 mg/L. Throughout the 1991 season the measured TPH concentration decreased to 45% of the initial value. By December 1992 the TPH concentration in the recovered groundwater decreased to 30% of the initial value.

The populations of hydrocarbon degrading-microorganisms in the recovered groundwater were generally below 10 CFU/mL during the 1991 season but reached 10^3 CFU/mL to 10^4 CFU/mL during the 1992 season.

SUMMARY

Construction and operation of the hydrocarbon containment system minimized the impact of the contamination on the waters of Great Slave Lake. The biological enhancement and site monitoring results presented herein indicate that bioremediation was successful in decreasing the concentrations of petroleum-derived constituents present in the groundwater despite the relatively cold groundwater temperatures. The concentration of TPH measured in groundwater samples decreased approximately 55% between August 1991 and December 1991. Between September 1992 and December 1992 the concentration of TPH decreased by an additional 15%. The concentrations of benzene, toluene, ethylbenzene, and xylenes, while already low, were generally reduced to nondetectable levels. The apparent consumption of oxygen during the operation of the recovery system suggests the theoretical mineralization of approximately 1,200 L of petroleum products.

MONITORING AN ABOVEGROUND BIOREACTOR AT A PETROLEUM REFINERY SITE USING RADIORESPIROMETRY AND GENE PROBES: EFFECTS OF WINTER CONDITIONS AND CLAYEY SOIL

R. Samson, C. W. Greer, T. Hawkes, R. Desrochers, C. H. Nelson, and M. St-Cyr

INTRODUCTION

Remediation of contaminated soil using an aboveground heap-pile bioreactor is applicable when the contaminants are known as biodegradable, when the soil offers a suitable structure (e.g., low clay content), and when the climate is not a detrimental factor. In Canada, bioremediation is still considered applicable only from April to October and with sandy or loamy soil. The objectives of this field study were to demonstrate that an aboveground bioreactor could be used to remediate a soil high in clay content to meet the stringent Quebec criterion B for oil and grease (below 1,000 mg/kg), to study the effect of winter conditions on microbial activity, and to validate the use of radiorespirometry and gene probes as tools to assess the biological activity. This technical note summarizes the results of a comprehensive study that was carried out from October 1991 to August 1992 at the Shell Canada refinery located in Montreal East, Quebec, Canada.

TEST SITE

The aboveground heap-pile bioreactor was installed at the Montreal East Shell Canada refinery (Quebec, Canada) and was operated by Groundwater Technology Inc. For this study, 1,500 m^3 of clay-loam soil (42% clay, 33% silt, and 25% sand) contaminated initially with an average concentration of 6,500 mg/kg of petroleum hydrocarbons with an average chain length of 12 to 24 carbons was treated in a cell (60 m × 20 m × 1.8 m) over a 10-month period beginning in October 1991. Before piling, the soil was screened to remove rocks. Nutrients, gypsum, and sawdust were then added to improve the nitrogen content, water-holding capacity, and soil structure, respectively. Continuous aeration of the pile was initiated only in May 1992 to avoid unnecessary cooling of the pile and O_2 depletion. Three 100-m^3 control piles also were assessed. The first control pile was untreated, the second had screened soil, and the third pile had screened and nutrient-amended soil without aeration.

MONITORING PROCEDURE

Sampling. Soil samples were collected using a split spoon every month for 10 months at eight locations and two depths (0.6 and 1.2 m) in the soil pile. For the three control piles, samples were collected at one location and two depths. Between each sampling, the equipment was washed with ethanol and acetone, and evaporated to dryness.

Chemical Characterization. The concentration of total petroleum hydrocarbon (TPH) was monitored using the oil and grease method derived from the American Public Health Association (APHA) method 5520 (1989), including methods 5520F (mineral oil separation) and 5520C (infrared measurement). The hydrocarbon composition and the concentration of two specific compounds (hexadecane and pristane) were evaluated using gas chromatography/mass spectrometry (GC/MS). The pH, percentage of humidity, concentration of NH_4+, NO_3^-, and PO_4^{-3}, and the C/N ratio (Control Equipment Corp CHN elementary analyzer, model 240XA) also were determined for each sample.

Respirometry and ^{14}C Mineralization Study. For each soil sample, biological performance was followed by monitoring the rate of mineralization of radiolabeled ^{14}C-hexadecane and ^{14}C-dotriacontane in soil microcosms, and by monitoring the oxygen uptake rate using an electrolytic respirometer (Bioscience Inc., model ER-100, Bethlehem, Pennsylvania). The soil samples (200 g) were placed in 1-L reactors incubated at the temperature measured during sampling without added substrate and the oxygen consumption was monitored for 10 days. To verify the activity of hydrocarbon-degrading microorganisms, mineralization of 1-^{14}C-hexadecane (100 mg/kg) and 16,17-^{14}C-dotriacontane (100 µg/kg) was studied in microcosms (100-mL serum bottles with KOH trap). The production of $^{14}CO_2$ was recorded for each soil sample at 5-day intervals for 20 days.

Gene Probes. Viable bacteria were determined by the spread plate method following dilution of the soil in sterile minimal salts medium (MSM) (Greer et al. 1990). An aliquot from the dilution series was then spread-plated in quadruplicate onto MSM-containing yeast extract, tryptone, and starch (250 mg/L each). The plates were incubated at three temperatures (4, 10, and 20°C) for at least 1 week before counting colonies. For gene probe assays, bacterial colonies were lifted onto nylon membranes, and colony hybridizations were performed according to Sambrook et al. (1989).

Gene probes were prepared by the polymerase chain reaction (PCR) using primers, generally 30 nucleotides (nt) in length, derived from published sequences. The *xylE* gene probe is an 834-nt probe from within the coding sequence of the *xylE* gene (encoding catechol 2,3-dioxygenase) from *Pseudomonas putida* mt-2

(ATCC 33015) carrying the Tol plasmid (Nakai et al. 1983). The *alkB* gene probe is an 870-nt probe from within the coding sequence of the *alkB* gene (encoding alkane hydroxylase) of *P. oleovorans* (ATCC 29347) (Kok et al. 1989). The resulting probe fragments were visualized by agarose gel electrophoresis, extracted and purified from the gel using the Geneclean kit (Bio101 Inc.), and labeled with [^{32}P]dATP using the Multiprime DNA labeling system (Amersham).

RESULTS AND DISCUSSION

During the winter, the average temperature measured within the soil never decreased below 7°C, even when the outside temperature was below −20°C (Figure 1). The soil pile never froze and for January and February the temperature close to the surface (0.6 m) was about 4°C. At the bottom (1.6 m) the temperature remained at 10°C. The highest concentration of hydrocarbon-degrading bacteria that were positive for the *xylE* and *alkB* gene probes were found to be mesophilic and therefore more sensitive to cold temperatures. Their number never exceeded 0.4% of the total heterotrophic population (Figure 1). The rate of mineralization at 4°C in February ranged from 0.5 to 1.5 mg/kg/day, depending on the sampling point. This indicated that hydrocarbon-degrading bacteria were quite active during the winter. However, the three control piles froze during the winter. The pile to which nutrients were added froze only in February, and its activity resumed more quickly in spring, suggesting that the addition of nutrients was an important factor to maintain microbial activity during the winter.

Although the microorganisms remained active in winter in the treatment soil pile, most of the biodegradation occurred when the soil temperature was above 15°C (Figure 1). The biodegradation of the petroleum hydrocarbons was achieved primarily during the first 2 months after soil processing. During this period, the concentration of oil and grease decreased from 3,400 to 2,000 mg/kg. The lag phase between December and April resulted in a poor removal of hydrocarbons. The biodegradation resumed in May when continuous aeration was initiated. The Quebec criterion B for mineral oil and grease (<1,000 mg/kg) was achieved in August 1992. The APHA method was not reliable in determining low concentrations (around 1,000 mg/kg) of oil and grease in clayey soil. Significant variations, associated with the silica gel cleaning procedure and the use of nonappropriate infrared standards, were observed from the data obtained from various commercial laboratories. By monitoring the disappearance of hexadecane and pristane using GC/MS, we were able to prove compliance with Quebec criterion B. Up to 50% of the hexadecane and 35% of the pristane were removed between June and August. This conclusion was also supported by the fact that the measured rate of mineralization of ^{14}C-hexadecane and ^{14}C-dotriacontane was high enough to allow the biodegradation of about 800 mg hydrocarbon/kg. The rate of oxygen consumption for the tested period (9,600 mg O_2/kg) was sufficient to enable the biodegradation of approximately 4100 mg of hydrocarbons/kg and the production of 3,200 mg of biomass/kg.

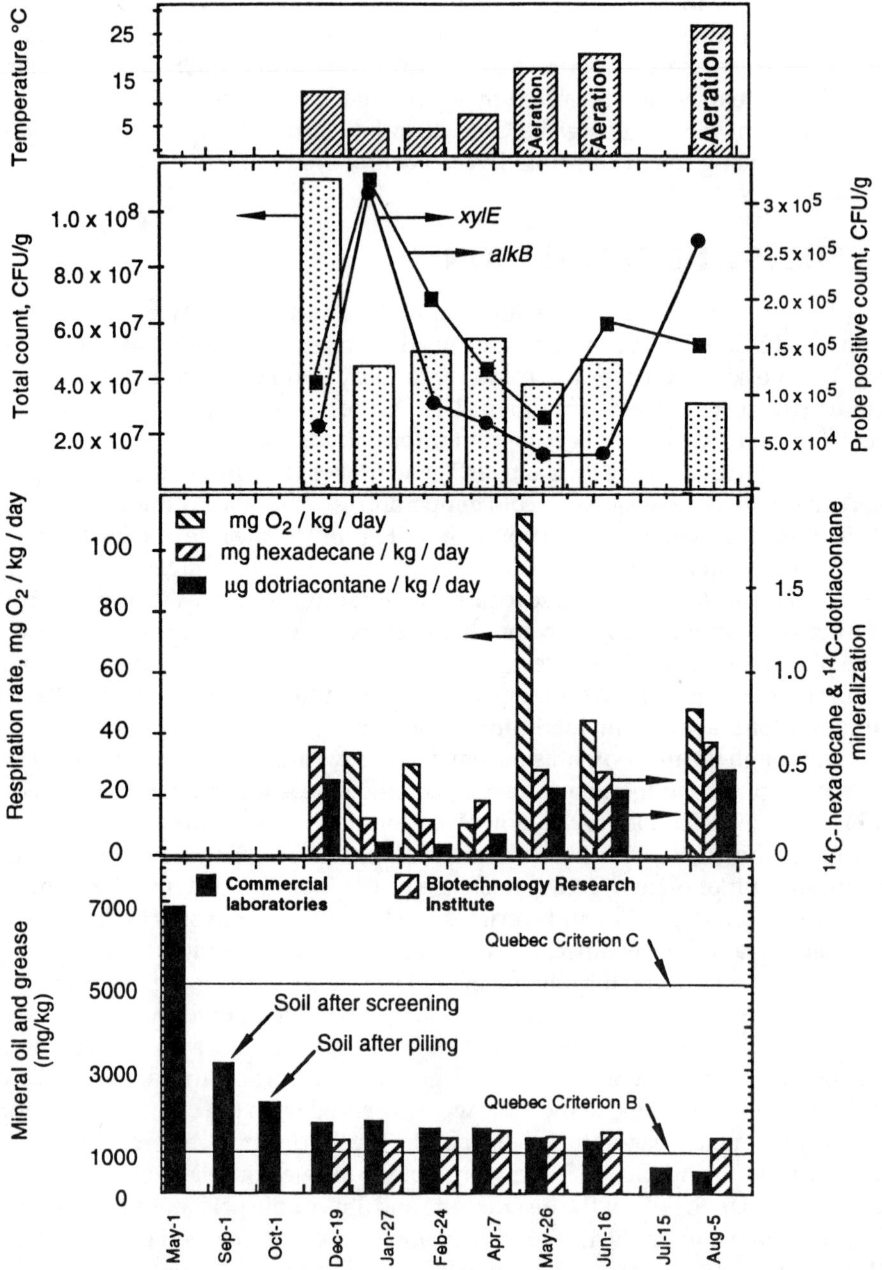

FIGURE 1. Summary of performance of the aboveground bioreactor treating hydrocarbon-contaminated soil (each data point represents the average of 16 samples).

The use of gene probes and radiorespirometry to monitor the microbial activity were quite instrumental in elucidating bioreactor performance. These techniques helped to visualize the chemical and biological heterogeneity within the pile. For example, it was observed that, at some locations within the pile, hydrocarbons were not biodegraded. A high residual concentration of hydrocarbons was clearly correlated to the low ^{14}C-hydrocarbon mineralization activity. A clear correlation was also obtained between an increase in the C/N ratio and the decrease in the mineralization activity, indicating that the distribution of nutrients during the soil preparation step is crucial. For example, the level of activity exhibited by hydrocarbon-degrading microorganisms depended on the local concentrations of NH_4^+ and NO_3^- ions. Because the organic matter was not entirely of petroleum origin (e.g., sawdust), respiration rates were not found to correlate well with the hydrocarbon-degrading activity, suggesting that the monitoring of CO_2 in the exhaust gas may not be a valid indicator of biodegradation activity.

CONCLUSION

Results from this study suggested that bioremediation of clayey soil in an aboveground heap pile bioreactor could be achieved (below the Quebec criterion B of 1,000 mg/kg) and that the pile remained active during the winter. However, a comprehensive monitoring program is required to properly assess the performance of the process. The combined use of tracer mineralization studies, respirometry, and gene probes allowed a good representation of biological activity within the pile and was instrumental in evaluating the state of the treatment.

ACKNOWLEDGMENTS

This study was funded by Shell Canada and Environment Canada (St-Lawrence Centre).

REFERENCES

Greer, C. W., J. Hawari, and R. Samson. 1990. "Influence of environmental factors on 2,4-dichlorophenoxyacetic acid degradation by *Pseudomonas cepacia* isolated from peat." *Arch. Microbiol. 154*: 317-322.

Kok, M., R. Oldenhuis, M. P. G. van der Linden, P. Raatjes, J. Kingma, P. H. van Lelyveld, and B. Witholt. 1989. "The *Pseudomonas oleovorans* alkane hydroxylase gene: Sequence and expression." *J. of Biol. Chem. 264*: 5435-5441.

Nakai, C., H. Kagamiyama, M. Nozaki, T. Nakazawa, S. Inouye, Y. Ebina, and A. Nakazawa. 1983. "Complete nucleotide sequence of the metapyrocatechase gene on the TOL plasmid of *Pseudomonas putida* mt-2." *J. Biol. Chem. 258*: 2923-2928.

Sambrook, J., E. F. Fritch, and T. Maniatas. 1989. *Molecular Cloning. A Laboratory Manual*, 2nd ed. Cold Spring Harbour Laboratory, Cold Spring Harbour, New York, NY.

BIOTREATMENT OF PETROLEUM HYDROCARBON-CONTAINING SLUDGES BY LAND APPLICATION: A CASE HISTORY AND PROSPECTS FOR FUTURE TREATMENT

N. A. Persson and T. G. Welander

INTRODUCTION

Petroleum hydrocarbon-containing wastes from a Scandinavian petrochemical facility (STATOIL Petrokemi AB, Stenungsund, Sweden) were disposed of by land application for a period of more than 10 years. The waste was comprised of filter residue (⅔ by volume) and waste-activated sludge (⅓ by volume) from the facility's wastewater treatment works. The application took place in spring each year. The stored sludge that accumulated between applications was stabilized with quicklime, principally as an odor control measure.

This study was undertaken to (1) evaluate the performance of the land application and (2) verify conclusions and investigate possible treatment approaches in laboratory microcosms. In addition, a laboratory-scale aerated slurry reactor was tested as an approach to stabilize the sludge prior to land application.

LAND APPLICATION CASE HISTORY

The summary is based on data from logbooks from chemical analyses performed once per year by independent laboratories. The land application was alternately conducted on three distinct parcels at the facility (A, B, and C). During 11 years, sludge (about 4×10^5 kg) was applied once or twice per year in April-June at a yearly load of around 40 kg/m^2. A total of approximately 4.5×10^6 kg stabilized sludge was disposed of on the 2.8×10^4 m^2 total area. The composition of the sludge varied during the period. The average dry content was 45 to 55%, and the average quicklime content was 30 to 45%. The average oil content as total extractable oil (TEO, according to Swedish standard SS 02 81 45, extraction with 1,1,2-trichloro-1,2,2-trifluoroethane and subsequent infrared (IR) analysis, the polar part identified as the part adsorbed to aluminum-oxide) was 1%, and of the TEO 5% was identified as polycyclic aromatic hydrocarbons (PAHs; gas chromatography/mass spectroscopy (GC-MS) with selective ion monitoring, internal standard added in order to compensate for losses during preparation).

Due to unequal distribution of the sludge and subsequent difficulties in representative sampling in the field, it was possible only to estimate the long-term landfarming performance on the basis of several years of cumulative site usage. Soil sampling at different depths across the site showed that the oil load essentially remained in the top 0.3-m layer of the soil. The interval from 0.3 to 0.6 m depth showed concentrations in the range of 5% of the top layer, and the interval from 0.6 to 0.9 m showed 0.5%, correspondingly. The land treatment areas were surrounded by ditches connected to ponds to enable surface runoff analysis and by wells (depth: 2 m) to permit site groundwater analyses.

Both the ponds and the wells frequently showed elevated pH values (>8). PAHs were not detected in any of the wells, and only once were detected (sum-identified at 40 μg/L) in a pond. TEO ranged from below detection limit (< 0.1 mg/L) to a maximum 9 mg/L in one of the wells. Surface water analysis showed a maximum TEO level of 8 mg/L in one of the ponds.

Figure 1 shows the analyses of TEO and PAHs, and calculated accumulated levels of TEO in the top 0.3 m of soil in area B. This is the area that was used for the longest period of time for land treatment. Areas A and C presented the same overall pattern (however the data are not shown); biotic (and/or abiotic) hydrocarbon reduction was insignificant. The pH value of the soil was elevated, ranging from 7.5 to 8.5 (occasionally above 9), compared to the reference soil (pH 6.5). This is most likely due to the addition of the quicklime as a sludge-stabilizing agent.

The content of heavy metals in the soil was low, but slightly elevated levels (background levels are commonly around 10 mg lead/kg soil [Fitchko 1989]),

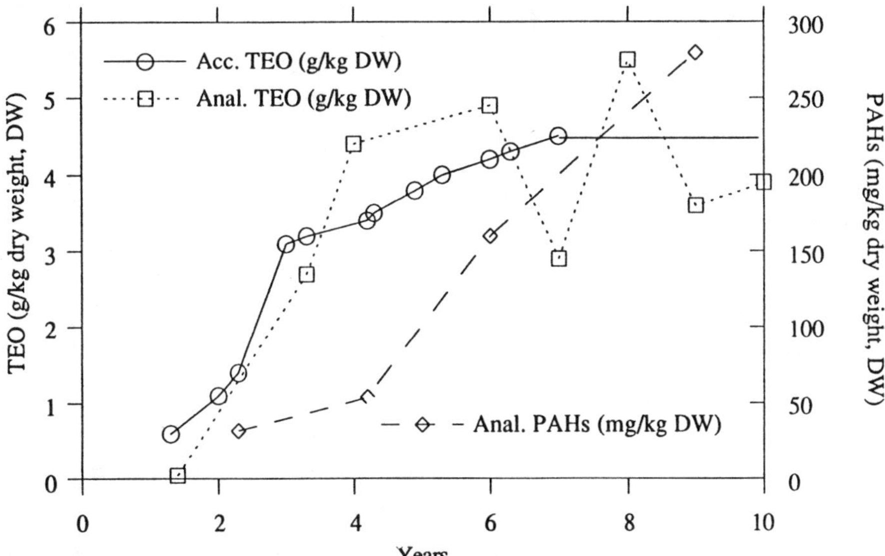

FIGURE 1. Accumulated (calculated) levels of TEO, analyzed levels of TEO, and sum of identified PAHs in the top 0.3 m of soil B.

up to 30 mg lead/kg soil, were shown. The applied sludge was the most probable source for this lead; nonstabilized sludge contained 140 mg lead/kg dry sludge.

LABORATORY MICROCOSM STUDIES

Laboratory studies were performed at room temperature in 35-kg soil microcosms (open pans of 0.3 m depth). TEO and PAH analyses were performed by independent laboratories. The first 30-week trials were designed to verify conclusions from field data. The microcosms were manipulated to try to stimulate the degradation of TEO (including PAHs) in the stabilized sludge. The sludge was mixed with topsoil from area C (the area presently in use) in proportion to full-scale treatment (45 g stabilized sludge/ kg top soil). The sludge/soil mixture was then added to the microcosms. To monitor the influence of pH adjustment (HCl), nutrient amendment (100 mg N-NH_4 and 20 mg P-PO_4 per week), oxygenation via H_2O_2 (3.5 g per week), grinding of soil (<0.5 cm), and inoculum (prepared every second week from circulated water enriched with glucose, NH_4Cl, and KH_2PO_4), various treatments were evaluated.

None of the above-mentioned factors significantly increased the degradation rate. A typical course of events is outlined in Figure 2. The noted increase in TEO might be dependent on a change in hydrocarbon availability for extraction during the trial.

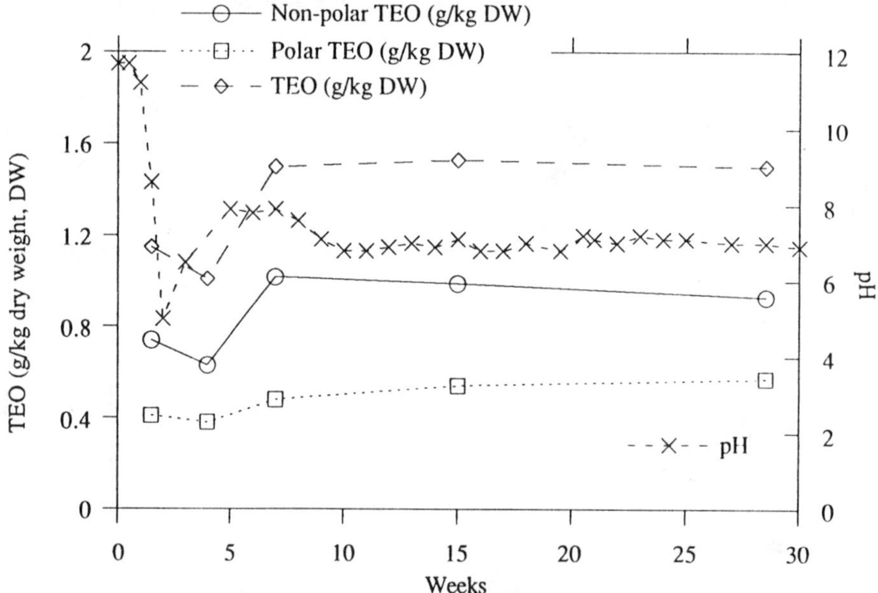

FIGURE 2. TEO and pH from microcosm with soil from area C mixed with stabilized sludge (pH control, nutrient amendment, oxygenation with H_2O_2, inoculum, and grinding applied).

A second set of 30-week trials in 35-kg microcosms focused on the exclusion of quicklime (nonstabilized sludge, 50 g/kg soil). Area C soil was compared to reference soil collected nearby area C and not previously used for land treatment. The effects of nutrient amendment and soil texture enhancement were then evaluated as shown in Table 1. The soil in each pan was tilled by hand every second week.

Final TEO reductions are outlined in Table 1. An overall reorganization to more polar TEO can be recognized. Nutrient addition to the reference soil increased the degradation rate, but the effects of soil texture enhancement seem more obscure. Decreases in old TEO residues were small, and the rate of TEO reduction in soil C was slower than in the reference soil. As seen from Figure 3, the process levels off but does not seem stationary after 30 weeks.

Changes in PAHs are summarized in Table 2. The sum of the identified PAHs, mainly contributed by 2- and 3-ring components, decreased to 10 mg/L in 30 weeks. More complex PAHs (present in low levels) seem more resistant to degradation, even though some reduction of benz(b+k)fluoranthene and benz(a)pyrene can be seen in soil from area C.

LABORATORY-SCALE AERATED SLURRY REACTOR

One part of nonstabilized sludge was diluted with nine parts of water and pH-adjusted to 7.0. Nutrients (50 mg/L P-PO_4 and 190 mg/L N-NH_4) were added to the slurry. The resulting slurry was introduced continuously to an aerated reactor of 500 mL working volume, equipped with a condenser. The trials were performed at room temperature.

Reduction of TEO at 3.5 days retention time in the slurry reactor was comparable to 30 weeks of land application, and an increased reactor retention time of 14 days did not significantly improve this. Reduction of the sum of identified PAHs is illustrated in Table 3, and reduction results were comparable to those given in Table 2. The reduction of "intermediate"-sized PAHs such as pyrene, benz(a)anthracene, and chrysene in the slurry was less than in an extended land treatment situation.

DISCUSSION

The climate in Scandinavia does not allow for year-round land application, so winter time sludge storage is necessary. Unfortunately, stabilizing sludge with quicklime decreases degradation in subsequent seasonal biotreatment.

Biotreatment in a closed system, such as a slurry reactor, prior to field application of sludge provides several advantages: requirements for time and space are minimized, waste treatment conditions are more easily controlled, and the escape of volatiles and odors can be avoided. It has been shown (Yare 1991) that slurry-phase bioremediation is potentially more efficient than soil-phase bioremediation. The slurry-phase treatment does not introduce factors inhibitory for seasonal land

TABLE 1. Influence of soil texture enhancement and nutrient addition on degradation of TEO in reference soil and soil from area C.

				Soil enh.[d]		TEO g/kg dry weight[e]		% Reduction	
Trial No.	Soil[a]	Sludge[b]	Nutr.[c]	5%	20%	Start	30 Weeks	Total	Nonpolar
1	Ref.	x				4.4 (80)	3.3 (58)	25	42
2	"	x	x			5.1 (80)	3.5 (58)	31	55
3	"	x	x	x		4.9 (79)	3.0 (51)	38	58
4	"	x	x		x	4.9 (82)	3.1 (55)	37	60
5	Soil C					1.2 (62)	0.9 (44)	25	47
6	"		x			1.2 (62)	1.0 (50)	17	32
7	"		x	x		1.3 (58)	1.3 (55)	0	5
8	"		x		x	1.2 (62)	0.9 (50)	25	40
9	Soil C	x				5.9 (77)	4.3 (65)	27	39
10	"	x	x			5.9 (76)	4.8 (63)	19	31
11	"	x	x	x		5.8 (75)	4.7 (57)	19	35
12	"	x	x		x	4.8 (73)	3.8 (60)	21	34

(a) Ref. = virgin (reference) soil collected near area C; soil C = topsoil from area C.
(b) Nonstabilized sludge.
(c) Weekly additions of 350 mg N, 100 mg P, and trace elements.
(d) Sand:bark 1:1 by volume; 5 or 20% of soil volume.
(e) Parentheses indicate percent nonpolar TEO.

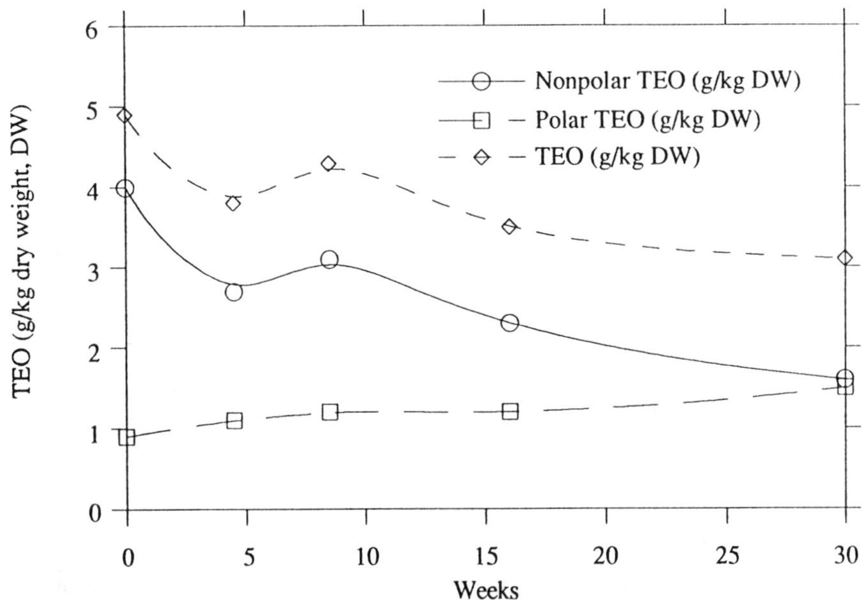

FIGURE 3. TEO from microcosm with reference soil mixed with nonstabilized sludge (nutrient amendment and soil enhancement applied, trial No. 4, Table 1).

treatment, and optimally it could replace land application. However capital and operating costs will be higher in a slurry reactor. Further investigations are needed to improve the slurry reactor results before this technology will be applicable to the sludge of interest in this study.

During land application, a residual concentration is always left in the soil. The amount remaining depends on nondegradable residuals or a lack of bioavailability (Harmsen 1991). Bioavailability of TEO-containing wastes can be increased by using detergents/emulsifiers. However, these substances may be better suited for use in the slurry phase because increased availability could also increase the toxic effects on the surroundings (Weissenfels et al. 1992).

The breakdown of PAHs in soil can be slow (Wild et al. 1991). The faster reduction of PAHs compared to removal in this study, therefore, could be surprising, but similar observations have already been described (Wang et al. 1990). Although 5- and 6-ring PAH components were at low levels initially and chromatograms showed an actual decrease, our observations could be explained by differences in analytical methods used. The nonspecific TEO analysis will include more metabolites than the specific PAH analysis. This was supported by the increase of polar TEO found during the treatment process.

TABLE 2. Reduction of identified PAHs in microcosm studies with nonstabilized sludge applied to reference soil (No. 4) and soil from area C (No. 12), and in microcosms with old residues alone (soil from area C, No. 8).

	mg/kg dry weight[b]					
	No. 4[a]		No. 8		No. 12	
Component	Start	30 Weeks	Start	30 Weeks	Start	30 Weeks
Naphthalene	<0.0	<0.05	0.1	0.1	0.2	0.2 (0)
Acenaphthalene	2.6	2.2 (15)	0.4	0.6	3.1	3.2 (0)
Acenaphthene	8.8	0.1 (99)	0.1	0.1	1.1	0.1 (91)
Fluorene	27	0.3 (99)	0.2	0.3	1.4	0.5 (65)
Phenanthrene	39	0.3 (99)	0.1	0.5	1.1	0.3 (73)
Anthracene	2.2	0.1 (95)	<.05	<.05	0.4	0.1 (75)
Fluoranthene	7.1	0.3 (96)	0.3	0.2	3.6	0.3 (91)
Pyrene	13	1.1 (92)	0.5	0.2	9.4	0.7 (92)
Benz(a)anthracene	4.0	0.4 (90)	0.2	0.2	2.3	0.7 (70)
Chrysene	2.4	0.4 (83)	0.2	0.2	1.7	0.4 (76)
Benz(b+k)fluoranthene	1.4	1.4 (0)	0.2	0.1	1.9	1.2 (26)
Benz(a)pyrene	1.3	1.6 (0)	0.1	0.1	1.9	1.5 (21)
Indeno(1,2,3-cd)pyrene and Dibenz(ah)anthracene	0.4	0.6 (0)	<.05	<.05	0.6	0.7 (0)
Benz(ghi)perylene	0.2	0.3 (0)	<.05	<.05	0.3	0.4 (0)
Sum of identified PAHs	109.4	9.1 (92)	2.4	2.6	29	10.3 (65)

(a) See Table 1.
(b) Parentheses indicate percent reduction.

TABLE 3. Reduction of identified PAHs in aerated slurry reactor fed with diluted, nonstabilized sludge.

		mg/L[(a)]		
		In	Out	
Component	Retention Time		3.5 days	14 days
Naphthalene		1.2	<.005 (>99)	<.005 (>99)
Acenaphthalene		7.0	0.22 (97)	0.20 (97)
Acenaphthene		2.7	0.82 (70)	1.6 (41)
Fluorene		33	1.7 (95)	0.07 (99.8)
Phenanthrene		33	0.24 (99.3)	0.07 (99.8)
Anthracene		1.9	0.16 (92)	0.03 (98)
Fluoranthene		2.0	0.82 (59)	0.20 (90)
Pyrene		2.4	1.4 (42)	1.3 (46)
Benz(a)anthracene		0.46	0.12 (74)	0.24 (48)
Chrysene		0.39	0.24 (38)	0.08 (79)
Benz(b+k)fluoranthene		0.20	0.16 (20)	0.16 (20)
Benz(a)pyrene		0.20	0.17 (15)	0.18 (10)
Indeno(1,2,3-cd)pyrene		0.04	0.05 (0)	0.04 (0)
Dibenz(ah)anthracene				
Benz(ghi)perylene		0.01	0.02 (-)	0.02 (-)
Sum of identified PAHs		84.5	6.1 (92)	4.2 (95)

(a) Parentheses indicate percent reduction.

ACKNOWLEDGMENT

The authors are very grateful to STATOIL Petrokemi AB, Stenungsund, Sweden (Jonny Andersson) for cooperating in the study and for giving us access to all available information.

REFERENCES

Fitchko, J. 1989. "Table 2.11: Background levels of selected elements in soil." In *Criteria for Contaminated Soil/Sediment Cleanup*. Pudvan Publishing Co., Inc. Northbrook, IL.

Harmsen, J. 1991. "Possibilities and limitations of landfarming for cleaning contaminated soils." In R. E. Hinchee and R. F. Olfenbuttel (Eds.), *On Site Bioreclamation: Processes for Xenobiotic and Hydrocarbon Treatment*, pp. 255-272. Butterworth-Heinemann, Stoneham, MA.

Wang, X., X. Yu, and R. Bartha. 1990. "Effect of bioremediation on polycyclic aromatic hydrocarbon residues in soil." *Environ. Sci. Technol.* 24:1086-1089.

Weissenfels, W., H.-J. Klewer, and J. Langhoff. 1992. "Adsorption of polycyclic aromatic compounds (PAH's) by soil particles: Influence on biodegradability and biotoxicity." *Appl. Microbiol. Biotechnol.* 36:689-696.

Wild, S. R., M. L. Berrow, and K. L. Jones. 1991. "The persistence of polynuclear aromatic hydrocarbons in sewage sludge amended agricultural soils." *Environ. Poll.* 72:141-157.

Yare, B. S. 1991. "A comparison of soil phase and slurry phase bioremediation of PNA-containing soils." In R. E. Hinchee and R. F. Olfenbuttel (Eds.), *On Site Bioreclamation: Processes for Xenobiotic and Hydrocarbon Treatment*, pp. 173-187. Butterworth-Heinemann, Stoneham, MA.

MATRIX EFFECTS ON THE ANALYTICAL TECHNIQUES USED TO MONITOR THE FULL-SCALE BIOLOGICAL LAND TREATMENT OF DIESEL FUEL-CONTAMINATED SOILS

M. A. Troy and D. E. Jerger

INTRODUCTION

Biological land treatment is the remediation technology in which contaminated soil is placed in an engineered treatment cell and managed to enhance indigenous microbiological activity and promote biodegradation of the constituents. Bioremediation of diesel fuel has been proven to be an effective treatment technology for the remediation of contaminated soils (Riser-Roberts 1992; Song et al. 1990). The assessment of the bioremediation progress of diesel fuel-contaminated soils, as well as the desired cleanup criteria, are often defined by analytical methods dictated by state regulations (Bell et al. 1991). Infrared (IR) spectroscopy and gas chromatography (GC) measurements of total petroleum hydrocarbons (TPH) are the current analytical methods of choice for the assessment/screening of diesel fuel-contaminated soils, as well as for determining site closure.

The TPH-IR (EPA 418.1 modified for soils; US EPA 1986, 1989) is the method where a soil sample is solvent extracted. The total mass of hydrocarbon dissolved in the solvent is then quantitated by comparing the infrared absorption of the extraction liquid against that of a defined hydrocarbon mixture.

Gas chromatographic methods for petroleum hydrocarbon analysis include the American Society for Testing and Materials (ASTM method 3328.78 (ASTM 1989) and EPA modified 8015 method (US EPA 1986, 1989). ASTM method 3328.78 is a GC method that uses flame ionization detection (FID) to measure petroleum hydrocarbons. EPA modified 8015 method is also a GC method that uses FID to measure petroleum hydrocarbons.

The following two scenarios discuss the remediation of diesel fuel and potential matrix interferences encountered when assessing bioremediation progress.

BIOLOGICAL LAND TREATMENT – SANDY LOAM

The remediation of the sandy loam soil was brought about as a result of the spillage of 8,400 gal (32,000 L) of diesel fuel (Troy et al. 1993). A synthetically lined treatment cell with a drainage system was constructed to treat approximately 1,200 cubic yards (920 m^3) of contaminated soil. Conditions in the cell were managed to ensure an optimum environment for microbial growth. Initial regulatory requirements indicated that bioremediation progress was to be monitored by measuring TPH concentrations by IR (EPA Method 418.1) and by GC (ASTM Method 3328.78). The results from the GC analysis were used as a qualitative measure of diesel fuel-contamination. The mean TPH-IR results for the entire treatment cell that were obtained for the first two months of operation indicated that the apparent concentration of the diesel stayed the same (1,000 to 1,500 mg/kg). However, observations made by comparing chromatograms produced by qualitative GC analysis over time indicated that the levels of the diesel had decreased (Troy et al. 1993).

The treatment system was shut down after 2 months of operation due to winter weather conditions. For startup in the next season, the regulatory requirements for monitoring bioremediation progress were allowed to be modified for the use of a GC method (EPA modified 8015 Method) that permitted specific quantitation of the diesel fuel. The modified 8015 Method allowed for accurate quantitation of the diesel fuel remediation progress. The treatment of the diesel-contaminated soil followed typical first-order kinetics (K_1 = 0.018 day^{-1}), and remediation to less than the designated cleanup level of 100 mg/kg was accomplished after 147 days of treatment.

BIOLOGICAL LAND TREATMENT – FILL MATERIAL

The biological land treatment of fill comprised of ash, cinders, crushed stone mixed with soil, and clay was required as a result of diesel spillage over a number of years at a railroad engine fueling yard (Troy et al. 1992). Approximately 3,500 yd^3 (2,700 m^3) of material were excavated and placed in an engineered biological land farm treatment cell. The soils were contaminated with diesel fuel at concentrations exceeding 3,000 mg/kg as measured by TPH (EPA modified 8015 Method). TPH measurements by GC (EPA modified 8015 Method) were accepted by the regulatory agency for use in monitoring bioremediation progress. The treatment cell was managed to ensure optimum conditions for microbial growth.

Initially, during the first three months of operation, diesel bioremediation progress followed typical first-order degradation kinetics. The site was then shut down over the winter months and started up again the following spring. During the second operational season very erratic results were obtained using the modified 8015 Method (Troy et al. 1992). Initially the difficulties were believed to be due to the analytical procedure and the extraction method used.

However, additional investigation indicated that the presence of coal and shale in the site matrix, caused interferences to the TPH-GC (EPA mod. 8015 Method) analytical method. Tripp et al. (1981) have indicated that unburned coal can be a significant source of environmental hydrocarbon levels. However, the chromatograms produced by the soil collected from the treatment cell area have indicated the disappearance of the straight-chain peaks normally associated with diesel fuel, as is illustrated in Figure 1. This suggested that the diesel concentration in the treatment cell soils was remediated. Observations at the site also supported this theory, as there was no evidence of staining or odors that would be associated with the reported high levels of diesel that were measured (260 to 5,100 mg/kg) at the end of the second operational season. The soils are currently being utilized as construction material for industrial development being performed at the site.

CONCLUSION

Comparison of the two projects shows that the matrix in which the diesel fuel bioremediation is being performed can have an affect on the currently accepted

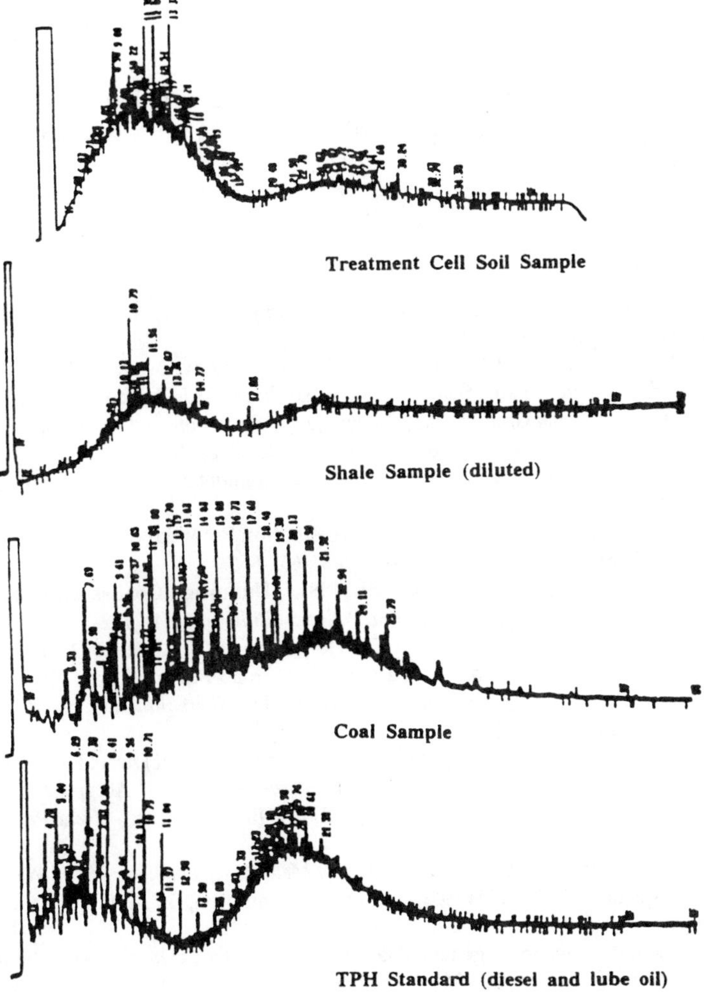

FIGURE 1. Representative chromatograms produced by the TPH-GC EPA modified 8015 Method.

analytical methods being used to monitor progress and determine closure. Although these matrix effects are site specific, the current analytical methods for assessing petroleum hydrocarbon remediation are not capable of overcoming these effects. Part of the difficulty is the use of a non-compound-specific measurement to assess progress. One way of overcoming this difficulty may be through the use of target or specific marker compounds. The study of bioremediation indicators requires further attention.

Biological land treatment is an effective technology for the remediation of diesel fuel-contaminated soils. However, it is important to be aware of potential matrix interferences with the presently accepted analytical techniques that are used to assess bioremediation progress.

ACKNOWLEDGMENTS

The authors would like to extend their gratitude to Paul Flathman, Bruce Demaine, Bruce Allen, and Pat Woodhull of OHM for discussions regarding activities at these sites.

REFERENCES

American Society for Testing and Materials. 1989. *Annual Book of ASTM Standards.* Volume 11.02. Philadelphia, PA.

Bell, C. E., P. T. Kostecki, and E. J. Calabrese. 1991. "Review of State Cleanup Levels for Hydrocarbon Contaminated Soils." In P. T. Kostecki and E. J. Calabrese (Eds)., *Hydrocarbon Contaminated Soils and Groundwater, Analysis, Fate, Environmental and Public Health Effects, Remediation,* Vol. 1, pp.77-89. Lewis Publishers, Chelsea, MI.

Riser-Roberts, E. 1992. *Bioremediation of Petroleum Contaminated Sites.* CRC Press, Inc., Boca Raton, FL.

Song, H., X. Wang, and R. Bartha. 1990. "Bioremediation Potential of Terrestrial Fuel Spills." *Applied and Environmental Microbiology.* 56(3): 652-656.

Tripp, B. W., J. W. Farrington, and J. M. Teal. 1981. "Unburned Coal as a Source of Hydrocarbons in Surface Sediments." *Marine Pollution Bulletin* 12(4): 122-126.

Troy, M. A., S. R. McGinn, B. P. Greenwald, D. E. Jerger, and L. B. Allen. 1992. "Bioremediation of Diesel Fuel Contaminated Soil at a Former Railroad Fueling Yard." In P. T. Kostecki and E. J. Calabrese (Eds), *Contaminated Soil: Diesel Fuel Contamination,* pp. 165-216. Lewis Publishers, Ann Arbor, MI.

Troy, M. A., S. W. Berry, and D. E. Jerger. 1993. "Bioremediation of Diesel Fuel Contaminated Soil; A Case Study From Emergency Response Through Closure." Accepted for Publication in P. Flathman, J. Exner, and D. E. Jerger (Eds.), *Bioremediation – Field Practice.* Lewis Publishers, Ann Arbor, MI.

U.S. Environmental Protection Agency. 1989. *Test Methods for Evaluating Solid Waste, Physical/ Chemical Methods.* 3rd ed., SW-846, Update 1. Washington, DC, Office of Solid Waste and Emergency Response.

U.S. Environmental Protection Agency. 1986. *Test Methods for Evaluating Solid Waste, Physical/ Chemical Methods.* 3rd ed., SW-846. Washington, DC, Office of Solid Waste and Emergency Response.

INITIAL RESULTS FROM A BIOVENTING SYSTEM WITH VAPOR RECIRCULATION

D. C. Downey, O. A. Awosika, and E. Staes

INTRODUCTION

The use of air injection or extraction to stimulate aerobic biodegradation of fuel hydrocarbons, or bioventing, has become a widely accepted low-cost method of soil remediation. In contrast to standard soil vapor extraction, the objective of bioventing is in situ bioremediation rather than volatilization and extraction. Although bioventing is most effective for less volatile hydrocarbons found in diesel and jet fuels, bioremediation of more volatile hydrocarbons found in gasoline can be enhanced with proper modification of the bioventing system. This technical note describes a combined soil gas extraction and air reinjection system that is being optimized to remediate gasoline-contaminated soils by maximizing in situ bioremediation and minimizing volatilization.

SITE HISTORY AND DESCRIPTION

The site is located at a gasoline service station located on Eglin Air Force Base in northwest Florida. The station has been in operation since 1955, and leaking pipelines have caused significant gasoline contamination beneath and downgradient of the station. Following several years of investigation and remedial design, a free product recovery and groundwater treatment system was installed on the site by the U.S. Army Corps of Engineers in 1987 and operated intermittently from 1988 to 1990. The system included six groundwater recovery wells, two free product recovery wells, an oil/water separator, and packed-column air-stripping system. In 1990, Engineering-Science, Inc. (ES) evaluated the remediation system and completed modifications to enhance free product recovery and groundwater treatment.

The surficial soils at the site are fine to medium-grained sands of the Lakeland Soil Association. Surficial sands extend to a depth of approximately 10 m where a 30-m-thick layer of Pensacola Clay is encountered (Geraghty and Miller 1985). Groundwater at the site occurs in the unconfined shallow sands at a depth of 1.5 to 2.1 m below ground surface (bgs). Groundwater elevations may vary by as much as 1.5 m as a result of heavy rainfall or drought. As a result, free product has impacted soils throughout an interval of 0.6 to 2.1 m bgs. Prior to initiating bioventing at the site, soil samples from the unsaturated zone indicated total recoverable petroleum hydrocarbon (TRPH) concentrations as high as 1,200 mg/kg and total benzene, toluene, ethylbenzene, and xylene (BTEX) concentrations of 500 mg/kg (ES 1992).

In 1991, the Air Force Center for Environmental Excellence (AFCEE) recommended that a bioventing system be established on the site to remediate these soil residuals and to remove the long-term source of groundwater contamination. Following initial pilot testing in September 1991, ES installed a full-scale bioventing system in May 1992. This technical note describes the results of initial testing and system optimization during the first 6 months of operation.

INITIAL PILOT TESTING

Pilot testing was completed to determine the potential radius of influence of each venting well and the potential for aerobic fuel biodegradation in contaminated soils. An initial soil gas survey was performed to determine the availability of oxygen in the contaminated area. Soil gas samples were extracted from existing monitoring wells and newly constructed soil gas monitoring points. Anaerobic conditions (O_2<2%) existed beneath the asphalt areas and in grass-covered areas in contaminated soils greater than one meter deep. The soil gas survey confirmed that the site would benefit from oxygen addition and the much higher rates of biodegradation afforded by aerobic conditions.

A soil gas permeability test was performed using an existing 2-inch (5 cm) monitoring well for air injection. A 1-hp blower provided a constant air injection rate of 280 L/min at a well-head pressure of 500-600 mbar, while pressure influence was measured with soil gas probes located as far as 14 m from the injection well. Steady-state, radial flow equations described by Johnson et al. (1990) were used to estimated soil gas permeability at approximately 2×10^{-7} cm^2 (20 darcy), which is typical for medium-grained sands.

The final phase of pilot testing was an in situ respiration test to estimate the rates of fuel biodegradation that could be achieved under aerobic conditions. The in situ respiration methods described by Hinchee et al. (1992) were used for this test. Oxygen utilization rates varied from 0.15 to 0.28 percent of O_2 consumed per hour. Based on a conservative ratio of 3.5 mg O_2 for every milligram of hydrocarbon degraded, a fuel biodegradation rate of approximately 3 to 5 mg fuel/kg soil per year day was measured during this test. These rates fall within the range of hydrocarbon biodegradation observed in similar soils and at soil temperatures of 16-23°C observed at Tyndall AFB, Florida (Miller et al. 1990).

FULL-SCALE BIOVENTING DESIGN

The full-scale bioventing system was designed to (1) control hydrocarbon vapor movement to eliminate any risk of vapor accumulation in the service station building; (2) provide sufficient oxygen to the vadose zone to promote in situ biodegradation of fuel hydrocarbons; and (3) minimize volatilization by returning extracted hydrocarbon vapors to biologically active soils. Figure 1 illustrates the

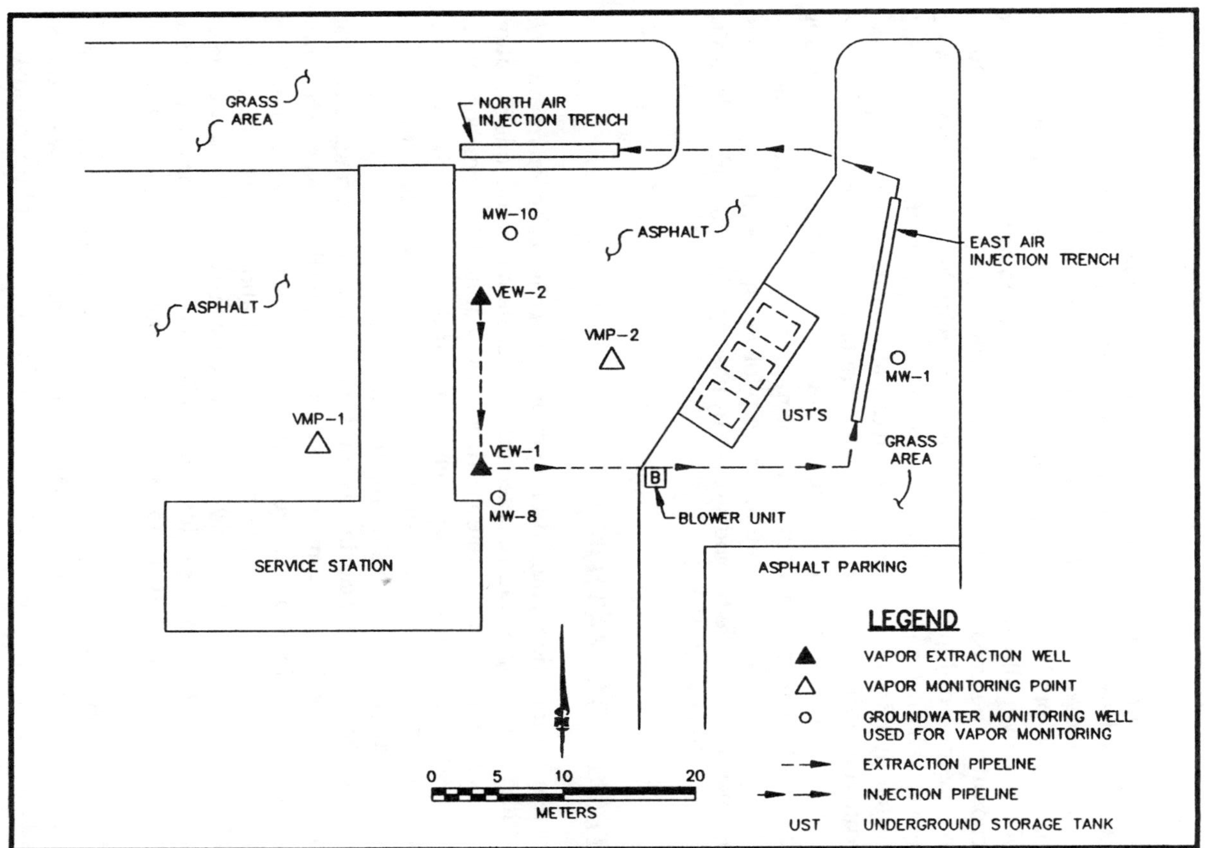

FIGURE 1. Eglin AFB, Florida, service station bioventing system.

design used to promote these objectives. Based on the results of pilot testing, two 4-inch (10.2-cm) PVC soil vapor extraction wells (VEWs), approximately 18 m apart near the center of the contaminated soil beneath the asphalt-covered service area, were screened over the 1.5-m contaminated interval from approximately 0.7 to 2.1 m bgs. The VEWs created a flow gradient away from the service station building and promoted the inward flow of oxygenated soil gas from the uncontaminated soils and atmosphere surrounding the site.

A 2.5-hp blower was installed but was later replaced with a 1-hp blower due to less-than-predicted operating vacuums. A dilution valve on the vacuum side of the blower allowed air at 20.8% O_2 to be added to the system prior to reinjection at the perimeter of the site. Two air injection trenches totalling 37 linear m were constructed along grassy areas near the edge of the asphalt-covered area. A 2-in-(5 cm-)diameter, slotted PVC pipe was placed at the bottom of each trench (approximately 1.2 m deep) to recirculate hydrocarbon vapors at the site perimeter and to provide additional opportunities for vapor biodegradation. A plastic liner was placed over each injection pipeline to reduce the potential for short circuiting and to extend the area of horizontal vapor migration. This concept had been employed at Tyndall AFB, Florida, where pilot testing results indicated that hydrocarbon vapors could be biodegraded by passing them through biologically active soils (Miller et al. 1990).

SYSTEM OPTIMIZATION

System optimization required a reduction in soil gas extraction from VEWs while still maintaining enough oxygen influx to sustain in situ biodegradation rates. A lower vapor extraction rate also decreases loading of hydrocarbon vapors to the reinjection trenches and vapor emissions to the atmosphere. However, too low of an extraction rate will reduce the radius of venting influence and could result in a portion of the site returning to anaerobic conditions.

Optimization began by closing the air dilution valve and extracting soil gas from the two VEWs at the maximum vacuum supplied by the blower. A flowrate of 700 L/min was achieved at a gauge vacuum of 850 mbar. Following 90 minutes of operation, the average soil gas oxygen concentration had increased from near 0% to over 4%, and a radius of vacuum influence in excess of 14 m was measured using vapor monitoring points. Initial hydrocarbon vapor levels in the extracted gas were measured at approximately 290 mg of hydrocarbons per liter of soil gas.

The 700-L/min flowrate produced a rapid increase in oxygen and more than adequate radial influence, so the extraction rate from the two VEWs was reduced by gradually opening the air dilution valve. At a flowrate of approximately 280 L/min through both VEWs, a slight vacuum could still be measured at the perimeter of the contaminated soil. The system was allowed to reach equilibrium over a 24-hour period. Soil-gas O_2 concentrations were then measured at various soil-gas monitoring points. O_2 levels began to stabilize at levels exceeding 10%, indicating aerobic conditions could be maintained at this reduced flow rate.

The reduction in soil gas extraction rates and addition of dilution air resulted in a lower hydrocarbon vapor concentrations (85 mg/L) the off-gas, but a higher flowrate of air (1,200 L/min) entering the injection trenches. During the first day of operation hydrocarbon loading to the air injection trenches was approximately 145 kg/day. Based on the average initial biodegradation rate of 4 mg TPH/kg soil/day, approximately 36,000,000 kg of biologically active soil would be required to degrade this initial hydrocarbon load. The volume of soil available for vapor treatment along the reinjection trench (assuming a 6-m horizontal influence on each side of the trench) is approximately 700 m^3 or 1,300,000 kg of soil. Clearly, the existing injection trenches were not large enough to biodegrade the high initial hydrocarbon vapor concentrations extracted from this site.

Because the extracted soil gas hydrocarbon concentrations were expected to decrease exponentially during the initial month of operation, a decision was made to continue to operate the system in this configuration but to replace the 2.5-ph blower with a smaller 1-hp blower to reduce the overall airflow of the system. After 35 days of operation, the hydrocarbon loading into the reinjection trenches had decreased by an order of magnitude to 14 kg/day. After 65 days, the loading was approximately 5 kg/day, which is approximately equal to the mass of fuel that theoretically can be biodegraded each day in the soil mass influenced by the injection trenches.

RESULTS TO DATE

Figure 2 summarizes the estimated biodegradation and volatilization that occurred on the site during the initial 180 days of operation. To date, volatilization accounts for approximately 35% and biodegradation approximately 65% of the total hydrocarbon removed from the site. However, it is important to note that since day 30 of system operation, biodegradation has exceeded volatilization as the primary mechanism of removal. Estimates of biodegradation are based on the

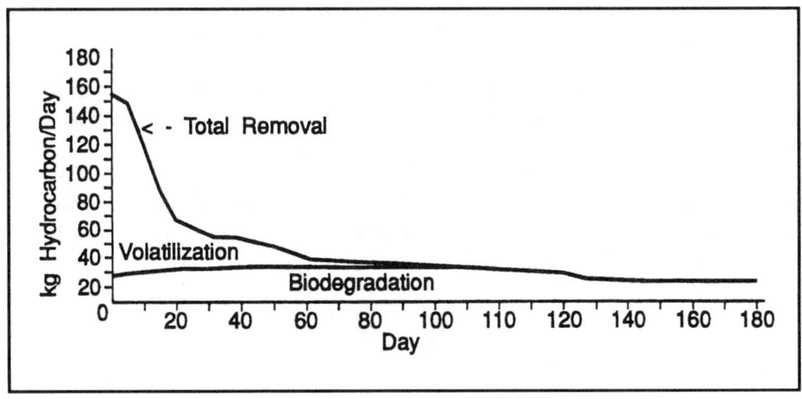

FIGURE 2. Biodegradation and volatilization vs. time.

initial average in situ respiration rates and a repeat respiration test conducted 90 days after system startup, when the biodegradation average rate had slightly increased to more than 5 mg TPH/kg soil/day. This increase may be the result of increasing soil temperatures during July and August and an expected increase in aerobic bacteria populations and biological activity in these previously anaerobic soils. The reduction in total biodegradation from day 120 to day 180 was the result of a rising water table which has temporarily submerged a third of the contaminated vadose zone, making it unavailable for bioventing.

RECOMMENDATIONS

Several methods of improving bioventing systems that treat gasoline-contaminated soil have been tested at this site, and other methods are under consideration. On this site, a smaller blower was used to reduce total air flow and volatile emissions. The use of vapor treatment systems such as an internal combustion engine or adsorbent bed during the first month of operation would greatly reduce the initial volatiles loading on injection trenches without a significant impact on total cost. The use of a pulsed system that would provide enough oxygen for biodegradation, and then shut down until the oxygen was depleted, could also reduce total air flow and volatilization. ES is currently designing bioventing systems for two additional gasoline sites which will incorporate these improvements.

REFERENCES

Engineering-Science, Inc. 1992. *Engineering Work Plan for Bioventing System, 7th Street BX Service Station.* Prepared for the Air Force Center for Environmental Excellence Brooks Air Force Base, TX.

Geraghty and Miller, Inc. 1985. *Environmental Assessment for the 7th Street BX Station.* Prepared for the Army Corps of Engineers Kansas City, MO.

Hinchee, R. E., S. K. Ong, R. N. Miller, D. C. Downey, and R. Frandt. 1992. *Test Plan and Technical Protocol for a Treatability Test for Bioventing.* Prepared for the Air Force Center for Environmental Excellence, Brooks Air Force Base, TX.

Johnson, P. C., M. W. Kemblowski, and J. D. Colhart. 1990. "Quantitative Analysis for the Cleanup of Hydrocarbon Contaminated Soils by In Situ Soil Venting." *Ground Water 28*(3): 413-429.

Miller R. N., R. E. Hinchee, C. M. Vogel, R. R. Dupont, and D. C. Downey. 1990. "A Field Scale Investigation of Enhanced Petroleum Hydrocarbon Biodegradation in the Vadose Zone." *Proceedings of NWWA/API Conference on Petroleum Hydrocarbons and Organic Chemicals in Ground Water*, pp. 339-351.

FIELD DEMONSTRATION OF NATURAL BIOLOGICAL ATTENUATION

H. S. Rifai and P. B. Bedient

INTRODUCTION

Natural aerobic biodegradation is the process by which indigenous microorganisms break down chemicals present in the soil or groundwater, using oxygen as an electron acceptor, into other products and sometimes to carbon dioxide and water. The basic requirements for aerobic biodegradation at the field scale include (1) the presence of indigenous microorganisms; (2) the availability of an energy source, a carbon source, oxygen, and inorganic nutrients; and (3) the prevalence of acceptable environmental conditions.

The challenge to scientists and engineers relates to the question of how to verify and quantify natural aerobic biodegradation at a field site. Commonly used procedures have involved conducting a series of laboratory studies of biodegradation and site-specific field characterization studies. Dissolved oxygen (DO) content measurements in groundwater, studies of numbers and activities of microorganisms, and material balances on contaminants have been used at several field sites to quantify the process (Borden & Bedient 1987, Barker et al. 1987, Chiang et al. 1989, Klecka et al. 1990, and Rifai et al. 1988). In addition, researchers have developed analytical and numerical models to simulate biodegradation and increase our understanding of the interaction between the different processes occurring in the subsurface (Borden & Bedient 1986, Srinivasan & Mercer 1988, Molz et al. 1986, and Rifai et al. 1988).

The need remains, however, to establish a standardized field protocol that would allow practitioners to estimate the rate of natural biodegradation at a field site and predict the impact of biodegradation on contamination levels at the site. Given that we are faced with the almost impossible task of actively remediating thousands of contaminated sites in the United States, it might be possible to view natural aerobic biodegradation as a viable remediation technology. The technology might be used for sites that are associated with lower risks to human health and the environment and for sites that cannot attain cleanup standards using other technologies.

PROPOSED FIELD MONITORING PROGRAM

Research aimed at developing a field protocol for assessing aerobic biodegradation of benzene, toluene, ethylbenzene, and xylenes (BTEX) at field sites is being conducted at Rice University. The conceptual framework for the project involves analyzing the soil and groundwater chemistry at a gasoline spill site undergoing aerobic biodegradation, as well as completing modeling studies to predict future site conditions. A various number of parameters were identified for field measurements, as shown in Table 1. The objective of the listed sampling program is to determine which of those parameters could be used as indicators of biodegradation, and as such provide information on the rate of the reactions.

PROJECT FIELD SITE

The site that is currently being studied in this research is an active gasoline station facility located at the intersection of Highway 290 and Antoine Road in Houston, Texas. The site contains two dispensing islands located northeast and southeast of the facility building (see Figure 1). An underground storage tank (UST) system consisting of three 8,000-gallon (30,283 L) steel tanks is located east of the station building near the eastern property boundary. The UST system is used

TABLE 1. Proposed parameters for field measurements.

Media	Parameter
Soil	Microbial counts/activity TOC BTEX
Ground Water	Temperature Dissolved oxygen Carbon dioxide Conductance pH, TDS Redox potential Ca, Mg, Na, Mn, Fe, SO_4, Cl Total alkalinity BOD, COD, TOC, BTEX NO_3, NO_2, NH_4, PO_4
Soil Vapor	O_2, CO_2, CH_4, H_2S

Notes: TOC = total organic carbon;
TDS = total dissolved solids;
BOD = biological oxygen demand;
COD = chemical oxygen demand.

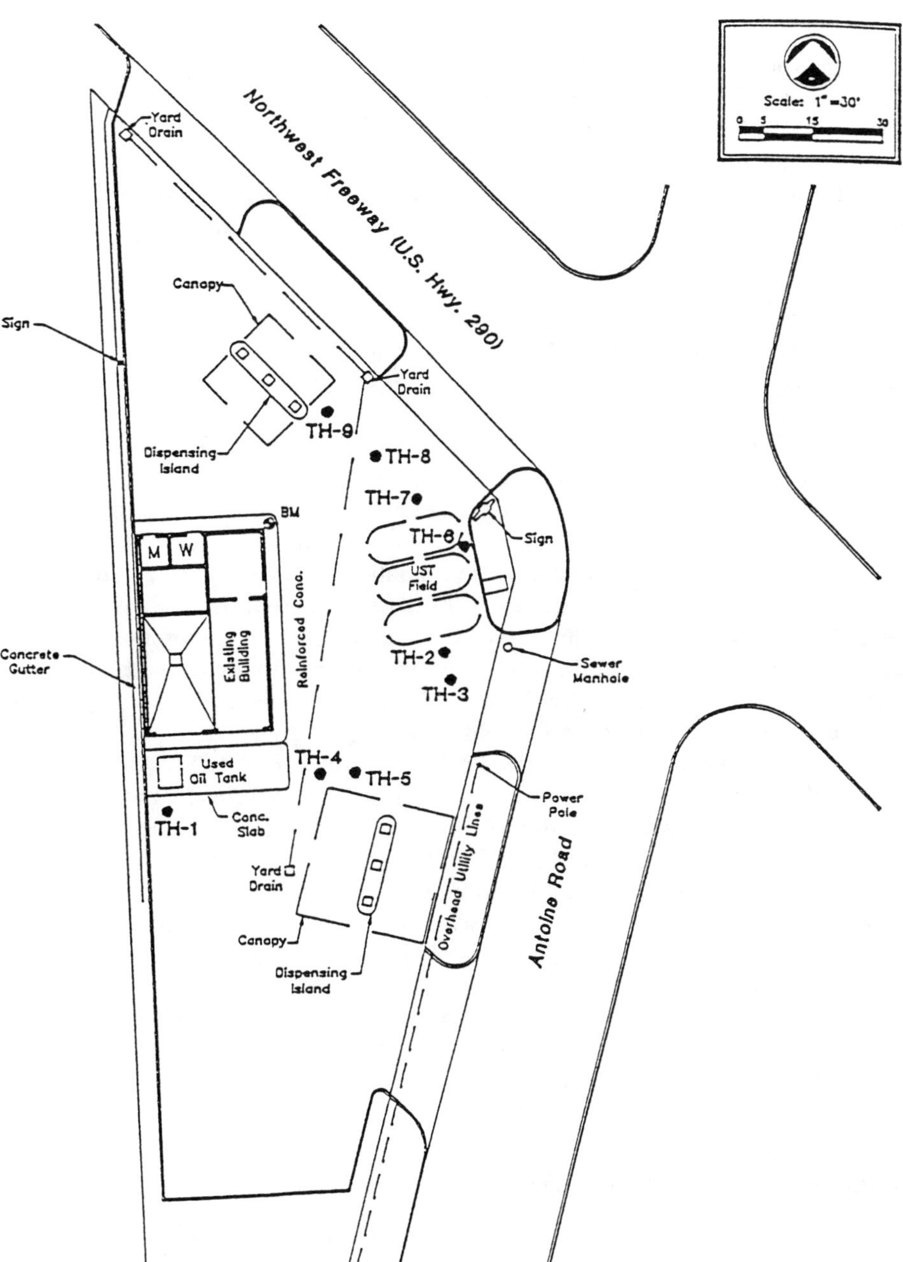

FIGURE 1. Site layout.

to store three grades of unleaded gasoline. A steel 550-gallon (2,082 L) underground used oil tank is located south of the building near the western property boundary.

Subsurface hydrocarbon contamination was detected at the site in early 1991 during a routine pre-UST removal/replacement investigation. Analyses of soil samples taken from borings around the dispenser islands (locations TH-1 through TH-9, Figure 1) indicated levels of total petroleum hydrocarbons (TPHs) up to 1,112 mg/kg, and benzene, toluene, ethylbenzene, and xylenes (BTEX) levels up to 174 mg/kg. The higher TPH and BTEX concentrations were found in soil samples obtained from exploratory holes TH-4 and TH-5 located in the northwest corner of the southern dispensing island area.

Two types of soils underlie the facility. The first is a tan to gray clay that is sandy to silty, stiff to very stiff, and varies in thickness from 9 to 9.6 ft (2.7 to 2.9 m). The second underlies the clay layer and is a light gray to tan sand. The sand is clayey to silty with interbedded clay seams, and it varies in thickness from 6 to 10.8 ft (1.8 to 3.3 m).

The USTs at the site were excavated and removed from the site in April 1991. The three tanks containing unleaded gasoline had no visible holes or damage. The used oil tank, however, exhibited small holes, which are some of the sources of contamination at the site. It is also likely that parts of the distribution system may have contributed to subsurface contamination as evidenced by hydrocarbon-contaminated soils located underneath the dispensing islands. Eight monitoring wells have been installed at the site since the tank removal operations (Figure 2). Groundwater levels and BTEX concentrations in the monitoring wells are measured on a quarterly basis (see Table 2). Additionally, several soil sampling events have taken place at the site (see Table 3 and Figure 2).

Groundwater at the site is located between approximately 17.5 to 18 ft (5.3 to 5.5 m) below ground surface. The general direction of groundwater flow is toward the west-northwest, with a gradient in the range of 0.002 ft/ft (0.002 m/m). The water quality data listed in Table 2 indicate relatively high levels of BTEX (exceeding 1 mg/L) at most of the monitoring wells, with the exception of wells MW-7 and MW-8. The data also indicate that there is a significant amount of fluctuation in the BTEX concentrations measured during the various sampling events. It is believed that this is due to seasonal variations in the groundwater levels at the site. The soil sampling data listed in Table 3 indicate the presence of residual hydrocarbons at most of the sampled locations between 12 and 17 ft (3.7 to 5.2 m) below ground surface.

BIODEGRADATION SAMPLING

A sampling program modified from that listed in Table 1 was conducted in July 1992 at the facility to screen the chemistry of the groundwater at the site from samples collected from monitoring wells MW-2, MW-5, and MW-7. The three wells were selected such that each of the wells would be representative of either a low, medium, or high level of contamination. The samples were analyzed for BTEX, dissolved oxygen, temperature, pH, Eh, ferrous iron, manganese, alkalinity, ammonia, nitrate, sulfate, total organic carbon (TOC), carbon dioxide, and hydrogen sulfide. Results from the July 92 analyses are shown in Table 4.

TABLE 2. Groundwater levels and water quality data at the field site.

	Water Table Levels (ft)							
	MW-1	MW-2	MW-3	MW-4	MW-5	MW-6	MW-7	MW-8
5/24/91	81.96	82.08	81.97					
6/14/91	81.97	82.06	81.95	81.94	81.94	82.01	82.02	
8/2/91	82.57	82.68	82.58	82.56	82.56	82.62	82.63	
11/6/91	81.33	81.32	81.35	81.29	81.29	81.41	81.40	
2/5/92	85.76	85.90	85.67	85.24	85.90	85.91	85.90	
5/18/92	87.66	87.52	87.96	87.66	87.87	87.77	87.90	
6/16/92	88.12	88.03	88.47	88.17	88.94	88.26	88.44	87.76
8/14/92	85.43	85.58	85.59	85.43	85.58	85.59	85.73	85.59
	BTEX Concentrations (mg/L)							
	MW-1	MW-2	MW-3	MW-4	MW-5	MW-6	MW-7	MW-8
5/24/91	17.01	0.13	2.31					
6/14/91				5.33	11.21	4.11	0.02	
8/2/91	1.61	1.63	1.31	3.13	3.31	3.66	0.02	
11/6/91	7.45	32.39	2.32	52.68	7.22	6.54	0.10	
2/5/92	1.31	20.95	18.14	0.46	3.55	24.93	0.03	
5/18/92	10.67	24.25	34.90	16.29	47.67	73.08	0.08	
6/16/92								0.00
8/14/92	27.78	10.18	17.63	6.83	17.97	58.84	0.00	0.00

The measured BTEX concentrations at the three wells ranged from less than 2 mg/L at well MW-7 to 23.5 mg/L at well MW-5. Dissolved oxygen concentrations ranged from 0.82 mg/L at MW-2 to 2.12 mg/L at MW-7. This is consistent with the commonly observed phenomenon of depressed oxygen levels in contaminated zones and background oxygen levels in uncontaminated zones. The redox conditions at the site decrease from a high of 72 mV at MW-7 to 39 mV at MW-2. Total organic carbon is the highest at MW-5 as would be expected, because it is the most contaminated well. Similarly, alkalinity was the highest in MW-5. Carbon dioxide is higher in wells MW-2 and MW-5 than in the less contaminated well MW-7. Sulfate concentrations decreased from a high of 55 mg/L at MW-7 to a low of 9 mg/L at MW-2.

Future plans for the site include monthly sampling of groundwater at the monitoring wells to verify the trends detected in the screening sampling event conducted in October 1992. Additional parameters listed in Table 1 also will be sampled. The data would then provide a more comprehensive framework for assessing and modeling the natural biodegradation at the site.

TABLE 3. Soil sampling data of field site.

SAMPLE LOCATION	DEPTH (ft)	DEPTH (m)	TOTAL BTEX (mg/kg)	TPH (mg/kg)
MW-1	15	4.6	78	282
	17.5	5.3	31	331
MW-2	15	4.6	0	ND
	17.5	5.3	53	149
MW-3	15	4.6	8	142
	17.5	5.3	334	729
TB-1	5	1.5	0	ND
	17.5	5.3	371	575
TB-2	12.5	3.8	212	1171
	17.5	5.3	1	23
TB-3	15	4.6	12	157
	17.5	5.3	0	28
TB-4	7.5	2.3	4	37
	17.5	5.3	42	166
TB-5	10	3	14	160
	17.5	5.3	3034	6718
MW-4	15	4.6	4	ND
	17.5	5.3	10	101
MW-5	17.5	5.3	10	13
	25	7.6	2	81
MW-6	7.5	2.3	9	74
	17.5	5.3	125	245
MW-7	15	4.6	0	ND
	17.5	5.3	0	ND
MW-8	16.5	5	ND	ND
	26.5	8.1	ND	ND

NOTES: ND indicates the constituent was not detected above the method detection limit.
Method detection limits: BTEX — 0.001 mg/kg
TPH — 5.0 mg/kg.

CONCLUSIONS

A proposed field protocol for assessing natural biodegradation is being tested at a field site in Texas. Initial results indicate that the chemistry of the groundwater at the site varies from uncontaminated to contaminated zone in a

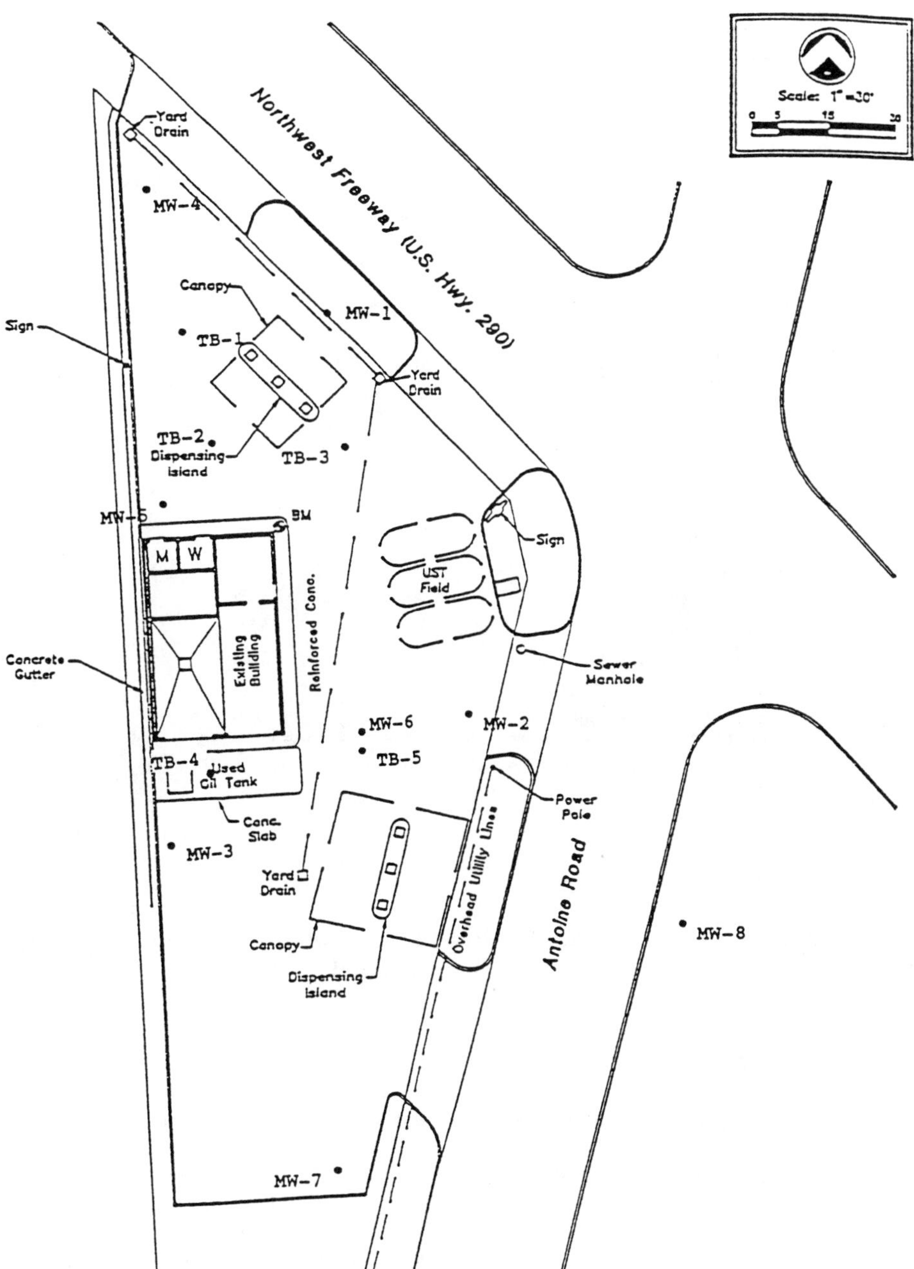

FIGURE 2. Locations of monitor wells and soil samples.

manner that could be attributed to some degree to biological processes. Additional sampling is needed before a more comprehensive geochemical model of biodegrading plumes can be developed.

TABLE 4. July sampling data at field site.

	Monitoring Well		
	MW-2	MW-5	MW-7
Total BTEX (µg/L)	14,070	23,400	187
Dissolved Oxygen (mg/L)	0.82	1.20	2.12
Water Table (ft)	85.58	85.64	85.72
Thickness of Free Product	na	na	na
Temperature (°C)	27.9	25.1	26.6
Conductivity (mmho/cm)	na	na	na
pH	6.51	6.51	6.46
Eh (mV)	39	41.00	72
Ferrous Iron (mg/L)	1.6	2.8	1.5
Manganese (mg/L)	0.9	1.24	1.9
Alkalinity (mg/L)	309	446	301
N-Ammonia (mg/L)	0.20	0.24	0.29
N-Nitrate (mg/L)	0.40	0.20	0.20
Sulfate (mg/L)	9	27	55
Total Organic Carbon (mg/L)	73	100	67
Carbon Dioxide (mg/L)	131	132	122
Hydrogen Sulfide (mg/L)	0.1	0.1	0.0

Notes: ns — not sampled; na — not applicable

ACKNOWLEDGMENTS

Funding from the American Petroleum Institute and the Shell Chair in Environmental Science is gratefully acknowledged.

REFERENCES

Barker, J. F., G. C. Patrick, and D. Major. 1987. "Natural Attenuation of Aromatic Hydrocarbons in a Shallow Sand Aquifer." *Ground Water Monitoring Review* 7(1):64-71.

Borden, R. C., and P. B. Bedient. 1986. "Transport of Dissolved Hydrocarbons Influenced By Oxygen-Limited Biodegradation: 1. Theoretical Development." *Water Resources Research 13*:1973-1982.

Borden, R. C., and P. B. Bedient. 1987. "In Situ Measurement of Adsorption and Biotransformation at a Hazardous Waste Site." *Water Resources Bulletin 4*:629-636, American Water Resources Association.

Chiang, C. Y., J. P. Salanitro, E. Y. Chai, J. D. Colthart, and C. L. Klein. 1989. "Aerobic Biodegradation of Benzene, Toluene, and Xylene in a Sandy Aquifer — Data Analysis and Computer Modeling." *Ground Water 6*:823-834.

Klecka, G. M., J. W. Davis, D. R. Gray, and S. S. Madsen. 1990. "Natural Bioremediation of Organic Contaminants in Ground Water: Cliffs-Dow Superfund Site." *Ground Water* 4:534-543.

Molz, F. J., M. A. Widdowson, and L. D. Benefield. 1986. "Simulation of Microbial Growth Dynamics Coupled to Nutrient and Oxygen Transport in Porous Media." *Water Resour. Res.* 22(8):1207-1216.

Rifai, H. S., P. B. Bedient, J. T. Wilson, K. M. Miller, and J. M. Armstrong. 1988. "Biodegradation Modeling at Aviation Fuel Spill Site." *Journal of Environmental Engineering* 5:1007-1029.

Srinivasan, P., and J. W. Mercer. 1988. "Simulation of Biodegradation and Sorption Processes in Ground Water." *Ground Water* 4: 475-487.

ASPECTS OF SOIL VENTING DESIGN

N. V. Mark-Brown

INTRODUCTION

Soil venting for bioremediation and vapor extraction processes both rely on the effective movement of air through contaminated zones of soil. Engineering issues relevant to the design and operation of venting systems are discussed. Data are described on the performance of several completed and currently operating projects.

The measured engineering performance parameters of several venting projects are summarized and compared with parameters predicted from theoretical mathematical relationships between soil airflow, applied suction pressure, and soil permeability. For most of the projects documented, the measured parameters diverged significantly from those predicted using theoretical relationships. The apparent explanations for this divergence, relating to the inferred physical patterns of airflow in the soil, are presented. The implications of these for the engineering design and performance prediction of venting systems are discussed. The discussion covers issues such as surface sealing and the design of cost-effective air extraction wells and trenches.

The methodology for designing irrigation systems for maintaining optimum soil moisture conditions and applying controlled doses of nutrients for bioventing systems is described.

PREDICTION OF AIRFLOW

Equations useful for predicting vapor flowrates through soil by way of extraction wells have been presented in the literature. The following equation is derived from the simplistic steady-state radial flow solution for compressible flow (Johnson et al. 1990):

$$\frac{Q}{H} = \pi \frac{K}{\mu} P_W \frac{[1-(P_{ATM}/P_W)^2]}{\ln (R_W/R_I)} \tag{1}$$

Where Q = air flowrate through a vapor extraction well
H = length of extraction well screen
K = soil air permeability
μ = viscosity of air
P_W = applied air pressure at extraction well (absolute)

P_{ATM} = ambient pressure (absolute)
R_W = radius of extraction well
R_I = radius of influence of extraction well

The requirements for input data for this equation are not onerous. The equation is relatively insensitive to large changes in the radius of influence. Soil air permeability can be measured or estimated from field investigations.

Values of the radius of influence, which depend on soil conditions, are reported to typically vary up to 30 m (Johnson et al. 1990). The measured radii of influence for a number of projects carried out by Woodward-Clyde in New Zealand and Hong Kong range from 2 to 3 m for 1-m depth of unsealed silt up to 12 m for 0.5- to 3-m depth of silt sealed with concrete pavement.

COMPARISON OF MEASURED AIRFLOW AND SUCTION PRESSURE WITH THEORETICAL PERFORMANCE PARAMETERS

Extraction well airflow and well vacuum pressure data from several projects have been collated and are presented in Figure 1.

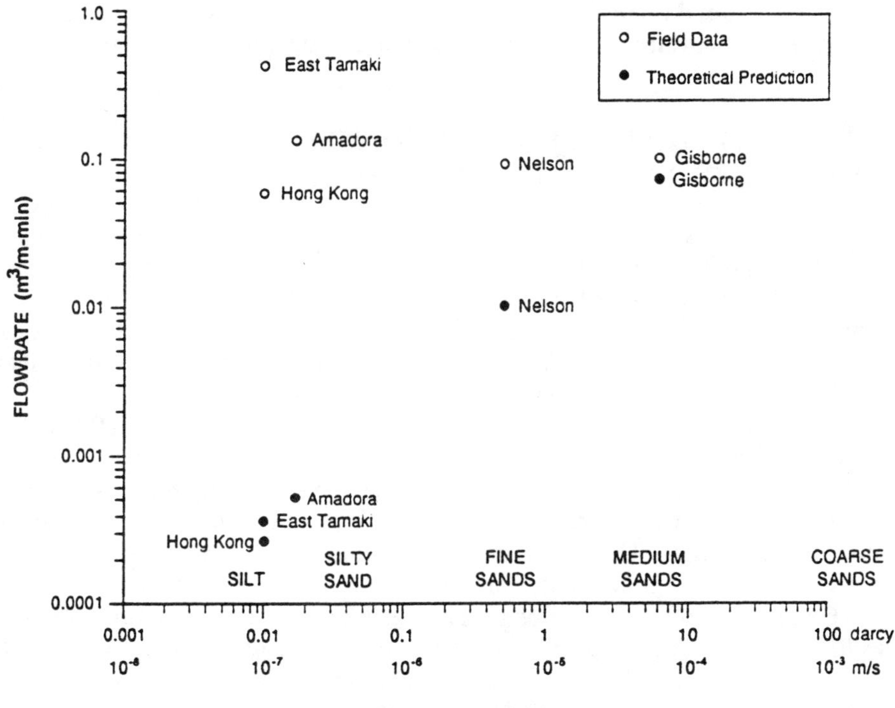

FIGURE 1. Steady-state flowrates (per unit well screen length) for a range of soil permeabilities and applied vacuums at different sites.

The data points on Figure 1 represent averaged data for each of the sites. Air was extracted at depths below ground level ranging from 0.5 m to 3 m. The value of soil permeability for the Gisborne site was assessed from sand grain size analysis. The values of soil permeability for the other remaining sites were assessed based on a visual assessment of samples from soil borings. The data points show a significant divergence from the theoretical prediction, with the exception of the Gisborne data. The divergence is greater for the sites where fine-grained soils are present, e.g., East Tamaki, Amadora, and Hong Kong.

The most likely reason for the divergence between measured and theoretical results is that the radial flow assumption of the predictor equation is not valid because of a significant vertical component of the flow occurring near the extraction wells. The influence of relatively high soil moisture content due to capillary rise above the water table in the fine-grained soils also may be significant. There was no evidence of the existence of preferred flow paths through the soil, due, for example, to cracking. Varying soil moisture levels due to seasonal rainfall also influenced results.

Leakage of airflow around the wellhead is not considered to be a major factor on these projects. At the sites studied, the potential for such leakage was minimized by the use of cement grout seals of at least 300 mm vertical thickness at the extraction wellheads. There was no evidence of direct leakage, for example, as might be indicated by air noise around the sides of the grout seals. Air leakage for the East Tamaki data in particular was very unlikely as water was ponded on the ground surface immediately adjacent to the wellhead during the test.

IMPLICATIONS FOR ENGINEERING DESIGN AND PERFORMANCE OF VENTING SYSTEMS

The occurrence of significant airflow in a vertical direction close to extraction wells is generally undesirable as it results in reduced airflows being available for venting of soil areas further away from the extraction well. In some conditions, sealing the ground adjacent to the wellhead will assist in increasing the effective radius of extraction wells. This, however, is not always a practical option. Figure 2(a) illustrates a situation of a site with concrete pavement.

The relatively permeable granular basecourse can act as a conduit for airflow passing through joints or cracks in the pavement. Modification to the surface sealing can be achieved only by sealing the joints or removing large areas of the existing pavement and relaying it with impermeable material replacing the granular basecourse. This is unacceptable in situations where the benefits of soil venting are its cost effectiveness and its lack of impact on site use.

One option to minimize this problem is to provide two separate extraction wells with screens at different levels as shown in Figure 2(b). This option allows separate operation of the upper and lower well inlets. This is an advantage on a site where the blower capacity is limited; groups of extraction wells can be operated in shifts to achieve appropriate cumulative airflows passing through the upper and lower soil zones.

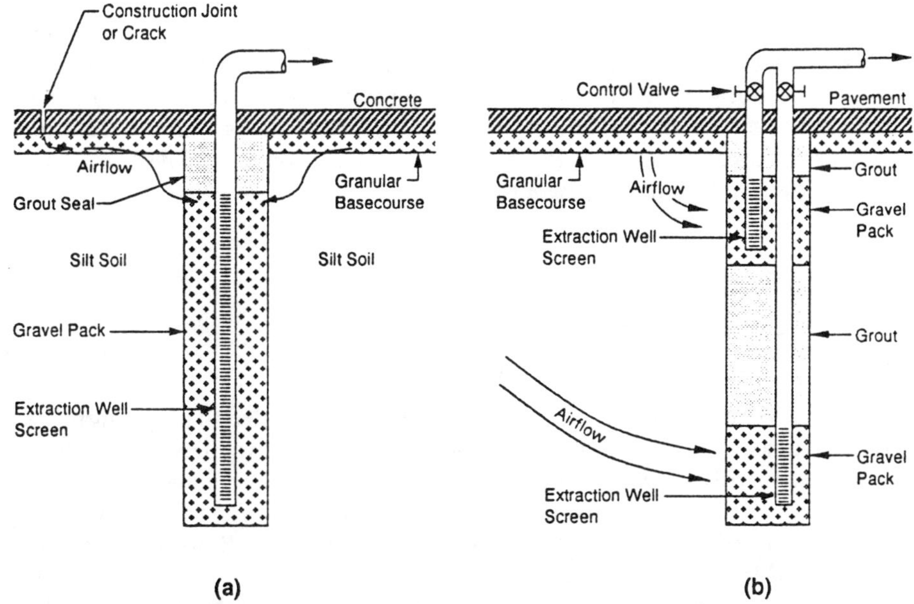

FIGURE 2. Extraction well design.

A similar design principle can be used for extraction trenches, as shown in Figure 3. The use of layers of compacted clay together with sprayed bitumen emulsion is a cost-effective method to minimize 'short circuiting' of extracted air vertically through the soil adjacent to the trench. One particular feature of the trench design shown on Figure 3 is that it allows the use of gravel basecourse as the upper foundation, rather than grout or concrete. Using compacted clay as the pavement foundation can be inappropriate due to the difficulty of compacting it to provide the necessary bearing strength for the foundation of heavily trafficked pavement.

Our experience with shallow venting (approximately 1.3 m deep) in sand with a clay surface showed that approximately 75% of the extracted air entered

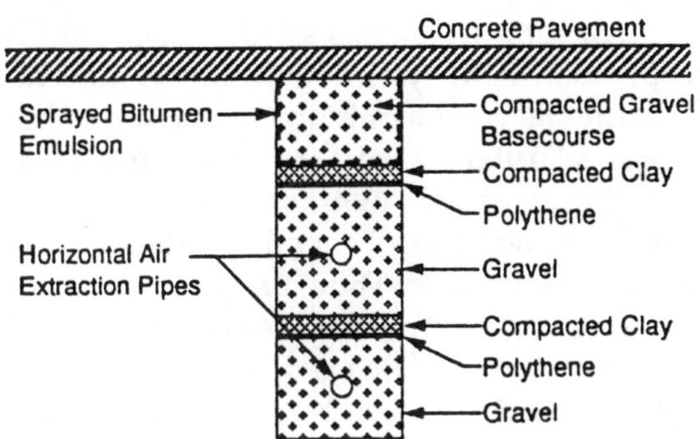

FIGURE 3. Extraction trench design.

through the ground surface and 25% entered through the inlet holes. This was determined by measuring airflow velocities in extraction wells and inlet wells. The implications from these results, together with the earlier discussion on the preferential vertical flow of air for shallow extraction systems, indicate that unless areas are completely sealed at the surface it is not necessary to provide inlet holes for shallow (up to 3 m deep) venting.

IRRIGATION FOR MAINTENANCE OF SOIL MOISTURE CONTENT AND ADDITION OF NUTRIENTS

For projects in which biological degradation is intended to play a significant role in the clean up process, it is advantageous to maintain soil moisture at levels that assist microbial activity. Soil moisture content can be adjusted and nutrients added by the application of irrigation water. Application rates of irrigation water containing nutrients need to be carefully matched to the climatic conditions at the site and the moisture content of the soil.

In dry weather, the application rate needs to be sufficient to maintain at least 10% (w/w) moisture content in the soil. This can be achieved by balancing the application rate against evapotranspiration and leaching losses from the soil. In wet weather, the application rate needs to be reduced to zero, if necessary, to avoid leaching of nutrients past the zone being remediated and to the underlying groundwater. Control can be achieved by ensuring that the soil moisture content does not exceed the field capacity of the soil.

For a bioventing project with approximately 1.3 m depth of diesel-contaminated sand it was decided to apply nutrients (nitrogen and phosphorus) in the field, based on results of bench-scale testing. The results of bench-scale testing, performed over a period of 110 days, indicated that aeration increased hydrocarbon degradation rates two-fold and that addition of nitrogen and phosphorus increased degradation rates an additional two-fold over aeration alone. Data on the nutrient status of the soil was not obtained. The design nutrient ratio adopted was 60 g N and 6 g P per kg of total recoverable petroleum hydrocarbons (TRPHs).

At this site, irrigation water was applied by proprietary buried irrigation drippers with operation of the system controlled by a solenoid valve and a timer. Soil moisture conditions were monitored by the use of buried gypsum blocks, and the timer was adjusted by local personnel to provide the required daily flowrate to match climatic and soil moisture conditions. The N and P were applied by way of a proprietary liquid fertilizer mix, made up in batches as required and injected into the irrigation water. The average reduction in TRPH concentration over the site was 423 g/m^3, and the total mass of nitrogen and phosphorus applied over the duration of site remediation was approximately 100 g N and 16 g P per kg TRPH reduction. The applied rate of N and P exceeded the design ratio to allow for inefficiency in application and for soil fixing of phosphorous. Control data were not collected during the project to measure how beneficial nutrient addition was in the field. However, in our opinion, nutrient addition was justified because of the positive results of the bench-scale testing and because it was relatively inexpensive to add nutrients to the soil via the irrigation system.

CONCLUSIONS

The equation derived from the simplistic steady-state radial flow solution for compressible flow is a useful guide to assist soil venting design at locations with relatively permeable soils to be vented and relatively impermeable surface conditions or where venting occurs at significant depth. The equation appears unsuitable for use at locations where venting is from a shallow depth and soils have low permeability. Care needs to be taken with the design of extraction wells and trenches, the provision of airflows, and the prediction of time for cleanup in systems where significant vertical airflow can be expected to occur.

Techniques are available for minimizing vertical flow near extraction wells and trenches and for cost-effective construction of extraction trenches.

Provision of irrigation water and nutrients at a small incremental cost to a project is considered worth-while in order to assist in achieving optimum conditions for biodegradation of organic compounds.

REFERENCE

Johnson, P. C., M. W. Kemblowski, J. D. Colthard, D. L. Byers, and C. C. Stanley. 1990. *A Practical Approach to the Design, Operation and Monitoring for In-Situ Soil Venting Systems.* Shell Development/Shell Oil Company, Westhollow Research Center, Houston, TX.

ADVANCES IN VACUUM HEAP BIOREMEDIATION AT A FIXED-SITE FACILITY: EMISSIONS AND BIOREMEDIATION RATE

G. R. Hater, J. S. Stark, R. E. Feltz, and A. Y. Li

INTRODUCTION

Since the late 1980s, vacuum heap bioremediation technology has been used at clients' facilities to comply with tightening air quality standards. The vacuum heap technique uses a slight vacuum to maintain aerobic conditions in large ex situ remediations (Hater 1988, von Wedel et al. 1990). The technique allows for remediation of thousands of tons of soil in a relatively small area and provides an economical, risk-free alternative to soil burning or landfilling.

In 1992 a joint venture of Chemical Waste Management, Inc., and Waste Management North America, Inc., opened a recycling center in Cincinnati, Ohio, at the ELDA Recycle and Disposal Facility. Because the facility is a fixed site, the state air board requires extensive air monitoring. The monitoring has resulted in proof that the technique does not emit significant emissions and that measured degradation rates are significant.

The ELDA Soil Center is located on a recompacted clay base that is approximately 210 × 440 ft (70 × 145 m). The facility is permitted to process non-RCRA petroleum-contaminated soils at a capacity of more than 391,071 tons/year. The soil contents must be reduced to the criteria noted in Table 1.

TABLE 1. Soil TPH and BTEX concentrations of remediated soil and criteria used by the Ohio Environmental Protection Agency (OEPA) to establish nonregulated soil.

Parameter	Typical Remediated Soil	OEPA Criteria
TPH	48 ppm	105 ppm
Benzene	<0.005 ppm	0.006 ppm
Toluene	<0.005 ppm	4.0 ppm
Ethylbenzene	<0.005 ppm	6.0 ppm
Xylene	<0.005 ppm	28.0 ppm

Current piles contain petroleum soil with total petroleum hydrocarbon (TPH) levels ranging from 1,100 to 3,300 ppm (mg/kg) and benzene ranging between 0.4 and 2 ppm (mg/kg). Permit weighted average values of TPH and benzene cannot exceed 5,000 ppm (mg/kg) and 50 ppm benzene (mg/kg), respectively. Air emissions cannot exceed 190.0 ppm, 1.1 pph (0.5 kg/hr), 26.4 ppd (12 kg/day), or 4.8 tpy.

Air emissions are controlled by two processes. As indicated in Figure 1, the vacuum airstream enters a biofilter followed by carbon "once through" or is partially recycled back to the soil pile.

Typical pretreatment vacuum heap analytical data are given as weighted means: TPH, 1,187 mg/kg; benzene, 0.42 mg/kg; toluene, 8.428 mg/kg; ethylbenzene, 0.548 mg/kg; and xylene, 34.54 mg/kg. The remediation initiated in November 1992 has soil piles that weigh as much as 16,000 tons. As time passes at the ELDA Soil Center, the tonnage will continue to increase.

AIR EMISSIONS

Air emissions are monitored daily using a Foxboro OVA 128 (organic vapor analyzer). The instrument has gas chromatographic (GC) capabilities for benzene, toluene, ethylbenzene, and xylene (BTEX) analyses. The instrument is zeroed

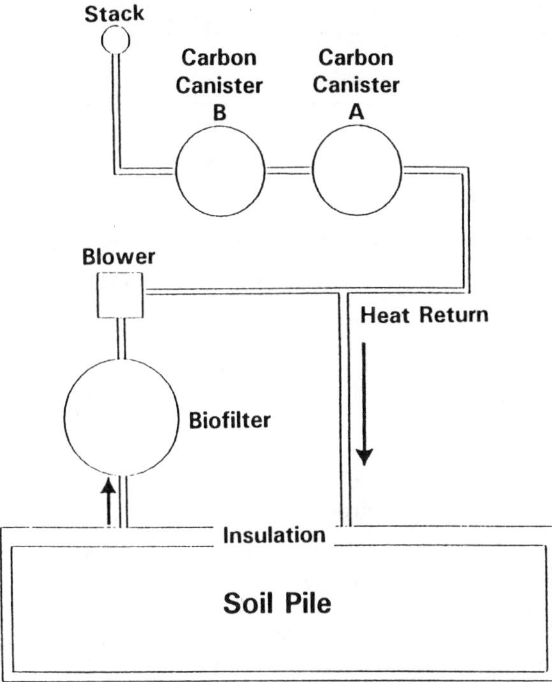

FIGURE 1. Operation schematic depicting "once through" emissions control vs. air recycle.

using grade "D" breathing air and calibrated with methane and BTEX standards produced by Alphagaz. Backup quality control data were produced by an independent laboratory.

Emissions testing was performed on a soil pile that began active remediation the day of the testing. Table 2 compares the "once through" emissions sequence. Volatile organic carbon (VOC) emissions were 0.021 lb/hr (0.0095 kg/hr) prior to treatment. Posttreatment maximum VOC emissions were 0.0067 lb/hr (0.0030 kg/hr).

Table 3 depicts the organic loading during the recycle mode of operation. The recycle of the air back to the soil pile creates less back pressure on the blower than the carbon canisters and results in a large increase in airflow from the soil pile. Without recycle airflow to the carbon averaged 430 dcfm, but with partial recycle back to the soil pile, airflow averaged 830 dcfm. VOC maximum loading to the prefilter is 0.067 lb/hr (0.003 kg/hr). The maximum emissions rate at the stack are 0.0044 lb/hr (0.019 kg/hr). Table 4 compares the permit vs. actual maximum emissions. The soil center typically is under the permitted levels by two orders of magnitude.

TABLE 2. Organic pretreatment loading and posttreatment stack emissions w/o use of return air line.

Sampling			Total Load (kg/hr)				
Type	Date	Time	Benzene	Toluene	Ethyl-benzene	Xylene	VOC
Pre	10/06/92	0905	<0.018	<0.014	<0.016	<0.016	0.009
Post	10/06/92	0905	<0.006	<0.006	<0.008	<0.008	<0.001
Pre	10/06/92	1702	0.0021	<0.013	<0.014	<0.014	0.006
Post	10/06/92	1702	<0.006	<0.006	<0.008	<0.008	<0.001
Pre	10/07/92	0855	<0.007	<0.013	<0.063	<0.063	0.006
Post	10/07/92	0855	<0.034	<0.006	<0.030	<0.030	<0.030
Pre	10/07/92	2000	<0.006	<0.012	<0.054	<0.056	<0.003
Post	10/07/92	2000	<0.032	<0.006	<0.029	<0.029	<0.001
Pre	10/08/92	0800	<0.006	<0.012	<0.054	<0.054	<0.003
Post	10/08/92	0815	<0.029	<0.005	<0.026	<0.026	<0.001
Pre	10/09/92	2000	<0.006	<0.013	<0.059	<0.059	<0.003
Post	10/09/92	2000	<0.0028	<0.005	<0.025	<0.025	<0.001
Pre	10/10/92	0900	<0.007	<0.013	<0.059	<0.059	<0.009
Post	10/10/92	0900	<0.0	<0.005	<0.026	<0.026	<0.003
Pre	10/11/92	0800	<0.006	<0.012	<0.054	<0.054	<0.002
Post	10/11/92	0800	<0.033	<0.006	<0.029	<0.029	<0.001

TABLE 3. Organic loading and stack emissions into soil center with use of return air line.

Sampling			Total Load (kg/hr)				
Type	Date	Time	Benzene	Toluene	Ethyl-benzene	Xylene	VOC
Pre	10/06/92	1129	0.206	<0.021	<0.024	<0.024	0.029
Post	10/06/92	1137	<0.003	<0.004	<0.004	<0.004	<0.001
Pre	10/07/92	1101	<0.013	<0.025	<0.113	<0.113	0.028
Post	10/07/92	1110	<0.002	<0.004	<0.017	<0.017	<0.001
Pre	10/08/92	0205	<0.014	<0.027	<0.122	<0.122	0.012
Post	10/08/92	0215	<0.002	<0.004	<0.048	<0.021	0.001
Pre	10/09/92	1130	<0.014	<0.026	<0.118	<0.118	0.012
Post	10/09/92	1140	<0.002	<0.004	<0.018	<0.018	<0.001
Pre	10/10/92	1145	<0.013	<0.025	<0.118	<0.118	0.006
Post	10/10/92	1145	<0.002	<0.004	<0.019	<0.019	0.001
Pre	10/11/92	0900	<0.014	<0.026	<0.122	<0.122	<0.006
Post	10/11/92	0910	<0.002	<0.004	<0.020	<0.020	0.001

BIOREMEDIATION RATE

The CO_2 content of the gas pulled from the pile was monitored. The CO_2 content of the gas above ambient CO_2 levels is representative of organic degradation by microbial activity.

Goldsmith and Balderson (1989) used the following equation to describe the microbial degradation of diesel fuel:

$$C_{13}H_{28} + 8.6O_2 + 2.4NH_4 \rightarrow 2.4C_5H_7O_2N + CO_2 + 10.4H_2O \qquad (1)$$

$C_{13}H_{28}$ is a weighted average of diesel fuel and $C_5H_7O_2N$ represents a typical cell. The stoichiometric cell yield computation found that 71% of the diesel fuel was used for cell yield. Assuming the system to be aerobic and that carbon not incorporated into cell growth goes to CO_2, equation 1 becomes:

$$C_{13}H_{28} + 11.232O_2 + 1.846NH_4 \rightarrow 1.846C_5H_7O_2N + 3.77CO_2 + 11.231H_2O \qquad (2)$$

Equation 2 predicts that the degradation of 15,500 pounds of TPH would result in 3,810 pounds of C as CO_2. The CO_2 concentration of air existing in the pile the first morning of the stack test was greater than 3,600 ppm. After 8 hours of operation the CO_2 concentration had dropped to 2,300 ppm. Monitoring of the

TABLE 4. Permit/actual data comparison.

Permit	Actual
Maximum Emissions	Maximum Emissions
190 ppm of VOC	2 ppm of VOC
1.1 lb/hr (0.5 kg/hr) VOC	0.0067 lb/hr (0.003 kg/hr) VOC
26.4 lb/day (12.0 kg/hr) of VOC	0.161 lb/day (0.073 kg/hr) VOC
4.8 ton/year VOC	0.029 ton/year VOC

system over the remainder of the remediation period demonstrated that the concentration remained relatively constant and averaged 2,000 ppm. All CO_2 concentrations are corrected for the background CO_2 levels at the soil center, which were found to average 350 ppm.

On average 3.10 lb (1.4 kg) of CO_2 per hour was generated from the pile. Over the 41 days of operation, the total carbon emitted as CO_2 would be 3,020 lb (1,370 kg) or 80.1% of the CO_2 evolved as predicted by equation 2.

The CO_2 evolved from the pile generally can be attributed to the degradation of TPH and BTEX. The soils and underground storage tank bedding materials that are remediated at the ELDA Soil Center have very small organic carbon contents (the soils are from subsurface horizons). Stoichiometrically fertilized soils that are wetted, as occurs before the soil is incorporated into the pile, experience a temporary increase in biological activity (Alexander 1977). The increase in activity lasts about 3 to 7 days before the readily oxidized organic carbon is consumed. Referring to Figure 2, there was an initial increase in CO_2 production that decreased rapidly the first day and then maintained a constant level over time. The biodegradation of the lower-molecular-weight organics and BTEX constituents occurs initially at a faster rate as simpler, more readily oxidized forms are consumed. Without the use of isotope labeling, it is not possible to determine the source of the CO_2-C.

SUMMARY AND CONCLUSIONS

The vacuum heap technology used at the ELDA Soil Center effectively bioremediates petroleum-impacted soils from multiple sites simultaneously. The technique creates an aerobic environment in the pile to maximize the aerobic decomposition rate. The vacuum established is designed to maintain an oxidizing environment and to prevent fugitive VOC emissions from the pile.

From measured CO_2 emissions from the pile, 80% of the reduction in the TPH content of the pile is accounted for using equation 2 (Goldsmith & Balderson 1989). Considering the experimental variables that could not be controlled during the study, the relatively close agreement of the actual and predicted CO_2 evolution indicates that aerobic degradation, as described by equation 2, is occurring with the vacuum heap technology.

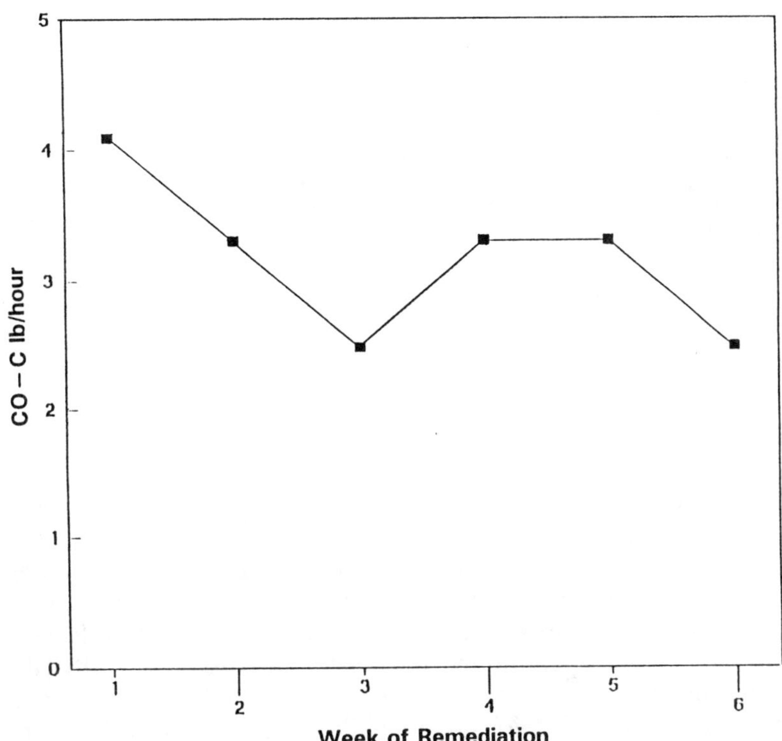

FIGURE 2. Average hourly CO_2 generation.

REFERENCES

Alexander, M. 1977. *Soil Microbiology*, 2nd ed. John Wiley & Sons, Inc., New York, NY.

Goldsmith, C. D., Jr., and R. K. Balderson. 1989. "Biokinetic Constants of a Mixed Microbial Culture with Model Diesel Fuel." *Hazardous Waste & Hazardous Materials*, pp. 145-152. Mary Ann Liebert, Inc., Publishers.

Hater, G. R. 1988. "Bioremediation of In Situ and Ex Situ Contaminated Soils." *Proceedings of the 4th Annual Hazardous Materials Management Conference West* (Haz Mat '88), pp. 234-240. Tower Conference Management Company, Glen Ellyn, IL.

von Wedel, R. J., G. R. Hater, R. Farrell, and C. D. Goldsmith. 1990. "Excavated Soil Bioremediation for Hydrocarbon Contaminations Using Recirculating Leachbed and Vacuum Heap Technologies." Paper presented at *Air & Waste Management Association and EPA Hazardous Waste Treatment of Contaminated Soil Symposium*. Cincinnati, OH.

BIODEGRADATION OF BTEX COMPOUNDS IN A BIOFILM SYSTEM UNDER NITRATE-REDUCING CONDITIONS

J. P. Arcangeli and E. Arvin

INTRODUCTION

Monoaromatic compounds are widely used in the industry as degreasers, as solvents, and as primary constituents in gasoline. It is not surprising that they have become a common soil and groundwater contaminant. Besides improper waste disposal, spills and leaking storage tanks are the main causes of contamination. The compounds are mostly benzene, toluene, ethylbenzene, and xylenes (BTEX).

Many published reports on anoxic biodegradation have shown that the BTEX family of compounds, except benzene, can be mineralized or transformed cometabolically (Flyvbjerg et al. 1991) under denitrifying conditions. However, most of the research has been carried out with suspended cultures, in aquifer columns (Zeyer et al. 1986), or in the field (Barbaro et al. 1992, Hutchins & Wilson 1991). The purpose of the present work was to study the BTEX biodegradation in a biofilm system under nitrate-reducing conditions, in particular the kinetics of degradation.

MATERIALS AND METHODS

Experimental Setup. The experimental system is depicted schematically in Figure 1. The biofilm reactor, called a "Biodrum system" (Kristensen & Jansen 1980), consists of a cylinder rotating inside another with the biomass growing on the surface of the rotator and the stator. The rotation and the recycling system ensure total mixing in the bulk liquid, and therefore give regular biofilm thickness. System specifications and operating conditions are summarized in Table 1.

The dissolved oxygen in the medium was removed continuously using a gas exchange chamber. The chamber is a 5-L reservoir containing a concentrated solution of Na_2SO_3 (100 g/L) and $CoCl_2$ (200 mg/L) covering 20 m of silicone tubing (wall thickness 1 mm; inner diameter 4 mm) through which the medium flowed (Figure 1, item 5). This chamber, reduced the oxygen concentration in the medium to below the detection limit (0.05 mg/L).

Analytical Techniques. Analysis of BTEX was performed with a Dani 8520 gas chromatograph equipped with a J. & W. Scientific column (DB5, 30 m, catalog

TABLE 1. Operational data for the biofilm reactor during the biofilm growth.

Condition	Unit	Value
Surface area	m^2	0.16
Reactor volume	m^3	0.96×10^{-3}
Average flow of water	L/h	4.5
Toluene infl. conc.	mg/L	5
Nitrate infl.	mg N/L	50-200[(a)]
Phosphate infl.	mg P/L	46.5
Ammonium infl.	mg N/L	10.4
Temperature	°C	20-21
Alkalinity	meq/L	20.8
pH in the reactor		6.9-7.1

(a) During kinetic experiments the nitrate concentration was raised in order to keep it in excess.

no. 125-1032, series 9416145) and a flame ionization detector (FID). The peak areas were determined by integration on a MAXIMA 820 Chromatographic Workstation (Millipore Corporation, Massachusetts). Extraction of BTEX was carried out in a 10 ml volumetric flask with 500 µL of pentane; the solution was shaken vigorously for 2 min; 1 µL of pentane was injected into the GC using heptane (6 mg/L) as an internal standard. Because of their identical retention times, *m*-xylene and *p*-xylene could not be separated with this column and were analyzed as a single compound. Nitrate and nitrite were measured on a Technicon Autoanalyser II by the hydrazine sulfate reduction procedure (Kamphake et al. 1967). The average biofilm thickness was calculated based on ten measurements. A typical standard deviation of ±10% has been found.

Experimental Procedure. Prior to the degradation studies, the experimental setup was sterilized with a solution of chloramine (2 g/L) and fed with a mixture of BTEX in order to estimate the absorption of these compounds in the reactor. A dissimilarity of approximately ±3% was observed in the liquid concentration between the inflow and the outflow of the reactor. Because the BTEX concentration in the bulk liquid could be analyzed with a standard deviation of ±2 to 3% errors due to absorption phenomena in the reactor can be neglected.

The reactor was inoculated with a culture originating from a denitrifying sewage sludge enriched with toluene for 16 generations. The biofilm was grown anaerobically with toluene as the sole carbon source for 6 weeks. Nitrate was supplied in excess, to prevent the reduction of nitrate to nitrogen. For several levels of biofilm thickness, kinetic experiments were performed with toluene and

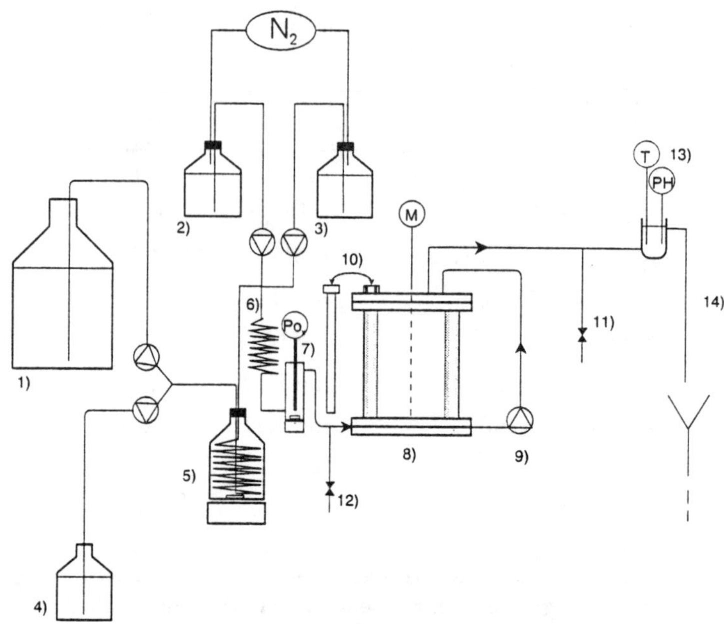

FIGURE 1. Experimental setup: (1) mineral medium; (2) toluene stock solution; (3) BTEX stock solution; (4) nitrate stock solution; (5) oxygen removal chamber; (6) mixing coil; (7) oxygen electrode; (8) biofilm reactor; (9) recirculation pump; (10) removable slides; (11) inlet sampling port; (12) outlet sampling port; (13) temperature and pH control; (14) outlet.

with a mixture of BTEX to find the stoichiometry of the reaction as well as different kinetic parameters. The experiments consisted of measuring the substrate removal rate per unit of surface area and the associated nitrite production rate versus the substrate concentrations in the reactor. Typically the BTEX feed concentration was varied in the range of 0 to 60 mg/L. The toluene feed range concentration is indicated in Figure 2.

RESULTS AND DISCUSSION

Toluene Biodegradation. The result from two kinetic experiments related to toluene biodegradation are shown in Figures 2A and 2B. These tests were performed on the same denitrifying biofilm at two different biofilm thicknesses. A first-order reaction controls the degradation for a bulk concentration below 1 to 1.3 mg toluene in the reactor (feed concentration: 2 to 3 mg/L), giving Monod constants (half saturation constants) of 0.4 and 0.85 mg/L, respectively. The degradation follows a zero-order kinetics for concentrations above 3 to 15 mg/L (feed concentration: 8 to 30 mg/L), depending on the amount of attached biomass. Figure 2B shows that toluene can be degraded even at a relatively high concentration.

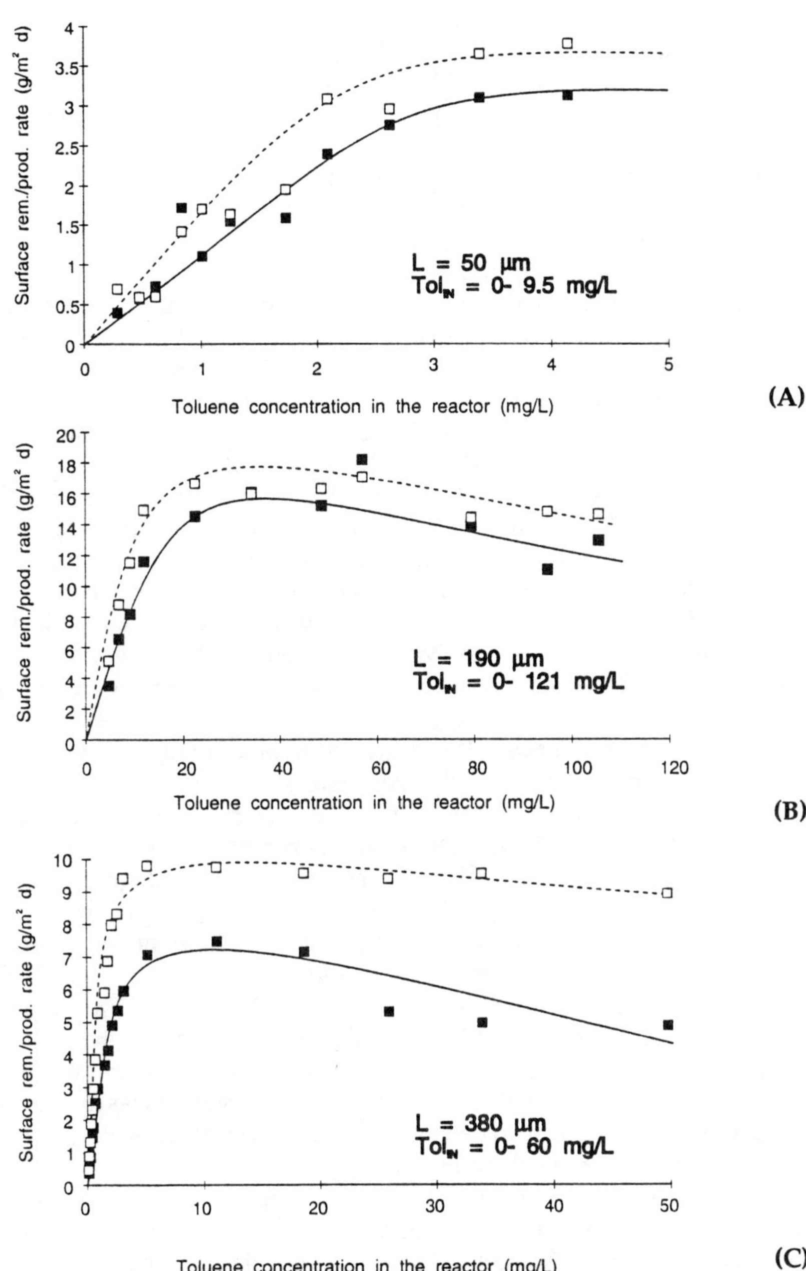

FIGURE 2A,B,C. Surface removal rate of toluene under anaerobic conditions: (■) toluene removal; (□) nitrite production; (L) biofilm thickness; (Tol_{in}) toluene feed concentration range. Fig. 2C shows toluene degradation in the presence of BEX.

However, a slight inhibition noticed for a substrate concentration above 50 to 60 mg/L (feed concentration: 70 to 80 mg/L) was corroborated with the nitrite production (Figure 2A and 2B). Maximum biomass yield coefficients were determined from the nitrite production per unit of toluene used (α_{NO2}) at a relatively high toluene concentration (i.e, from 2 to 3 mg/L). The calculation was based on the assumption that degradation products of toluene were not present, using an empirical cell formula of $C_5H_7NO_2$. A specific nitrite production of 1.2 and 1.3 mg N-NO_2/mg toluene degraded was found for two kinetic experiments giving a maximum yield coefficient of 1.1 and 1.0 mg biomass/mg toluene degraded. The presence of the other aromatic compounds, "BEX," seemed to inhibit the toluene removal. This is shown in Figure 2C where the BEX inflow concentration was maintained at 2.8 mg/L, whereas the toluene increased up to a concentration of 60 mg/L. The maximum toluene removal rate dropped from 15.4 g m^{-2} d^{-1} (Figure 2B) to 7.5 g m^{-2} d^{-1}. Furthermore, the toluene inhibition threshold was reduced from 40 mg/L (Figure 2B) to 10 to 15 mg/L when the other aromatics were present. Numerous experimental runs confirmed the inhibition of BEX toward the primary substrate, toluene (data not shown). Degradation of toluene under denitrifying condition has been investigated (Jørgensen 1992) with the same microbial consortium, but in a batch system. He found a Monod constant of 0.4 ±0.2 mg/L, and an average α_{NO2} of 1.14 ±0.05 mg N-NO_2/mg toluene, leading to a yield coefficient of 1.15 mg biomass/mg toluene degraded. Besides, he found a toxic level for toluene of approximately 120 mg/L.

BTEX Biodegradation. The degradation of a mixture of benzene, toluene, *m*/*p*-xylene, *o*-xylene, and ethylbenzene was studied in the concentration range from 0 to 10 mg/L for each compound. The composition of this mixture at the reactor inlet was (%): 37, 23, 20, 10, 10 for benzene, toluene, *m*/*p*-xylene, *o*-xylene, and ethylbenzene, respectively. The benzene was not degraded; preliminary experiments revealed a very small removal rate that might be due to absorption in the reactor or in the biomass. The total TEX (toluene, ethylbenzene, and xylenes), was treated as a single compound (benzene not included). Figure 3A displays the surface removal rate of separate TEX and total TEX versus total TEX concentration in the reactor; individual TEX concentrations are shown in Figure 3B. It appears that, among the TEX, toluene is the most biodegradable aromatic in the mixture. On average, 80% of the total BTEX converted was due to the toluene biodegradation and 12% to *o*-xylene; the other compounds were more difficult to degrade (5% for *m*/*p*-xylene and 3% for ethylbenzene). When toluene was used as a sole carbon source, the specific production of nitrite was constant at 1.2 to 1.3 mg NO_2-N per mg of toluene degraded. However, as the reactor was supplied with a mixture of BTEX, the specific production of nitrite increased with the TEX concentration in the reactor from 0.9 to 1.6 mg NO_2-N/mg TEX degraded. The reason for that has not been identified.

Cometabolism. The evidence of cometabolism of xylenes and ethylbenzene is supported by Figure 4, which demonstrates that their removal were strongly

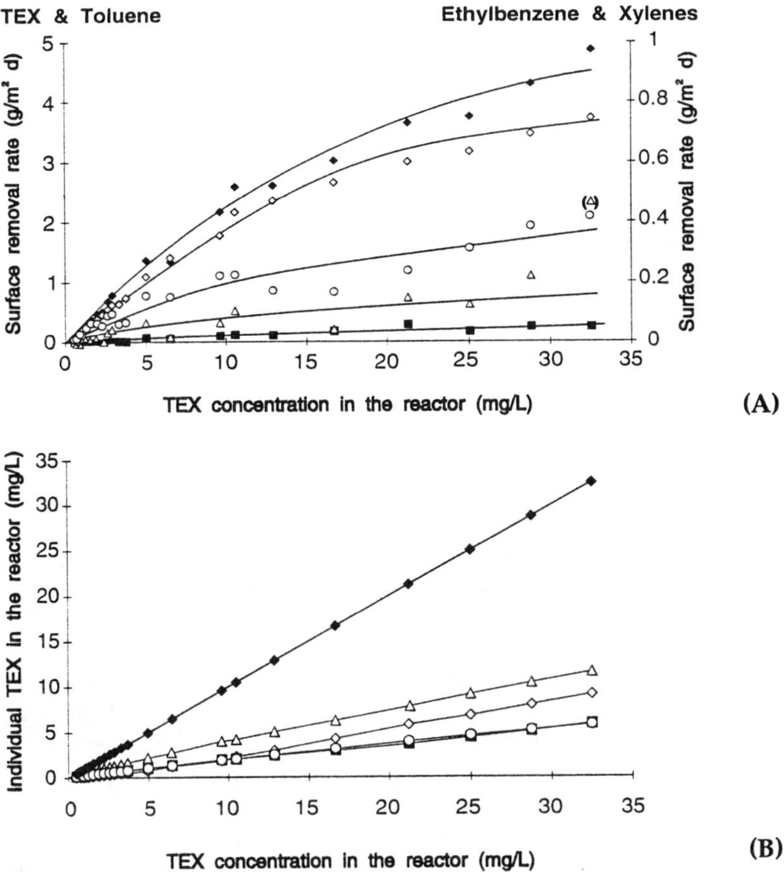

FIGURE 3. (A) Surface removal rate of total TEX; (B) individual TEX under anaerobic conditions. (♦) TEX; (◊) toluene; (○) *o*-xylene; (△) *m*/*p*-xylene; (■) ethylbenzene.

dependent on toluene in the reactor. This is consistent with previous observations that ethylbenzene and xylenes are degraded by cometabolism (Jensen et al. 1988, Jørgensen 1992). In that experiment, the TEX feed concentration was 4, 1, 0.8, and 0.5 mg/L for toluene, *m*/*p*-xylene, *o*-xylene, and ethylbenzene, respectively.

Competitive Inhibition. The removal of *o*-xylene was governed by competitive inhibition. Figures 4 and 5 show that a lack of toluene cease the degradation of *o*-xylene, but an excess of toluene inhibits its removal. The maximum removal rate for *o*-xylene was obtained with a toluene concentration in the reactor from 0.3 to 0.8 mg/L. In this experiment, the *o*-xylene concentration in the reactor inlet was 0.6 mg/L while the toluene feed concentration was varied in the range from 0 to 60 mg/L. Competitive inhibition has also been observed with *m*/*p*-xylene and

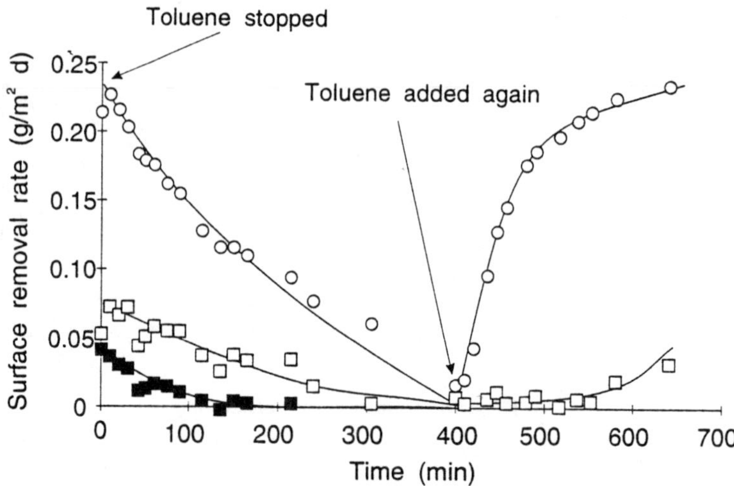

FIGURE 4. Evidence of cometabolism of (○) *o*-xylene; (□) *m/p*-xylene; (■) ethylbenzene. NB: When toluene was added again, the ethylbenzene could not be quantified due to an analytical problem.

ethylbenzene (data not shown). Another interaction phenomenon between the secondary substrates seems to exist: as toluene was added again to the reactor, *m/p*-xylene degradation did not start before the *o*-xylene removal rate reached its previous value (Figure 4). One probable explanation is competitive inhibition between the xylenes: the concentration of the most degradable secondary substrate (*o*-xylene) has to be below a certain level before bacteria start to attack the less degradable one (*m/p*-xylene). In the present experiment this critical *o*-xylene concentration was 0.35 mg/L in the reactor.

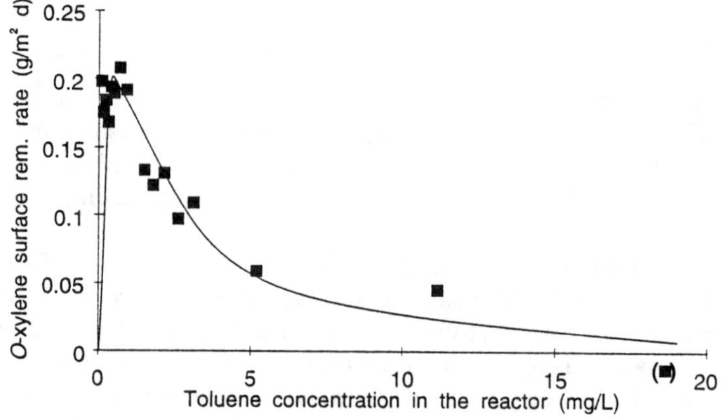

FIGURE 5. Effect of toluene concentration on *o*-xylene removal.

CONCLUSION

In this study, a biofilm was grown with toluene as the carbon source and nitrate as the terminal electron acceptor. The toluene was easily biodegraded with only slight inhibition at a toluene concentration higher than 60 mg/L. The average yield coefficient was 1.0 to 1.1 mg biomass/mg toluene degraded, and the Monod constant was 0.4 to 0.85 mg/L. When a mixture of BTEX was fed into the reactor, the toluene was the most degradable among all the compounds. Nevertheless, its degradation rate was reduced because of the presence of BEX in the reactor. The benzene was not degraded. The xylenes and ethylbenzene were degraded cometabolically using toluene as the primary carbon source; their removal was influenced by competitive inhibition with toluene. This study confirms that nitrate can be used to enhance in situ TEX biodegradation of a contaminated aquifer. However, the small removal rate observed for EX and the absence of benzene degradation suggests that in a practical situation the use of oxygen always will be required after an anoxic treatment in order to eliminate pollutants that are nonbiodegradable under denitrifying condition.

ACKNOWLEDGMENTS

The authors would like to thank the Groundwater Research Centre at the Technical University of Denmark for its financial support.

REFERENCES

Barbaro J. R., J. F. Baker, L. A. Lemon, and C. I. Mayfield. 1992. "Biotransformation of BTEX under anaerobic, denitrifying condition: Field and laboratory observations." *Journal of Contaminant Hydrology*, 11, 245-272.

Flyvbjerg, J., E. Arvin, B. K. Jensen, and S. K. Olsen. 1991. "Biodegradation of oil- and creosote-related aromatic compounds under nitrate-reducing conditions." In R. E. Hinchee and R. F. Olfenbuttel (Eds.), *In Situ Bioreclamation: Applications and Investigations for Hydrocarbon and Contaminated Site Remediation*, pp. 471-479. Butterworth-Heinemann, Stoneham, MA.

Hutchins, S. R., and J. T. Wilson. 1991. "Laboratory and field studies on BTEX biodegradation in a fuel-contaminated aquifer under denitrifying condition." In R. E. Hinchee and R. F. Olfenbuttel (Eds.), *In Situ Bioreclamation: Applications and Investigations for Hydrocarbon and Contaminated Site Remediation*, pp. 157-172. Butterworth-Heinemann, Stoneham, MA.

Jensen, B. K., E. Arvin, and A. T. Gundersen. 1988. "Biodegradation of phenolic and monoaromatic hydrocarbons by a mixed wastewater culture under denitrifying conditions." Presented at *Cost 691/681 Workshop on Organic Contaminants in Wastewater, Sludge & Sediments: Occurrence, Fate and Disposal*. Brussels, Oct. 26-27, 1988.

Jørgensen, C. 1992. "Anaerobic microbial degradation of aromatic hydrocarbons" (In Danish). Ph.D. Thesis. Department of Environmental Engineering, The Technical University of Denmark, Denmark.

Kamphake, L. J., S. A. Hannah, and J. M. Cohen. 1967. *Water Resour. Res.* 1: 205-216.

Kristensen, G. H., and J. C. Jansen. 1980. *Fixed film kinetics. Description of laboratory equipment.* Department of Environmental Engineering, Technical University of Denmark, Lyngby, Denmark.

Zeyer, J., E. P. Kuhn, and R. P. Schwarzenbach. 1986. "Rapid microbial mineralization of toluene and 1,3-dimethylbenzene in the absence of molecular oxygen." *Appl. Environ. Microbiol.* 52 (4): 944-947.

BIOREMEDIATION OF WASTE OIL-CONTAMINATED GRAVELS VIA SLURRY REACTOR TECHNOLOGY

M. A. Wilson, A. G. Saberiyan, J. S. Andrilenas, R. S. Miller, C. T. Esler, G. H. Kise, and P. DeSantis

The objective of this study was to evaluate the most feasible, efficient, and cost-effective means for bioremediating waste oil-contaminated gravels from several sites. The goal was to develop a mobile system that could be used repeatedly for various applications. The contaminated material used in this study consisted of gravel fill and some native sandy silts. Analysis by EPA Method 418.1 (U.S. EPA 1983) indicated total petroleum hydrocarbon (TPH) concentrations ranging up to 100,000 mg/kg. An on-site treatment method was desired since disposal of these soils in a landfill would be expensive and offered no solution for future environmental liabilities. A bench-scale biofeasibility study was performed to assess the efficiency of degrading waste oil in a slurry bioreactor under aerobic conditions. The high removal efficiency of the bench-scale slurry reactor indicated that the process was a feasible remedial approach for field-scale use. Full-scale bioremediation was implemented at the subject site and is continuing at the time of this report preparation.

BIOFEASIBILITY STUDY

A study was performed to determine the feasibility of remediating waste oil-impacted gravels via slurry reactor technology. The biofeasibility study involved an assessment of the indigenous microbial populations and an analysis of the ability of the microorganisms to degrade the waste oil contaminant under oxygen- and nutrient-rich slurry conditions. The native populations were found to include bacteria (i.e., *Pseudomonas* spp.) capable of degrading a wide range of petroleum hydrocarbons, including waste oil (Riser-Roberts 1992). Waste oil typically has both linear and branched carbon chains with lengths in excess of C_{22} (Speight 1991). During the course of the feasibility study, CO_2 and gravel TPH concentrations were periodically monitored. Gravel contaminant concentrations were reduced from approximately 34,000 mg/kg to 1,400 mg/kg over a 14-week time period.

BIOREMEDIATION SYSTEM DESIGN

The field-scale bioreactors were designed based on results obtained from the feasibility study, understanding the limitations of the laboratory scale, and incorporating future use requirements. The bioreactors were constructed from specially fabricated, watertight, steel drop boxes (Figure 1). This fabrication ensured durability for repeated use, security for the treatment operation, and the ability to transport the two 20-yd^3 (15.29-m^3) units with standard drop box lift trucks. The rate of airflow into each reactor unit, determined by scaling up from the laboratory model, was approximately 200 ft^3/min (5.66 m^3/min). Uniform air introduction into the base of each unit was achieved using horizontally placed galvanized well screen along the bottom of the reactors.

A water recirculation system was used in the reactors to aid in the distribution of nutrients and reduce the likelihood of "dead" spaces or areas not reached by the injected air. Further, the water recirculation system allowed for the monitoring of pH and temperature in a flowthrough system, instead of discrete points within the reactor vessel. The pH monitoring system measured general trends in pH over time, and data were collected using an electronic data logger unit. Recirculating water was collected from the base of the bioreactors through an intake

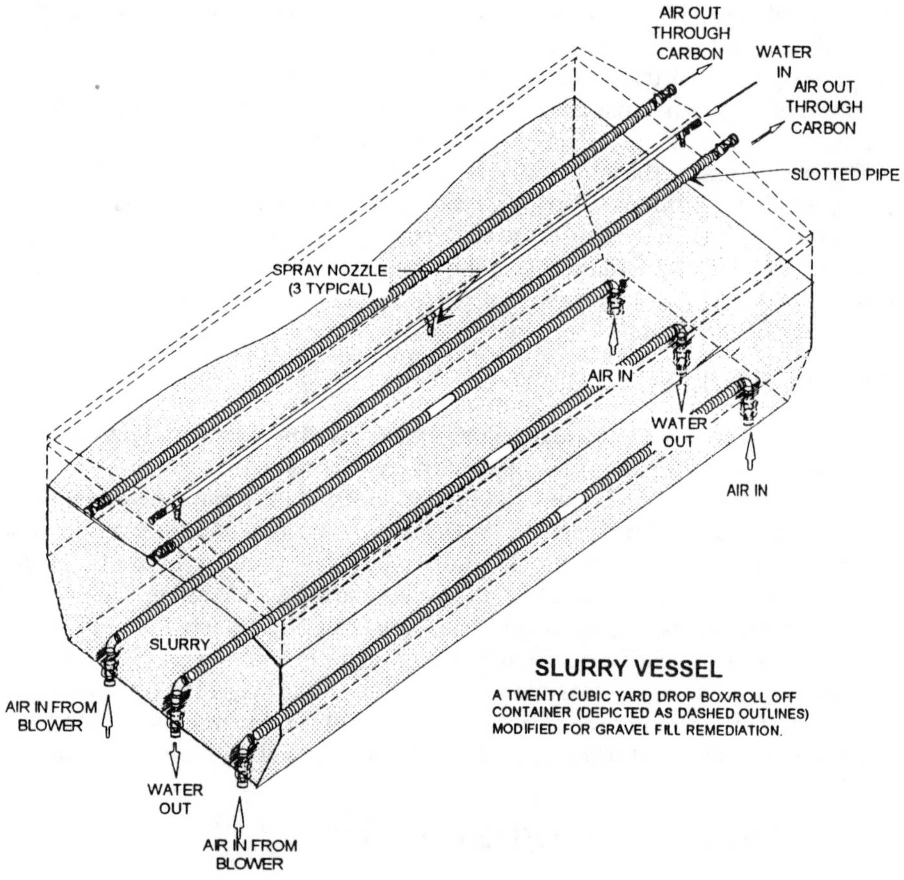

FIGURE 1. Slurry reactor design.

screen that ran the length of the units. The recirculation system utilized a flexible-impeller pump. The water, filtered through a 50-micron cartridge, was reintroduced to the top of the slurry reactors through three irrigation spray nozzles. The air-injection and water-recirculation systems were positioned to induce a circulation pattern within each reactor vessel.

Due to the odors expected from a biosystem, the reactors were operated under negative pressure, and off-gases from the units required treatment. The covers, equipped with pressure and vacuum relief valves, were fitted with off-gas recovery piping in addition to the water reintroduction spray fittings. Off-gases were collected through two vent lines that ran down the central spine of the covers. The off-gas collection system required balancing with the air injection system to maintain the lowest possible pressure differential. The collected off-gases were filtered through activated carbon units to remove any offensive smells from the bioreactors.

System Operation. Operation of the field-scale units began in May of 1992. The gravels had been stored on site in a VisQueen™-lined pile structure, the integrity of which was protected using hay bales. The 40 yds^3 (30.59 m^3) of contaminated gravels were loaded into the reactors by backhoe; approximately 20 yds^3 (15.29 m^3) were placed in each unit. Unfortunately, the hay bales were loaded into the reactors in addition to the gravels; this posed future problems with the aeration system, making even dispersement of oxygen difficult. After the gravels were in place, river water was added to sufficiently submerge all gravels, nutrients (urea [CH_4N_2O] and ammonium phosphate [$(NH_4)_2HPO_4$]) were added to maintain an optimum C:N:P ratio, and the system was activated.

In actual field operation, it became obvious that the air supply from the blowers was sufficient to keep some of the material suspended, but occasional mixing by backhoe would be necessary to ensure efficient transfer of the waste oil to the aqueous phase. Also contributing to the need for mechanical mixing was the ineffectiveness of the water recirculation system, caused by the addition of hay bales to the bottom of the reactors. Nutrient additions were timed to coincide with mixing events to ensure thorough and uniform nutrient dispersement. Additional maintenance and operational concerns involved weekly monitoring and sampling of the system. Due to the vigorous aeration system and hot summer temperatures, net loss due to evaporation required frequent additions of water to the system to keep the gravels submerged.

Performance Summary. Sampling of the gravels and water in each reactor indicates that the reactors are operating as expected; the trends follow those established during the biofeasibility study. Initial TPH concentrations of up to 100,000 mg/kg in the gravel media have been reduced to approximately 10,550 mg/kg in one reactor and 8,500 mg/kg in the other (Figure 2). The water TPH concentrations temporarily increase following each mixing event, demonstrating the efficiency of releasing the waste oil from the gravels in this manner. The TPH of the water then drops in the following weeks, back down to levels below 200 mg/kg (Figure 2). When the contaminant concentrations in the water decrease, the mixing is repeated.

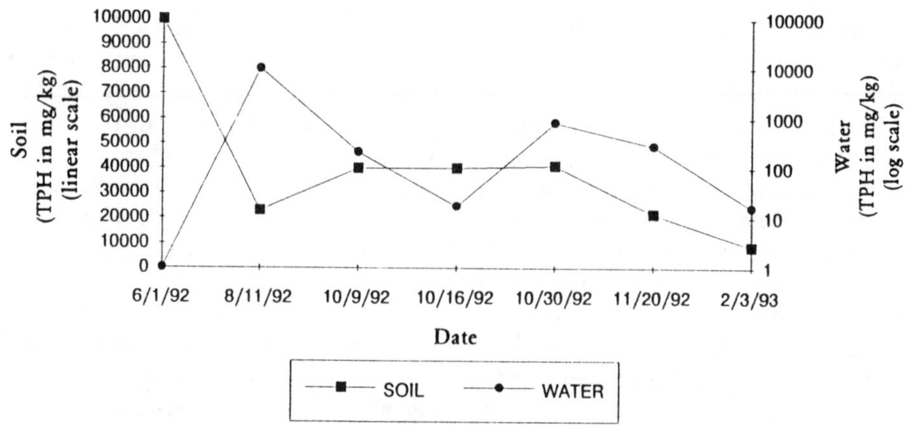

FIGURE 2. Combined gravel and water TPH data for field-scale slurry reactor.

Throughout the summer operating months, the water temperature in each reactor averaged 22°C and the pH remained between 6.5 and 7.0. Samples are collected on a monthly basis for bacterial analysis; repeated examination reveals a consistently healthy population of microorganisms. Population sizes to date have fluctuated between 4.0×10^6 CFU/mL and 9.0×10^7 CFU/mL for each reactor.

CONCLUSIONS AND RECOMMENDATIONS

The first field test of the gravel slurry reactors was ongoing at the time of this paper preparation. Progress to date reflects over a 90% reduction in TPH concentrations from the gravel media. Cleanup is estimated to be completed in the second quarter of 1993, although a decrease in microbial activity has been noted during the winter months. The use of a laboratory-scale model, emulating actual conditions of treatment, has proven to be a useful design step. It provided sufficient data for not only the construction of the field units, but also as a model for interpreting operational parameters during the field trials.

Optimization efforts for operation of the slurry reactors currently are centered on solving the problems encountered with the water recirculation system and the need for constant gravel mixing throughout the remediation phase. These appear to be limiting factors in the expedient completion of the treatment. Following completion of this project, the slurry reactors are to be modified with a gravel mixing system and a slurry pump for high solids fluid recirculation prior to being moved to another waste oil-impacted site.

ACKNOWLEDGMENT

We thank BP Oil Company (Environmental Resource Management) for funding for this project and for technical assistance in the development of the mobile slurry reactor design.

REFERENCES

Riser-Roberts, E. 1992. "Microbial Degradation and Transformation of Petroleum Constituents and Related Elements." *Bioremediation of Petroleum Contaminated Sites*, pp. A-73. CRC Press, Inc., Boca Raton, FL.

Speight, J. G. 1991. *The Chemistry and Technology of Petroleum*, pp. 700. Marcel Dekker, Inc., New York, NY.

U.S. EPA (Environmental Protection Agency). 1983. *Manual of Methods for Chemical Analysis of Water and Wastes*. EPA 600/4-79-022, STORET NO. 45501.

EVALUATION OF AN IN SITU BIOREMEDIATION USING HYDROGEN PEROXIDE

J. Heersche, J. Verheul, and H. Schwarzer

INTRODUCTION

For several reasons, in situ biorestoration (ISB) is not a generally established technique: (1) the technique still needs a great deal of practical experience to guide it out of its infancy; (2) to date only little knowledge and experience have been gained; and (3) with the use of biorestoration no guarantee can be made as to whether or not the reference-value will be obtained (Dutch A-level for mineral oil 50 mg/kg soil). In 1990, DHV started an ISB of a petrol station based on a literature survey, laboratory studies (van den Berg et al. 1987, Verheul et al. 1988, Hinchee et al. 1991), visits to sites abroad, and a cost-benefit analysis. The practical execution of an ISB by consultants was seen as the only way to build up the knowledge and experience needed as a base for consulting activities in this field. At this moment we can conclude that the groundwater cleanup has been successful. The soil cleanup has been partly successful. For a highly polluted spot that remained, excavation seemed the best solution. Although the soil remediation was not a complete success, the knowledge gained is of great value.

DESCRIPTION OF THE SITE VELSEN-NOORD

The location is a petrol station in Velsen-Noord, The Netherlands. The site contains an automatic car wash, a warehouse, and a shop that was renovated some years ago when a number of new petrol pump islands were added. The soil underneath the old part is contaminated. All pump islands are now provided with liquid-impermeable paving. Locally, the soil consists mostly of medium-fine to medium-coarse sand. Horizontal and vertical permeabilities are 5 and 1 m/day, respectively. The depth of the groundwater is about 1.2 m, and the groundwater flow is to the south-southeast. The pollution consisted mostly of petrol, caused by spillage and probably a leaking supply pipe. The soil has been polluted to a depth of 3 to 4 m.

The highest concentrations of mineral oil measured during a detailed study of a college office (De Ruiter Bv) were 2,690 $mg.kg^{-1}$, and more than 1,200 $mg.kg^{-1}$ for benzene, toluene, ethylbenzene, and xylenes (BTEX). The contaminated area totals around 250 m^2. This amounts to about 1,500 tons (density 1,500 kg/m^3) of polluted soil. The highest concentrations measured in the groundwater were

16,000 $\mu g.L^{-1}$ for mineral oil and 65,000 $\mu g.L^{-1}$ for BTEX. Both contents were found near the pump island close to the warehouse. The groundwater was polluted up to a depth of 4 to 5 m. The total surface area of contaminated groundwater was around 1,200 m^2, and the total volume around 1,600 m^3, with a layer thickness of 4 m (porosity 30%). This extended volume is due to the relative high mobility of contaminants. There was no free floating layer of contaminant that had to be removed.

METHOD OF REMEDIATION

The contaminated soil from the unsaturated zone beneath ground level (–gr.l.) (0 to 1.5 m –gr.l.) was excavated and treated elsewhere. The remaining part of the soil and groundwater was selected to function as an underground biological reactor. Within this reactor, biodegradation was stimulated to create optimal conditions for microorganisms by adding oxygen and nutrients to the infiltrated groundwater. This was studied in a previous laboratory study (see Figure 1). In view of the concentrations in the soil and the relatively limited thickness of the contaminated layer, it was decided to use vertical infiltration and withdrawal wells, flushing both the polluted soil and the groundwater. In designing a remediation measure, it is of utmost importance to prevent the pollution from spreading any further. This is done by geohydrologic isolation. The total infiltration flowrate is around 11.3 m^3/hour and the withdrawal flowrate exceeds the infiltration flow rate by 10% (12.5 m^3/hour)(see cross-section of the installation, Figure 2). The surplus withdrawn groundwater (1.2 m^3/hour) is discharged to the sewage system after treatment in the groundwater purification plant (GPP). The designed system consists of three parallel infiltration strings, 16 m apart. Each infiltration string consists of eight infiltration filters (1 to 5 m –gr.l.). Withdrawal strings — also with eight filters — have been placed in between and parallel to the infiltration strings, at 1 to 5 m –gr.l. The distance from one filter to the next one within each string is 5 m (see Figure 3).

Groundwater Purification Plant (GPP). The GPP functions to (1) remove iron and floating particles; (2) remove CO_2 that is formed during the biodegradation to avoid disturbance of biodegradation; and (3) purify the pumped-up water for discharge into the sewage system.

Dosing Installation for Nutrients. The nutrients were dosed from a central storage tank, which held ammonium nitrate (NH_4NO_3) and monosodium dihydrogen phosphate (NaH_2PO_4) dissolved in tap water. The concentrations were ±2.0 mg/L ammonium, ±3.5 mg/L nitrate, and ±12.0 mg/L phosphate.

Dosing Installation for Peroxide. Hydrogen peroxide was dosed from a storage tank with concentrations increasing from 10 to 100 mg/L. Because peroxide can decompose in pipelines, a study of the stability of hydrogen peroxide was undertaken by monitoring the concentration in the central dosing system.

The aim of this study was to measure the (possible) decomposition in the H_2O_2 in the pipelines.

RESULTS OF THE REMEDIATION

The regulatory limit agreed upon with regard to remediation of both soil and groundwater was the Dutch B-level, with regard to mineral oil as well as volatile aromatics (BTEX) (see Table 1).

In view of the behavior of these substances in the soil, the contaminant level in groundwater after remediation is the crucial parameter. If B-level is to be realized in the groundwater, the soil will have to be cleaned until at or even below A-value. The sampling and analysis of the monitoring filters took place every

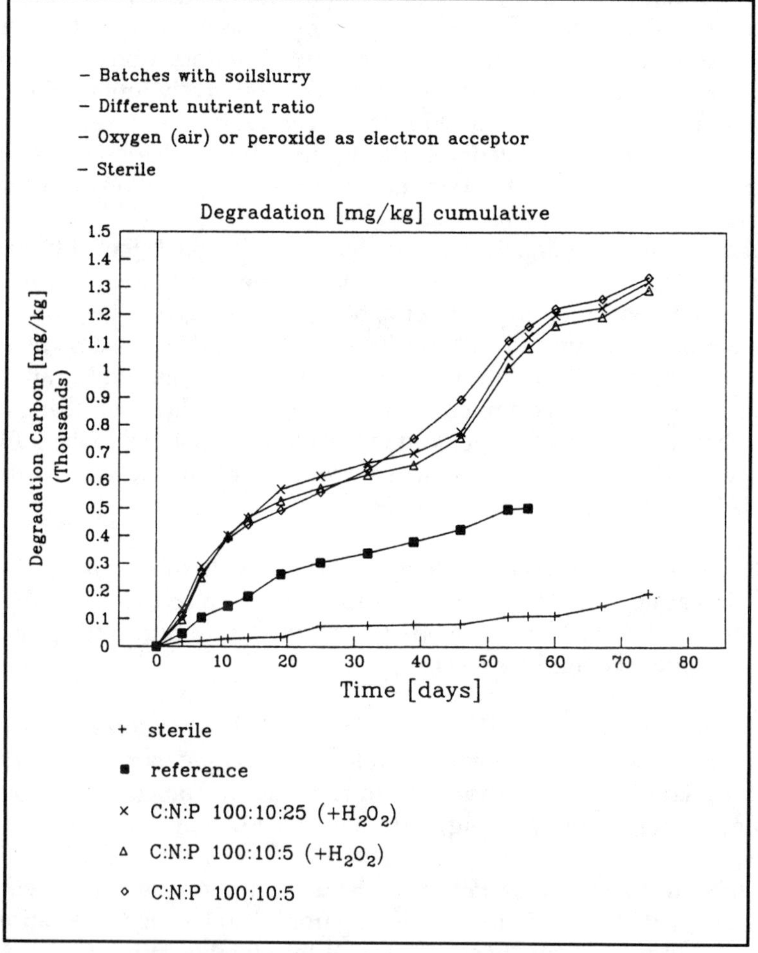

FIGURE 1. Laboratory/research.

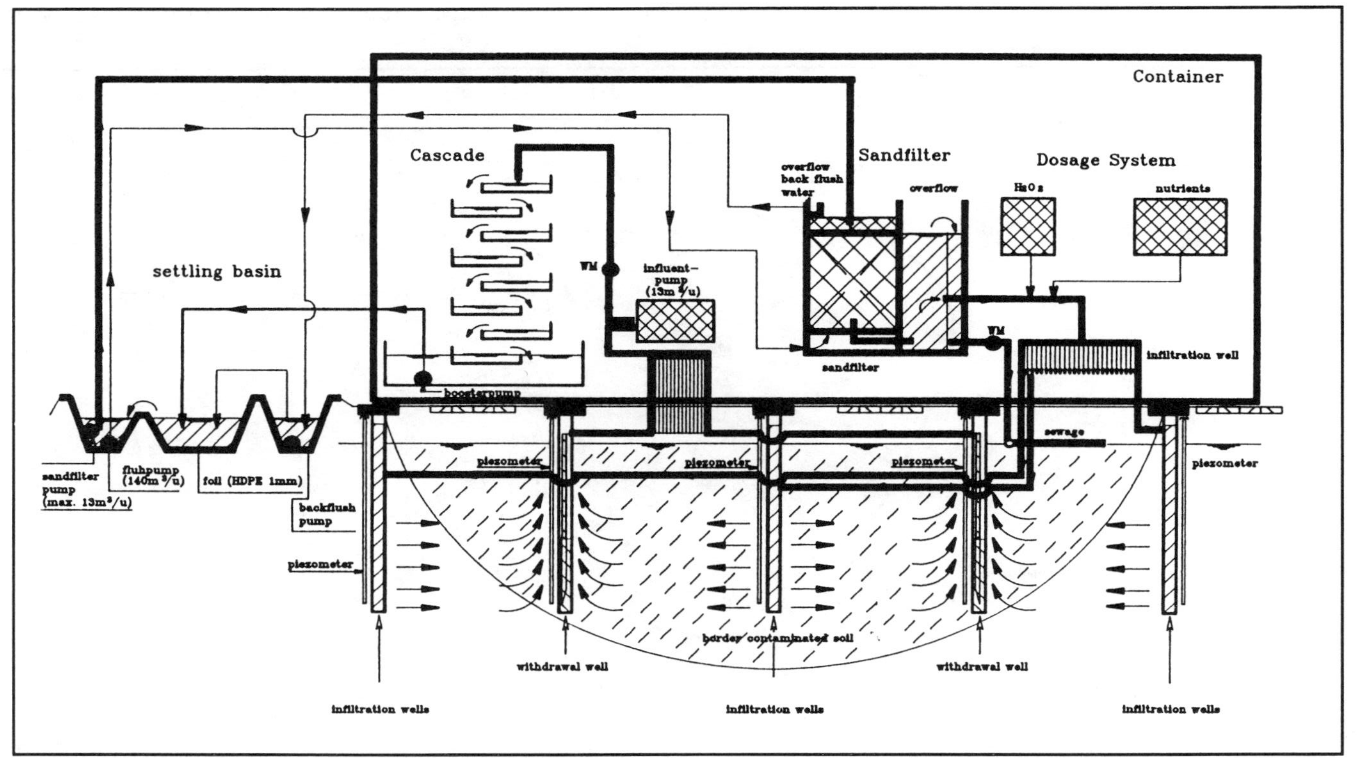

FIGURE 2. Cross-section of groundwater purification plant and infiltration and withdrawal wells.

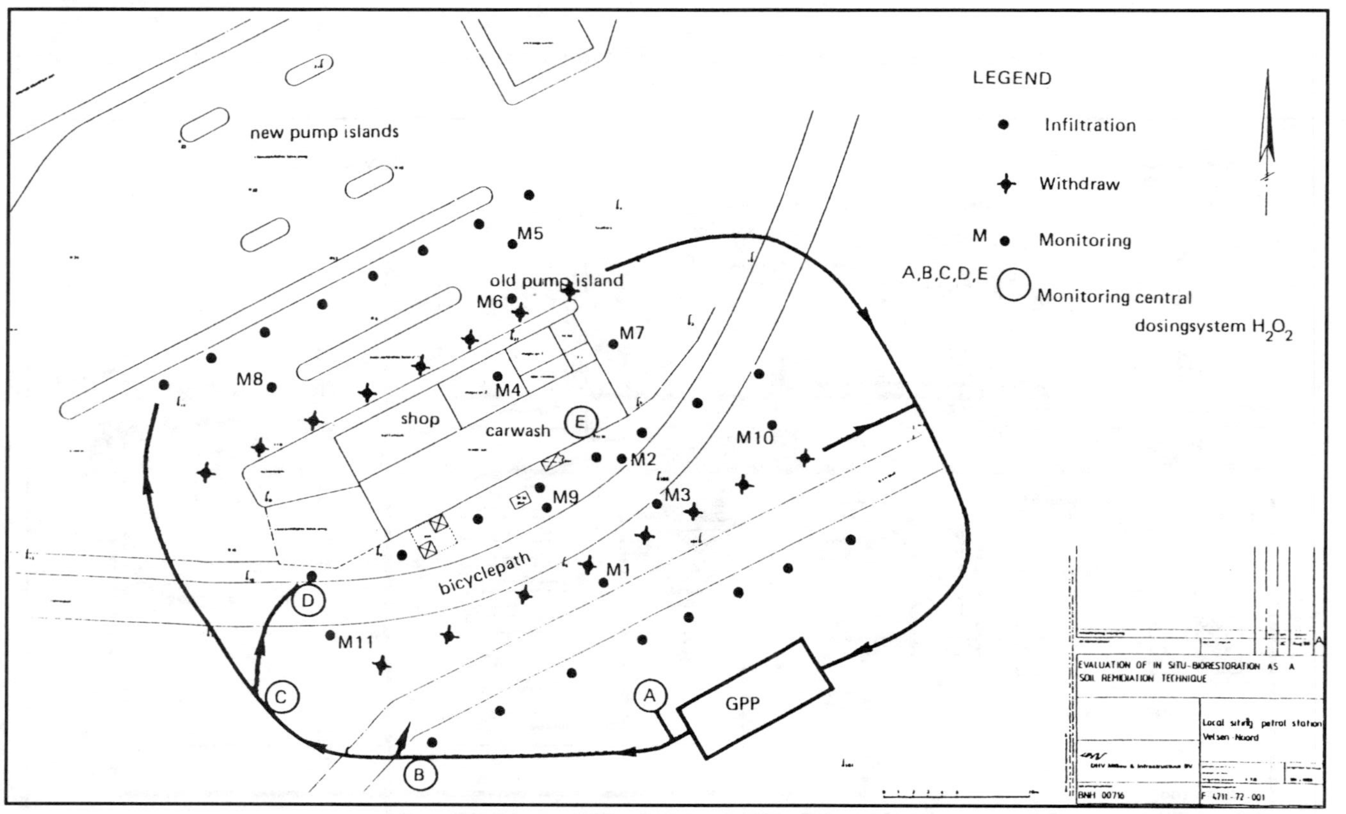

FIGURE 3. Location of filters and monitoring.

TABLE 1. Dutch regulatory limits for soil and groundwater.

		Dutch A level	Dutch B level	Dutch C level
soil (mg/kg)	mineral oil	50	1,000	5,000
	BTEX	detection limit	7	70
groundwater (µg/L)	mineral oil	50	200	600
	BTEX	detection limit	30	100

month. The course of the concentration in aromatics as well as in mineral oil is shown in Figures 4 and 5. The remediation of the Velsen-Noord site was started in November 1990. One year later the operation seemed a success except for the soil and groundwater underneath the pump island. An intensive monitoring program of this area showed that the soil was far more seriously contaminated than assumed at the start. The decision was made to halt the operation so that causes could be analyzed and solutions to the problems could be generated. Limited

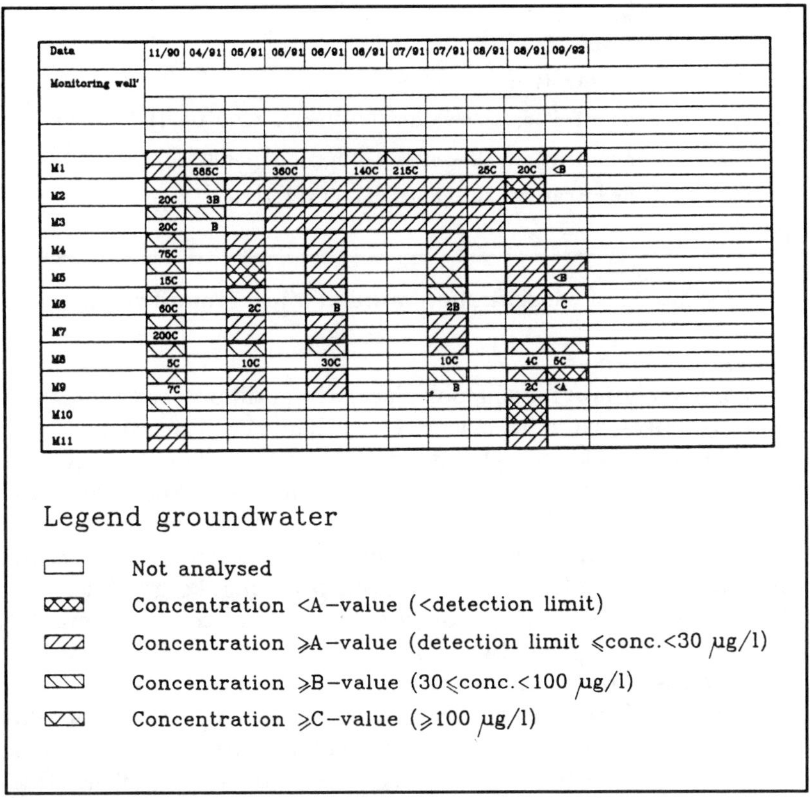

FIGURE 4. Course of concentration aromatics in groundwater.

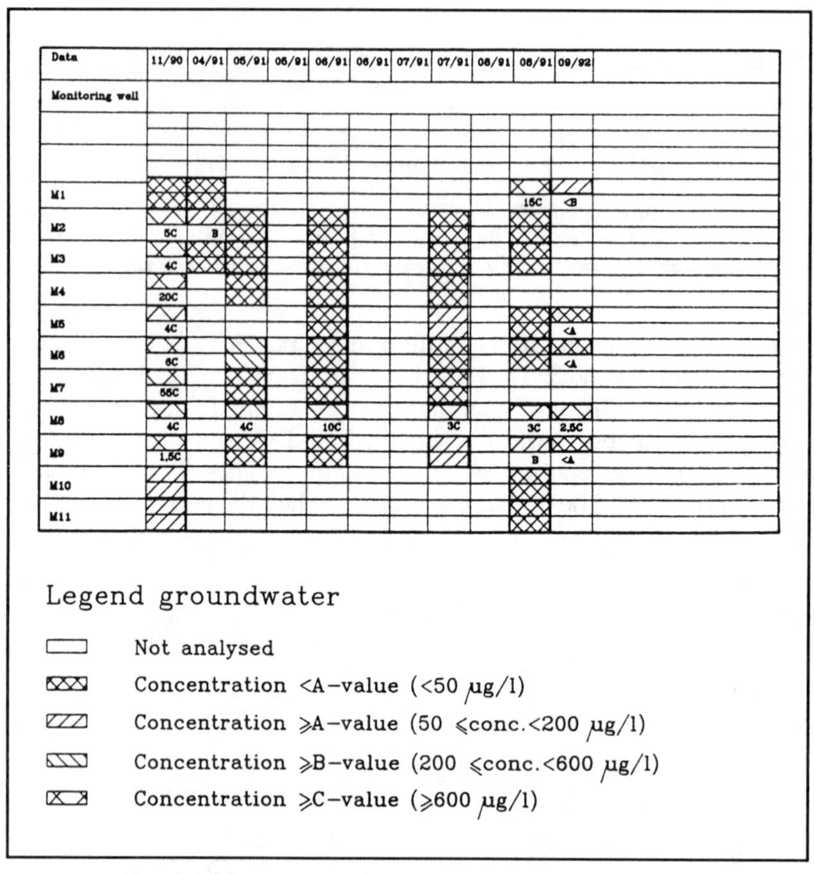

FIGURE 5. Course of concentration mineral oil in groundwater.

sampling and analysis of the soil took place. The additional sampling under the pump island shows that in the layer between 2 and 4 m –gr.l. remediation was successful. But degradation in the layer between 1.5 to 2 m –gr.l. seemed to be limited. Detailed analysis showed that, measuring from the point of infiltration, the first 1 to 2 m of the soil are clean (contents below A-level). In the remaining parts beneath the island, with the measuring starting at 2 m from the infiltration, remarkably high contents were found (up to 5,000 mg/kg gasoline).

The stagnation in the soil remediation is caused by a combination of (1) area-wide lowering of the groundwater level (since April 1990 the groundwater level has decreased from ±1.2 m to 1.6 m –gr.l.), resulting in a slightly flushed layer of ±40 cm; and (2) design not being based on the higher initial concentrations. Since installation, the subsoil has been flushed about 35 times with nutrients and peroxide, with a residence time of 7 days. Around 4,000 kg of peroxide have been introduced into the soil beneath the pump island. This amount corresponds to around 2,000 kg of oxygen, which is capable of degrading around 800 kg of hydrocarbons.

Geohydrology Results. The flowrates calculated previously were obtained. After several months, some infiltration filters had to be regenerated, because the flowrate kept dropping. In the period from July to August 1991 the withdrawal flowrate dropped sharply, from 13 m^3/hour to about 7 m^3/hour, due to iron hydroxide precipitation in the withdrawal system. After regeneration, the flowrate regained its original level.

Groundwater Purification Plant Results. The highest concentrations in the influent were measured during the first 3 months of the withdrawal, with concentrations of aromates varying from 1.2 mg/L to 3.5 mg/L. The aromatic concentrations in the effluent were normally reduced to a point below discharge standards (30 μg/L for total aromatics). Posttreatment with activated carbon was necessary only for 1 month. So it may be concluded that the GPP performed well.

Hydrogen Peroxide Dosage Results. Comparing the measurements of the peroxide concentration in the dosing system showed that hardly any decomposition occurred inside the pipelines (see Figure 6 and location of filters and monitoring in Figure 3). At the moment of infiltration (point E), the peroxide decomposes immediately up to approximately 50% of the influent concentration. The occurrence of an oxygen breakthrough usually indicates that the biodegradation process is nearing its end. Judging by the results of this investigation, this does not seem to have been the case beneath the pump island. The explanation is that oxygen-enriched water was transported through the clean soil which has a better permeability.

COST EVALUATION

The costs of the project Velsen-Noord are shown in Table 2. The first column states the budget drawn up in November 1989, and the second one the expenses as of March 1993. Operating costs covers the checking, maintenance, meetings, as well as the costs of chemicals (around $9,500). This budget was overrun due to a much longer adjustment period for the installations. Total remediation costs amounted to $517,000, for which 1,500 tons of soil as well as 2,000 m^3 of groundwater were cleaned up. This amounts to the rounded sum of $345 per ton, including remediation and on-surface purification of the groundwater. When comparing the above costs with a conventional remediation, the financial requirements appear to be higher for the in situ method. It should be noted that by complete excavation the costs for demolishing and rebuilding and temporary closing should be taken into account. Although the costs for the design stayed within the budget, these costs are relatively high for an installation of this type and size. Future installations could be designed with a budget reduction of approximately 25%.

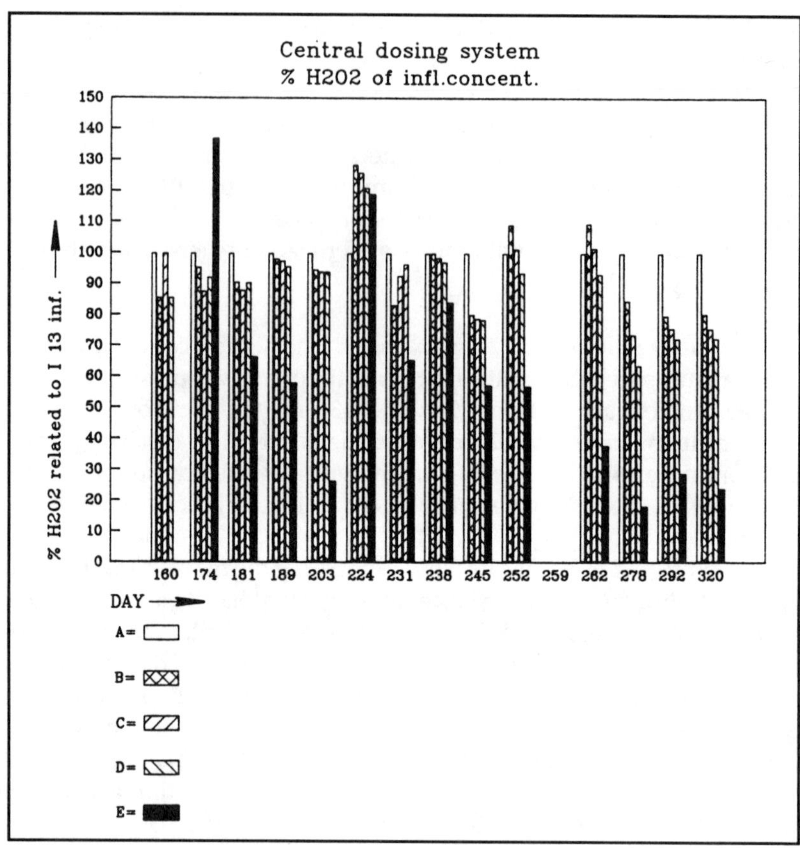

FIGURE 6. Stability of hydrogen peroxide.

CONCLUSIONS

From the evaluation of in situ biorestoration at Velsen-Noord, the following conclusions can be drawn:

1. The infiltration and withdrawal system functioned according to the design.
2. The withdrawn water was purified very well with a simple GPP.
3. Hydrogen peroxide is a satisfactory oxygen source to support biodegradation.
4. There is no need for a separate dosing system as hardly any decomposition in the pipeline took place.
5. The groundwater cleanup was a success; even after a standstill there was no new pollution.
6. The cleanup of the polluted soil was partly successful due to a combination of lowered groundwater level and higher initial concentrations.

TABLE 2. Summary budgeted costs and expenditures, ISB location Velsen-Noord (U.S. $$).

Entry	Budgeted costs (November 1989)	Expenditures (March 1993)
Preparation	8,000	24,000
Design/construction	102,000	104,000
Investment costs	121,000	149,000
Operating costs	51,000	139,000
Monitoring/analysis	38,000	28,000
Reporting	11,000	23,000
Additional costs for excavation	—	50,000
Total	331,000	517,000

REFERENCES

Balba, M. T., A. C. Ying, and T. G. McNeice. 1991. "Bioremediation of Contaminated Land: Bench Scale to Field Applications." In R. E. Hinchee, and R. F. Olfenbuttel (Eds.), *On-Site Bioreclamation*, pp. 464-476. Butterworth-Heinemann, Boston, MA.

De Ruiter Milieutechnologie BV. 1989. *Rapport Saneringsonderzoek Velsen Noord*. RP/IO/A890207, February.

Hinchee, R. E., and M. F. Arthur. 1991. "Bench Scale Studies of the Soil Aerating Process for Bioremediation of Hydrocarbons Contaminated Soil." *Applied Biochem. and Biotech. 28/29*: 902.

van den Berg, R., D. H. Eikelboom, and J. H. A. M. Verheul. 1987. "In situ biorestauratie van een met olie verontreinigde bodem. Resultaten van het laboratorium onderzoek." RIVM, Bilthoven.

Verheul, J. H. A. M., R. van den Berg, and D. H. Eikelboom. 1988. "In situ biorestoration of a subsoil contaminated with gasoline." In K. J. Wolf, J. van der Brink, and F. J. Colon (Eds.), *Contaminated Soil '88*, pp. 705-715. Kluwer Academic Publishers.

BIOREMEDIATION OF HYDROCARBON-CONTAMINATED SOIL

D. S. Hogg, M. R. Piotrowski, R. P. Masterson, M. R. Jorgensen, and C. Frey

BIOREMEDIATION PROGRAM

An on-site aboveground bioremediation program was developed to remediate 45,875 m^3 of petroleum hydrocarbon-contaminated soil. The soil, contaminated with gasoline, diesel fuel, and oil, was encountered during removal of underground storage tanks (USTs) during preparation for site development. Oil-affected soil also was encountered in a washwater recirculation pond located adjacent to oil-bearing machinery and within areas used to steam-clean or store equipment. The remedial criteria for the contaminated soil were 100 mg/kg total petroleum hydrocarbons (TPH, by EPA modified Method 8015) for the gasoline- and diesel fuel-affected soil and 100 mg/kg total recoverable petroleum hydrocarbons (TRPH, EPA Method 418.1) for the oil-affected soil. The approach involved excavating contaminated soil from the site source areas and placing it on 13 clay-lined biological treatment pads (biopads). Biodegradation of petroleum hydrocarbon compounds was stimulated by introducing oxygen to the indigenous soil microorganisms through periodic tilling of the contaminated soil and maintaining soil moisture and nutrient levels within limits necessary for enhanced microbial activity.

Design, Implementation, and Operations. The 13 biopads, covering approximately 16.4 acres (6.65 hectares), were constructed on site. The pads were built on a level portion of the site that consisted of low-permeability, compacted fill extending to a depth of at least 15 ft (4.6 m) below grade. An additional layer of low-permeability fill material 1 to 2 ft (30 to 60 cm) thick was placed as a liner for the base of each biopad and compacted to 90% to produce a permeability range of 1.5×10^{-7} to 4.8×10^{-8} cm s^{-1}.

Contaminated soil generated during remedial excavation was placed in the biopads. Soil contaminated mainly with turbine oil was placed on biopads 1, 3, and 4 through 9. Gasoline- and diesel-affected soil was placed on biopads 2 and 10, and spent oil-affected soil was placed on biopads 11, 12, and 13 (Figure 1). Once placed, the soil was leveled to a thickness ranging between 1 to 3 ft (30 to 90 cm) on each biopad. The biopad perimeters were bermed and trenched to control surface water runon and runoff.

Typically, tilling was conducted biweekly to both mix the soil mass and introduce oxygen to the contaminated soil. The soil moisture content was monitored periodically, and when it was observed to be less than 10% by weight, the

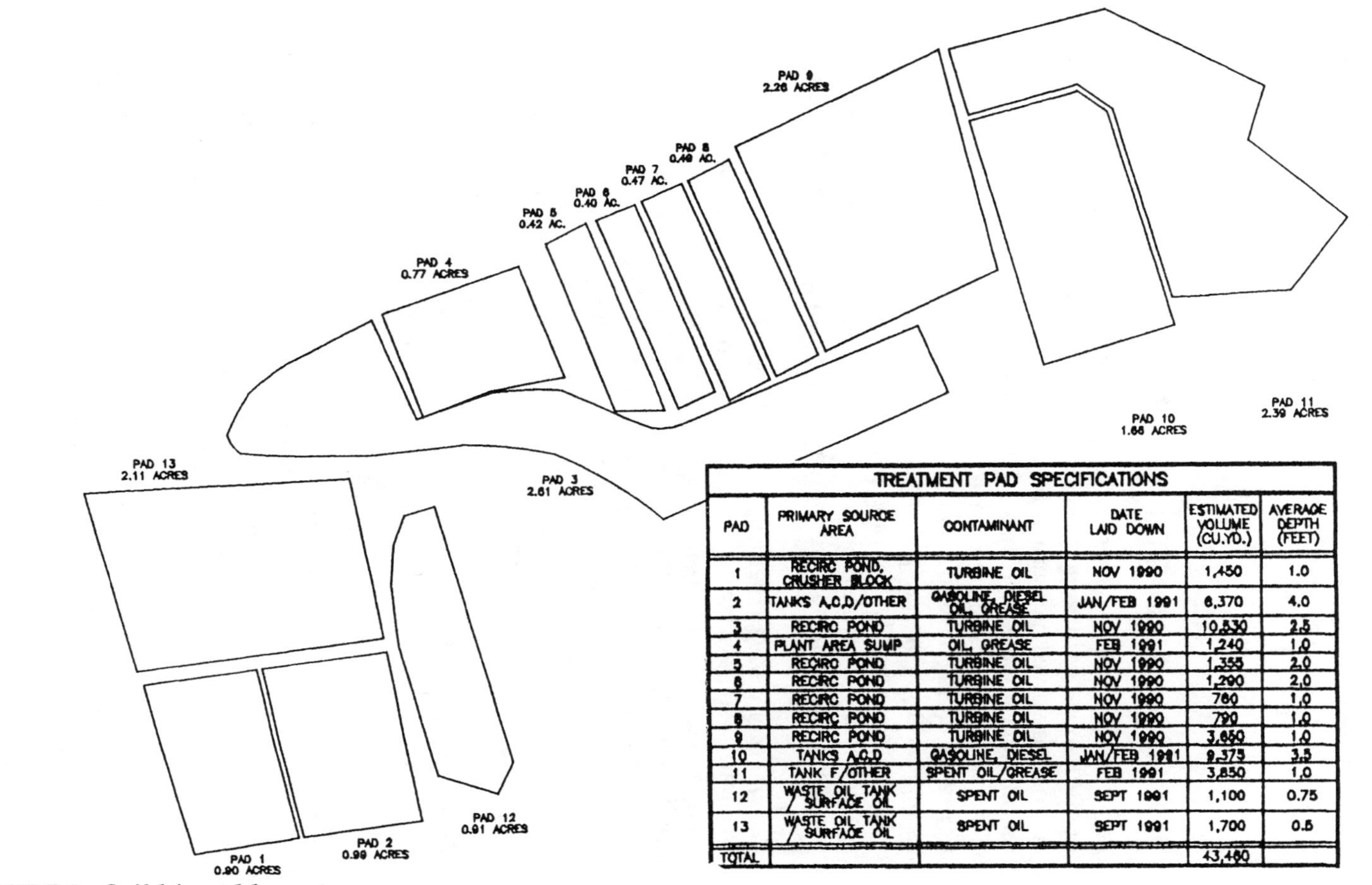

TREATMENT PAD SPECIFICATIONS

PAD	PRIMARY SOURCE AREA	CONTAMINANT	DATE LAID DOWN	ESTIMATED VOLUME (CU.YD.)	AVERAGE DEPTH (FEET)
1	RECIRC POND, CRUSHER BLOCK	TURBINE OIL	NOV 1990	1,450	1.0
2	TANKS A,C,D/OTHER	GASOLINE, DIESEL OIL, GREASE	JAN/FEB 1991	6,370	4.0
3	RECIRC POND	TURBINE OIL	NOV 1990	10,530	2.5
4	PLANT AREA SUMP	OIL, GREASE	FEB 1991	1,240	1.0
5	RECIRC POND	TURBINE OIL	NOV 1990	1,355	2.0
6	RECIRC POND	TURBINE OIL	NOV 1990	1,290	2.0
7	RECIRC POND	TURBINE OIL	NOV 1990	760	1.0
8	RECIRC POND	TURBINE OIL	NOV 1990	790	1.0
9	RECIRC POND	TURBINE OIL	NOV 1990	3,650	1.0
10	TANKS A,C,D	GASOLINE, DIESEL	JAN/FEB 1991	9,375	3.5
11	TANK F/OTHER	SPENT OIL/GREASE	FEB 1991	3,850	1.0
12	WASTE OIL TANK / SURFACE OIL	SPENT OIL	SEPT 1991	1,100	0.75
13	WASTE OIL TANK / SURFACE OIL	SPENT OIL	SEPT 1991	1,700	0.5
TOTAL				43,480	

FIGURE 1. Soil biopad layout.

biopads were irrigated with chlorine-free water. On September 4, 1991, an application of a wetting agent/surfactant called Bio-7SR (BD Chemicals, Denver, Colorado) and an inorganic nutrient mixture called Bio-D Nutrients (Medina Soil Agricultural Products, Inc., Hondo, Texas) was applied to biopads 1, 3, 4, 9, and 11 to further enhance contaminant biodegradation in these pads.

PERFORMANCE-MONITORING PROGRAM

Performance of the bioremediation program was evaluated over time by monitoring the change in petroleum hydrocarbon concentrations in each biopad. The approach involved measuring the petroleum hydrocarbon concentration in several soil samples in each biopad, the number of which varied with the size of the biopad (see Table 1), and when the average concentration was less than 100 mg/kg (including the upper 95% confidence interval), soil in this biopad was not resampled during the next performance monitoring event. Although biopads were deleted sequentially from the performance-monitoring program as the average petroleum hydrocarbon concentrations declined below the remedial criterion, the bioremediation program operations continued for all biopads.

Baseline Monitoring. The analytical results for soil samples collected during the preliminary site assessment and during the remedial excavation activities were considered representative of the initial concentrations of petroleum hydrocarbons in each biopad. Therefore, the values presented are order-of-magnitude estimates and have been used for comparison with results from the performance-monitoring phase. Table 1 presents the initial average petroleum hydrocarbon concentrations (number of samples analyzed and ±standard error of the mean) for soil in each biopad. The initial average petroleum hydrocarbon concentrations ranged from 284 mg/kg to 1,747 mg/kg.

Performance Monitoring. Performance monitoring was conducted on seven dates: November 19 and December 28, 1990; February 22, April 16, June 21, September 9, and December 23, 1991. Each biopad was divided into 10 ft by 10 ft (3 m by 3 m) cells, and sample locations among the cells in a biopad were randomly selected using X and Y coordinates. The number of samples collected and analyzed from each biopad are provided in Table 1. Samples from biopads 2 and 10 were analyzed for TPH using EPA Method 8015 (modified). Samples from the remaining biopads were analyzed for TRPH using EPA Method 418.1. The analytical results for each sample were used to calculate an average concentration for each biopad.

Results of the Performance-Monitoring Program. The average petroleum hydrocarbon concentrations and upper 95% confidence intervals for the biopads during each performance-monitoring sampling are presented in Table 1. Biopads 12 and 13 were sampled only once on December 23, 1991, and the average soil petroleum hydrocarbon concentrations [number of samples analyzed] and (the

TABLE 1. Concentrations of petroleum hydrocarbons in soil during performance monitoring program.

Sampling Date	Average Petroleum Hydrocarbon Concentration[a] [number of samples analyzed] (upper 95% confidence interval) (mg/kg)										
	Biopad 1	Biopad 2	Biopad 3	Biopad 4	Biopad 5	Biopad 6	Biopad 7	Biopad 8	Biopad 9	Biopad 10	Biopad 11
Baseline[b]	837[88] (±250)	300[160] (±80)	1078[68] (±317)	284[15] (±143)	1078[68] (±317)	1078[68] (±317)	1078[68] (±317)	1078[68] (±317)	1078[68] (±317)	300[160] (±80)	1747[101] (±494)
11-19-90	NS		NS	250[3] (325)	42[1] (NA)	56[2] (65)	373[8] (543)	579[7] (1258)	NS	NS	NS
12-28-90	NS	NS	NS	NS	NS	NS	NS	158[4] (214)	NS	NS	NS
2-22-91	NS	NS	NS	NS	NS	NS	114[9] (144)	120[6] (197)	NS	NS	NS
4-16-91	720[4] (1218)	32[4] (37)	92[5] (117)	272[4] (404)	33[3] (47)	30[3] (39)	145[3] (179)	83[3] (123)	117[5] (146)	113[4] (146)	276[5] (313)
6-9-91	99[5] (132)	NS	90[6] (118)	208[4] (395)	NS	NS	65[4] (89)	68[4] (100)	157[6] (285)	<5[5] (<5)	367[5] (472)
9/9/91	36[5] (49)	NS	25[6] (32)	34[4] (39)	NS	NS	NS	NS	80[6] (117)	NS	56[5] (80)
12/23/91	NS	NS	NS	NS	NS	NS	NS	NS	46[6] (58)	NS	NS

(a) Results indicate average concentrations [number of samples analyzed] (upper 95% confidence interval).
(b) Results indicate average concentrations (± standard error).
NA = Not applicable, only one sample collected.
NS = Not sampled.

upper 95% confidence interval) were 26 mg/kg [5], (36 mg/kg) and 30 mg/kg [5], (45 mg/kg) for each biopad, respectively. The results shown in Table 1 indicated that average petroleum hydrocarbon concentrations (including the upper 95% confidence interval) in soil from all 13 biopads were less than the remedial goal of 100 mg/kg by December 1991. Therefore, the verification sampling program was implemented to demonstrate that the excavated soil did not contain average petroleum hydrocarbon concentrations of greater than 100 mg/kg.

VERIFICATION SAMPLING PROGRAM

A statistically based verification sampling program (VSP) was developed to demonstrate that the excavated soil and biopad liner material did not contain average petroleum hydrocarbon concentrations greater than 100 mg/kg. Because the area of the biopads and volume of soil placed on each biopad varied, sampling was conducted at a ratio of one sample per 0.1 acre (0.04 hectare) of biopad area and one sample per 1 ft (30 cm) of soil thickness. Five to six samples were collected from the liner material beneath each biopad to assess whether any downward migration of the contaminants occurred during soil treatment.

Sampling and Collection Procedure. The VSP was implemented on December 20, 1991 for biopad 11, and the remaining biopads were subjected to the VSP over the period January 12 through 21, 1992. The biopads were divided into 25 ft by 25 ft (7.6 m by 7.6 m) cells. Each sample location was randomly selected within each treatment pad using an X, Y, and Z coordinate system. A discrete soil sample (113 g) was collected from each sample location. Samples from biopad 11 were analyzed for TRPH by EPA Method 418.1, using an off-site laboratory. Samples from biopads 2 and 10 were analyzed for TPH using EPA Method 8015 (modified) using an off-site laboratory. Samples from the remaining biopads were analyzed for TRPH using EPA Method 418.1 using an on-site, mobile laboratory.

Results of the Verification Sampling Program. The average petroleum hydrocarbon concentrations and 95% confidence intervals for soil in the biopads and sampled of the biopad liner material are presented in Table 2.

Except for three soil samples, TRPH concentrations in biopads 1 and 3 through 9 were less than the analytical detection limit of 10 mg/kg. One sample collected from biopad 4 contained 21 mg/kg TRPH, and two samples from biopad 7 contained 16 and 33 mg/kg TRPH, respectively. The concentrations of TRPH in soil from biopad 11 ranged from 6 to 230 mg/kg, with an average concentration of 51.3 mg/kg. Except for one sample which contained 27 mg/kg TPH (as diesel fuel), the concentrations of TPH in soil samples from biopads 2 and 10 were less than the analytical detection limit of 5 mg/kg. When this sample was subsequently reanalyzed, it contained a TPH concentration that was less than 5 mg/kg.

The concentrations of TRPH in liner soil from biopad 11 ranged from 7 to 61 mg/kg, with an average concentration of 22.7 mg/kg. The liner samples from the remaining biopads contained less than the respective analytical detection limits (5 mg/kg TPH or 10 mg/kg TRPH).

TABLE 2. Concentrations of petroleum hydrocarbons in soil from biopads and biopad liner material during verification sampling program.

				Concentration (mg/kg)		
Pad	Location	Number of Samples	Analyte	Average	Standard Error	95% Confidence Interval
1	Biopad	28	TRPH	<10	0	NA
	Liner	5	TRPH	<10	0	NA
2	Biopad	39	TPH	<5	0	NA
	Liner	5	TPH	<5	0	NA
3	Biopad	50	TRPH	<10	0	NA
	Liner	7	TRPH	<10	0	NA
4	Biopad	27	TRPH	10.4	0.40	9.9 - 10.9
	Liner	6	TRPH	<10	0	NA
5	Biopad	28	TRPH	<10	0	NA
	Liner	5	TRPH	<10	0	NA
6	Biopad	28	TRPH	<10	0	NA
	Liner	5	TRPH	<10	0	NA
7	Biopad	28	TRPH	11.04	0.83	10.0 - 12.1
	Liner	5	TRPH	<10	0	NA
8	Biopad	27	TRPH	<10	0	NA
	Liner	6	TRPH	<10	0	NA
9	Biopad	29	TRPH	<10	0	NA
	Liner	4	TRPH	<10	0	NA
10	Biopad	38	TPH	5.6	0.56	4.8 - 6.3
	Liner	5	TPH	<5	0	NA
11	Biopad	27	TRPH	51.3	9.8	38.4 - 64.2
	Liner	6	TRPH	22.7	7.7	11.5 - 33.8
12	Biopad	28	TRPH	<10	0	NA
	Liner	5	TRPH	<10	0	NA
13	Biopad	28	TRPH	<10	0	NA
	Liner	5	TRPH	<10	0	NA

SUMMARY

The analytical results from the VSP indicated that the remedial criteria had been achieved and that the soil had been remediated to a satisfactory level. There was no evidence of significant vertical migration of the contamination to the biopad liner during soil remediation. After agency approval, the remediated soil was placed on site in a layer at least 10 ft (3 m) (10 ft) below the ground surface and at least 3 m (10 ft) above the groundwater level. Site redevelopment activities were resumed shortly thereafter. The cost for site investigation, excavation, handling, and treatment of the soil was \$13.55 yd^3 of soil.

THE USE OF MULTIPLE OXYGEN SOURCES AND NUTRIENT DELIVERY SYSTEMS TO EFFECT IN SITU BIOREMEDIATION OF SATURATED AND UNSATURATED SOILS

R. D. Norris, K. Dowd, and C. Maudlin

INTRODUCTION

The removal of tanks and product lines following the sale of a retail service station in southern California revealed that subsurface soils were contaminated with gasoline (Norris & Dowd 1993). A soil boring and well installation program delineated the area of contamination as shown in Figures 1 and 2. Total fuel hydrocarbons in the largely sandy unsaturated zone soils ranged from below the detection limit to 5,200 mg/kg extending over an area of 27 m by 22 m to a maximum depth of 17 to 18 m. The saturated soils, which are largely clay and silt, were less contaminated with hydrocarbon concentrations ranging from below the detection limit to 32 mg/kg as benzene, toluene, and xylene (BTX). Maximum groundwater BTX levels were 2,700 µg/L for benzene, 6,600 µg/L for toluene, 4,100 µg/L for xylene, and 45,000 µg/L for total petroleum hydrocarbons (TPH). The aquifer contamination extended over an area of 18 m by 18 m directly under the former tank area. A smaller, lesser contaminated area was located on the eastern portion of the property.

The depth to water, the sandy unsaturated soils, and the location of the site adjacent to three streets and a commercial building precluded any significant excavation of contaminated soils. Pump-and-treat technology was considered too slow to implement at a property that was undergoing ownership transfer. The biodegradability and volatility of the contaminant source, gasoline, indicated that some combination of bioremediation and vapor extraction should be evaluated for remediation of the site. These technologies also were attractive based on their compatibility with the intended use of the site as a restaurant.

Generally, in situ bioremediation of aquifers is thought to be applicable for more permeable aquifers than the one underling this site because of the difficulty in delivering sufficient quantities of electron acceptor over an acceptable period of time (Thomas & Ward 1989). However, the total hydrocarbon burden to be addressed by the aquifer remediation system was estimated to consist of 110 kg of fuel hydrocarbons adsorbed to the soils, with an additional 2 to 3 kg of dissolved hydrocarbons. Based on approximate stoichiometry, approximately 3 kg of

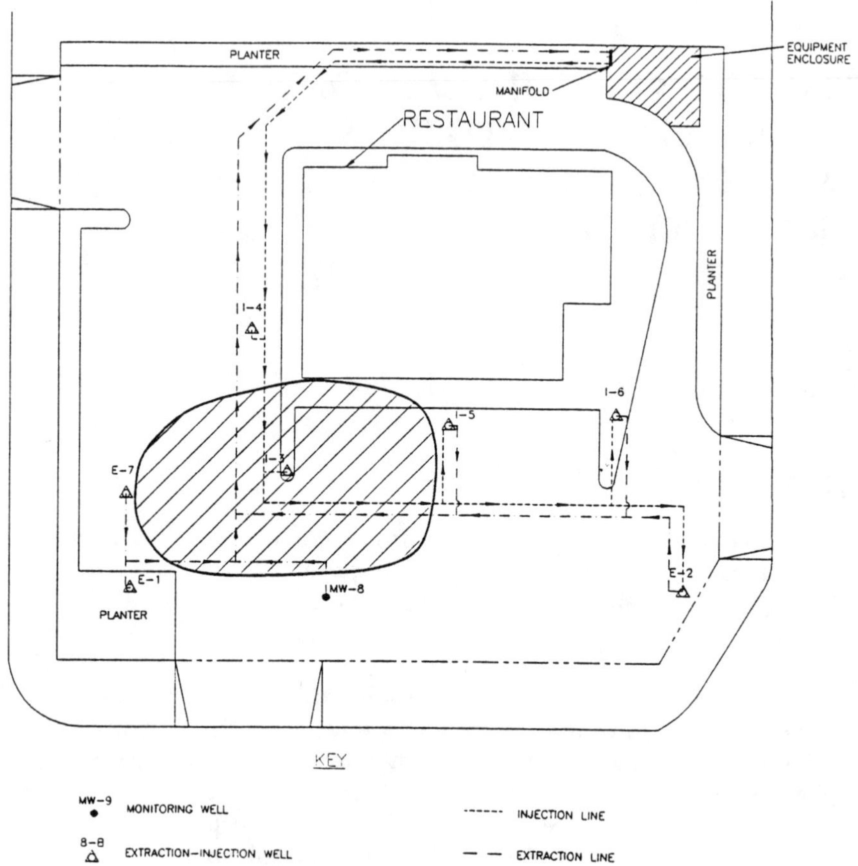

FIGURE 1. Groundwater injection/extraction system.

oxygen are required to biodegrade 1 kg of hydrocarbon, and 2 kg of hydrogen peroxide are required to generate 1 kg of oxygen. Thus, nearly 360 kg of oxygen or 720 kg of hydrogen peroxide would be needed to meet the oxygen requirements for biodegradation of hydrocarbons within the saturated zone, provided complete mineralization of the hydrocarbons occurred and that the mass of hydrocarbons was correctly estimated. At an injection concentration of 500 mg/L of hydrogen peroxide, this would require approximately 1.4 million L of groundwater to be recovered, amended, and reinjected if distribution were 100% efficient.

Gasoline consists essentially of volatile components. Thus, if sufficient air is passed through gasoline-impacted unsaturated soils, the gasoline should be entirely removed. In addition, gasoline components will undergo biodegradation in soils containing hydrocarbon-degrading indigenous bacteria, provided sufficient nutrients, oxygen, and moisture are present. Vapor extraction systems typically are operated at rates that provide far more than stoichiometric amounts of oxygen over the lifetime of their operation (Dupont 1992). Thus, the operation of a vapor recovery system was expected to address the gasoline contamination in the unsaturated soils by two mechanisms, physical removal and biodegradation. Additionally, by providing oxygen to the capillary zone, the demand on the aquifer remediation system would be diminished, thus expediting remediation.

SYSTEM DESIGN

The system for treating the aquifer consisted of three recovery wells and four injection wells (Figure 1). The recovered groundwater was amended with nutrients (batch additions) and hydrogen peroxide (continuous) and reinjected without aboveground treatment. All water and vapor recovery lines were installed prior to property development as a restaurant and were run underground to a small walled treatment compound so that the system was hidden from the view. At no time was the restaurant operation impacted as a result of the remediation efforts.

The aquifer system was designed to operate at 40 to 60 L/min. However, the falling head tests that were conducted to evaluate the aquifer characteristics overestimated the potential of the wells, largely because the injection wells could not accept water at the same rate as the recovery wells could produce it. Further, the aquifer test may have been conducted in the more permeable portion of the site and,

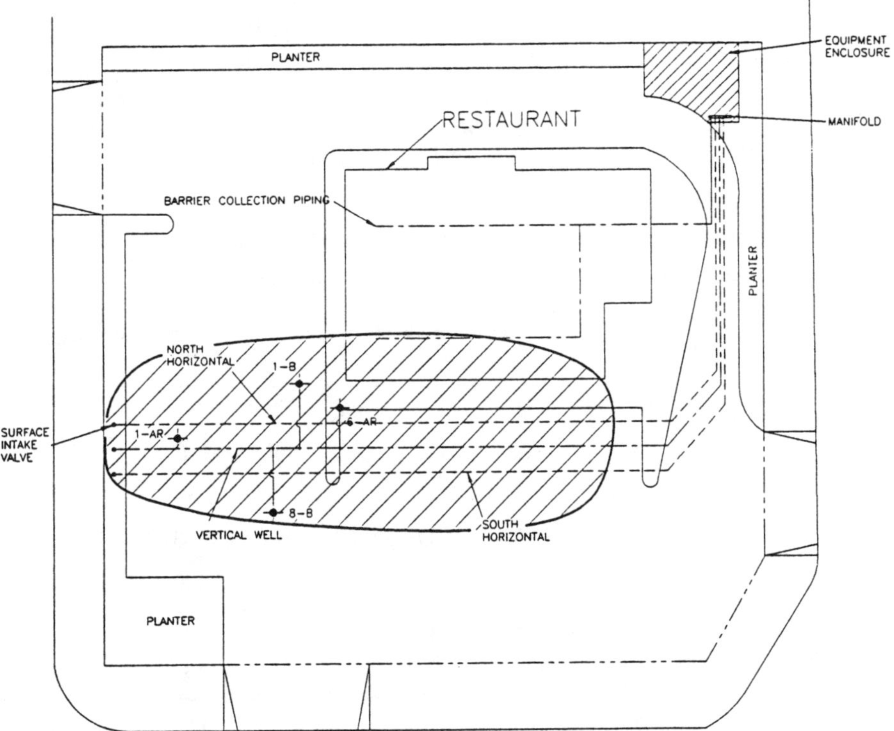

FIGURE 2. Soil vapor extraction system.

as is commonly the case, well performance decreased over time requiring frequent redevelopment. The final operating total flowrate was typically 4 to 8 L/min.

The performance of the system was monitored by regular sampling of both injection and recovery wells. The injection wells were sampled by stopping injection and removing several bore volumes of water prior to sampling. Sampled water was analyzed for dissolved oxygen, pH, ammonium ion, chloride, phosphate, nitrate, carbon dioxide, and contaminant parameters.

The vapor recovery system was designed by George Hoag of the University of Connecticut and consisted of two horizontal recovery wells installed at a depth of 1.5 m and four vertical wells screened from 1.5 m to 15 m below the surface (Figure 2). Air was recovered at an average flow rate of 2.2 m^3/min and treated with a catalytic oxidizer. The influent and effluent of the catalytic oxidizer were monitored using a flame ionization detector (FID) hydrocarbon analyzer.

The system design included horizontal wells for greater induced air flow at a depth approximate to the bottom of the former tank pit where the highest levels of contamination existed. At the same time, the vertical wells would induce air flow through the deeper, less contaminated soils and possibly provide some oxygen to the capillary fringe which would facilitate treatment of the aquifer. It was recognized at the time of the design that biodegradation also would contribute to the treatment of the unsaturated soils, although no effort was made to quantify this contribution.

RESULTS

As shown in Table 1, total BTX and TPH levels increased substantially in the recovery wells over the first several months before dropping below the detection

TABLE 1. Groundwater contamination during remediation.[a]

Date	Total BTX (µg/L)[b]							Maximum TPH[c]
	E1	E2	E7	I	I4	I5	I6	
9/87	ND	ND	ND	29	46	21	23	1,500
10/87	ND	4.0	5,800	3,000	ND	ND	ND	20,000
12/87	21	NE	2.0	ND	200	ND	1,500	400
1/88	ND	2.0	26,000	ND	300	800	250	220,000
2/88	2.0	ND	800	400	600	400	400	1,300
3/88	ND	2.0	4,200	ND	ND	ND	ND	61,000
5/88-9/90	ND	ND	ND	ND	ND	ND	ND	ND

(a) E1, 2, etc. are extraction wells, and I4, 5, etc. are injection wells.
(b) Measured by EPA Method 602.
(c) Measured by EPA Method 8015.

limit after 10 months, at which time samples obtained from the injection areas were also below the detection limit. The large fluctuations in TPH concentrations reflect initial mobilization due to increased groundwater flow and possible increased solubilization from biosurfactants followed by biodegradation. Following an initial acclimation period, hydrogen peroxide concentrations in the injected water ranged from 500 to 1,000 mg/L. A total of 1,350 kg of hydrogen peroxide was introduced directly into the aquifer over 10 months. This amount is roughly two times the calculated requirement based on the estimated mass of hydrocarbons in the saturated zone. The apparent excess hydrogen peroxide consumption might be attributed to inefficiencies in distribution, continued addition to areas where hydrocarbons were already biodegraded, or underestimation of the contaminant mass.

Nitrate levels in the groundwater are of interest because nitrate can serve as an alternative electron source (Hutchins et al. 1991) and could possibly be produced biologically from ammonium and oxygen. Nitrate concentrations were consistently lower in the treatment zone than in the samples from the upgradient monitoring well, which ranged from 80 to 236 mg/L as NO_3. Carbonate levels in groundwater are of interest because carbon dioxide is a product of biodegradation. During the aquifer remediation period, aqueous carbonate/bicarbonate levels increased from an average of 102 mg/L to 186 mg/L and then decreased in concert with the change in hydrocarbon levels to an average of 103 mg/L.

The vapor recovery system physically removed approximately 5,600 liters of hydrocarbons and probably induced biological degradation of an undetermined mass of hydrocarbons over a period of 22 months. Following a decrease in hydrocarbon removal rates to an average of 2 kg/day, bioflood operation was initiated by introducing nutrient- and hydrogen peroxide-amended groundwater through the horizontal and vertical vapor recovery piping. The bioflood process was initiated because of the low recovery rate from the vapor recovery system, the delivery system was in place, and biodegradation was successful in the saturated zone. After an additional 15 months, during which groundwater hydrocarbon levels first increased and then decreased, both soils and groundwater were determined to be below the levels set by the Los Angeles Regional Water Quality Control Board leading to site closure.

The successful treatment of this aquifer was accomplished by designing and adjusting the system operations based on the properties of the contaminant and soils and the performance of the system. By making use of site conditions rather than force-fitting a solution, the site was successfully remediated.

REFERENCES

Dupont, R. R. 1992. "Application of Bioremediation Fundamentals to the Design and Evaluation of In Situ Soil Bioventing Systems." Paper presented at 85th Annual Meeting and Exhibition of the Air & Waste Management Association, Kansas City, MO. Paper 92-30.03.

Hutchins, S. R., W. C. Downs, G. B. Smith, et al. 1991. *Nitrate for Biorestoration of an Aquifer Contaminated with Jet Fuel.* U.S. Environmental Protection Agency Technical Report, EPA 600/2-91/009, Robert S. Kerr Environmental Research Laboratory, Ada, OK.

Norris, R. D., and K. Dowd. 1993. "Successful In Situ Bioremediation in a Low Permeability Aquifer." In P. E. Flathman and J. Exner (Eds.), *Bioremediation: Field Experiences*, Lewis Publishers, Ann Arbor, MI.

Thomas, J. M., and C. H. Ward. 1989. "In Situ Biorestoration of Organic Contaminants in the Subsurface." *Environmental Science & Technology* 23:760-766.

AROMATIC HYDROCARBON DEGRADATION SPECIFICITY OF AN ENRICHED DENITRIFYING MIXED CULTURE

B. K. Jensen and E. Arvin

INTRODUCTION

A growing number of studies have reported biodegradation of monoaromatic hydrocarbons under denitrifying conditions (Barbaro et al. 1992, Evans et al. 1991, Hutchins 1991, Jensen et al. 1988, Jørgensen et al. 1991, Kuhn et al. 1988, Major et al. 1988). Many of these are laboratory studies using one or two compounds as model substrates and as such give few indications of substrate interactions in complex mixtures. In the cases with more complex mixtures, interaction phenomena, for instance sequential or preferential degradation, are not addressed. Substrate or population interactions in the biodegradation have been shown to occur under other redox conditions (Alvarez & Vogel 1991, Arvin et al. 1989, Hwang et al. 1989). In environments with limited flux of electron acceptors and a high load of organic compounds, the electron acceptor might become depleted. In such cases, the degradation sequence in a complex contaminant mixture and the ability of the microbial populations to adapt and shift to other compounds when the preferred compounds are depleted are of importance. The aim of this study was to identify, in complex mixtures of coal tar or creosote contaminants, the biodegradation interactions that could have significance for the use of bioremediation with nitrate as the electron acceptor.

MATERIALS AND METHODS

Mixed Substrate Experiment. The batch experiments were performed with suspended toluene-enriched mixed cultures in 5-L reagent bottles with glass tube stoppers and equipped with a three-way stopcock for sample collection. The inoculum was prepared by toluene enrichment of a mixed wastewater sample. The suspensions were stirred magnetically, and the mixture of organic substances was added from prepared basis solutions to the batches to give the final concentrations ranging from 0.2 to 0.5 mg/L for the different compounds. The following compounds were added as a complex mixture: benzene, toluene, the xylenes, ethylbenzene, 1,4-diethylbenzene, butylbenzene, propylbenzene, naphthalene, 1,4-dimethylnaphthalene, 1-ethylnaphthalene, biphenyl, phenanthrene, furan, fluorenone, benzothiophene, phenol, the cresols, 2,3-dimethylphenol, 2,4-dimethylphenol, 2,5-dimethylphenol, 2,6-dimethylphenol, 3,5-dimethylphenol, and 2,4,6-trimethylphenol.

Factorial Design Experiment. A factorial design experiment including benzene, toluene, ethylbenzene, *o*-xylene, and a mixture of the following compounds: cumol, 2,4,6-trimethylbenzene, ethyltoluene, butylbenzene, naphthalene, 1-methylnaphthalene, biphenyl, 1-ethylnaphthalene, 1,4-dimethylnaphthalene, furane, fluorenone, benzothiophene, and phenanthrene was set up using batches of 0.12 L. Adenosine triphosphate (ATP) content reflecting growth was the dependent variable. As indicated in Table 1, following substrate additions were included as independent variables: benzene, toluene, ethylbenzene, *o*-xylene, and hydrocarbon mixture. The experiment was designed as a confounded $\frac{1}{2} \cdot 2^5$ experiment with 16 treatments plus blanks. Benzene, toluene, ethylbenzene, and *o*-xylene were added in concentrations of 1 mg/L each, and the concentration of the individual compounds in the mixture was 0.15 to 0.2 mg/L. ATP contents were measured by applying the bioluminescence technique slightly modified after Leach (1981). A nutrient medium prepared according to Arvin et al. (1989)

TABLE 1. Factorial design experiment. Growth of biomass in terms of ATP production versus addition of aromatics: –, compound not added; +, compound added.

BATCH	BEN	TOL	EBEN	OXYL	HCM	ATP (ng/L)
1	–	–	–	–	–	213
2	+	–	–	–	+	150
3	–	+	–	–	+	1,200
4	+	+	–	–	–	4,400
5	–	–	+	–	+	161
6	+	–	+	–	–	247
7	–	+	+	–	–	8,400
8	+	+	+	–	+	4,800
9	–	–	–	+	+	247
10	+	–	–	+	–	155
11	–	+	–	+	–	1,700
12	+	+	–	+	+	598
13	–	–	+	+	–	196
14	+	–	+	+	+	160
15	–	+	+	+	+	2,800
16	+	+	+	+	–	5,400

BEN: benzene; TOL: toluene; EBEN: ethylbenzene; OXYL: *o*-xylene; HCM: hydrocarbon mixture.

was used. Anaerobic conditions in the batches and basis solutions were maintained by purging with N_2. All experiments were run at 20°C for 2 weeks in the dark. Batches poisoned with $HgCl_2$ served as controls. Samples from the bottles were collected by applying an N_2 pressure in the batch, thus forcing out the sample.

Chemical Analysis. Quantification of the organics was performed by gas chromatography (GC) analysis according to the technique described by Arvin et al. (1989) and Flyvbjerg et al. (1993). NO_3^--N and NO_2^--N were measured on a Technicon Autoanalyzer II by the hydrazine sulfate reduction procedure (Kamphake et al. 1967).

RESULTS

Nitrate and nitrite were monitored throughout the experiment. Normally, the acetylene block approach is applied to verify denitrification, but it cannot be used in this case, as it has been shown to interfere with the biodegradation of the aromatics (Hutchins et al. 1992, Jørgensen et al. 1992). It was not possible to make precise electron balances, because mineralization and incorporation data were not provided. However, electron balances for this culture have been calculated with toluene as the only carbon and energy source in a previous study showing a good agreement between nitrate reduction and toluene degradation (Jørgensen et al. 1992).

In the mixed substrate experiments, the compounds could be divided up in four groups according to their degradation response. The compounds in group one — toluene, phenol, 2,4-dimethylphenol, *m*-cresol, and *p*-cresol — were degraded without any observed lag phase. The compounds in a second group — *o*-xylene, butylbenzene, propylbenzene, cumol, and *o*-cresol — were dependent on a primary substrate, toluene, ethylbenzene, or phenol, to be degraded. The compounds 3,5-dimethylphenol and 2,4,6-trimethylphenol, the third group, seemed to lose degradability after an initial slow degradation. The compounds in the last group — benzene, 1,4-diethylbenzene, naphthalene, 1-methylnaphthalene, 1,4-methylnaphthalene, biphenyl, phenanthrene, 2,5-dimethylphenol, 2,6-dimethylphenol, and the NSO-substituted compounds furane, fluorenone, and benzothiophene — were not degraded (Table 2).

The dependency of a primary substrate observed for some of the compounds was either a concomitant degradation, as in the case of toluene and *o*-xylene, or the degradation of the primary substrate before the secondary substrate could be attacked, see Figure 1.

The factorial design experiment with the same mixture of compounds except the phenols was set up to identify the compounds that could support growth measured as ATP content. It confirmed the findings of the first batch tests in terms of degradability of the different compounds. The experiment proved that only degradation of ethylbenzene and toluene resulted in an increase of ATP content. The factorial experiment also showed that the addition of *o*-xylene and

TABLE 2. Aromatic compounds degraded by the mixed culture.

Compound	Removal rate (%/days)	Lag phase (days)	Degradation Interactions
benzene	0/210		
toluene	100/6	2	
o-xylene	100/66	2	Cometabolism with toluene
m-xylene	100/9	3	
p-xylene	100/9	3	
ethylbenzene	100/13	7	
1,4-diethylbenzene	0/210		
butylbenzene	100/26	6	Dependent on toluene
propylbenzene	100/19	12	Initially dependent on ethylbenzene
phenol	100/5	2	
o-cresol	62/13	2	Initially dependent on phenol
m-cresol	100/8	3	
p-cresol	100/8	3	
2,4-dimethylphenol	100/16	4	
2,5-dimethylphenol	0/210		
2,6-dimethylphenol	0/210		
3,5-dimethylphenol	78/45	2	Degradation stopped
2,4,6-trimethylphenol	33/24	2	Degradation stopped
naphthalene	0/210		
1-methylnaphthalene	0/210		
1,4-methylnaphthalene	0/210		
biphenyl	0/210		
phenanthrene	0/210		
furane	0/210		
fluorenone	0/210		
benzothiophene	0/210		

the hydrocarbon mixture inhibited ATP production. Benzene did not influence ATP production.

DISCUSSION

The mixed culture tested in this study could degrade a wide range of aromatic hydrocarbons and phenols with nitrate as electron acceptor. Different compounds, however, were not degraded simultaneously, and the culture exhibited a clear preference for the compounds in the above-mentioned group one. Not surprisingly, toluene was quite easily degraded, which could be due to the fact that the culture had been enriched on toluene but it also confirmed earlier studies showing that toluene is more readily biodegraded than the other BTEXs (Evans et al. 1991, Kuhn et al. 1998). Among the aromatic hydrocarbons, toluene and ethylbenzene were the only compounds that could support growth of the mixed culture. Availability

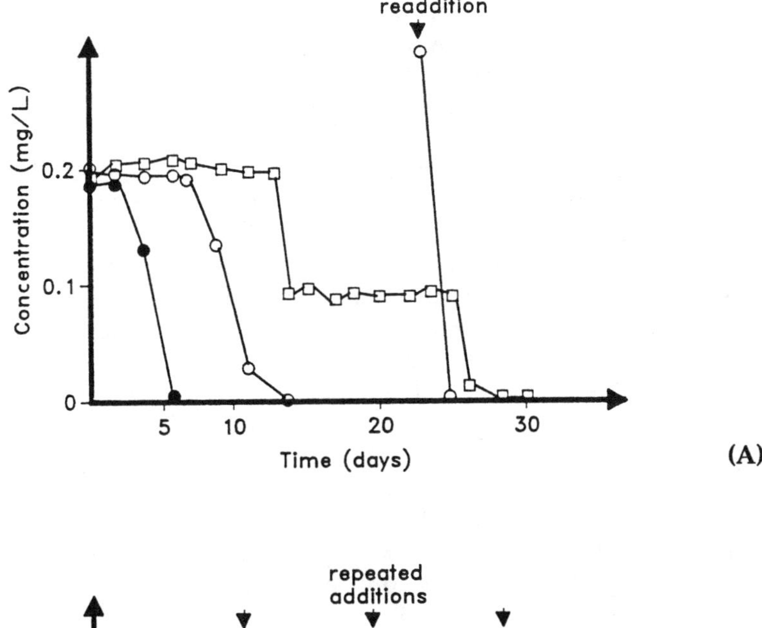

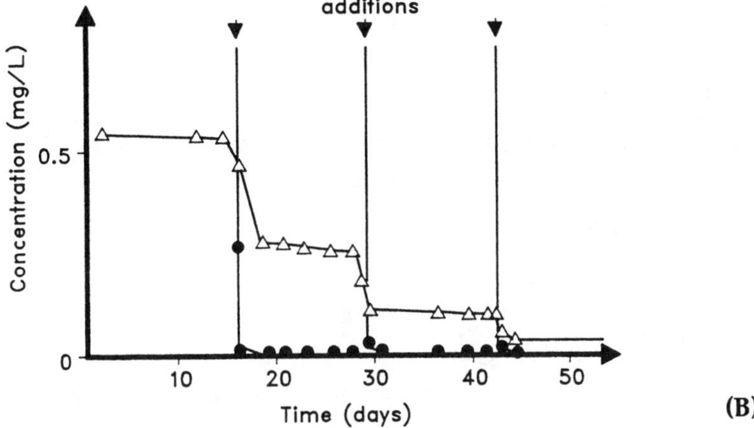

FIGURE 1. Types of primary substrate dependency. A: Preferential degradation of compounds combined with primary substrate dependency. B: Cometabolism. Toluene ●, ethylbenzene O, propylbenzene □, and *o*-xylene ▵.

of these compounds was required for the others to be degraded. In some cases, they served as primary substrates in a cometabolic relationship, i.e., with *o*-xylene (Figure 1A), a phenomenon that has been shown previously (Evans et al. 1991, Jensen et al. 1988). In other cases, these compounds had to be degraded first before other compounds were attacked, and a distinct sequence of degradation could be observed (Figure 1B). The explanation for this phenomenon might be that only the primary substrates are able to induce the production of the proper enzymes, and at the same time, they have a much higher affinity for these enzymes.

Of the dimethylphenols, only 2,4-dimethylphenol and 3,5-dimethylphenol could be attacked by this culture. These results are only partly in agreement with the findings of Flyvbjerg et al. (1993). They found that only the dimethylphenols with the methyl group in the para-position could be degraded, and they did not observe any loss of degradation ability.

Although most of the aromatics could be degraded by the mixed culture, benzene, some of the methylated phenols, and a number of NSO-compounds will persist in the anaerobic environment. The compounds that were shown to be able to support growth of the specific degrading microorganisms also were the most rapidly degraded. When these compounds are depleted, no substrates will be available to sustain the specific degrading population and thereby ensure degradation of the more persistent compounds.

The findings on preferential degradation of the contaminants lead to the conclusion that nitrate limitation also might result in persistence of compounds that themselves are degraded under denitrifying conditions.

Although it is not always justified to extrapolate from laboratory studies to field performances, the study showed that mixtures of creosote or coal for contaminants cannot be treated with bioremediation only on the basis of anaerobic degradation with nitrate as the only electron acceptor.

REFERENCES

Alvarez, P. J., and T. M. Vogel. 1991. "Substrate interactions of benzene, toluene and para-xylene during microbial degradation by pure cultures and mixed culture aquifer slurries." *Appl. Environ. Microbiol. 57*: 2981.

Arvin, E., B. K. Jensen, and A. T. Gundersen. 1989. "Substrate interactions during aerobic biodegradation of aromatic hydrocarbons and pyrrole." *Appl. Environ. Microbiol. 55*: 3221.

Barbaro, J. R., J. F. Barker, L. A. Lemon, and C. I. Mayfield. 1992. "Biotransformation of BTEX under anaerobic, denitrifying conditions: field and laboratory observations." *J. Contam. Hydrol.* (In press).

Evans, P. J., D. T. Mang, and L. Y. Young. 1991. "Degradation of toluene and *m*-xylene and transformation of *o*-xylene by denitrifying enrichment cultures." *Appl. Environ. Microbiol. 57*: 450.

Flyvbjerg, J., E. Arvin, B. K. Jensen, and S. K. Olsen. 1993. "Microbial degradation of phenols and aromatic hydrocarbons in creosote-contaminated groundwater under nitrate-reducing conditions." *J. Contam. Hydrol. 12*: 133.

Hutchins, S. R. 1991. "Biodegradation of monoaromatic hydrocarbons by aquifer microorganisms using oxygen, nitrate, or nitrous oxide as the terminal electron acceptor." *Appl. Environ. Microbiol.* 57: 2403.

Hutchins, S. R. 1992. "Inhibition of alkylbenzene biodegradation under denitrifying conditions by using the acetylene block technique." *Appl. Environ. Microbiol. 58*: 3395.

Hwang, H.-M., R. E. Hodson, and D. L. Lewis. 1989. "Assessing interactions of organic compounds during biodegradation of complex waste mixtures by naturally occurring bacterial assemblages." *Environ. Toxicol. Chem. 8*: 209.

Jensen, B. K., E. Arvin, and A. T. Gundersen. 1988. "Biodegradation of phenolic compounds and monoaromatic hydrocarbons by a mixed wastewater culture under denitrifying conditions." In *Proceedings from the COST 641/681 Workshop on Organic Contamination in*

Wastewater, Sludge and Sediments: Occurrence, Fate and Disposal, Brussels, Belgium, October 26-27, 1988.

Jørgensen, C., J. Flyvbjerg, E. Arvin, and B. K. Jensen. 1992. "Stoichiometry and kinetics of microbial toluene degradation under denitrifying conditions." Submitted for publication.

Jørgensen, C., E. Mortensen, B. K. Jensen, and E. Arvin. 1991. "Biodegradation of toluene by a denitrifying enrichment culture." In R. E. Hinchee and R. F. Olfenbuttel (Eds.), *In Situ Bioreclamation*, pp. 480-486. Butterworth-Heinemann, Stoneham, MA.

Kamphake, L. J., S. A. Hannah, and J. M. Cohen. 1967. "Automated analysis for nitrate by hydrazine reduction." *Water Res. 1*: 205.

Kuhn, E. P., J. Zeyer, P. Eicher, and R. P. Schwarzenbach. 1988. "Anaerobic degradation of alkylated benzenes in denitrifying laboratory aquifer columns." *Appl. Environ. Microbiol. 54*: 490.

Leach, F. 1981. "ATP determination with firefly luciferase." *J. Appl. Biochem. 3*: 473.

Major, D. W., C. I. Mayfield, and J. F. Barker. 1988. "Biotransformation of benzene by denitrification in aquifer sand." *Ground Water 26*: 8.

BACTERIAL DEGRADATION UNDER IRON-REDUCING CONDITIONS

H.-J. Albrechtsen

INTRODUCTION

The iron content of aerobic aquifer sediment is often relatively high (200 to 4,500 mg Fe/kg sediment). Iron oxides are not very soluble and are therefore associated with the sediment. Solid iron oxides cannot be transported by water and therefore differ significantly from other inorganic electron acceptors (oxygen, nitrate, sulfate), which can all be transported by water. Although only minimal attention has been paid to iron reduction in aquifer sediments, the high iron oxide content might prove to be very important as a redox buffer and electron acceptor for microbial processes if the sediment is polluted by organic compounds. Iron oxides also might be important electron acceptors for microbial degradation of xenobiotic compounds. Lovley et al. (1989) reported the bacterial degradation of toluene, phenol, and *p*-cresol by a pure culture (GS-15) isolated from a river sediment.

In an aquifer polluted by landfill leachate (Vejen, Denmark), a zone has been characterized by high content of soluble iron (Fe[II]) and no content of oxygen, nitrate, sulfide, or methane (Lyngkilde & Christensen 1992a). In this zone the concentration of organic matter and organic contaminants decreased rapidly and is therefore considered as very important for the degradation of the leachate from the landfill (Lyngkilde & Christensen 1992b).

The purposes of this investigation were to verify that the observed iron reduction in a landfill leachate plume was microbiologically mediated and to investigate the importance of pH and the type of carbon source and iron oxides present.

MATERIALS AND METHODS

All water and sediment samples were collected from the aquifer downstream from the Vejen landfill (Lyngkilde & Christensen 1992a). A long-duration experiment was set up in reactors (10 L) with a suspension of water collected from the leachate plume (sterilized by filtering) and inoculated with the fines fraction (clay and silt) extracted from sediment from the iron reducing zone. Synthetic amorphous iron oxides, prepared as described by Lovley & Phillips (1986) (final concentration: approx. 220 mg/L Fe[III]), were added to serve as the electron acceptor. An abiotic control (added 2 g/L of sodium azide) and a control without groundwater but with Millipore-filtered water were run in duplicate. The experiment has as of June 1992 run for more than 450 days (for further details see Albrechtsen et al. 1994). The concentrations of specific organic compounds (benzene, ethylbenzene, 1,3,5-trimethylbenzene, 1-propenylbenzene, toluene, 2-ethyltoluene, *p/m*-xylene, *o*-xylene, naphthalene, 2-methylnaphthalene, 1-methylnaphthalene, camphor = (1,7,7-trimethylbicyclo-(2,2,1)-heptane-2-one)) actually present in the leachate were determined by gas chromatographic/mass spectroscopic (GC/MS) analysis as described by Lyngkilde & Christensen (1992b) during the experimental period.

For use in the pH experiment, an artificial oligotrophic anaerobic medium (OAM) containing (g/L) yeast extract (0.25), Bacto Tryptone (Difco) (0.25), NH_4Cl (1.0), $K_2HPO_4{\cdot}3H_2O$ (0.04), $MgCl_2{\cdot}6H_2O$ (0.1), $CaCl_2{\cdot}2H_2O$ (0.05), vitamins, and trace metals was prepared and supplied with synthetic amorphous iron oxides (approx. 140 mg/L Fe[III]). The pH was buffered by a carbonate buffer and adjusted to pH of 5.0, 5.5, 6.0, 6.5, and 7.0, all in triplicate. The bottles were inoculated with mixed cultures of iron reducing bacteria obtained from the Vejen aquifer.

All manipulations were carried out using anaerobic and aseptic techniques, and all incubations were carried out at the actual groundwater temperature (10°C) with a headspace in the bottles of N_2/CO_2 (80/20%). For determination of Fe[II], subsamples of approx. 1 mL were collected with a syringe, immediately filtered (0.45 µm, Minisart SRP 15, Sartorius), and diluted in HNO_3 solution (25 mL conc. NHO_3 was dissolved in 2.5 L water after addition of 50 mL of 26.6% w/v LaCl [BDH Chemicals Ltd. Poole, England] with 32 g/L of CsCl). Fe[II] was quantified by a Perkin-Elmer 370® atomic absorption spectrophotometer at 247.9 nm.

RESULTS AND DISCUSSION

Ferric iron is found as many different oxides in aquifer sediments, such as amorphous iron, hematite (α-Fe_2O_3), and goethite (α-FeOOH). The stability of the different oxides varies, and the availability for microbial iron reduction is expected to be related to the stability of the iron oxide. Synthetic amorphous iron oxides (Figure 1), which in our case consist mainly of akageneite, and aquifer sediment rich in ferric iron (colored red) were tested as electron acceptor sources, and significant iron reduction was observed with both types of iron.

Because pH is important for the solubility of ferric iron, the importance of pH in microbial iron reduction was tested. The onset of the Fe[II]-evolution started earlier, and the rate was faster around pH 5.5 to 6.0, indicating that the optimum pH for iron reduction was in this range (Figure 2). This optimum pH range also has been observed in two other experiments, inoculated with other cultures. At the end of the experiment, nearly all the iron oxide in the bottles was reduced to Fe[II]. One bottle from each condition was sacrificed, and the final pH was measured. The final pH in all the bottles was approximately 6.0, which also indicates that this was the optimum pH. The solubility of Fe[III] increases with decreasing pH, and if the availability of Fe[III] depends of the solute concentration, an optimum iron reduction rate was expected at low pH. The fact that the optimum was

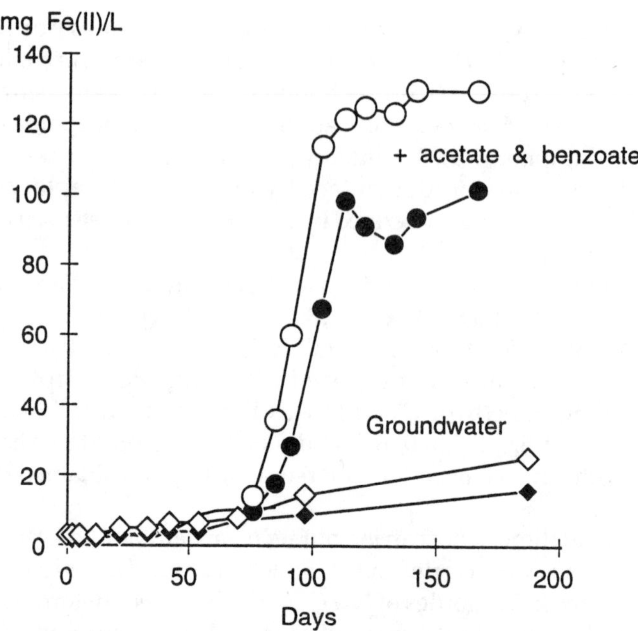

FIGURE 1. Iron reduction (measured as Fe[II] evolution) in a suspension of leachate-polluted groundwater, the fine fraction of the sediment from the iron-reducing zone, and amorphous iron oxides, shown for two replicates (◊, ♦). During the long-duration experiment, acetate and benzoate were added to subsamples (O, ●).

observed at a higher pH indicates that not only the iron reduction, but also the solubilization of precipitated Fe[III] is microbiologically mediated. The pH of the aquifer in the iron-reducing zone (measured in water samples) was between 4.9 and 7.9, with an average of 6.38 (Lyngkilde & Christensen 1992a).

By investigating different carbon sources, it was shown that landfill leachate-polluted groundwater could be used for iron reduction (with a rate of 0.13 mg Fe/L·day) (Figure 1). However, when subsamples were collected from the reactors, amended with acetate and benzoate, and incubated separately, the iron reduction was much faster (e.g., 0.36 mg Fe/L·day). No effect was observed the first two times this experiment was carried out. When cultures were transferred to OAM-medium (with tryptone/yeast extract), an even higher activity was obtained (e.g., 0.84 mg Fe/L·day). The increased iron reduction rate could be explained partly by enrichment and the addition of nutrients, but it also suggests that the iron reduction could be limited by lack of adequate carbon and nutrient sources in the leachate plume.

Microbial iron reduction was demonstrated (Figure 1) by long-duration incubations (up to 450 days) of leachate-polluted groundwater enriched with amorphous iron oxides and without oxygen or nitrate. Under these conditions, toluene also was degraded (Table 1). After lag periods of 10 to 130 days, toluene

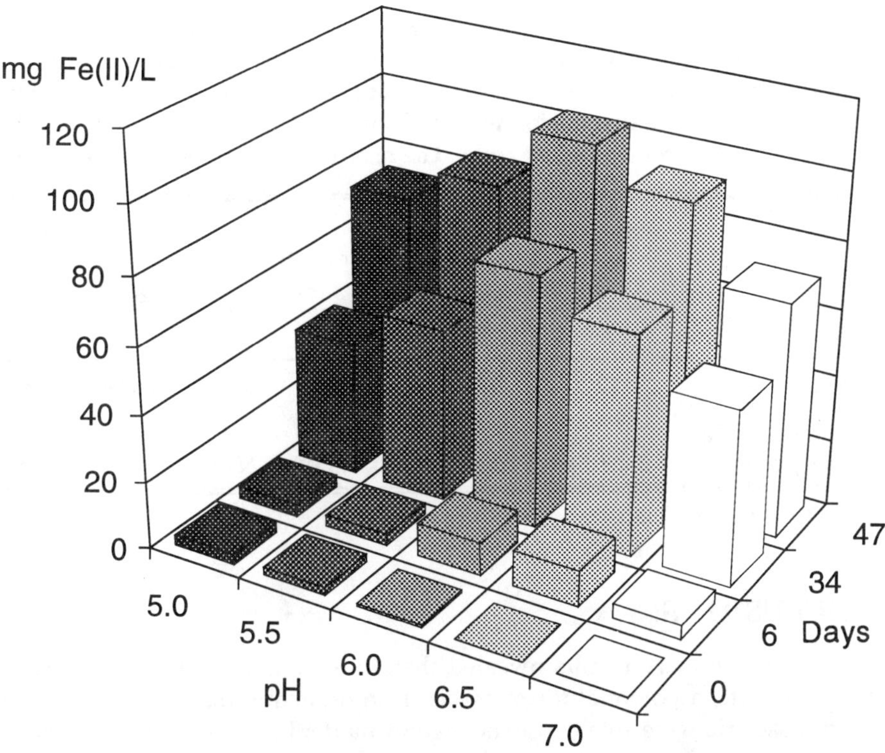

FIGURE 2. Effect of pH on the iron reduction (measured as Fe[II] evolution) in artificial OAM-medium with synthetic amorphous iron oxides as electron acceptor. The medium was inoculated with a mixed culture obtained from an iron-reducing aquifer sediment. The values are averages of triplicate.

(200 μg/L) was completely degraded during 70 to 100 days at a rate of 3.4 to 4.2 μg/(L·day) of toluene. The long lag period may indicate sequential degradation in which other carbon sources are used before the toluene is degraded. In the controls (sodium azide added), the toluene concentration remained constant. None of the other investigated 11 xenobiotic organic compounds observed in the leachate was degraded.

In the controls using artificial groundwater with no other dissolved organic carbon than that dissolved from the sediment particles, no iron reduction was observed, and the toluene concentrations remained constant. Addition of acetate and benzoate to subsamples did not result in any iron reduction potential. It is possible that the bacteria must colonize the particles of amorphous iron oxides before they are able to use the insoluble iron oxides as electron acceptors, and that they require some dissolved organic carbon for growth during this colonization. Although the bacteria may be able to use the organic matter attached to the surfaces of the sediment particles, these surface environments may be poor in available iron oxides.

TABLE 1. Long-duration experiment with degradation of toluene (the initial concentration was approx. 180 µg/L) in leachate-polluted groundwater under iron-reducing conditions with synthetic amorphous iron oxides as the electron acceptor. The suspension was inoculated with the fine fraction of sediment from the iron-reducing zone. The results are given for two replicates.

	Groundwater		Groundwater + azide		Artificial C-free groundwater	
	A	B	A	B	A	B
Lag-period, days	12	126	>380	>380	>380	>380
Maximum degradation rate, µg/(L·day)	3.4 ±0.1	4.2 ±0.7	N.O.[(a)]	N.O.	N.O.	N.O.

(a) N.O. = Not observed.

CONCLUSIONS

These preliminary results revealed that, in an aquifer contaminated with landfill leachate, a potential for microbial iron reduction may exist. The iron reducing bacteria were able to use the organic matter in the leachate as a carbon source and the iron oxides present in the aquifer as an electron acceptor to degrade toluene. It also seems that other carbon sources are more easily degradable than the organic matter present in the leachate. These experiments constitute the first phase in a research program investigating the significance of microbial degradation of organic pollutants in iron reducing aquifer zones.

ACKNOWLEDGMENTS

Mona Refstrup, Lajla Olsen, and Gitte Brandt are gratefully acknowledged for their technical assistance. This work was a part of a major research program focusing on the effects of waste disposal on groundwater. The program is funded by the Danish Technical Research Council, the Technical University of Denmark, and the Commission of the European Communities.

REFERENCES

Albrechtsen, H.-J., J. Lyngkilde, C. Grøn, and T. H. Christensen. 1994. "Landfill leachate-polluted groundwater evaluated as substrate for microbial degradation under different redox conditions." In R. E. Hinchee, D. B. Anderson, F. B. Metting, Jr., and G. D. Sayles (Eds.), *Applied Biotechnology for Site Remediation.* Lewis Publishers, Ann Arbor, MI.

Lovley, D. R., M. J. Baedecker, D. J. Lonergan, I. M. Cozzarelli, E. J. P. Phillips, and D. I. Siegel. 1989. "Oxidation of aromatic contaminants coupled to microbial iron reduction." *Nature 339:* 297-301.

Lovley, D. R., and E. J. P. Phillips. 1986. "Availability of ferric iron for microbially reducible ferric iron in bottom sediments of the freshwater tidal Potomac River." *Applied Environmental Microbiology 52:* 751-757.

Lyngkilde, J., and T. H. Christensen. 1992a. "Redox zones of a landfill leachate pollution plume (Vejen, Denmark)." *Journal of Contaminant Hydrology 10:* 273-289.

Lyngkilde, J., and T. H. Christensen. 1992b. "Fate of organic contaminants in the redox zones of a landfill leachate pollution plume (Vejen, Denmark)." *Journal of Contaminant Hydrology 10:* 291-307.

LABORATORY FEASIBILITY STUDY FOR BIOREMEDIATION OF A KEROSENE-CONTAMINATED SITE

M. E. Watwood and D. L. Carr

INTRODUCTION

A kerosene leak from a 40-year-old underground storage tank was discovered at a business site in Albuquerque, New Mexico. The underlying vadose zone was contaminated with up to 21,000 mg/kg total petroleum hydrocarbon (TPH). Subsurface material consisted primarily of medium to coarse-grained sand. Depth to groundwater was approximately 12 m. Subsurface oxygen levels were moderate to low, but not depleted. The State Environmental Improvement Division (EID) decided that the vadose zone should be remediated; 50 mg/kg TPH was approved as the cleanup target.

A laboratory study was conducted to assess the feasibility of bioremediating this site. Kerosene, which has a chromatographic profile similar to that of diesel fuel (Hazard et al. 1991), was considered to be a likely candidate for bioremediation if site-specific nutrient and peroxide recommendations could be made (Lapinskas 1989, Leahy & Colwell 1990, Nyer & Skladany 1989).

Soil samples were collected for microbial analysis from three depths within the contaminated zone; a sample from a nearby pristine site was analyzed for comparative purposes. Samples were placed in sterile gastight containers and transported to the laboratory, where they were maintained in the dark at 4°C prior to analysis.

EXPERIMENTAL DESIGN

Plating was used to estimate the size of aerobic heterotrophic and kerosene-degrading microbial populations in the vadose zone material at the site. The effects of nutrient and hydrogen peroxide additions on these bacterial population sizes also were assessed. For counting heterotrophic microorganisms, standard nutrient agar (NA) and a soil extract agar (SE) were used. NA is a relatively rich medium on which many subsurface microorganisms grow readily. SE, prepared from site subsurface material, contains nutrients in relatively low concentrations and mimics oligotrophic subsurface conditions.

To enumerate kerosene degraders, a mineral salts-kerosene agar (MSK) was used. Mineral salts agar contains all of the essential macro- and micronutrients without any organic carbon source. Inoculated plates were incubated in the presence of kerosene fumes. The kerosene fumes provide the sole organic carbon source; bacteria that form colonies on this medium are capable of degrading the contaminant as the sole carbon source for growth. Control plates were incubated without exposure to kerosene fumes.

To assess the effects of nutrient amendments on heterotrophic and kerosene-degrading populations, nutrients (final concentration 0.2%, weight/volume) were added to sterile soil extract broth. Nutrients included (1) ammonium nitrate, (2) mono- and di-basic phosphate salts, (3) ammonium nitrate/phosphate salts, and (4) ammonium sulfate/ammonium nitrate/phosphate salts. The flasks were inoculated with subsurface material and shaken at 15°C for 48 hours. After incubation, aliquots were counted by plating onto triplicate NA and MSK plates. (NA was the sole heterotrophic medium because preliminary results showed no difference between NA and SE counts.)

To assess the effects of peroxide on microbial populations, experiments were conducted as described above, substituting hydrogen peroxide for nutrients. In these experiments a preincubation period of 48 hours with the optimum nutrient (0.2% ammonium nitrate, mono- and di-basic phosphate salts, and ammonium sulfate) was performed prior to peroxide addition to increase the robustness of the bacterial population. Peroxide concentrations of 50, 100, 500, and 1,000 mg/mL were tested.

Miniature column experiments were used to determine the effects of nutrient and peroxide additions on kerosene removal from contaminated subsurface material. Composite material from the contaminated subsurface was packed into 40-mL screw top vials to approximately the same density found in the field. Nutrient and peroxide solutions (5 mL) were injected into the butts of the tubes, which were then inverted. The liquid was allowed to drain downward through the tubes and was collected as effluent. The injections were repeated for 6 consecutive days. Following treatment, tubes were resealed and tested for TPH by EPA Method 8015 by Analytical Technologies, Inc. Effluent samples were also analyzed for TPH.

FEASIBILITY STUDY RESULTS

Approximately 5×10^7 colony-forming units (CFUs)/g total heterotrophs and 4×10^5 CFUs/g kerosene degraders were cultured from the contaminated material. Standard errors calculated for various sample depths were large and ranged from 22 to 58% of the mean for heterotrophs and from 18 to 52% of the mean for kerosene degraders (data not shown). Heterotrophic counts on NA and SE agars did not differ, and counts from contaminated soil were not different from those obtained from pristine samples.

Counts of heterotrophs and kerosene degraders following batch culture exposure to various nutrient amendments are shown in Table 1. Data shown represent average values calculated by pooling data derived from each of the three sample depths. Unamended counts were derived from batch cultures that received no extra nutrients. For total heterotrophs as well as kerosene degraders, amendment with combined nitrogen/phosphate (or) nitrogen/phosphate/sulfate

TABLE 1. Heterotrophic and kerosene-degrading microorganisms cultured on nutrient agar following supplementation with nutrient amendments.

		Amendments			
	Unamended	Mono- and di-basic phosphate salts	Ammonium nitrate	Ammonium nitrate + phosphate salts	Ammonium nitrate + phosphate salts + ammonium sulfate
Population/sample			CFU/g (percent error)		
Heterotrophs/ contaminated	3.91×10^9 (30.7 %)	4.95×10^9 (20.4 %)	5.46×10^9 (21.8 %)	9.28×10^9 (20.0 %)	7.20×10^9 (27.6 %)
Heterotrophs/ pristine	4.90×10^9 (27.7 %)	4.88×10^9 (21.8 %)	3.65×10^9 (28.3 %)	9.33×10^9 (15.6 %)	1.91×10^{10} (15.7 %)
Kerosene degrad-ers/contaminated	4.16×10^6 (23.8 %)	3.89×10^6 (41.5 %)	8.30×10^6 (36.9 %)	1.16×10^7 (35.6 %)	1.30×10^7 (28.1 %)
Kerosene degrad-ers/pristine	2.74×10^7 (8.8 %)	1.05×10^7 (15.6 %)	2.91×10^7 (5.4 %)	2.40×10^7 (19.6 %)	1.59×10^7 (26.2 %)

resulted in the largest population increases. Amendment with nitrogen or phosphate alone had no significant effect on population sizes.

Results from treatment with hydrogen peroxide (Table 2) indicate that heterotrophs were able to tolerate peroxide concentrations up to 1,000 mg/L without suffering significant population decreases. However, kerosene-degrading populations exhibited a significant decrease in response to any peroxide addition. There was no difference in the response to varying peroxide concentrations; a decrease of approximately an order of magnitude occurred in response to peroxide for all concentrations tested.

Results from the miniature column experiments are shown in Table 3. Treatments included a high (1.0%) and low (0.2%) nitrogen/phosphate nutrient amendment with peroxide treatments of 100 and 500 mg/L. The difference between the untreated sample and the control sample treated with deionized water represents kerosene loss due to volatilization and biodegradation without nutrient amendment. This effect would be simply in response to soil wetting in the field. Negligible TPH was measured in effluent samples, indicating that leaching did not contribute to kerosene removal. A mass balance based on these controls indicated that TPH loss could be attributed to volatilization or biodegradation. The difference between results obtained after injecting treatment solutions and those obtained after injecting deionized water represents kerosene loss due to biodegradation enhanced by treatment. Results from the samples treated with nutrients and/or peroxide indicated that nutrient treatment alone at either concentration or peroxide alone at 100 mg/L resulted in the largest removal of kerosene from the subsurface material. Higher concentration of peroxide or any combination of peroxide and nutrient resulted in significantly less kerosene removal.

SUMMARY AND RECOMMENDATIONS

There are substantial populations of both general heterotrophs and kerosene-degrading microorganisms present within the contaminated soil at this site. Both populations exhibit similar levels in contaminated and pristine samples. The trend of the plating experiment results is that both general heterotrophs and kerosene-degrading populations respond positively to nutrient additions combining nitrogen and phosphate. While general heterotrophs tolerate treatment with moderate levels of hydrogen peroxide, kerosene degraders are inhibited by peroxide application, even at relatively low concentration. Column study results indicate that nutrient application alone is more effective than a nutrient/peroxide combination at stimulating degradation of kerosene.

It was recommended that a solution of approximately 1.0% (weight/volume) of nitrogen/phosphate be applied as a nutrient amendment and that hydrogen peroxide not be used to treat the contaminated soil. If oxygen depletion in the treated material had become a serious problem, an alternative means of oxygen delivery would have been considered.

TABLE 2. Effect of hydrogen peroxide treatment on heterotrophic and kerosene-degrading microbial populations.

Population/sample	Untreated	50 mg/L H_2O_2	100 mg/L H_2O_2	500 mg/L H_2O_2	1,000 mg/L H_2O_2
		Treatments			
	CFU/g (% error)				
Heterotrophs/contaminated	4.39×10^7 (25.6 %)	2.79×10^7 (31.5 %)	5.05×10^7 (26.5 %)	3.50×10^7 (31.3 %)	3.86×10^7 (23.7 %)
Heterotrophic/pristine	2.91×10^7 (24.4 %)	1.89×10^8 (20.2 %)	2.77×10^7 (48.0 %)	1.25×10^8 (6.4 %)	3.30×10^7 (20.2 %)
Kerosene degraders/contaminated	4.28×10^5 (18.9 %)	1.94×10^4 (31.0 %)	4.86×10^4 (20.1 %)	2.03×10^4 (32.0 %)	6.56×10^3 (47.8 %)
Kerosene degraders/pristine	6.57×10^5 (22.5 %)	1.73×10^4 (71.2 %)	2.80×10^4 (20.6 %)	2.20×10^4 (50.1 %)	1.00×10^4 (52.9 %)

TABLE 3. Total petroleum hydrocarbon (TPH) in contaminated soil (depth composite) following treatment in miniature column.

Treatment	Average TPH (mg/kg)	Std Dev	Std Error	% Error	% Initial TPH
Untreated	446.7	15.3	8.8	2	100
Deionized Water	306.7	51.3	29.6	10	69
0.2% Nutrient	108.7	27.2	15.7	14	24
1.0% Nutrient	86.7	20.6	11.9	14	19
100 mg/L H_2O_2	145.3	68.9	39.8	27	33
500 mg/L H_2O_2	346.7	51.3	29.6	9	78
0.2% Nutrient + 100 mg/L H_2O_2	413.3	56.9	32.8	8	93
0.2% Nutrient + 500 mg/L H_2O_2	330.0	104.4	60.3	18	74
1.0% Nutrient + 100 mg/L H_2O_2	293.3	5.8	3.3	1	66
1.0% Nutrient + 500 mg/L H_2O_2	410.7	75.5	43.6	10	92

SOIL BIOREMEDIATION

An infiltration gallery type biotreatment system was designed to deliver nutrients within the contaminated vadose zone. Water was pumped to the surface from a downgradient monitor well and was passed through a series of filters to remove solids and hydrocarbon. Nutrients were added to filtered water as recommended by the laboratory feasibility study. The nutrient solution moved into the treatment zone via the infiltration gallery at 1.9 L/min. Treatment involved three separate applications (1,520 L of solution per application) separated by 6-month intervals.

Following the last application, soil borings were made within the formerly contaminated zone. The borings were drilled and continuously sampled to a depth of 11.6 m. Representative samples were collected and analyzed (method 8015) for TPH and BTEX. None of the samples exceeded the regulatory goal of 50 mg/kg TPH; in most cases the levels were substantially lower than 10 mg/kg. Based on these results, site closure was approved by the EID.

REFERENCES

Hazard, S. A., J. L. Brown, and W. R. Betz. 1991. "Extraction and Analysis of Hydrocarbons Associated with Leaking Underground Storage Tanks." *LC-GC* 9:38-44.

Lapinskas, J. 1989. "Bacterial Degradation of Hydrocarbon Contamination in Soil and Groundwater." *Chemistry and Industry December*:784-789.

Leahy, J. G., and R. R. Colwell. 1990. "Microbial Degradation of Hydrocarbons in the Environment." *Microbiological Reviews 54*:305-315.

Nyer, E. K., and G. J. Skladany. 1989. "Relating the Physical and Chemical Properties of Petroleum Hydrocarbons to Soil and Aquifer Remediation." *Ground Water Monitoring Review Winter*:54-60.

DISTRIBUTION OF NONAQUEOUS PHASE LIQUID IN A LAYERED SANDY AQUIFER

C. D. Johnston and B. M. Patterson

INTRODUCTION

A site contaminated by diesel and kerosene was investigated to anticipate the effects of heterogeneous subsurface conditions on the movement and degradation of residual nonaqueous phase liquid (NAPL) during in situ bioremediation. Contamination has resulted from spills of diesel and kerosene onto the ground and subsequent infiltration over a period of many years. Of particular interest is the variability and layering within the soil profile in relation to the NAPL distribution. The entrapment of NAPL in coarse-textured layers shown by Wilson et al. (1990) and Dawe et al. (1992) suggests that layering is important in understanding the vertical distribution of NAPL at a site where there are significant seasonal water table variations. Vertical mobility and probable entrapment of NAPL in these coarser-textured layers is also of concern where bioremediation involves manipulation of the water table. We present the results of the field sampling program, showing the distribution of NAPL over the field site in relation to subsurface stratigraphy. Laboratory measurements also were made on core samples to characterize water and NAPL retention and movement in the soil materials.

METHODS

The depth distribution of NAPL was determined by coring at 25 sites within a 0.7-ha area down gradient of the source of contamination (Figure 1). Core samples were collected with a hollow-stem auger drilling rig. Relatively undisturbed cores were collected by pushing a thin-walled, 50-mm-diameter aluminum tube, adapted as a piston sampler, ahead of the augers. The total NAPL content of subsamples of the cores was measured by extracting the wet soil with dichloromethane-acetone and analyzing the extract by gas chromatography with flame ionization detector (GC-FID). Water contents also were determined. NAPL contents are reported as mass of total petroleum hydrocarbons per dry mass of soil. Saturated hydraulic conductivity, K_{sat}, was measured on selected cores by applying a constant head and determining the head gradient along the core. Water-air saturation-pressure (S-P) relations were determined on 40-mm-long undisturbed cores placed in retention cells. Residual NAPL contents were determined by displacing water from a 40-mm, water-saturated core sample with NAPL and then

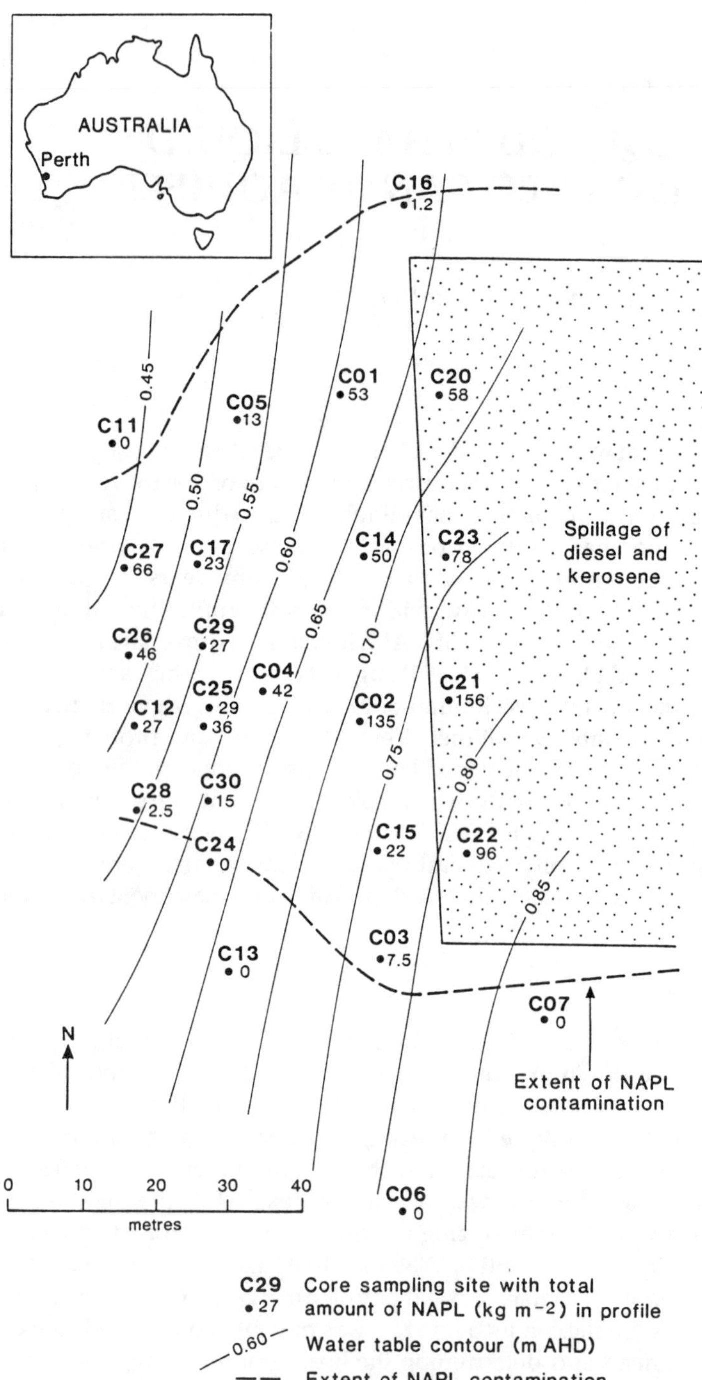

FIGURE 1. Location of sampling sites showing total amount of NAPL in profiles.

displacing the NAPL with water following the technique used by Wilson et al. (1990).

RESULTS AND DISCUSSION

The general stratigraphy at the Perth site is illustrated in Figure 2. Unconsolidated creamy tan dune sands (Safety Bay Sands) overlie the grey, calcareous Becher Sands (Semeniuk & Searle 1985), which in turn overlie the Tamala limestone. The ground surface is relatively level at an elevation of 4.6 m Australian Height Datum (AHD), although the source area of NAPL contamination is 0.1 to 0.4 m lower than the rest of the site. The top of the Becher Sands is generally at an elevation of 0.7 to 0.8 m AHD, although ranging from -0.27 to 1.13 m AHD, and is highest in the center of the contaminated area. Typically, the water table is encountered just below the top of the Becher Sands (Figure 2). The layering of fine and coarse sands in the vicinity of the water table is an important feature of the site. The depth distribution of particle sizes in the profile at site C07 (Figure 2) is typical. The layers are well differentiated and may be only a few tens of millimeters thick. Hydraulic properties such as K_{sat} vary markedly at scales of 50 to 100 mm. Measured K_{sat} values lie in the range 5 to 80 m/day for the Safety Bay Sands and 5 to 25 m/day for the Becher Sands. Air-water S-P relationships show a corresponding range of behavior. S-P relations for a range of materials are shown in Figure 3.

NAPL has moved at least 40 m west of the source of contamination (Figure 1). The spread of NAPL is generally consistent with the west-northwest direction of groundwater flow inferred from water table contours (Figure 1). However, the absence of NAPL at C11 suggests a more westerly migration of NAPL. The reason for this may be the extensive cementation within the Becher Sands. A partly cemented layer, 0.2 to 0.3 m thick, is found over most of the site at around 0.2 m AHD, and more extensive cementation is found below this layer at sites C11 and to a lesser extent C05, C26, and C27. Water table contours, however, do not indicate any change in aquifer conditions in the vicinity of C11.

Outside the area of spillage, NAPL thickness varied from 0.1 m to 1.3 m in the profile with a median value of 0.6 m. Most NAPL distributions have a well-defined peak within the top 0.1 to 0.2 m of the contaminated interval (see Figure 4). Maximum NAPL contents were mostly in the range 0.09 to 0.12 g/g, but three sites (C02, C26, and C27) had maxima in the range 0.23 to 0.26 g/g. C02 was the only site outside the area of spillage to have a measurable free NAPL layer in the borehole, but this layer has not exceeded 60 mm in thickness. NAPL contents in most of the profiles were 0.01 to 0.05 g/g over the lower portion of the contaminated interval (Figure 4). The interval of NAPL contamination coincided with the interval of annual water table fluctuations (annual amplitude about 0.5 m). The peaks in the NAPL distributions were at or immediately above the water table at the time of sampling. The distance of the peak above the water table is consistent with the S-P relationships appropriately scaled for interfacial tensions between NAPL, water, and air. In most cases the maximum NAPL content was at the same

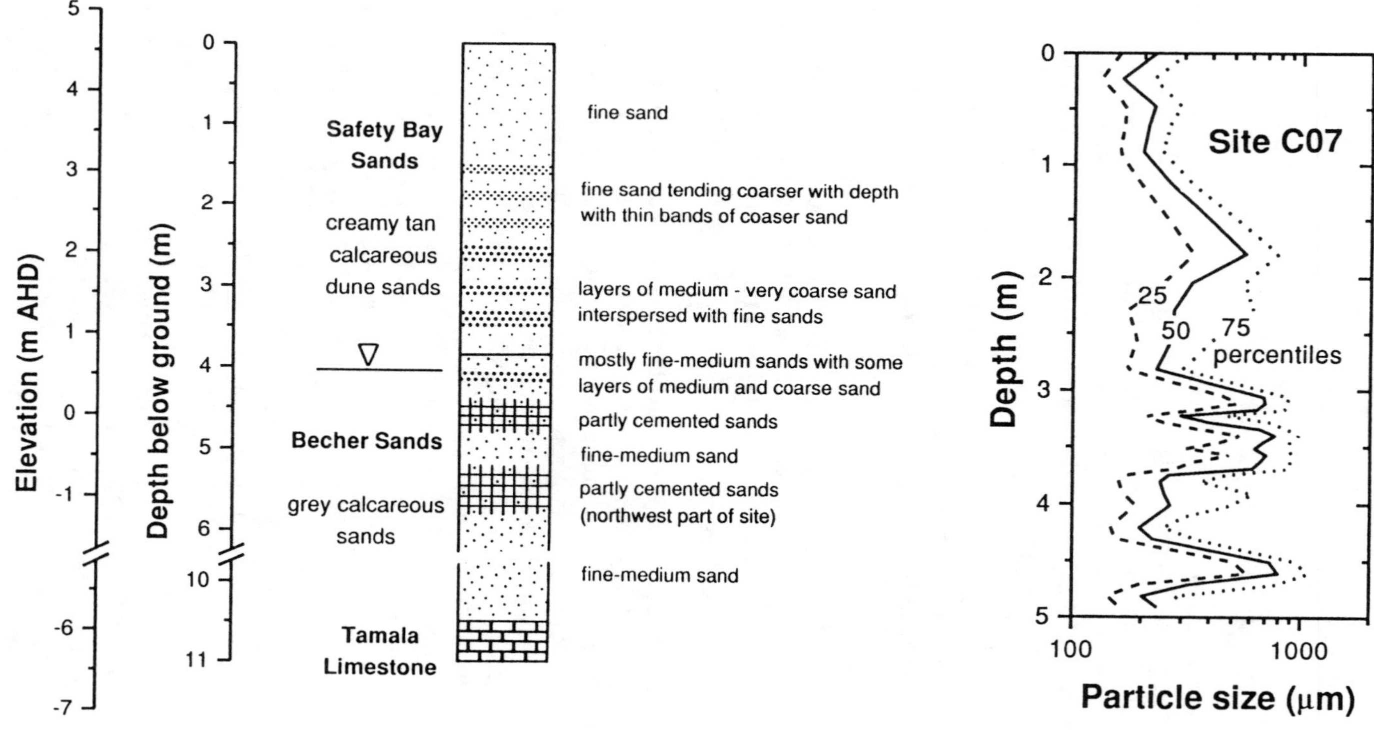

FIGURE 2. Generalized stratigraphy and particle-size distributions in the profile.

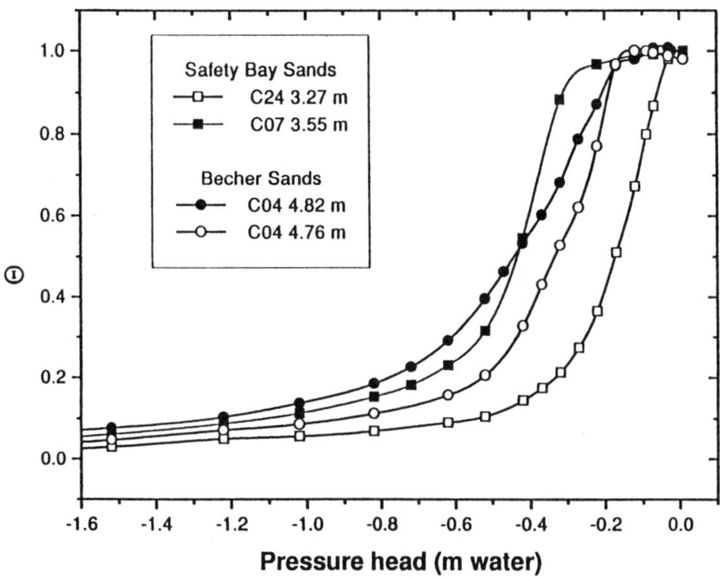

FIGURE 3. The main drainage water-air saturation-pressure relationship for samples of Safety Bay Sands and Becher Sands. Saturation given as Q = $(q-q_r)/(q_s-q_r)$, where q is volumetric water content, q_r is residual water content, and q_s is saturated water content.

elevation as the maximum water table. Hysteresis in the S-P relations (not shown in Figure 3) under rising water table conditions would account for the peak being at the maximum level of the water table. At several of the sites, the location and magnitude of the peak suggests NAPL may be locally mobile. NAPL distributions within the area of spillage are similar in the vicinity of the water table but additionally have NAPL throughout the unsaturated zone. Average NAPL content down to 3 m ranged from 0.007 g/g at C23 to 0.021 g/g at C21.

There was no clear association between NAPL contents and particle sizes within the profile. In about 70% of cases, the top of the NAPL interval was in fine-textured sands, whereas the lower part of the NAPL interval was almost always variably textured, characteristically with interbedded fine and coarse sand textured materials. Figure 4 shows the particle size distributions at four sites, illustrating complete NAPL saturation of a coarse layer at the water table (C02), coarse layers immediately above the NAPL interval (C04, C12, and C17), and coarse layers within the NAPL interval (C12 and C17). The NAPL was always found above the cemented layer in the Becher Sands. There was no evidence of high NAPL contents below the water table suggestive of entrapment in coarse layers. NAPL contents below the peak also are in agreement with the 0.02 to 0.05 g/g residual NAPL contents determined in the laboratory for two-phase water-NAPL systems. These are lower than the 0.05 to 0.09 g/g residual contents measured by Wilson et al. (1990). The lower contents seen in field profiles compared to

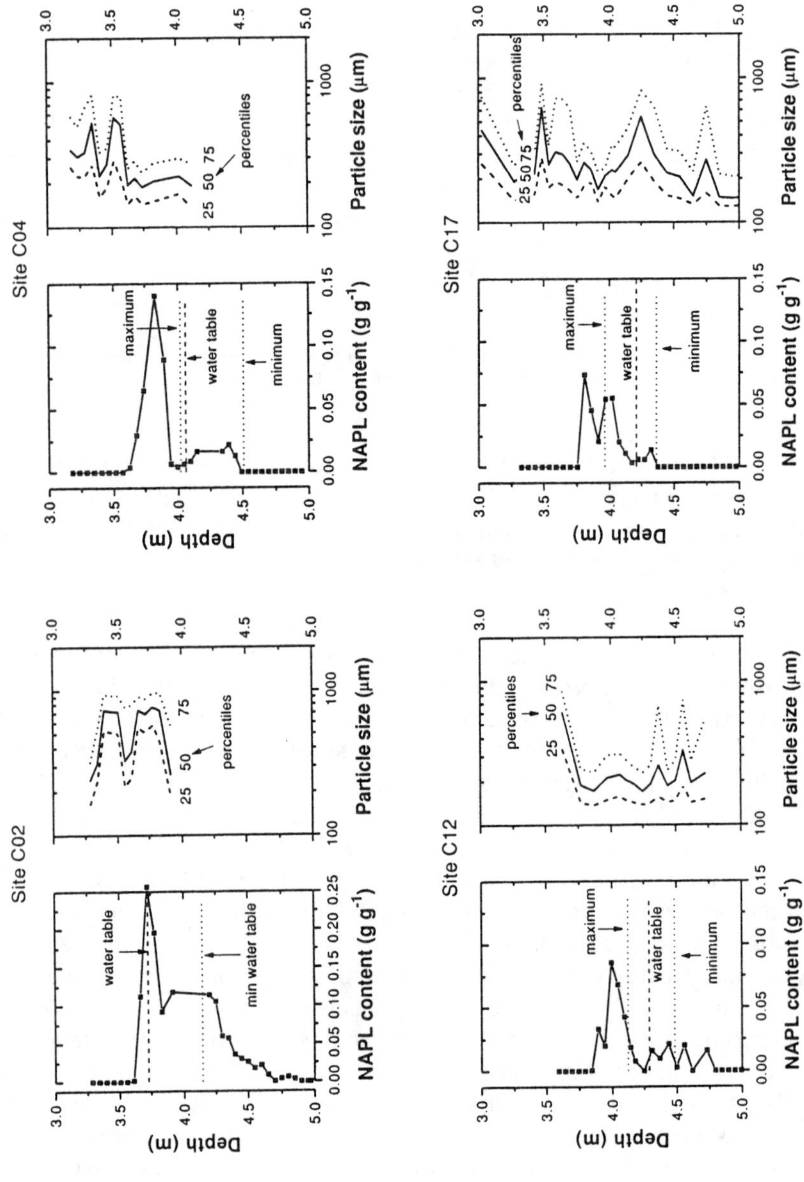

FIGURE 4. Vertical distributions of NAPL in relation to the water table and particle-size distributions.

measured residual contents may be because of natural degradation of NAPL or because continual movement of the water table does not allow the NAPL to come to an equilibrium distribution. The lack of evidence for entrapment in coarse layers may be explained because vertical displacement of NAPL at the field site is under three-phase conditions compared to the water-NAPL systems of Wilson et al. (1990) and Dawe et al. (1992).

CONCLUSIONS

Previous laboratory work has indicated that subsurface layering may lead to entrapment of NAPL in the coarser layers during manipulation of the water table or during natural annual variations. Residual NAPL occurred above and below the water table at the field site, but there was no evidence that NAPL was entrapped at higher contents in coarser layers above or below the water table. The position of the water table and S-P relation have a greater influence on the NAPL distribution. Future work should be aimed at investigating whether the entrapment demonstrated for water-NAPL systems occurs for water-NAPL-air conditions at the water table. Determination of entrapped NAPL under water-NAPL and water-NAPL-air conditions for the profile materials is required to understand whether there will be horizontal or vertical movement of NAPL at the field site.

ACKNOWLEDGMENTS

This work was partly funded by a cooperative research agreement with Broken Hill Pty Co. Ltd. The assistance of David Briegel, Michael Lambert, and Tracy MacLachlan is appreciated.

REFERENCES

Dawe, R. A., M. R. Wheat, and M. S. Bidner. 1992. "Experimental Investigation of Capillary Pressure Effects on Immiscible Displacement in Lensed and Layered Porous Media." *Transport in Porous Media 7*: 83-101.

Semeniuk, V., and D. J. Searle. 1985. "The Becher Sand, a New Stratigraphic Unit for the Holocene of the Perth Basin." *J. Royal Society Western Australia 67*: 109-115.

Wilson, J. L., S. H. Conrad, W. R. Mason, W. Peplinski, and E. Hagan. 1990. *Laboratory Investigation of Residual Liquid Organics from Spills, Leaks and the Disposal of Hazardous Wastes in Groundwater*. U.S. Environmental Protection Agency Report, EPA/600/6-90/004, R.S. Kerr Environmental Research Laboratory, Ada, OK.

REVIEW OF BIOREMEDIATION EXPERIENCE IN ALASKA

B. L. Kellems, A. Leeson, and R. E. Hinchee

INTRODUCTION

Recent experience in Alaska has shown that bioremediation is an effective technology for cleaning up petroleum hydrocarbon contamination in a cold region environment. Although many projects currently are under way, documentation in the scientific and engineering literature is limited. This paper summarizes the current state of bioremediation technology being practiced in Alaska. The engineered controls used to improve bioremediation performance in a cold region environment are identified.

BIOREMEDIATION PROJECTS IN ALASKA

A list of pilot- and full-scale bioremediation projects that have been completed or currently are under way is presented in Table 1. The list has been compiled to provide an indication of current practice and is not intended to be comprehensive.

As shown in Table 1, a wide range of media have been, or currently are being, remediated through biological means, including soils, gravel pads, tundra, beach sand and cobbles, pond sediments, and oilfield drilling waste. The contaminants being remediated in all cases are of petroleum origin.

Although numerous methods and process options currently are being utilized, at least five general bioremediation technologies have been used in Alaska:

1. *In Situ Surface Bioremediation*: Microbial enhancement through seasonal treatment of surface soils, sediments, or tundra. This may include addition of nutrients, tilling to mix and aerate the soil, watering or dewatering to control soil moisture, bioaugmentation through microbial inoculation, and/or addition of organic matter.
2. *Ex Situ Cell Bioremediation*: Soils are excavated and placed on a geosynthetic liner in one or more lifts. Distribution piping typically is installed to provide active aeration and leachate control. The presence of the bottom liner minimizes the potential for further environmental release of contaminants compared to in situ methods. A top liner may be installed to provide further temperature control. This technology is equivalent to landfarming with the addition of a liner. Individual process options are similar to those for in situ surface bioremediation (1).

TABLE 1. Summary of bioremediation projects in Alaska.

Location/Site	Media	Contaminants	Technology	Mode of Operation	Scale	Status	Reference
Barrow/Arctic Lake	Pond sediments	Leaded gasoline	In situ surface (fertilization)	Seasonal	Full-scale	Completed (1977)	Horowitz & Atlas, 1978
Brooks Range/Village of Anaktuvuk Pass	Soil	#2 diesel fuel, waste oil	Ex situ recirculation leach bed (fertilization, aeration, bioaugmentation)	Seasonal	Full-scale (500 cy)	Completed	Gilkinson & Travis, 1991
Deadhorse/various sites	Sandy gravel	Petroleum hydrocarbons	Ex situ cell (fertilization, watering, tilling)	Seasonal	Full-scale	Completed	Galloway, 1991
Fairbanks/Fairbanks International Airport	Soil, groundwater	Jet fuel	In situ groundwater, ex situ soil landfarming	Seasonal	Full-scale (3,700 cy)	Operational	Reynolds et al., 1994
Fairbanks/Fairbanks International Airport	Soil	Jet fuel	Ex situ recirculation leach bed (fertilization, aeration)	Seasonal	Pilot-scale (150 cy)	Ongoing	Reynolds et al., 1994
Fairbanks/Eielson AFB	Silty sand	JP-4 jet fuel	In situ bioventing (aeration, heating)	Year-round	Pilot-scale	Ongoing (2nd season)	Leeson et al., 1993
Fairbanks/Eielson AFB	Soil	Jet fuel	Ex situ composting (fertilization, aeration, organic addition)	70-day test	Pilot-scale (3 cells, 3.5 cy ea.)	Completed (1991)	Simpkin et al., 1991
	Silty sand	JP-4 jet fuel	In situ bioventing	Year-round	Pilot-scale	Ongoing	Ong et al., 1994
	Silty sand	JP-4 jet fuel	In situ bioventing	Year-round	Pilot-scale	Ongoing	Ong et al., 1994
Kenai Peninsula/Poppy Lane Gravel Pit	Oil exploration and production solid waste	Crude oil, gas condensate	Ex situ cell (fertilization, moisture control, tilling)	Seasonal	Full-scale (60,000 tons)	Operational (1st year)	Cox et al., 1992
Kenai Peninsula/Soldotna UST Site	Soil	Diesel fuel	Ex situ cell (fertilization, watering, tilling, organic addition)	Seasonal	Full-scale (1,000 cy)	Operational (1st season)	Duffey & Authier, 1992

TABLE 1. (Continued)

Location/Site	Media	Contaminants	Technology	Mode of Operation	Scale	Status	Reference
Kenai Peninsula/Swanson River Oilfield/3-9 Tank Setting	Sandy silt	Crude oil	In situ bioventing (aeration, heating)	Year-round	Full-scale (10,000 cy)	Operational (1st year)	Kellems et al., 1991
Kuparuk Oilfield/Drill Site 2U	Tundra	Crude oil, produced water	In situ surface (fertilization, tilling)	Seasonal	Full-scale (1.4 ac)	Operational (3rd season)	Jorgenson et al., 1991
North Slope/Service City Gravel Pad	Sandy gravel	Diesel fuel	In situ bioventing (aeration, fertilization, watering, bioaugmentation)	10-week test	Pilot-scale	Completed (1992)	Walden, 1993
Prince William Sound/*Exxon Valdez* Oil Spill	Beach sand and cobbles	Crude oil	In situ surface (fertilization)	Seasonal	Full-scale	Completed	U.S. Congress, 1991
Prudhoe Bay/S. E. Eileen Well Site	Tundra	Crude oil	In situ surface (fertilization, tilling)	Seasonal	Full-scale	Operational (2nd season)	Jorgenson & Cater, 1991
Prudhoe Bay/Gravel Pads	Sandy gravel	Diesel fuel	In situ bioventing (aeration)	Seasonal	Pilot-scale	Completed (1992)	Simpkin et al., 1992
50 mi. east of Prudhoe Bay/ Point Thomson No. 1 Pad	Sandy gravel	Diesel fuel	In situ surface (fertilization, watering)	Seasonal	Full-scale	Operational	Liddell et al., 1991
Prudhoe Bay/Surfcote Pad	Sand gravel	Diesel fuel	In situ surface (fertilization, watering, tilling)	Seasonal	Full-scale	Operational	Evans et al., 1991
Prudhoe Bay/various relief pits	Tundra	Crude oil	In situ surface (fertilization, tilling)	Seasonal	Pilot-scale (6 pits)	Ongoing (2nd season)	Bledsoe et al., 1992
Trans-Alaska Pipeline/ Check Valve 23	Tundra	Crude oil	In situ surface (fertilization, tilling)	Seasonal	Full-scale	Completed (1985)	Brendal & Eschenbach, 1985
Trans-Alaska Pipeline/ Pump Station #9	Sandy gravel	Crude oil	Ex situ recirculation leach bed (fertilization, aeration, organic addition)	Seasonal	Full-scale (800 cy)	Operational (1st season)	Sweeney et al., 1992

3. *Ex Situ Recirculation Leach Bed*: Similar to ex situ cell bioremediation (2), except that water is forced through the soil, usually in the upflow direction, and all of the material can be processed at one time rather than in lifts.
4. *Ex Situ Composting*: Similar to ex situ cell bioremediation (2), except that biodegradation rates are enhanced through the addition of organic matter. The process usually is conducted in a closed reactor to conserve the additional heat produced through biological activity.
5. *In Situ Bioventing*: This technology uses forced aeration to stimulate indigenous microorganisms to aerobically degrade petroleum hydrocarbons. Volatile hydrocarbons also may be removed due to soil venting during this process; however, volatilization of petroleum hydrocarbons is minimized during bioventing, while biodegradation is optimized.

COLD-REGION PROCESS OPTIONS

By far, the most prevalent method for implementing bioremediation in Alaska is through seasonal operation. Active bioremediation typically is limited to the summer months and little or no removal takes place during the colder portions of the year. Almost all of the full-scale applications of bioremediation in Alaska to date have been in situ under ambient conditions. Whether remediating beaches in Prince William Sound or tundra on the North Slope, in situ bioremediation under ambient conditions has proven to be an effective technology in Alaska.

For in situ bioremediation under engineered conditions and ex situ bioremediation applications, process options are available to extend active bioremediation throughout the colder months of the year. These process options have consisted of heating the soil either directly or indirectly. The soil has been heated directly using heating elements or a jacketed reactor, or by simply covering the soil to utilize solar heating. The soil has been heated indirectly through injecting heated air and heated water. The use of active process options to heat the soil has seen only limited application in Alaska to date, primarily in pilot studies.

CONCLUSION

Successful bioremediation applications currently under way in Alaska have shown that year round bioremediation in cold regions is possible. However, temperature remains an important variable. Several process options for heating the soil are available to extend active bioremediation throughout the colder months of the year.

REFERENCES

Bledsoe, K., B. Elder, R. Hoffman, B. Collver, and B. Gerken. 1992. "Full Scale Investigation of Bioremediation of Tundra Contaminated with Crude Oil." Presented at the BP Exploration (Alaska) Inc. Soil Remediation Conference, Anchorage, Alaska, Nov. 12-13.

Brendel, J. E., and T. G. Eschenbach. 1985. "Check Valve 23 Revegetation Study." In M. L. Lewis (Ed.), *Northern Hydrocarbon Development Environmental Problem Solving*, Proceedings, 8th Ann. Meet. Int. Soc. Petrol. Ind. Biol., Banff, Alberta, Sept. 24-26, pp. 230-239.

Cox, B., B. Penn, and D. A. Vossler. 1992. "Bioremediation of Oilfield Wastes Excavated from the Poppy Lane Gravel Pit, Kenai Peninsula, Alaska." Presented at the BP Exploration (Alaska) Inc. Soil Remediation Conference, Anchorage, Alaska, Nov. 12-13.

Duffey, A. M., and B. D. Authier. 1992. "Case Study: Use of Site Assessment Results to Design and Implement a Bioremediation Alternative." Presented at the BP Exploration (Alaska) Inc. Soil Remediation Conference, Anchorage, Alaska, Nov. 12-13.

Evans, D., B. Elder, and R. Hoffman. 1991. "Bioremediation of Diesel Contamination at the Prudhoe Bay Unit E.O.A. Surfcote Pad." Presented at the BP Exploration (Alaska) Inc. Soil Remediation Workshop, Anchorage, Alaska, Nov. 19-20.

Galloway, G. 1991. "Appropriate Hydrocarbon Cleanup Levels and Successful Bioremediation on the North Slope." Presented at the BP Exploration (Alaska) Inc. Soil Remediation Workshop, Anchorage, Alaska, Nov. 19-20.

Gilkinson, B., and M. Travis. 1991. "Bioremediation of Diesel and Waste Oil Contaminated Soils at Anaktuvuk Pass, Alaska." Presented at the BP Exploration (Alaska) Inc. Soil Remediation Workshop, Anchorage, Alaska, Nov. 19-20.

Horowitz, A., and R. M. Atlas. 1978. "Hydrocarbons and Microbial Activities in Sediment of an Arctic Lake One Year After Contamination with Leaded Gasoline." *Arctic 31*:180-191.

Jorgenson, M. T., and T. C. Cater. 1991. *Cleanup, Bioremediation and Tundra Restoration After a Crude-Oil Spill, S. E. Eileen Exploratory Well Site, Prudhoe Bay, Alaska.* 1991 Annual Report, prepared for ARCO Alaska, Inc., Anchorage, Alaska by Alaska Biological Research, Inc., Fairbanks, AK.

Jorgenson, M. T., T. C. Cater, and K. Kielland. 1991. *Bioremediation and Tundra Restoration After a Crude Oil Spill Near Drill Site 2U, Kuparuk Oilfield, Alaska.* 1991 Annual Report, prepared for ARCO Alaska, Inc., Anchorage, Alaska by Alaska Biological Research, Inc., Fairbanks, AK.

Kellems, B. L., R. E. Hinchee, and G. L. Upson. 1991. "Feasibility of In-Situ Bioventing for Remediation of Hydrocarbon Contaminated Soils in Cold Regions." Presented at the BP Exploration (Alaska) Inc. Soil Remediation Workshop, Anchorage, Alaska, Nov. 19-20.

Leeson, A., R. E. Hinchee, J. Kittel, G. D. Sayles, C. M. Vogel, and R. N. Miller. 1993. "Optimizing Bioventing in Shallow Vadose Zones and Cold Climates." *Hydrological Sciences Journal 38*(4) (in press).

Liddell, B. V., D. R. Smallbeck, and P. C. Ramert. 1991. "Arctic Bioremediation: A Case Study." Proceedings, International Arctic Technology Conf., Soc. Petrol. Engrs., Anchorage, Alaska, May 29-31, pp. 659-670.

Ong, S. K., A. Leeson, R. E. Hinchee, J. Kittel, C. M. Vogel, G. D. Sayles, and R. N. Miller. 1994. "Cold Climate Applications of Bioventing." In R. E. Hinchee, B. C. Alleman, R. E. Hoeppel, and R. N. Miller (Eds.), *Hydrocarbon Bioremediation*. Lewis Publishers, Ann Arbor, MI.

Reynolds, C. M., M. Travis, and W. A. Braley. 1994. "Applying Field-Expedient Bioreactors and Landfarming in Alaskan Climates." In R. E. Hinchee, B. C. Alleman, R. E. Hoeppel, and R. N. Miller (Eds.), *Hydrocarbon Bioremediation*. Lewis Publishers, Ann Arbor, MI.

Simpkin, T. 1991. "Treatment of Fuel Product Contaminated Soil in a Cold Climate Using Enhanced Composting." Presented at the BP Exploration (Alaska) Inc. Soil Remediation Workshop, Anchorage, Alaska, Nov. 19-20.

Simpkin, T., K. Bledsoe, R. Hoffman, B. Elder, and R. Hinchee. 1992. "Potential for Bioventing of Gravel Pads on the North Slope." Presented at the BP Exploration (Alaska) Inc. Soil Remediation Conference, Anchorage, Alaska, Nov. 12-13.

Sweeney, J., B. D. Authier, and G. Galloway. 1992. "Remediation of Diesel-Impacted Soils in a Thick-Layer Biocell at Pump Station #9." Presented at the BP Exploration (Alaska) Inc. Soil Remediation Conference, Anchorage, Alaska, Nov. 12-13.

U.S. Congress, Office of Technology Assessment. 1991. Bioremediation for Marine Spills — Background Paper. OTA-BP-0-70. U.S. Government Printing Office, Washington, DC.

Walden, T. 1993. "Bioventing Field Trials with Amendment Comparison in a Cold Climate Setting." Presented at In Situ and On-Site Bioreclamation, The Second International Symposium, San Diego, California, April 5-8.

COLD CLIMATE APPLICATIONS OF BIOVENTING

S. K. Ong, A. Leeson, R. E. Hinchee, J. Kittel, C. M. Vogel, G. D. Sayles, and R. N. Miller

INTRODUCTION

At most petroleum-contaminated sites, oxygen usually is the limiting nutrient for hydrocarbon biodegradation (Hinchee & Ong 1992, Miller et al. 1991). Bioventing is a simple technology whereby air (oxygen) is introduced into the unsaturated zone to enhance mineralization of hydrocarbons by indigenous bacteria. Air can be introduced into the subsurface by extracting air through the contaminated zone from the surface or through passive vent wells or by injecting air through vent wells into the contaminated zone. Bioventing is similar to soil vacuum extraction or soil venting with one key difference. The main objective of bioventing is to supply sufficient oxygen to maintain biodegradation at its optimum level, but with minimum volatilization of low-molecular-weight compounds.

Although bioventing is classified as an emerging technology, successful completion of several full-scale applications of bioventing has been documented (Dupont et al. 1991). The majority of bioventing sites are located in temperate or subtropical regions of the United States, whereas the application of bioventing at sites in arctic or subarctic regions has not been demonstrated. Several questions must be addressed before bioventing can be applied at these sites. The more important questions are: (1) Is there microbial activity in soil at temperatures close to 0°C? (2) If microbial activity is present, are microbial respiration rates sufficient to warrant implementation of bioventing? (3) Which factors control mineralization of organics in soils? and (4) What approaches can be taken to optimize bioventing?

This paper presents three case studies demonstrating the presence of microbial activity at hydrocarbon-contaminated sites in cold regions. Microbial respiration rates and data from pilot-scale bioventing systems are presented. Case Study 1 was conducted in early spring when part of the soil was still frozen. Case Study 2 was conducted over an 18-month period throughout the winter months. Also, in Case Study 2, three forms of soil warming were tested as a means of increasing biodegradation of hydrocarbons. Case Study 3 involves a total of five sites and was initiated during the summer months.

SITE DESCRIPTIONS

Case Study 1: Northern Alaska. Case Study 1 is located in Northern Alaska. This site was contaminated with diesel and motor oils from refueling and maintenance of motor vehicles. The soil at the test site consists of sandy gravel with silt. The gravel size may be as large as 1.5 inches. The depth to the water table is approximately 4 feet. Preliminary results from soil borings indicated that contamination to a depth of 2 feet was as high as 22,000 mg total petroleum hydrocarbons (TPH) per kg of soil.

Case Study 2: Eielson Air Force Base (AFB), Alaska. Eielson AFB is an active base located in the Alaskan Interior region approximately 25 miles southeast of Fairbanks, Alaska. The climate is characterized as subarctic with low annual precipitation and an average annual temperature near 0°C. Temperatures in the region cover a broad range with winter lows falling below –30°C and summer highs above +30°C. Soils consist primarily of glaciofluvial deposits derived from glacial outwash from the Alaskan Mountain Range. Surface soils consisting of interbedded layers of loose sand and gravel with silt concentration increasing with depth to approximately 6 ft. Groundwater typically is encountered at 6 to 7 ft. Soil is contaminated at the site with JP-4 jet fuel. Soil concentrations of the jet fuel have been detected at concentrations as high as 1,000 mg TPH per kg soil.

Case Study 3: Galena Air Force Station (AFS) and Eielson AFB, Alaska. Case Study 3 involves a total of five sites that are part of the U.S. Air Force Center for Environmental Excellence's Bioventing Initiative. Three of the sites are located at the Galena AFS in northwestern Alaska and two are located at Eielson AFB near Fairbanks, Alaska. The sites are contaminated with a variety of petroleum products including JP-4 jet fuel and diesel and motor oils. The depth to the water table at the sites is relatively shallow with an average depth of 6 to 8 feet.

METHODS AND MATERIALS

Case Study 1: Northern Alaska. The respiration rates of indigenous microorganisms in hydrocarbon-contaminated soils were measured using the rapid in situ respiration test method as described by Hinchee and Ong (1992). The in situ respiration test consisted of aerating the contaminated soil with air. Microbial activity was monitored by measuring the change of oxygen and carbon dioxide over time after the air had been turned off. A 1-cm inner diameter hollow stainless steel tube with a slotted well point assembly on one end or a 2-inch outer diameter polyvinyl chloride (PVC) pipe with a screen at one end was used as a venting well. The slotted end of the tube or the screen was placed in the zone of contamination. Figure 1a shows a schematic diagram of the in situ respiration test setup. Air mixed with 2 to 4% helium was injected into the contaminated zone over a 24-hour period. After the air was turned off, the oxygen, carbon

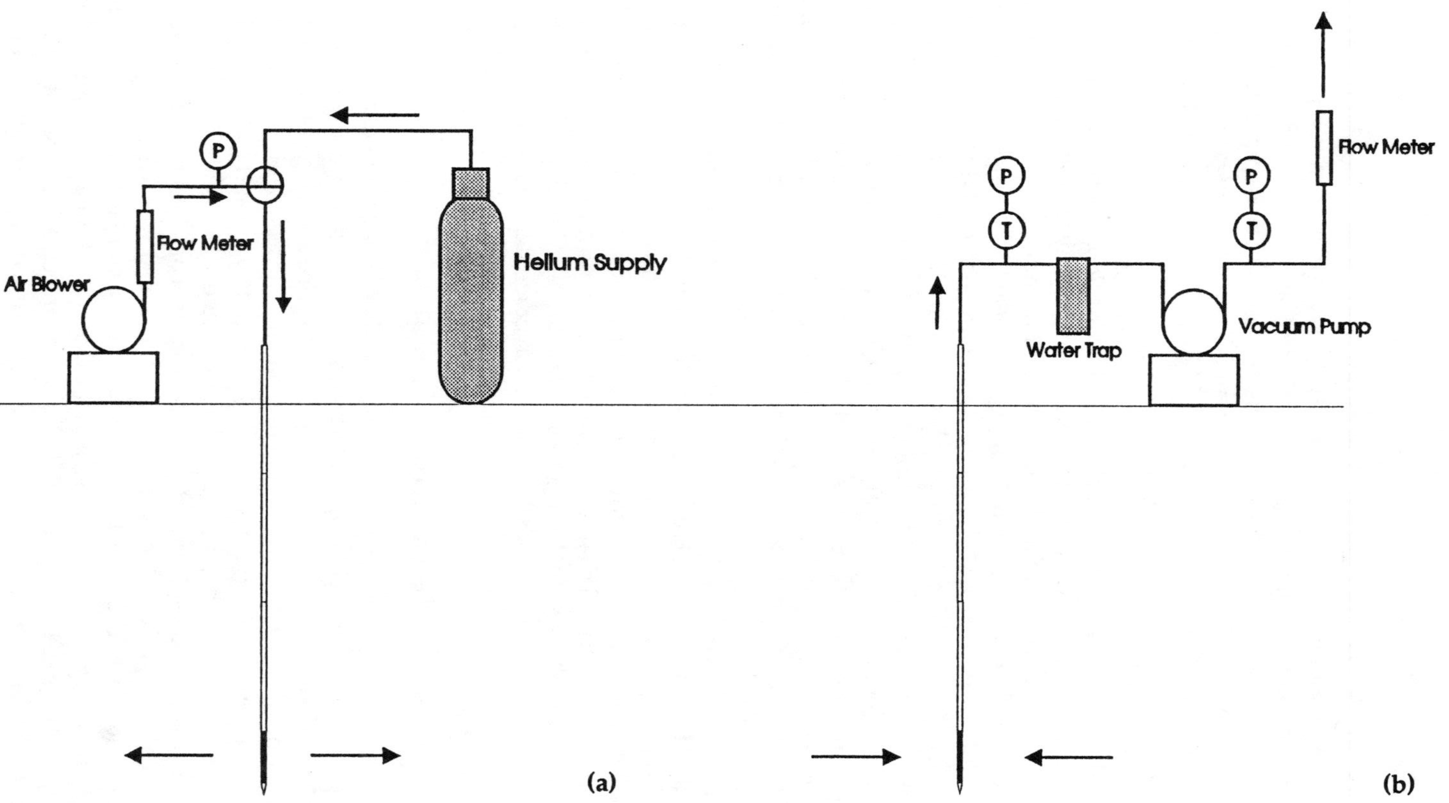

FIGURE 1. (a) Typical setup for conducting an in situ respiration test. (b) Typical setup for a bioventing system using air extraction.

dioxide, and helium concentrations were measured periodically over time. Oxygen and carbon dioxide were measured using a Gastech Model 35250X, and helium was measured using a Marks Helium Detector Model 9821. The respiration test was terminated when the oxygen level was less than 5%.

In addition to the in situ respiration test, a pilot-scale bioventing system using the air extraction approach was operated over a 2-week test period. The pilot-scale bioventing is illustrated in Figure 1b. A water trap was provided on the inlet line to trap moisture that might have been extracted from the soil. Oxygen, carbon dioxide, and TPH concentrations in the off-gas of the bioventing system were measured. Airflow rates were maintained at 1.5 ft^3/min throughout the test period.

Case Study 2: Eielson AFB, Alaska. Four test plots, 50 ft by 50 ft, were installed at the site. Each test plot contained air injection wells, groundwater monitoring wells (except the surface warming test plot), soil gas sampling probes, and thermocouples. An uncontaminated area was used as a background location for measurement of natural microbial respiration rates, and one air injection well and one groundwater monitoring well were installed in this area.

Four 2-inch-diameter air injection wells were installed in each plot at a depth of 6 feet, with 3 ft of 10-slot screen and 4 ft of schedule 40 PVC casing finished 1 ft above grade. One 2-inch-diameter air injection well was installed at the center of each test plot to 13 ft, with 6.5 ft of 10 slot screen and 7.5 ft of schedule 40 PVC casing finished 1 ft above grade. Air blowers were plumbed to the air injection wells and airflow was set at approximately 10 ft^3/min.

Tri-level soil gas monitoring points were installed in each of the test plots and in the background area. Each probe was constructed of either ¼-inch diameter schedule 80 PVC or of ¼-inch polyvinyl tubing with a 6-inch screened area. The screened area was surrounded with a sand filter pack, and a bentonite seal was used to fill the space between probes.

In three of the test plots (active, passive, and surface warming), a form of soil warming is being tested. The fourth test plot (control test plot) receives air injection but no soil warming. In the active warming test plot, the soil is warmed by circulating heated groundwater through soaker hoses in the soil. Styrofoam™ insulation covers the ground surface. In the passive warming test plot, the soil is warmed with solar warming. Plastic sheeting was placed over the ground surface of the test plot during the spring and summer months to capture solar heat and passively warm the soil. During the late fall and winter, the plot was covered with insulation to help retain heat during the dark winter months. In the surface warming test plot, the soil is warmed with heat tape buried at a depth of 3 feet. This test plot was installed in October 1992 and was covered with insulation to retain heat.

Operation of the bioventing system involves using blowers to introduce oxygen into the vadose zone by injecting atmospheric air into the contaminated subsurface. Samples taken from soil gas monitoring points are analyzed for oxygen, carbon dioxide, and TPH concentration. Air has been injected at the site for approximately 18 months. Biodegradation rates are measured quarterly through in situ respiration tests as described for Case Study 1.

Case Study 3: Galena AFS and Eielson AFB, Alaska. As part of the Bioventing Initiative, short-term bioventing tests were conducted at sites at Galena AFS and Eielson AFB. An initial in situ respiration test was conducted at each site by the method described by Hinchee and Ong (1992). After conducting the in situ respiration tests, pilot-scale bioventing systems were installed for 1 year of operation and currently are operating. This technical note presents results from the initial in situ respiration testing.

RESULTS AND DISCUSSION

Case Study 1: Northern Alaska. The temperature of the soil at this site during the in situ respiration test was between 4 to 5°C. The change in oxygen and carbon dioxide concentrations in the soil gas over time is shown in Figure 2. The increase in carbon dioxide shows that mineralization of hydrocarbons is occurring. Helium data presented in Figure 2 show a slow but steady decline over time. The helium data indicated minimum diffusion of oxygen from ground level into the test area. The respiration rate based on oxygen utilization was 13.2% oxygen per day.

Carbon dioxide and oxygen concentrations in the off-gas of the pilot-scale bioventing system over a 12-day demonstration period are presented in Figure 3. Oxygen concentrations were consistently below atmospheric oxygen level (20.9%), indicating that oxygen had been utilized for respiration. The carbon dioxide concentrations also were consistently above the atmospheric carbon dioxide level of 0.05%. Based on the field data, and from the radius of influence found at the

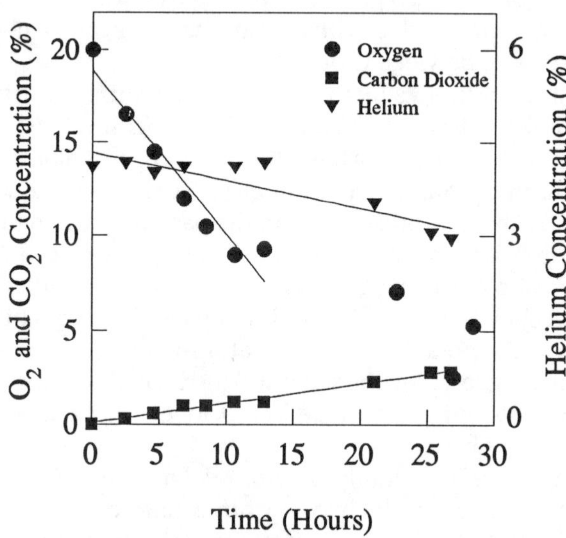

FIGURE 2. Oxygen, carbon dioxide, and helium concentrations during the in situ respiration test.

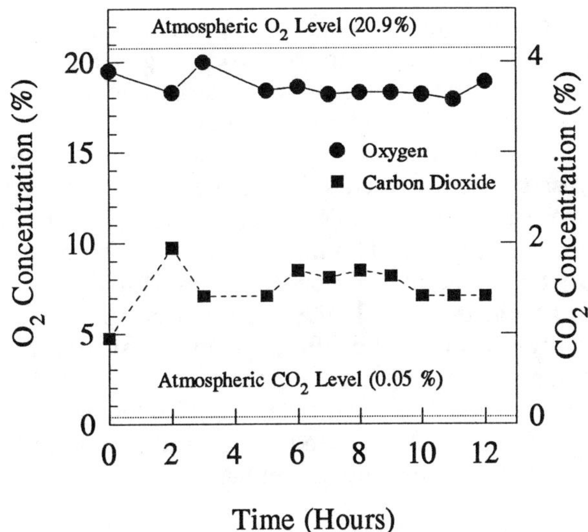

FIGURE 3. Oxygen and carbon dioxide concentrations during in off-gas during pilot-scale bioventing demonstration.

site, the respiration rate under dynamic conditions was computed with the following equation:

$$K_o = \frac{Q\left(C_o - C_f\right)}{V_p} \tag{1}$$

where K_o = oxygen utilization rate (% oxygen/day); C_o = initial concentration (oxygen – 20.9%); C_f = off-gas oxygen concentration (%); Q = airflow rate (m^3/day); V = volume of soil (m^3); and p = air-filled porosity (assumed to be 0.2).

In order to relate respiration rates and resulting biodegradation rates to active bioventing measurements and other sites in the literature, a stoichiometric relationship of the oxidation of the fuel is required. Hexane (C_6H_{14}) was used as the representative hydrocarbon for the jet fuel for the purpose of comparing the carbon dioxide and oxygen rates. The stoichiometric relationship is given by:

$$C_6H_{14} + 9.5O_2 \rightarrow 6CO_2 + 7H_2O \tag{2}$$

Based on oxygen utilization rates (percent/day), biodegradation rates in terms of mg/kg/day can be estimated by assuming a soil bulk density of 1,440 kg/m^3 and an air-filled porosity of 0.2 and using the following equation:

$$K_\beta = \frac{-K_o A D_o C}{100\%} \tag{3}$$

where K_B = biodegradation rate (mg/kg/day); A = volume of air/kg of soil (L/kg), in this case 300/1,440 = 0.21; D_o = density of oxygen gas (mg/L), assumed to be 1,330 mg/L; and C = mass ratio of hydrocarbon to oxygen required for mineralization, assumed to be 1:3.5 from Equation 2.

The radius of influence at this site was found to be approximately 4.6 m (15 ft) from soil gas pressure measurements. The average rate of oxygen use over the 12-day period was calculated with Equation 1 to be 7.1%/day.

The respiration rates obtained for this site were comparable to those for other bioventing sites in temperate and subtropical regions (Table 1). The rates obtained by the in situ respiration test were slightly less than twice the rate obtained during the bioventing demonstration. The differences in the rates can be accounted for by the assumptions made in Equation 1 for the computation of biodegradation rates under dynamic conditions. Furthermore, the biodegradation rates obtained by the in situ respiration test are based on the initial change in oxygen concentration. The initial rates usually are higher.

Case Study 2: Eielson AFB, Alaska. In situ respiration tests were conducted during October 1991, January 1992, August 1992, October-November 1992, and January 1993 (Figure 4). Soil temperature measurements taken during this time are shown in Figure 5.

A comparison of the biodegradation rates observed throughout the year in the active warming test plot generally has shown similar rates between in situ respiration tests, with the exception of a decrease during January 1992 (Figure 4). Biodegradation rates in the active warming test plot consistently have been higher than those found in the other test plots, corresponding to the higher temperature in this test plot.

The in situ respiration tests conducted during the winter months produced interesting results in the other test plots. Despite low temperatures in the passive warming and control test plots, microbial activity was observed at most sampling points. In general, biodegradation rates dropped slightly during the winter months in these test plots. However, some biodegradation rates were equivalent or slightly higher than those observed during the warmer months, and actual reductions in biodegradation rates were not as great as might be expected.

Soil temperatures in the passive warming test plot in November 1992 and January 1993 were a few degrees higher overall than at the same time in 1991 due to the plastic mulch that covered the test plot during the summer. Slightly higher biodegradation rates also were observed during November 1992 and January 1993, with an average biodegradation rate of 1.1 and 1.0 mg/kg/day, respectively, compared to an average rate of 0.44 mg/kg/day during January 1992.

Biodegradation rates measured in the control test plot during August 1992 were nearly as high as those measured in the passive warming test plot. It would be expected that due to the differences in soil temperature, respiration rates would be higher in the passive warming test plot; however, respiration rates in the control plot generally have been comparable to or higher than those measured in the passive warming test plot. Differences in the level of contamination between the two test plots may cause the differences in respiration rates.

TABLE 1. Comparison of respiration rates from cold regions with rates from warmer regions.

Location		Soil Temperature (°C)	Oxygen Utilization Rate from Respiration Test (% O_2/day)	Biodegradation Rate (mg/kg/day)	Oxygen Utilization Rate from Bioventing Demonstration (%O_2/day)	Biodegradation Rate (mg/kg/day)
Case Study 1		4.0 - 5.0	13.2	10.5	7.7	4.1
Case Study 2		1.0	0.7 ± 0.4	0.5 ± 0.3	NA	NA
Case Study 3	Eielson AFB, ST-10	16	6.9 ± 0.70	5.5 ± 0.56	NA	NA
	Eielson AFB, Site 48	8.0	4.2 ± 2.6	3.3 ± 2.1	NA	NA
	Galena AFS, Saddle Tank Farm	8.5	25 ± 13	20 ± 10	NA	NA
	Galena AFS, Power Plant	12	34 ± 23	27 ± 18	NA	NA
	Campion POL Tank Site	7.0	24 ± 13	19 ± 10	NA	NA
Florida[(a)]		25	10 ± 0.5	8.0 ± 0.4	NA	NA
Maryland[(a)]		21	3.0 ± 0.2	2.4 ± 0.2	NA	NA
Nevada[(a)]		21	6.0 ± 0.2	4.8 ± 0.2	NA	NA
Oklahoma[(a)]		17	4.0 ± 0.5	3.2 ± 0.4	NA	NA

NA Not applicable
(a) Hinchee and Ong, 1992.

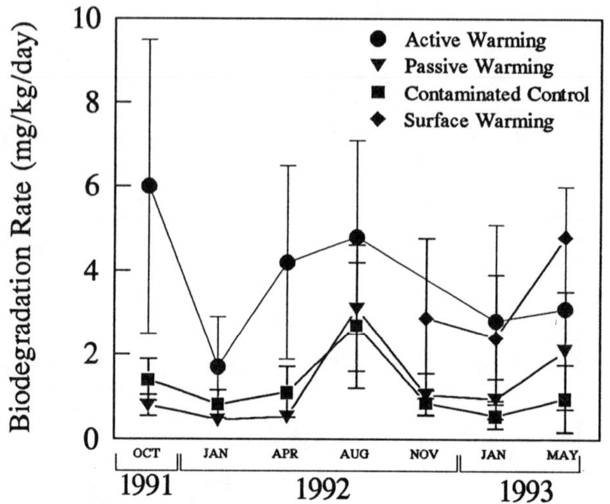

FIGURE 4. Average biodegradation rates in the four test plots.

Biodegradation rates in the surface warming test plot were measured for the first time in November 1992. Biodegradation rates ranged from 0.84 up to 6.9 mg/kg/day, with an average biodegradation rate of 2.9 mg/kg/day. Biodegradation rates measured during January 1993 changed little from November, with average rates of 2.4 mg/kg/day.

Case Study 3: Galena AFS and Eielson AFB, Alaska. The in situ respiration tests conducted at Galena AFS and Eielson AFB as part of the Bioventing Initiative

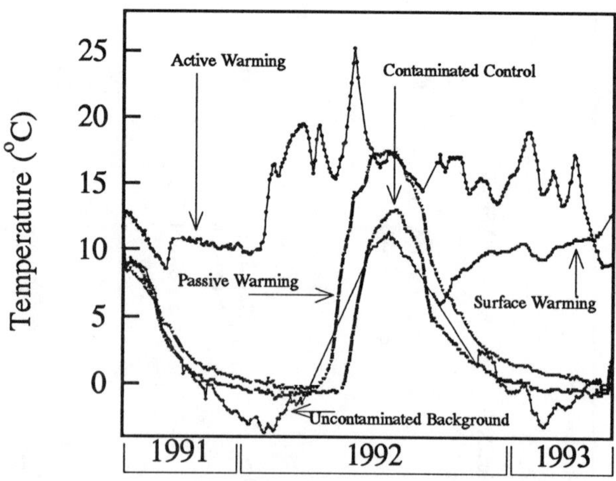

FIGURE 5. Soil temperature in four test plots and background area.

were conducted in July and August, 1992. Soil temperatures at this time were close to the warmest that the soil will achieve during the calendar year. Soil temperatures at Galena AFS were relatively low, with average temperatures ranging from 7.0 to 12°C. Soil temperatures at Eielson AFB were similar to those seen at Galena AFS, with average temperatures of 8.0 and 16°C at the two sites (Table 1). Respiration rates were relatively high at Galena AFS and Eielson AFB with average rates ranging from 19 to 27 mg/kg/day at Galena AFS and from 3.3 to 5.5 at Eielson AFB (Table 1).

SUMMARY AND CONCLUSIONS

Introduction of air into the subsurface to promote biodegradation of hydrocarbons in contaminated soils has been recognized as an economical and efficient method to remediate hydrocarbon-contaminated soils. To date there are limited applications of the bioventing technology for remediation of hydrocarbon-contaminated soils in cold regions. We have presented results from two test sites in cold regions that demonstrate biodegradation rates for the mineralization of hydrocarbons were similar to rates in temperate and subtropical regions. In Case Study 1, biodegradation rates ranged from 6.7 to 10.9 mg/kg/day when soil temperatures were approximately 4° to 5°C. In Case Study 2, biodegradation rates during January 1993 were found to be between 0.25 to 1.1 mg/kg/day in unheated areas (soil temperature of approximately 1°C), whereas biodegradation rates ranged from 0.23 to 6.9 mg/kg/day in actively heated test plots. In Case Study 3, average biodegradation rates at the five sites ranged from 3.3 up to 27 mg/kg/day. Based on these results, it is therefore feasible to apply the bioventing technology to simulate biodegradation for hydrocarbon-contaminated soils in cold regions.

REFERENCES

Dupont, R. R., W. Doucette, and R. E. Hinchee. 1991. "Assessment of In Situ Bioremediation Potential and the Application of Bioventing at a Fuel-Contaminated Site." In R. E. Hinchee and R. F. Olfenbuttel (Eds.), *In Situ and On-Site Bioreclamation*, Butterworth-Heinemann, Stoneham, MA. pp. 262-282.

Hinchee, R. E., and S. K. Ong. 1992. "A Rapid In Situ Respiration Test for Measuring Aerobic Biodegradation Rates of Hydrocarbons in Soil." *J. Air and Waste Manage. Assoc.* 42(10): 1305-1312.

Miller, R. N., R. E. Hinchee, and C. Vogel. 1991. "A Field-Scale Investigation of Petroleum Hydrocarbon Biodegradation in the Vadose Zone Enhanced by Soil Venting at Tyndall AFB, Florida." In R. E. Hinchee and R. F. Olfenbuttel (Eds.), *In Situ and On-Site Bioreclamation*, Vol. 1. Butterworth-Heinemann, Stoneham, MA.

A SIMPLE METHOD FOR DETERMINING DEFICIENCY OF OXYGEN DURING SOIL REMEDIATION

H. Würdemann, M. Wittmaier, U. Rinkel, and H. H. Hanert

INTRODUCTION

Two prerequisites for a successful aerobic biological remediation of contaminated soil are comprehensive monitoring and optimization of the processes involved in biological degradation, particularly the occurrence and correction of oxygen deficiency. In practice, monitoring generally is carried out by chemical analyses of soil or groundwater contaminants. Although chemical analysis will always remain an essential part of monitoring programs, the suitability of analysis for monitoring is limited, because the data it provides on the progress of biological degradation always involve a delay of several weeks and the heterogeneity of soil reduces the validity of the results, especially for in situ remediations.

A very quick and effective way to monitor and optimize biological processes is to determine the respiration rate, which is closely associated with the biodegradation of the contaminants (e.g., Lund et al. 1991, Miller et al. 1991, Wittmaier et al. 1992, Gudehus et al. 1993). An oxygen deficiency in the soil air is indicated by a decrease in O_2 consumption and CO_2 production. The oxygen content might be influenced by several biological processes, and it is sometimes difficult to determine CO_2 production because of the balance of lime and carbon. Therefore, the initiation of anaerobic processes will be a helpful indicator of oxygen deficiency. For this purpose measurements of N_2O, as an intermediate of denitrification, were carried out.

The aim of the investigations presented here was to develop and test a sensitive monitoring system for optimizing biological, on-site remediation of a former gasworks site. The silty-sandy soil, containing about 3,000 mg/kg of hydrocarbons and 180 mg/kg of PAH, was mixed with fertilizer and placed in bio piles. Because the piles were about 2 m high, a venting system was installed.

METHOD

Gas sampling in the coarse pore system of the soil was carried out with the help of a vacuum pump via soil-air probes. The samples were kept in airtight

Hungate tubes. Soil gases (N_2, O_2, CO_2, N_2O) were determined by a gas chromatographic (GC) thermal conductivity detector.

LABORATORY EXPERIMENTS

In order to establish the principal correlations between incipient oxygen deficiency and the production of CO_2 and N_2O for soil from the contaminated site, preliminary experiments were conducted using 120-L soil reactors. Figure 1 shows the curve for O_2 consumption and the associated curves for CO_2 and N_2O production measured in the soil reactors. Until an O_2 concentration of 1 to 2% was reached, there was a linear decrease in O_2 and a linear increase in CO_2. Because of low biological activity during the experiment, the limitation of biodegradation as a result of oxygen deficiency became noticeable only at very low O_2 concentrations in the soil atmosphere. The simultaneous drop in CO_2 production and the start of N_2O production give a very clear indication of when decreasing oxygen concentration becomes a limiting factor. Because N_2 production begins where O_2 concentration starts to become a limiting factor for aerobic degradation, it is a particularly suitable indicator of oxygen deficiency, where nitrate is available for denitrification.

The point of oxygen limitation is not fixed to a particular O_2 concentration, but depends on various factors. The control of O_2 and CO_2 during remediation via monitoring is possible only when a number of measurements are taken until the limiting concentration is reached (respiration kinetics). By monitoring N_2O

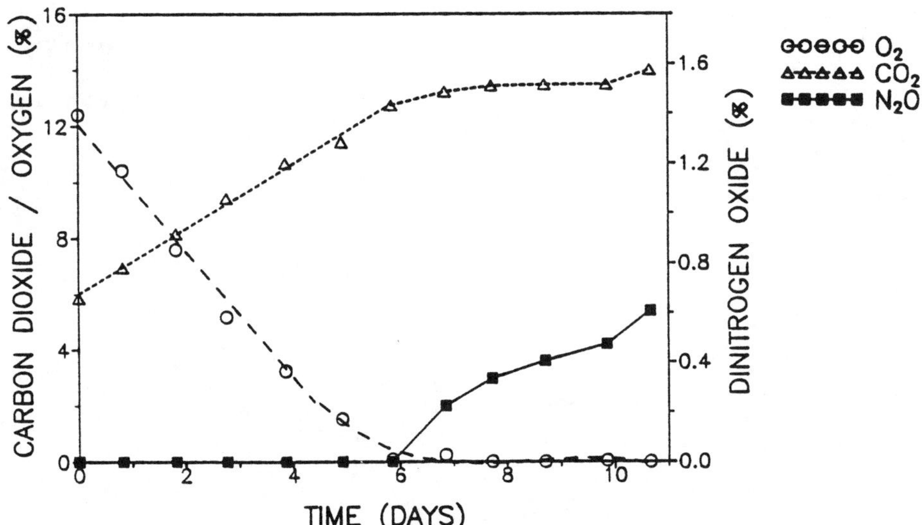

FIGURE 1. Monitoring of soil gases in soil reactor: CO_2, O_2, and N_2O concentrations over time without venting. After depletion of O_2, N_2O increases rapidly.

in the soil atmosphere, on the other hand, it is possible to establish with a single measurement whether or not the process is being limited by a shortage of oxygen.

FIELD MEASUREMENTS IN BIO PILES

During biological on-site remediation, investigations on the efficiency of the aeration were conducted. To optimize the process it was necessary to find out at which O_2 concentration the aerobic degradation starts to be limited. After deposition of piles, oxygen, carbon dioxide, and nitrogen oxide concentrations at the bottom of the pile were measured (Figure 2). Immediately after deposition, the oxygen concentration in the soil fell to 1% as a result of aerobic degradation of organics and oxidation of inorganic substances, e.g., sulfides. Subsequently, the CO_2 production rate decreased significantly. At the same time there was an increase of N_2O, which indicated denitrifying processes.

To evaluate the soil gas measurements it is necessary to consider that a soil sampling technique can be representative only for the coarse pore system (Greenwood & Goodman 1967, Flühler et al. 1976). Although the applied gas sampling technique allowed measurements of gas concentration only in the coarse pore system of the soil, where some O_2 still remained, the presence of N_2O in the soil gases indicated that at least parts of the fine pore system must be anaerobic. Due to the diffusion of oxygen from the top of the pile, the O_2 concentration remained at 2%. Because of both aerobic and anaerobic degradation there was a further slow increase in CO_2.

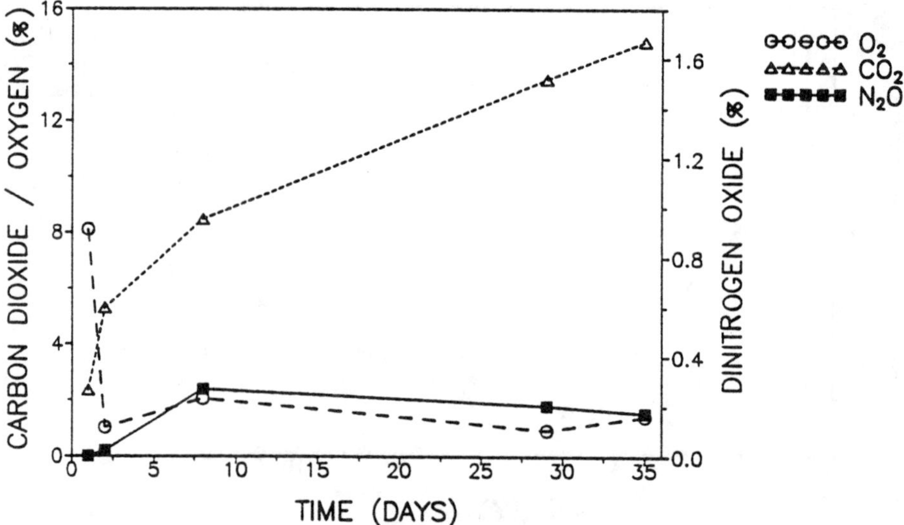

FIGURE 2. Monitoring of soil gases in the bio pile directly after deposition: CO_2, O_2, and N_2O concentrations over time without venting.

Once venting began, the O_2 concentration in the pile increased quickly to nearly 21%. For measuring respiration rates, ventilation was shut off and the changes in soil gases were monitored. Over a period of 80 h there was a linear increase in CO_2 and a linear decrease in O_2. Expecting a decrease in the O_2 consumption rate and CO_2 production rate because of limiting O_2 concentration, several additional respiration kinetics were determined. Because of the relatively low degradation activity and the corresponding long time interval needed to deplete O_2 in the soil air, it was impossible to distinguish the decrease of O_2 consumption and the corresponding CO_2 production by O_2 limitation from the influence of diffusion (loss of CO_2 and influx of O_2). Therefore, it was impossible to determine the exact point where O_2 starts to become a limiting factor for aerobic degradation in the pile.

Because of the difficulty in excluding the influence of diffusion in a bio pile, the correlation between O_2, CO_2, and N_2O concentration was investigated. The data obtained from soil gas monitoring before starting ventilation are presented in Figures 3a and 3b. At an O_2 concentration of 5%, N_2O increased significantly. This increase correlated with the further increase of CO_2. At a concentration of 5% O_2 in the coarse pore system, in many soil micropores the O_2 already was consumed and denitrification was in progress.

Some months later the investigations were repeated. On account of lower moisture content, lower degradation activity, and higher soil temperature, the conditions for diffusion had been improved. Therefore denitrification started at measured O_2 concentrations of 2%.

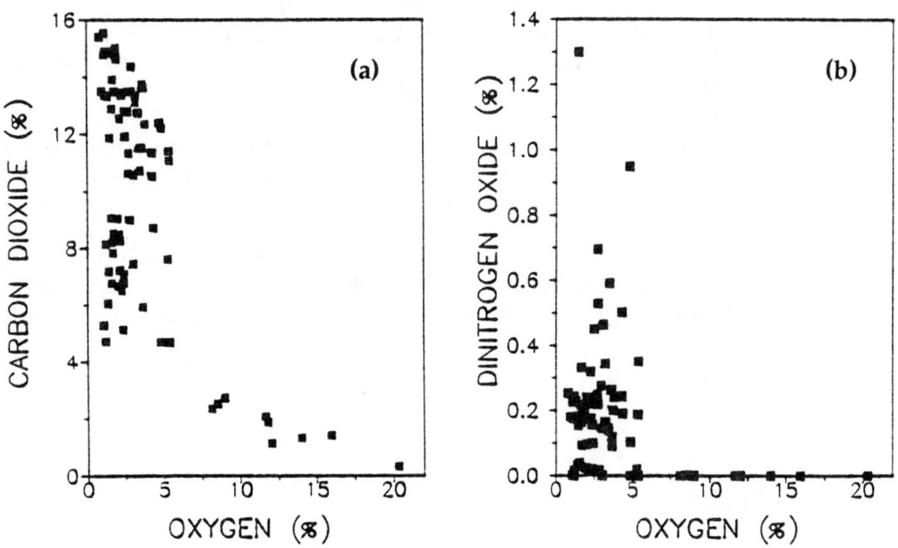

FIGURE 3. Correlation of CO_2 and N_2O concentration and the O_2 concentration in soil air. (a) CO_2 versus O_2; (b) N_2O versus O_2.

CONCLUSIONS

Depending on respiration activity and conditions for diffusion in soil, degradation was limited at different O_2 concentrations in the investigated coarse pore system. Even in silty sand, oxygen concentrations of 5% were limiting. To optimize remediation it is necessary to take additional criteria into consideration beyond the O_2 content. In combination with O_2 and CO_2 in situ respiration kinetics, the determination of N_2O is helpful for optimizing the venting process. In particular, if CO_2 production is difficult to measure because of a high pH value, N_2O will be an important factor in the control of ventilation. Reducing the air flow to a minimum will decrease emissions of gases significantly. When the soil is contaminated with highly volatile substances, reducing the air flow will improve cost effectiveness considerably by reducing or eliminating the need for cleaning off-gases. On the other hand, if aeration is reduced too much, both the degradation will be slower, and a huge amount of fertilizer (nitrogen) will be lost to denitrification.

REFERENCES

Flühler, H., L. H. Stolzy, and M. S. Ardakani. 1976. "A Statistical Approach to Define Soil Aeration in Respect to denitrification." *Soil Science 122*: 115-123.

Greenwood, D. J., and D. J. Goodman. 1967. "Direct Measurement of Distribution of Oxygen in Soil Aggregates and Columns of Fine Soil Crumbs." *Journal Soil Science 18*: 182-196.

Gudehus, G., J. Swiniansky, and H. Würdemann. 1993. "Biological In-Situ Remediation of Sandy Gravelly Gasworks Subsoils." Fourth International KfK/TNO Conference on Contaminated Soil, May 1993. Kluwer Academic Publishers, Berlin, Germany.

Lund, N. C., H. Würdemann, J. Swinianski, and G. Gudehus. 1991. "Experiences from a Field Test for a Biological In-Situ Remediation of Contaminated Gasworks Subsoils." BIOMEC/COMETT-Conference: Soil and Groundwater Cleaning. Nov. 1991. Hindsgavl Castle, Denmark.

Miller, R. N., R. E. Hinchee, and C. Vogel. 1991. "A Field-Scale Investigation of Petroleum Hydrocarbon Biodegradation in the Vadose Zone Enhanced by Soil Venting at Tyndall AFB, Florida." In R. E. Hinchee and R. F. Olfenbuttel (Eds.), *In Situ and On-Site Bioreclamation*, Vol. 1. Butterworth-Heinemann, Stoneham, MA.

Wittmaier, M., P. Harborth, and H. H. Hanert. 1992. "Biological Activity, Pollutant Degradation and Detoxification in Soil Highly Contaminated with Tar Oil." International Symposium: Soil Decontamination Using Biological Processes. Dec. 1992. Dechema, Karlsruhe, Germany.

AUTHOR LIST

C. M. Aelion
University of South Carolina
Environmental Health Sciences
Columbia, SC 29208 USA

J. G. Agar
O'Connor Associates Environmental, Inc.
201, 1144 - 29 Avenue N.E.
Calgary, Alberta
CANADA T2E 7P1

H.-J. Albrechtsen
Technical University of Denmark
Department of Environmental Engineering, Bldg. 115
DK-2800 Lyngby
DENMARK

R. M. Allen-King
Department of Geology
Washington State University
Pullman, WA 99164-2812 USA

S. D. Andrews
Groundwater Technology, Inc.
7346 South Alton Way, Suite A
Englewood, CO 80112 USA

J. S. Andrilenas
RZA-AGRA Engineering and Environmental Services
7477 SW Tech Center Drive
Portland, OR 97223 USA

J. P. Arcangeli
Technical University of Denmark
Department of Environmental Engineering, Bldg. 115
DK-2800 Lyngby
DENMARK

J. Armstrong
The Traverse Group, Inc.
3772 Plaza Drive, Suite 5
Ann Arbor, MI 48108 USA

E. Arvin
Technical University of Denmark
Department of Environmental Engineering, Bldg. 115
DK-2800 Lyngby
DENMARK

R. M. Atlas
University of Louisville
Louisville, KY 40292 USA

O. A. Awosika
57 Executive Park South, N.F., Suite 590
Atlanta, GA 30329-2265 USA

R. S. Baker
ENSR Consulting and Engineering
35 Nagog Park
Acton, MA 01720 USA

D. Ballerini
Institut Français du Pétrole
1/4 Avenue de Bois-Préau
92500 Rueil Malmaison FRANCE

R. E. Bare
Exxon Research and Engineering
Clinton Township, Route 22E
Annandale, NJ 08801 USA

J. F. Barker
Waterloo Centre for Groundwater Research
BFG, University of Waterloo
Waterloo, Ontario
CANADA N2L 3G1

P. J. Barter
Kinnetic Laboratories Inc.
Anchorage, AK 99501 USA

G. Battermann
DVGW-Forschungsstelle
Universität Karlsruhe (TH)
Richard-Willstatter-Allee 5
D-7500 Karlsruhe 1 GERMANY

C. Bech
SINTEF Applied Chemistry
N-7034 Trondheim
NORWAY

M. T. Becker
Dames & Moore, Inc.
12 Commerce Drive
Cranford, NJ 07016 USA

P. B. Bedient
Rice University
Environmental Science and Engineering
P.O. Box 1892
Houston, TX 77251 USA

C. Bocard
Institut Français du Pétrole
1/4 Avenue de Bois-Préau
92500 Rueil Malmaison FRANCE

R. Boni
Eniricerche S.p.A. Environmental Sciences
Via E. Ramarini, 32
00016 Monterotondo Scalo
Roma ITALY

H. Bonin
Department Environment BRGM
Avenue de Concyr
B.P. 6009
45060 Orléans Cedex 2 FRANCE

R. C. Borden
North Carolina State University
Department of Civil Engineering
Box 7908
Raleigh, NC 27695 USA

J. P. Bowman
Center for Environmental Biotechnology
University of Tennessee
10515 Research Drive, Suite 100
Knoxville, TN 37932 USA

J. F. Braddock
University of Alaska
Fairbanks, AK 99775 USA

J. R. Bragg
Exxon Production Research Company
Houston, TX 77098 USA

W. A. Braley
Fairbanks International Airport
P.O. Box 60396
Fairbanks, AK 99706 USA

E. J. Brown
University of Northern Iowa
Cedar Falls, IA 50614 USA

K. L. Brown
IT Corporation
312 Directors Drive
Knoxville, TN 37923 USA

R. A. Brown
Groundwater Technology, Inc.
310 Horizon Center Drive
Trenton, NJ 08691 USA

W. de Bruin
Agricultural University Wageningen
H. van Suchtelenweg 4
6703 CT Wageningen
THE NETHERLANDS

E. L. Butler
ENSR Consulting and Engineering
35 Nagog Park
Acton, MA 01720 USA

D. L. Carr
Department of Biology
Texas Tech University
Lubbock, TX 79423 USA

P. Carrera
Eniricerche S.p.A. Environmental Services
Via Maritano, 26
20097 San Donato Milanese
Milano ITALY

J. G. Carss
O'Conner Associates Environmental, Inc.
201, 1144 - 29 Avenue N.E.
Calgary, Alberta
CANADA T2E 7P1

T. H. Christensen
Technical University of Denmark
Department of Environmental Engineering, Bldg. 115
DK-2800 Lyngby
DENMARK

J. R. Clark
Exxon Biomedical Sciences Inc.
East Millstone, NJ 08875 USA

J. W. Costerton
University of Calgary
Department of Biological Sciences
2500 University Drive NW
Calgary, Alberta
CANADA T2N 1N4

P. DeSantis
RZA-AGRA Engineering and Environmental Services
7477 SW Tech Center Drive
Portland, OR 97223-8024 USA

R. Desrochers
Groundwater Technology Canada Ltd.
10200 L. H. Lafontaine Street
Anjou, Quebec
CANADA H1J 2T3

J. S. Devinny
Environmental Engineering Program KAP224D
University of Southern California
Los Angeles, CA 90089-2531 USA

C. Di Leo
Eniricerche S.p.A. Environmental Sciences
Via E. Ramarini, 32
00016 Monterotondo Scalo
Roma ITALY

G. S. Douglas
Battelle Ocean Sciences
397 Washington Street
Duxbury, MA 02332 USA

K. Dowd
W. W. Irwin, Inc.
2750 Signal Parkway
Long Beach, CA 90806 USA

D. C. Downey
Engineering-Science, Inc.
1700 Broadway, Suite 900
Denver, CO 80290 USA

J. Ducreux
Institut Français du Pétrole
1/4 Avenue de Bois-Préau
92500 Rueil Malmaison
FRANCE

C. T. Esler
RZA-AGRA Engineering and Environmental Services
7477 SW Tech Center Drive
Portland, OR 97223-8024 USA

L. G. Faksness
SINTEF Applied Chemistry
N-7034 Trondheim NORWAY

P. M. Fedorak
University of Alberta
Department of Microbiology
Edmonton, Alberta
CANADA T5J 4E6

R. E. Feltz
Chemical Waste Mangement, Inc.
6316 State Route 128
P.O. Box 626
Miamitown, OH 45041-0626 USA

C. Frey
Woodward-Clyde Consultants
1550 Hotel Circle North
San Diego, CA 92108 USA

R. Fried
DVGW-Forschungsstelle
Universität Karlsruhe (TH)
Richard-Willstatter-Allee 5
D-7500 Karlsruhe 1 GERMANY

C. E. Furlong
Genetics and Medicine
Division of Medical Genetics
University of Washington
Seattle, WA 98195 USA

J. Ghaemghami
Plant and Soil Sciences
Stockbridge Hall
University of Massachusetts
Amherst, MA 01003 USA

M. M. Ghosh
Department of Civil Engineering
University of Tennessee
Knoxville, TN 37996 USA

R. W. Gillham
Waterloo Centre for Groundwater Research
Department of Earth Sciences
University of Waterloo
Waterloo, Ontario
CANADA N2L 3G1

C. A. Gomez
Research Assistant
Civil Engineering Department
North Carolina State University
Box 7908
Raleigh, NC 27695 USA

C. W. Greer
Biotechnology Research Institute
National Research Council Canada
6100 Royalmount Avenue
Montreal, Quebec
CANADA H4P 2R2

M. J. Grossman
Exxon Research and Engineering
Clinton Township, Route 22E
Annandale, NJ 08801 USA

M. Guillerme
Department Environment BRGM
Avenue de Concyr, B.P. 6009
4500 Orléans Cedex 2 FRANCE

I. Guo
Alberta Environmental Centre
Bag 4000
Vegreville, Alberta
CANADA T9C 1T4

H. H. Hanert
Department of Soil Mechanics and Foundation Engineering
University of Karlsruhe, Postbox 6980
D-76128 Karlsruhe GERMANY

E. J. Harner
West Virginia University
Morgantown, WV 26506 USA

G. R. Hater
Chemical Waste Management, Inc.
6316 State Route 128, P.O. Box 626
Miamitown, OH 45014-0626 USA

T. Hawkes
Groundwater Technology Canada Ltd.
10200 L. H. Lafontaine Street
Anjou, Quebec
CANADA H1J 2T3

J. Heersche
DHV Environment & Infrastructure
Post Office Box 1076
3800 BB Amersfoort
THE NETHERLANDS

D. C. Herman
University of Calgary
Department of Biological Sciences
2500 University Drive NW
Calgary, Alberta
CANADA T2N 1N4

G. Heron
Technical University of DK
Department of Environmental
Engineering, Bldg. 115
DK-2800 Lyngby
DENMARK

R. T. Herrington
U.S. Environmental Protection Agency
Risk Reduction Engineering Laboratory
26 West Martin Luther King Drive
Cincinnati, OH 45268 USA

R. J. Hicks
Groundwater Technology, Inc.
310 Horizon Center Drive
Trenton, NJ 08691 USA

R. E. Hinchee
Battelle
505 King Avenue
Columbus, OH 43201-2693 USA

D. S. Hodge
Environmental Engineering Program
KAP210
University of Southern California
Los Angeles, CA 90089-2531 USA

D. S. Hogg
Woodward-Clyde Consultants
1550 Hotel Circle North
San Diego, CA 92108 USA

P. Hubbard, Jr.
Integrated Science and Technology, Inc.
1349 Old Highway 41, Suite 225
Marietta, GA 30060 USA

M. H. Huesemann
Shell Development Company
P.O. Box 1380
Houston, TX 77251-1380 USA

W. P. Hunt
Department of Civil Engineering
University of Tennessee
Knoxville, TN 37996 USA

S. R. Hutchins
U.S. Environmental Protection Agency
Robert S. Kerr Environmental
Laboratory
P.O. Box 1198
Ada, OK 74820 USA

B. K. Jensen
Microbiological Department
The Water Quality Institute
Agern Allé 11
DK-2800 Hørsholm
DENMARK

D. E. Jerger
ONM Remediation Services
Corporation
16406 U.S. State Route 224 East
Findlay, OH 45840 USA

R. L. Johnson
Alberta Environmental Centre
Bag 4000
Vegreville, Alberta
CANADA T9C 1T4

C. D. Johnston
CSIRO Division of Water Resources
Private Bag, Post Office
Wembley, WA 6014
AUSTRALIA

E. F. Johnstone
Exxon Company
5601 77 Center Drive, Suite 200
Charlotte, NC 28217 USA

M. R. Jorgensen
Hargis and Associates
2223 Avenida De La Playa, Suite 300
San Diego, CA 92037 USA

D. H. Kampbell
U.S. Environmental Protection Agency
Robert S. Kerr Environmental
Laboratory
P.O. Box 1198
Ada, OK 74820 USA

C.-M. Kao
Department of Civil Engineering
North Carolina State University
Box 7908
Raleigh, NC 27695 USA

B. L. Kellems
Consulting Engineers, Inc.
4000 Credit Union Drive, Suite 600
Anchorage, AL 99503 USA

J. M. Kennedy
Kinnetic Laboratories Inc.
Anchorage, AK 99501 USA

G. H. Kise
RZA-AGRA Engineering and
Environmental Services
7477 SW Tech Center Drive
Portland, OR 97223-8024 USA

J. Kittel
Battelle
505 King Avenue
Columbus, OH 43201-2693 USA

M. E. Leavitt
IT Corporation
312 Directors Drive
Knoxville, TN 37923 USA

P. Lecomte
Department Environment BRGM
Avenue de Concyr
B.P. 6009
45060 Orléans Cedex 2 FRANCE

Z. M. Lees
International Centre for Waste
Technology (Africa)
University of Natal
Post Office Box 375
Pietermaritzburg, 3200
SOUTH AFRICA

A. Leeson
Battelle
505 King Avenue
Columbus, OH 43201-2693 USA

A. Y. Li
Chemical Waste Management, Inc.
1950 South Batavia Avenue
Geneva, IL 60134 USA

J. E. Lindstrom
University of Alaska
Fairbanks, AK 99775 USA

S. C. Long
University of South Carolina
Environmental Health Sciences
Columbia, SC 29208 USA

C. J. Lu
Department of Environmental
Engineering
National Chung Hsing University
Taichung 40227 TAIWAN

G. Lucchese
Eniricerche S.p.A. Environmental
Sciences
Via E. Ramarini, 32
00016 Monterotondo Scalo
Roma ITALY

M. D. MacKinnon
Syncrude Canada Ltd.
Edmonton, Alberta
CANADA T5J 4E6

L. M. Mallory
Department of Plant and Soil Sciences
Stockbridge Hall
University of Massachusetts
Amherst, MA 01003 USA

M. Manfredi
Department Environment BRGM
Avenue de Concyr
B.P. 6009
45060 Orléans Cedex 2 FRANCE

N. V. Mark-Brown
Woodward-Clyde (N.Z.) Ltd.
Post Office Box 37547
Auckland
NEW ZEALAND

R. P. Masterson
Woodward-Clyde Federal Services
Stanford Place 3, Suite 600
4582 South Ulster Street
Denver, CO 80237-2637 USA

C. Maudlin
W. W. Irwin, Inc.
2750 Signal Parkway
Long Beach, CA 90806 USA

D. H. McNabb
Alberta Environmental Centre
Bag 4000
Vegreville, Alberta
CANADA T9C 1T4

V. F. Medina
Environmental Engineering Program
KAP210
University of Southern California
Los Angeles, CA 90089-2531 USA

M. Meier-Löhr
DVGW-Forschungsstelle
Universität Karlsruhe (TH)
Richard-Willstatter-Allee 5
D-7500 Karlsruhe 1
GERMANY

R. N. Miller
U.S. Air Force
Center for Environmental Excellence
Brooks Air Force Base, TX 78235 USA

R. S. Miller
RZA-AGRA Engineering and
Environmental Services
7477 SW Tech Center Drive
Portland, OR 97223 USA

K. O. Moore
Shell Development Company
P.O. Box 1380
Houston, TX 77251 USA

C. H. Nelson
Groundwater Technology, Inc.
7346 South Alton Way, Suite A
Englewood, CO 80112 USA

R. D. Norris
Eckenfelder, Inc.
227 French Landing Drive
Nashville, TN 37228 USA

K. E. O'Leary
355 Burrard Street, Suite 935
Vancouver, British Columbia
CANADA V6C 2G8

S. K. Ong
Polytechnic University
Civil and Environmental Engineering
Six Metrotech Center
Brooklyn, NY 11201 USA

P. R. Parrish
Society of Environmental Toxicology
and Chemistry
1010 North 12th Avenue
Pensacola, FL 32501-3307 USA

B. M. Patterson
CSIRO Division of Water Resources
Private Bag, Post Office
Wembley W.A. 6014 AUSTRALIA

N. A. Persson
Anox AB
Ideon Research Village
S-223 70 Lund SWEDEN

M. R. Piotrowski
Biotransformations, Inc.
1670 Newport Center Road, Suite 300
Colorado Springs, CO 80916 USA

R. C. Prince
Exxon Research and Engineering Co.
Clinton Township, Route 22E
Annandale, NJ 08801 USA

S. Ramstad
SINTEF Applied Chemistry
N-7034 Trondheim NORWAY

H. J. Reisinger
Integrated Science & Technology, Inc.
1349 Old Highway 41, Suite 225
Marietta, GA 30060 USA

P. E. Reith
RZA-AGRA Engineering and Environmental Services
7477 SW Tech Center Drive
Portland, OR 97223-8024 USA

C. M. Reynolds
U.S. Army Cold Regions Research and Engineering Laboratory
72 Lyme Road
Hanover, NH 03755 USA

R. J. Richter
Departments of Genetics and Medicine
Division of Medical Genetics
University of Washington
Seattle, WA 98195 USA

H. S. Rifai
Rice University
Environmental Science and Engineering
P.O. Box 1892
Houston, TX 77251 USA

U. Rinkel
Department of Soil Mechanics and Foundation Engineering
University of Karlsruhe, Postbox 6980
D-76128 Karlsruhe
GERMANY

A. Rinzema
Agricultural University Wageningen
H. van Suchtelenweg 4
6703 CT Wageningen
THE NETHERLANDS

A. Robertiello
Eniricerche S.p.A. Environmental Sciences
Via E. Ramarini, 32
00016 Monterotondo Scalo
Roma
ITALY

K. G. Robinson
Department of Civil Engineering
University of Tennessee
Knoxville, TN 37996 USA

E. O. Roe
RZA-AGRA Engineering and Environmental Services
7477 SW Tech Center Drive
Portland, OR 97223-8024 USA

A. G. Saberiyan
RZA-AGRA Engineering and Environmental Services
7477 SW Tech Center Drive
Portland, OR 97223 USA

R. Samson
Biotechnology Research Institute
National Research Council Canada
6100 Royalmount Avenue
Montreal, Quebec
CANADA H4P 2R2

G. S. Sayler
Center for Environmental Biotechnology
University of Tennessee
10515 Research Drive, Suite 100
Knoxville, TN 37932 USA

G. D. Sayles
U.S. Environmental Protection Agency
Risk Reduction Engineering Laboratory
26 West Martin Luther King Drive
Cincinnati, OH 45268 USA

R. J. Scholze
U.S. Army Construction Engineering Research Laboratory
P.O. Box 9005
Champaign, IL 61826-9005 USA

G. Schraa
Agricultural University Wageningen
H. van Suchtelenweg 4
6703 CT Wageningen
THE NETHERLANDS

H. Schwarzer
Peroxid-Chemie GmbH
8023 Höllriegelskreuth bei München
dr. Gustav-Adolph-Strasse 3
GERMANY

E. Senior
International Centre for Waste Technology
University of Natal
Post Office Box 375
Pietermaritzburg, 3200
SOUTH AFRICA

S. Simkins
Department of Plant and Soil Sciences
Stockbridge Hall
University of Massachusetts
Amherst, MA 01003 USA

M. St-Cyr
Shell Canada Products Limited
10501 Sherbrooke East
Montreal, Quebec
CANADA H1B 1B3

E. Staes
57 Executive Park South, N.F., Suite 590
Atlanta, GA 30329-2265 USA

J. S. Stark
Waste Management of North America
5701 Este Avenue
Cincinnati, OH 45232 USA

W. G. Steinhauer
Battelle Ocean Sciences
397 Washington Street
Duxbury, MA 02332-0601 USA

G. E. Surbey
Transport Canada
Canada Place, 12th Floor
9700 Jasper Avenue
Edmonton, Alberta
CANADA T5J 4E6

P. Sveum
SINTEF Applied Chemistry
N-7034 Trondheim NORWAY

J. C. Tjell
Technical University of DK
Department of Environmental Engineering, Bldg. 115
DK-2800 Lyngby DENMARK

M. D. Travis
RZA-AGRA Engineering and Environmental Services
711 H Street, Suite 450
Anchorage, AK 99501-3442 USA

M. A. Troy
OHM Remediation Services Corporation
200 Horizon Center Drive
Trenton, NJ 08691 USA

J. van Eyk
Delft Geotechnics
Post Office Box 69
2600 AB Delft
THE NETHERLANDS

A. D. Venosa
U.S. Environmental Protection Agency
Risk Reduction Engineering Laboratory
26 West Martin Luther King Drive
Cincinnati, OH 45268 USA

J. H. Verheul
DHV Environment & Infrastructure
Post Office Box 1076
3800 BB Amersfoort
THE NETHERLANDS

P. Vis
Agricultural University Wageningen
H. van Suchtelenweg 4
6703 CT Wageningen
THE NETHERLANDS

C. M. Vogel
U.S. Air Force
Environmental Quality Directorate of the Armstrong Laboratory
139 Barnes Drive
Tyndall Air Force Base, FL 32403 USA

E. A. Voudrias
Georgia Tech
Environmental Engineering
School of Civil Engineering, MC0512
200 Bobby Dodd Way
Atlanta, GA 30332-0355 USA

M. E. Watwood
Idaho State University
Biological Sciences
Campus Box 8007
Pocatello, ID 83209 USA

T. G. Welander
Anox AB and University of Lund
Department of Biotechnology
Post Box 124
S-221 00 Lund SWEDEN

P. Werner
DVGW-Forschungsstelle
Universität Karlsruhe (TH)
Richard-Wallstatter-Allee 5
D-7500 Karlsruhe 1 GERMANY

J. T. Wilson
U.S. Environmental Protection Agency
Robert S. Kerr Environmental Laboratory
P.O. Box 1198
Ada, OK 74820 USA

M. A. Wilson
RZA-AGRA Engineering and Environmental Services
7477 SW Tech Center Drive
Portland, OR 97223 USA

G. Winter
Alaska Department of Environmental Conservation
Anchorage, AK 99503 USA

M. Wittmaier
Department of Soil Mechanics and Foundation Engineering
University of Karlsruhe
Postbox 6980
D-76128 Karlsruhe
GERMANY

H. Würdemann
Lehrstuhl für Bodenmechanik und Grundbau
Universität (TH) Fridericiana Karlsruhe
Universität K'HE Postfach 6980
D-76128 Karlsruhe
GERMANY

M.-F. Yeh
26157 Georgia Tech Station
Atlanta, GA 30332 USA

INDEX

aboveground bioreactor 329, 332
accessibility 237, 242
accidental spill 133, 134
acenaphthalene 340, 341
acenaphthene 340, 341
activated carbon 12, 14, 19, 43, 44, 130, 136, 175, 385, 395
activated sludge 72, 75, 334
activity 2, 3, 7, 15, 33, 35, 37, 60, 61, 63, 64, 73, 107, 108, 114, 119, 133, 134, 136-138, 140, 141, 157, 163, 170, 171, 178, 181, 182, 186, 192, 193, 195, 198, 199, 220, 246, 267-272, 275, 276, 288, 300, 311, 329-331, 333, 343, 352, 366, 371, 372, 386, 398, 420, 441, 444, 445, 450, 455, 457, 458
adsorption 2, 13, 14, 18, 21, 38, 43, 44, 48, 128, 142, 237
aeration 21, 22, 59, 75, 137, 140, 149, 150-152, 154, 157, 158, 160, 301, 329-331, 366, 385, 438, 441, 456, 458
aerobic 45, 80, 81, 93, 94, 128, 144, 157, 177, 186, 201, 202, 218, 237, 255, 262, 278, 279, 281-283, 287, 300, 348, 350, 368, 371, 372, 383, 416, 418, 424, 453-457
  aerobic bacteria 151, 352
  aerobic biodegradation 46, 60, 149, 218, 347, 353, 354
aggregation 296, 298, 299
air biofilter 1-4
airflow 3, 5-7, 21, 22, 32, 35, 42, 47, 130, 178, 186, 351, 362-364, 366, 367, 370, 384, 447, 449
air injection 40, 347, 348, 350, 351, 385, 447
air sparging 148, 154, 158-160
air stripping 2, 5, 75, 175
akageneite 280, 419
Alaska 100, 101, 105-109, 115, 118, 122, 311, 438, 439, 441, 445, 447, 448, 450, 452
Alcaligenes faecalis 137
aliphatic hydrocarbons 224, 229
alkylbenzene(s) 80, 84, 85, 90, 132, 201, 208, 215, 217, 416
amendment(s) 14, 38, 47, 88, 90, 101, 158, 167, 188, 189, 296, 297, 298-301, 303, 336, 337, 339, 425, 426, 427
anaerobic 81, 84, 85, 138, 171, 216-218, 279-281, 283, 348, 350, 352, 377, 379, 381, 413, 416, 419, 454, 456
  anaerobic bioremediation 80, 88
  anaerobic degradation 21, 217, 416, 456
analog 283
anoxic 218, 374, 381
anthracene 149, 337, 340, 341
AODC (acridine orange direct count) 195, 197
aquifer(s) 38, 73, 80, 81, 84-88, 90, 91, 126, 128, 142, 152, 157, 158, 160, 192, 193, 197-199, 202, 204, 206, 210, 212, 217, 237, 252, 255, 256, 258, 259, 262, 263, 278-283, 290, 294, 374, 405, 407-409, 418-422, 431, 433
Arabian light crude 305, 306
arctic 100, 164, 313, 323, 444
aromatic hydrocarbon(s) 21-24, 62, 73, 84, 96, 118, 131, 176, 187, 220, 224, 230, 232, 318, 334, 411, 414
attenuation 84, 202, 207, 210, 212, 213, 214-217, 287, 353
augmentation 52, 79

basin(s) 154, 163-166, 168-173, 437
batch culture(s) 163, 164, 166-171, 267, 268, 271, 288, 425
bentonite 447
benz(a)anthracene 337, 340, 341
benz(a)pyrene 337, 340, 341
benz(b+k)fluoranthene 337, 340, 341
benz(ghi)perylene 340, 341

benzene 3, 20, 26, 32-34, 74, 80, 85, 86, 88, 90, 93, 126, 131, 140-142, 144-146, 152, 160, 175, 176, 178, 181, 184, 187, 192, 201, 202, 205, 206, 208, 210, 211, 213-218, 244, 245, 262, 290-294, 323, 328, 348, 353, 356, 368-371, 374, 378, 381, 388, 405, 411-414, 416, 419
bioaugmentation 72, 73, 75, 76, 78, 79, 438
bioavailability 134, 237, 241, 242, 318, 339
biodegradation
  kinetics 13, 58-60, 63-66, 68-70, 138
  rates 58, 59, 65, 69, 70, 100, 152, 188, 241, 305, 350, 441, 447, 449, 450, 452, 453
biofilm 2-4, 11, 20-22, 151, 374-377, 381
  reactor 374-376
biofilter(s) 1-4, 7, 8, 10, 11-15, 17-19, 45, 369
biofiltration 1, 3, 5, 11, 12, 19, 43
bioflood 409
biological oxygen demand 75, 166
bioluminescence 412
biomass 2, 3, 15, 18, 21, 22, 26, 30, 38, 45, 59, 95, 116, 117, 125, 163-165, 171, 185, 189, 232, 233, 242, 269, 271, 305, 307, 309, 331, 374, 376, 378, 381, 412
biomineralization 318, 319, 321
biopad(s) 398-404
biosurfactant(s) 99, 137, 138, 139, 311, 318-322, 409
biotransformation 145, 175, 185, 202, 218, 381, 416
biotreatment 32, 105, 267, 271, 334, 337, 429
bioventing 32-34, 40, 41, 45, 47, 52-57, 148, 152, 154, 243, 244, 347-349, 352, 362, 366, 441, 444-453
biphenyl 226, 411-414
brine 296-301
BTEX 20, 26, 27, 33, 80, 81, 83-91, 93, 126, 127, 131, 152, 154, 156, 160, 175, 176, 178, 179, 181, 182, 186-189, 192-194, 196, 197, 201, 202, 206, 209-211, 213-215, 218, 248, 249, 262-266, 290-294, 323, 348, 353, 356, 357, 368-370, 372, 374-376, 378, 381, 388-390, 393, 429
  aromates 94
bulking agents 76
2-butoxyethanol 108
n-butyric acid 140-144, 146

carbon dioxide (CO_2) 2, 33, 35, 36, 43, 44, 58-61, 63, 66, 98, 115, 125, 149, 158, 166, 170, 171, 177, 178, 181, 182, 204, 244-248, 269, 292, 294, 305, 311, 333, 353, 356, 357, 371, 372, 373, 383, 389, 408, 409, 419, 445, 447-449, 454-458
  production 62, 63, 182, 244-248, 372, 454-458
carbons 329
casein 303-310
catalyst 43
catalytic oxidation 43
catechol 330
cations 320
CH_4, *see* methane
chloride 48, 85, 101, 117, 126, 135, 194, 203, 220, 223, 225, 228, 232, 262, 305, 408, 445
chlorinated 126, 127, 267
  ethenes 149
chloroform 177, 178, 182, 183, 319
chrysene(s) 117, 121, 228, 233, 234, 337, 340, 341
clogging 15, 19
CO_2, *see* carbon dioxide
coal tar 411
coarse-textured layers 431
cold 100, 101, 105, 201, 327, 331, 333, 438, 441-444, 451, 453
column(s) 12, 13, 17, 18, 22, 25, 26, 32-39, 60, 73, 84, 108, 118, 119, 128, 135, 140, 141-146, 151, 175-183, 185, 186, 187, 188, 218, 220, 223, 224, 232, 237, 238, 254, 262, 264, 286, 287, 304, 305, 312, 347, 374, 375, 395, 425, 427, 429
cometabolism 149, 272, 378-380, 414, 415
commercial bacteria 58-60, 69
competition 271
competitive inhibition 379-381
compost 2, 12, 14-17, 45, 59
composting 15, 441
compression 296, 299, 300
concentrate 141

consumption 2, 10, 58, 59, 61-70, 85, 96, 98, 113, 130, 186, 244, 246, 248, 280, 307, 309, 328, 330, 331, 409, 454, 455, 457
containment system 323-325, 327
copper 267
creosote 381, 411, 416
p-cresol 413, 414, 418
crude oil 58-65, 67-74, 76, 107, 116-118, 163-170, 217, 219, 225, 226, 228-234, 274, 296, 304-306, 309, 311, 312, 313-315
cyclohexane carboxylic acid 276
cyclopentane carboxylic acid 276

DCE (ethylene di-chloride) 48
decontamination 19, 99, 139, 252, 458
deionized water 279, 427, 429
demonstration 76, 88, 90, 100, 149, 271, 301, 353, 448-451
project 243
denitrification 413, 454, 455, 457, 458
denitrifying 80, 81, 84, 85, 88, 90, 93, 94, 98, 99, 137, 138, 160, 374, 375, 376, 378, 381, 411, 416, 456
desorption 88, 130, 242
dibenz(ah)anthracene 340, 341
dichloromethane 238, 431
2,4-dichlorophenoxyacetic acid 333
diesel-contaminated 100, 344, 366
diesel fuel 48, 54, 59, 74, 133, 134, 228, 237, 343-346, 371, 373, 398, 402, 424
diffusion 2, 4, 13, 21, 42, 181, 182, 186, 262, 275, 279, 301, 309, 311, 448, 456-458
2,4-dimethylphenol 412-414, 416
dioxygenase 267, 330
dissolution 96, 107, 258, 259, 261
dissolved oxygen (DO) 41, 94, 95, 101, 110, 112, 128, 130, 131, 140, 142, 143, 145, 146, 151, 160, 167, 186, 198, 217, 219, 255, 256, 264, 293, 294, 296, 301, 324, 326, 353, 356, 357, 374, 408, 433, 450
DNA 331
DO *see* dissolved oxygen

EDTA, *see* ethylenediaminetetraacetic acid
electron acceptor(s) 33, 45, 46, 52, 80, 81, 88, 90, 93-95, 131, 132, 148, 158, 160, 175, 189, 202, 207, 211, 212, 293, 294, 353, 381, 405, 411, 414, 416, 418, 419, 421, 422
electron microscopy 286
emissions 1, 2, 11, 19, 44, 244, 350, 352, 368-372, 458
encapsulated 303-307, 309
enhanced biodegradation 58, 68-70, 73, 99, 116
enrichments 81
entrapment 431, 435, 437
enumeration 193, 195, 199, 246, 247
enzymes 157, 165, 267, 415
EPA, *see* U.S. Environmental Protection Agency
epifluorescence counting 246
ethane 31
ethanol 12, 15, 16, 19, 166, 296, 330
ethenes 149
ethylbenzene 3, 20, 22, 26, 27, 32-35, 80, 85, 93, 126, 152, 175, 176, 178, 181, 192, 201, 202, 205, 216, 217, 262, 290-294, 323, 328, 348, 353, 356, 368-371, 374, 378, 379, 380, 381, 388, 411-415, 419
ethylenediaminetetraacetic acid (EDTA) 61, 279-281, 283
European 422
evolution 35, 36, 148, 149, 195, 238, 255, 256, 259, 305, 372, 419-421
ex situ 368, 373, 438, 441
external carbon 165
extraction well(s) 43, 154, 175, 350, 362-367, 408
Exxon Valdez oil spill 73, 79, 107

fate and transport 176, 233
feasibility study 2, 3, 21, 70, 75, 128, 383, 424, 425, 429
ferrihydrite 280
field study 290, 329
fish meal 166-171
fixed-film 45
fluoranthene 337, 340, 341
fluorene 117, 226, 233, 340, 341
fluorescence 120
fractured bedrock 154, 324
freshwater 113, 164, 277
fuel carbon 194

fuel oil 101, 228
full-scale biological land treatment 343
fungi 45, 288

gas blockage 148, 156
gasoline 1, 3, 4, 6, 11, 12, 17-19, 41, 44, 46, 48, 49, 54, 80, 81, 94, 125, 133, 134, 148, 151, 154, 156, 175, 176, 178, 179, 201, 202-206, 208, 215, 243, 258, 262, 263, 290, 294, 347, 352, 354-356, 374, 394, 398, 405, 406, 407
gene probe(s) 74, 329-331, 333
geochemistry 207, 294
glucose 281, 336
goethite 280, 419
granular activated carbon 12, 14, 43, 44
gravels 110, 126, 201, 383, 385
groundwater remediation 266
growth on HC 246

HC oxidation 246
HCl 36, 280, 282, 283, 336
heavy metals 285, 287, 335
hematite 280, 419
hexadecane 113-115, 276, 330, 331
n-hexadecane 113, 286-288
Hg (mercury) 48
horizontal wells 156, 408
humic acid 281
humus 68
hydrocarbon degraders 46, 58, 60, 61, 69, 74
hydrocarbon-degrading bacteria 2, 3, 5, 151, 331
hydrogen peroxide 46, 72, 73, 80, 93, 95, 126, 128, 130-132, 140, 141, 142-148, 151, 154, 156, 157, 160, 252, 253, 255, 256, 324, 388, 389, 395, 396, 406, 407, 409, 424, 425, 427, 428
hydrogen sulfide 356
hydroquinone 281

immobilization 164, 165, 167, 171
indeno(1,2,3-cd)pyrene 340, 341
indigenous bacteria 73, 133, 149, 219, 406, 444
infrared 111, 120, 166, 204, 330, 331, 334, 343
inhibition 38, 58, 378-381
Inipol™ EAP22 108, 110, 113, 163, 166, 167, 168, 170, 171, 311, 312
injection wells 80, 81, 90, 151, 152, 154, 156, 252, 255, 407, 408, 447
inoculation 6, 22, 27, 69, 70, 74, 76, 310, 438
iron 24, 80, 84, 95, 96, 99, 156-158, 209-212, 269, 271, 278, 279-283, 292, 294, 356, 389, 395, 418-422
  reduction 201, 278, 418-422
irrigation 51, 76, 181, 301, 362, 366, 367, 385

jet fuel 33, 84, 192, 196, 199, 237, 409, 445, 449
JP-4 33, 84, 192-194, 196, 197, 199, 445

kerosene 48, 424-428, 431
kinetic model 176
kinetic studies 189
kinetics 2, 13, 15, 58-60, 63-70, 94, 112, 134, 138, 175, 176, 189, 215, 258, 261, 344, 374, 376, 455, 457, 458
KNO_3, *see* potassium nitrate

laboratory
  column 175-180, 182, 183, 186, 187, 188
  simulation 133, 135
  study 95, 389, 424
landfarm, landfarming 70, 100-105, 139, 335, 341, 438
landfill(s) 44, 141, 143, 258, 279, 281-283, 383, 418, 420, 422
landfill leachate 418, 420, 422
land treatment 72, 73, 76, 335, 337, 343, 344, 346
large-scale operation 93
layering 177, 431, 433, 437
leach 101, 266, 296, 412, 441
leachate 101, 141, 143, 418-422, 438
leachbed 100, 102, 104, 373
leaching 74, 85, 101, 139, 213, 278, 366, 427
liposome 303-305, 307, 309

LNAPL (light, nonaqueous-phase liquid) 93, 94
low temperature 105
lube oil 248

macronutrients 45, 46, 52
magnetite 280, 281
manganese 24, 95, 96, 99, 156, 157, 278, 279-282, 356
marine 107, 119-121, 149, 164, 283, 303, 305, 310, 311, 317, 346
mass transfer 21, 40, 43, 51
matrix effects 343, 346
metabolites 238, 339
methane (CH_4) 203, 209-211, 213, 214, 216, 267, 268-272, 279, 292, 294, 370, 418
  monooxygenase (MMO) 267
methanogenesis 201, 211, 213, 279, 294
methanogenic 80, 88, 201, 215, 217, 279, 281, 282, 293
methanol 37, 267
methanotroph(s) 267, 268, 269, 271
methanotrophic 267, 271
1-methylnaphthalene 412-414, 419
2-methylnaphthalene 419
Methylosinus trichosporium 267-270, 272
microbial
  activity 2, 7, 33, 37, 108, 114, 140, 141, 192, 193, 199, 246, 275, 276, 329, 331, 333, 366, 371, 386, 398, 444, 445, 450
  community 33, 193, 195
  enumerations 74, 195
  populations 59, 107, 108, 110, 113, 164, 383, 411, 424, 425, 428
microcosm(s) 45, 81, 85-88, 90, 91, 175-177, 182, 184, 185, 188, 189, 215, 217, 286, 287, 288, 330, 334, 336, 337, 339, 340
micronutrients 324, 424
Microtox™ 274-276
mineral oil 248, 250, 287, 305, 330, 331, 388-390, 393, 394
mineralization 33, 35, 37, 38, 40, 46, 63, 67, 98, 113-116, 119, 165, 267, 275, 294, 307, 319, 320, 322, 328, 330, 331, 333, 406, 413, 444, 448, 450, 453
mixed culture(s) 17, 163, 306, 411, 414, 416, 419, 421
mobility 311, 389, 431
modeling 14, 19, 90, 128, 132, 189, 354, 357
monitoring 3, 5, 33, 89, 90, 107, 108, 111, 113, 117-122, 125-126, 130, 131, 133, 163, 178, 193, 202, 204, 205-208, 212-215, 217, 219, 224, 226, 228, 243, 246, 248, 252, 255, 263, 264, 266, 274, 285, 290, 291-294, 323, 324, 327, 329, 330, 331, 333, 334, 344, 348, 350, 353, 356, 357, 368, 371, 384, 385, 389, 390, 392, 393, 395, 397, 400, 401, 409, 447, 454, 455-457
monocyclic aromatic hydrocarbons 21
most probable number 62, 68, 113, 199
MTBE (methyl tertiary butyl ether) 48

naphthalene 21-23, 27, 29, 38, 74, 75, 79, 117, 149, 226, 233, 267, 319, 340, 341, 411-414, 419
naphthenic acid(s) 274, 276, 277
NAPL(s), *see* nonaqueous-phase liquid(s)
natural
  annual variations 437
  arctic shoreline conditions 313
  (bio)attenuation 84, 202, 203, 353
  (bio)degradation 122, 202, 353, 357, 358, 437
  bioreclamation 201
  (bio)remediation 189, 290, 294
  cleansing 107
  column of soil 178
  environment 255
  gas 43
  logarithm 215
  microbial respiration rates 447
  open systems 164
  organics 69
  origin 212
  processes 274
  rate of degradation 108
  reaction 136
  resources 217
  soil 47, 69, 175-177, 180, 188
  surface water 275
  systems 46
necromass 165, 167, 171

nitrate(s) 35, 38, 80, 81, 84-96, 98, 101, 108, 111-113, 120, 126, 131, 132, 137, 148, 158, 160, 181, 187, 201, 209, 210-212, 238, 267, 269-271, 279, 292, 294, 303, 305, 324, 327, 356, 374, 375, 376, 381, 389, 408, 409, 411, 413, 414, 416-418, 420, 425, 426, 455
nitrate-reducing 374, 381, 416
nitrite 85, 96, 112, 126, 375-378, 413
nitrogen 6, 32-38, 45, 52, 59, 60, 77, 87, 88, 90, 101, 108, 110, 112, 113, 118, 119, 121, 131, 135, 142, 149, 163, 164-167, 171, 175, 179, 180, 185, 187-189, 220, 223, 252, 270, 287, 303-307, 309, 324, 329, 366, 375, 425, 427, 456, 458
nonaqueous-phase liquid(s) (NAPL) 93, 258, 261, 290, 431-433, 435-437
nonionic surfactant 237, 238
nonstabilized sludge 336-341
nutrient(s) 2, 5, 6, 14, 17, 24, 26, 33, 38, 45, 52, 58, 59, 64, 69, 72, 74, 76-78, 80, 84, 85, 88, 94, 101, 103, 107, 108, 110-114, 119, 121, 122, 125, 126, 127, 128, 130, 131, 134, 135, 141, 149, 151, 152, 154, 156-158, 163-165, 167, 171, 175, 178, 181, 184-187, 189, 195, 198, 199, 219, 237, 238, 252, 253, 258, 261, 271, 286, 287, 303-305, 309, 313, 316, 324, 327, 329, 330, 331, 333, 337, 338, 353, 362, 366, 367, 383, 384, 385, 389, 394, 398, 400, 405, 406, 407, 409, 412, 420, 424, 425, 426, 427, 429, 438, 444
  amendment 188, 336, 337, 339, 427
  injection 130, 131, 156

O_2, *see* oxygen
off-gas treatment 40, 43, 45, 51, 52, 54, 56
oil(s) 11, 46, 58-65, 67-74, 76, 78, 94, 101, 104, 107-110, 113, 114, 116, 117, 118-128, 130, 149, 163-171, 195, 217, 219, 220, 223, 225-238, 240-242, 248, 250, 252, 253, 258, 274, 285, 286-288, 296-307, 309, 311-317, 329-331, 334, 335, 347, 355, 356, 383, 385, 386, 388-390, 393, 394, 398, 445
oil (*continued*)
  bioremediation 58
  sands tailings 274, 276, 277
  spill(s) 73, 107, 121, 122, 163, 192, 228, 233, 234, 235, 303, 304, 309, 311, 316, 317
open system 167, 168
organic
  amendments 298
  carbon content 34, 180
  compound(s) 1, 2, 10, 32, 33, 56, 69, 74, 152, 203, 214, 216, 254, 255, 318, 367, 411, 416, 418, 419, 421
oxidation 24, 38, 43, 63, 131, 142, 149, 246, 267, 269, 270, 272, 276, 278, 280, 281, 283, 449, 456
  capacity 278, 279, 281-283
oxides 278, 283, 418-422
oxygen (O_2) 2, 21, 22, 25, 27, 29, 33, 40, 45, 46, 47, 52, 58-62, 65, 66, 70, 72, 80, 81, 84, 86, 88, 90, 93-95, 98, 110, 112, 125, 126-128, 130-132, 134, 138, 140, 141-149, 151, 152, 156-158, 160, 175, 177, 178, 182-187, 189, 201, 203, 204, 209, 210-212, 219, 237, 238, 241, 248, 252, 255, 256, 258, 261, 262, 263-266, 279, 280, 283, 292, 293, 294, 304, 307, 316, 324, 326, 328, 331, 348, 350, 352, 353, 356, 357, 374, 376, 381, 383, 385, 389, 395, 396, 398, 405-409, 418, 420, 424, 427, 444, 445, 447-451, 454-456, 458
  consumption 58, 59, 61-70, 113, 244, 246, 248, 307, 309, 330, 331, 454, 455, 457
  demand 47, 75, 80, 81, 166
  uptake 62, 304, 330
oxygenation 204, 253, 255-257, 336
PAH(s), *see* polycyclic aromatic hydrocarbons
parent compounds 233
PC, *see* phosphatidylcholine
PCE, *see* tetrachloroethene
permafrost 323
permeability (-ies) 51, 52, 94, 128, 156, 192, 237, 242, 248, 285, 287, 297, 318, 348, 362-364, 367, 388, 395, 398
peroxidase 157

petroleum 17, 33, 41, 54, 56, 72-74, 76, 80, 101, 108, 144, 149, 190, 219, 220, 224, 226, 228, 232, 233, 235, 243, 252, 262, 290, 294, 301, 323, 327-329, 333, 360, 368, 369, 372, 438, 444, 445
hydrocarbon(s) (*see also* total petroleum hydrocarbon) 45, 59, 64, 66, 68, 69, 70, 72, 73, 80, 81, 85, 101, 108, 111, 125, 151, 176, 192, 193, 199, 206, 217, 223, 232, 303, 329, 331, 334, 343, 346, 383, 398, 400, 401-403, 424, 438, 441
pH 2, 3, 5, 14, 27, 33, 34, 36, 45, 59-62, 74, 76, 78, 110, 113, 120, 126, 134, 137, 149, 156, 164, 201, 264, 266, 269, 271, 274, 279, 280, 286, 288, 292-294, 298, 305, 306, 319, 320, 330, 335-337, 351, 356, 375, 376, 381, 384, 386, 408, 418, 419-421, 458
phenanthrene 113-116, 149, 233, 242, 318, 319-322, 340, 341, 411-414
phenol(s) 412-414, 416, 418
phenolic compounds 416
phosphatidylcholine (PC) 304, 305, 306
phosphorous 158, 163, 171, 366
phosphorus 6, 45, 52, 59, 60, 70, 77, 85, 101, 108, 110, 121, 135, 149, 163, 164, 166, 167, 171, 180, 252, 303, 306, 309, 324, 366
pilot 3-5, 12, 17, 18, 27, 28, 30, 31, 33, 54, 72, 75, 76, 81, 93, 95-99, 252-255, 257, 348, 350, 438, 444, 447-449
study (-ies) 10, 20, 21, 26, 29, 73, 93, 99, 441
plasmid 331, 333
plugging 5, 80, 81
polar compounds 62, 65, 118, 220, 223, 232, 233, 238, 240-242
polycyclic aromatic hydrocarbon(s) (PAH) 73, 84, 118, 121, 131, 221, 224, 226-228, 230, 231, 233, 237, 318, 334-337, 339-340, 454
pool 259, 260, 278, 279
pore volume 90, 94
porous media 242, 258, 437
potassium nitrate (KNO_3) 304-308
precipitation 51, 81, 84, 95, 157, 158, 193, 199, 206, 237, 395, 445
pristane 118, 163, 166, 167, 169, 170, 172, 173, 223, 226, 233, 286-288, 313-315, 330, 331
propionic acid 140-142, 144, 146
Pseudomonas 1, 3, 5, 318-322, 330, 333, 383
P. aeruginosa 318-322
pyrene 131, 337, 340, 341
pyrolusite 280

quicklime 334, 335, 337

Raoult's Law 41
rates 5, 6, 8, 9, 11-13, 18, 19, 24, 38, 45, 58, 59, 63, 65, 68-72, 85-88, 100, 108, 116, 119, 128, 130, 152, 157, 164, 175, 188, 215, 216, 219, 233-235, 241, 244, 247, 270, 271, 288, 293, 294, 303, 305, 307, 311, 316, 333, 348, 350-352, 366, 368, 406, 409, 441, 444, 445, 447, 449-453, 457
RCRA (Resource Conservation and Recovery Act) 368
reactor(s) 12, 17, 18, 20-27, 29-31, 45, 74, 75, 102, 140-142, 144, 145, 271, 330, 334, 337, 339, 341, 374-376, 378-381, 383-386, 389, 418, 420, 441, 455
recirculating 6, 27, 100-102, 104, 373, 384
recovery system 158, 324, 328, 407, 408, 409
redox 96, 99, 208, 211, 264, 278-280, 282-283, 292, 293, 357, 411, 418, 422
pools 282
reduction(s) 3, 6, 9, 26, 51, 72, 75, 78, 95, 137, 149, 152, 154, 156, 163, 201, 210, 211, 212, 214, 218, 244, 257, 270, 271, 274, 276, 278-281, 283, 319, 335, 337-341, 350-352, 366, 372, 375, 386, 395, 413, 418-422, 450
refinery 93, 139, 257, 329
regulation 286
reinjection gallery 323, 324
residual concentration 22, 333, 339
respirometry 58, 59, 64, 66, 68, 69, 330, 333
retail gasoline station 243, 251
river sediment 418

S-P relations, relationships 433, 435
salinity 33, 75, 110, 113
sand and clay fractions 196, 199
scale-up 149
scanning electron microscopy (SEM) 286, 288
seawater 163-166, 168, 304-306, 312
sediment(s) 33, 107, 108, 109, 110, 113-120, 127, 128, 130, 165, 166, 170, 192-199, 218, 220, 223, 228, 233, 255, 257, 278-283, 312, 313, 314-317, 418-422, 438
SEM, *see* scanning electron microscopy
serum bottles 330
shoreline(s) 33, 107-110, 111, 113, 116, 121, 163-165, 170, 171, 173, 311-316, 323
simulated groundwater system 140, 141, 144, 146
sludge 70, 72-75, 139, 277, 334-341, 375, 381
slurry (-ies) 32, 38, 62, 70, 74, 77, 102, 176, 195, 274, 334, 337, 339, 341, 375, 383, 386, 416
  bioreactor 383
  -phase 337
  reactor 334, 337, 339, 341, 383, 384, 386
sMMO, *see* soluble methane monooxygenase
soil
  biopad 399
  bioventing 244, 409
  organic matter 58, 63, 66, 68-70
  respirometry 58, 59, 64, 68, 69
  sampling 60, 132, 248, 254, 335, 356, 358, 456
  type 58-60, 67, 128
  vapor extraction (SVE) 12, 17, 19, 33, 41, 128, 130, 152, 154, 158, 160, 347, 350, 407
  venting 40-43, 45, 47, 48, 51-57, 243, 251, 352, 362, 364, 367, 441, 444, 453, 458
solubility (-ies) 46, 48, 52, 73, 94, 140, 142, 233, 258, 259, 260, 298, 315, 318-322, 419
soluble methane monooxygenase (sMMO) 267-272
solvent(s) 34, 37, 79, 110, 201, 220, 223, 261, 287, 343, 374
sorption 85, 182, 316
soybean oil 166-169
spectroscopy 84, 120, 334, 343
spent oil 398
stabilized sludge 334, 336
stoichiometry 63, 67, 90, 145, 246, 376, 405
substrate(s) 13, 33, 64, 68, 164, 175, 184, 185, 188, 189, 267, 269, 307, 318, 330, 376, 378, 380, 411-413, 415, 416
subsurface bacteria 81, 99, 247
subsurface sediment heterogeneity 193
sulfate 77, 158, 184, 185, 209-212, 220, 292, 294, 356, 357, 418, 425, 426
  reduction 201, 211, 279, 375, 413
sulfide(s) 209, 356, 418, 456
Superfund 75, 158
surface area 2, 3, 17, 29, 84, 117, 252, 375, 376, 389
surfactant(s) (*see also* biosurfactant) 73, 158, 237, 238, 240, 241, 242, 253-255, 257, 286, 288, 318, 319, 320, 400
survival 119, 275
SVE, *see* soil vapor extraction

tar 411, 458
T_c, *see* transformation capacity
TCE (*see also* trichloroethylene *and* trichloroethene) 20, 267, 268, 270, 271
temperature(s) 2, 3, 5, 7, 22, 33, 43, 45, 59, 60, 61, 64, 94, 96, 98, 103, 105, 110, 113, 149, 165, 182, 219, 223, 224, 245, 258, 269, 286, 287, 304, 305, 312, 313, 319, 324, 326, 328, 330, 331, 336, 337, 348, 352, 356, 375, 376, 384, 385, 386, 419, 438, 441, 444, 445, 448, 450, 451, 452, 453, 457
TEO, *see* total extractable oil
tetrachloroethene (PCE) 20, 21, 31
tetrachloroethylene 201, 202
thermal incineration 43
Ti^{3+} 279-281, 283
TOL plasmid 331, 333

toluene 3, 20, 22-27, 32-38, 80, 85, 93, 126, 152, 175-179, 181-187, 189, 192, 201, 202, 205, 206, 212, 213, 215-217, 244, 258, 259, 260, 262, 290-294, 296, 323, 328, 348, 353, 356, 368, 369, 370, 371, 374-381, 388, 405, 411, 412-415, 418-422
total extractable oil (TEO) 228, 334-339
total petroleum hydrocarbon(s) (TPH) 61, 62, 66, 67, 74, 80, 83, 84, 87, 100, 101-104, 111, 126, 127, 131, 192, 203, 206, 214, 223, 232, 235, 305, 306-309, 323, 327, 328, 330, 343, 344, 345, 351, 352, 356, 368, 369, 371, 372, 383, 385, 386, 398, 400, 402, 403, 405, 408, 409, 424, 425, 427, 429, 431, 445, 447
total recoverable petroleum hydrocarbons (TRPHs) 78, 347, 366, 398, 400, 402, 403
toxicity 59, 73, 74, 108, 111, 119, 120, 157, 158, 168, 271, 274-276, 286, 303
TPAH (total PAH) 118, 119
TPH, *see* total petroleum hydrocarbon(s)
tracer 34, 35, 96, 98, 178, 182, 202, 333
transformation capacity (T_c) 24, 323
treatability 32, 128
  study (-ies) 38, 80, 81, 85, 86, 126
  test 32
trichloroethene (TCE) 20
trichloroethylene (TCE) 201, 202, 267, 272
TRPHs, *see* total recoverable petroleum hydrocarbons
turbidity 286
turbine oil 398

U.S. Environmental Protection Agency (EPA) 41, 59, 73, 80, 122, 125, 131, 132, 160, 192, 194, 199, 220, 223, 224, 226, 228, 232, 266, 305, 310, 322, 343-345, 383, 398, 400, 402, 408, 425

vacuum heap 368, 369, 372, 373
vadose zone 32-34, 40-42, 46-48, 51, 52, 56, 84, 125, 126, 152, 158, 199, 348, 352, 424, 429, 447
vapor
  extraction 2, 12, 17, 19, 33, 38, 41, 48, 56, 125, 126, 128, 130, 152, 154, 347, 350, 362, 405, 406, 407
  pressure 42, 47, 48, 52
  treatment 19, 351, 352
variability 77, 100-103, 187, 212, 219, 235, 431
venting 38, 40-45, 47, 48, 51-57, 156, 237, 243-245, 248-251, 348, 350, 352, 362, 364-367, 441, 444, 445, 453-458
vertical distribution 193, 206, 208-211, 215, 315, 431
VOC(s), *see* volatile organic compound(s)
volatile organic compound(s) (VOC) 1-3, 6, 7, 9, 10, 11, 19, 56, 152, 203, 370-372
volatile organics 45
volatilization 21, 22, 24, 27, 29, 32, 33, 36, 40, 42, 47, 48, 75, 142, 181, 182, 186, 291, 347, 348, 351, 352, 427, 441, 444

waste oil 127, 383, 385, 386
wastewater 2, 70, 72, 74, 75, 334, 381, 411, 416
water retention 296, 297, 301
water table variations 431

xenobiotic 11, 70, 341, 418, 421
xylene(s) 3, 20, 22-27, 29, 33, 35, 80, 85, 93, 126, 152, 175, 176, 178, 181, 184, 192, 201, 202, 205, 215, 216, 217, 244, 262, 290-294, 323, 328, 348, 353, 356, 368, 369-371, 374, 375, 378-381, 388, 405, 411, 412, 413-416, 419
m-xylene 85, 178, 181, 201, 216, 375, 414, 416, 419
o-xylene 25, 85, 178, 181, 201, 205, 216, 217, 262, 291-294, 378-380, 412-416, 419
p/*m*-xylene 419
p-xylene 85, 181, 184, 201, 205, 216, 291, 292-294, 375, 378-380, 414